DATE DUE

WITHDRAWN

DEMCO 38-297

D1452160

ELECTRONIC ASPECTS OF ORGANIC PHOTOCHEMISTRY

ELECTRONIC ASPECTS OF ORGANIC PHOTOCHEMISTRY

Josef Michl
Center for Structure and Reactivity
Department of Chemistry
University of Texas
Austin, Texas

Vlasta Bonačić-Koutecký
Freie Universität Berlin
Institut für Physikalische Chemie
und Theoretische Chemie
Berlin, West Germany

A WILEY-INTERSCIENCE PUBLICATION
John Wiley & Sons, Inc.
NEW YORK / CHICHESTER / BRISBANE / TORONTO / SINGAPORE

Copyright © 1990 by John Wiley & Sons, Inc.

All rights reserved. Published simultaneously in Canada.

Reproduction or translation of any part of this work beyond that permitted by Section 107 or 108 of the 1976 United States Copyright Act without the permission of the copyright owner is unlawful. Requests for permission or further information should be addressed to the Permissions Department, John Wiley & Sons, Inc.

Library of Congress Cataloging in Publication Data:

Michl, Josef, 1939–
  Electronic aspects of organic photochemistry/Josef Michl, Vlasta Bonačić-Koutecký.
   p. cm.
  "A Wiley-Interscience publication."
  Includes bibliographical references.
  ISBN 0-471-89626-8
  1. Organic photochemistry. I. Bonačić-Koutecký, Vlasta.
II. Title.
QD708.2.M53 1990
547.1'35—dc20   89-39500
                   CIP

Printed in the United States of America

10 9 8 7 6 5 4 3 2 1

Vlasta Bonačić-Koutecký: To Jaroslav Koutecký

Josef Michl: To my father, Josef, and to the memory of my mother, Věra

PREFACE

While the experimental, synthetic, and mechanistic aspects of organic photochemistry are well covered in quite a few recent textbooks and monographs, to our knowledge no book has been dedicated to the discussion of its theoretical aspects, that is, to the analysis of the detailed electronic nature of the elementary photochemical reaction steps. Much information on the subject is scattered throughout the journal literature, and the present text represents an attempt to organize it at a level suitable for graduate students. We assume that the reader is already familiar with the phenomenology of organic photochemistry and give little information about experimental results, which can be readily found elsewhere (see Section 1.4 for leading references). While it would have been ideal to combine theory and experiment in a single volume, it would have been of prohibitive size.

Parts of the text have been used in graduate courses for organic chemists at Yale University, Stanford University, California Institute of Technology, the University of Utah, and the University of Texas at Austin in the United States, at the Technion (Haifa) in Israel, and at the University of Münster and the Free University, Berlin in Germany, and we are grateful for the critical comments provided by the students.

It is hoped that in addition to graduate students, the book will also be of interest to practicing photochemists and possibly to physical and theoretical chemists seeking inspiration and looking for problems to which to apply their newly improved tools. Surely, the theory of organic photochemical reactions lags far behind the theory of organic thermal reactions, and a fresh influx of researchers and ideas would be most welcome. If this book helps to bring about such a turn of events, an important goal will have been accomplished.

The origin of the book may be traced back to a series of articles on photochemical mechanisms published in 1972 in *Molecular Photochemistry*, in 1974 in *Topics of Current Chemistry*, and subsequently mostly in the *Journal of the American Chemical Society,* by one of the authors, and to a series of articles on *ab initio* excited state wave functions and their implications for photochemistry, mostly in the *Journal of the American Chemical Society,* by the other author, starting with a 1975 paper on "sudden polarization" in *Angewandte Chemie* with Lionel Salem and several others. Over the years, our views on the subject developed, and the present text represents an outgrowth of several years of collaborative effort, highlighted by a 1987 review article in *Angewandte Chemie*.

The collaboration was made possible by the granting of an Alexander von Humboldt Senior U.S. Scientist award and a J. S. Guggenheim Fellowship award to J.M., and these are gratefully acknowledged. J.M. is grateful to Professor Jaroslav Koutecký at the Free University, Berlin, Professor Albert Weller at the Max-Planck-Institute for Biophysical Chemistry, Göttingen, Professor Horst Kramer at the University of Stuttgart, and Professor Yitzhak Appeloig at the Technion, Haifa,

Israel, for their warm hospitality during work on the manuscript. He is further much obliged to his friends in the following Departments of Chemistry, where much of the text was written during extended visits: at California Institute of Technology, at Stanford University, at the University of Chicago, where he held a Morris S. Kharasch visiting professorship, at Yale University, where he held a Treat B. Johnson visiting professorship, and at the Technion, where he held a Manson visiting professorship. Finally, we are thankful to the U.S. National Science Foundation and to the Deutsche Forschungsgemeinschaft and Fond der Chemischen Industrie, who have supported our work in photochemistry for many years.

The programs used in our computations were originally provided by Professors Robert Buenker and Sigrid Peyerimhoff. Drs. John Downing, Jutta Köhler, Daniella Papierowska-Kaminski, Maurizio Persico, and Klaus Schöffel participated in some of the calculations and computer art production. Professor Piercarlo Fantucci provided advice on valence-bond theory. Numerous authors sent us reprints of their work. Many kindly read parts or all of the text and suggested improvements: Professors Gerhard Closs, Marye-Anne Fox, Karl Jug, Martin Klessinger, Horst Kramer, J. Michael McBride, Lisa McElwee-White, Gabriella Poggi, Lionel Salem, Jack Saltiel, and Jacob Wirz. Dr. Claas Zachariasse has kindly provided unpublished results, and Dr. Regai Makar translated a page from Arabic. We are indebted to all of these, to our illustrator, Mr. Alexis Kelner, for producing art out of the simple sketches we provided, and to Mr. Philip Willden, Ms. Rebecca Cunningham, Ms. Katheryn Clayton, and Ms. Angie Watson for producing a superbly typed manuscript from our dictation and notes.

A special note of thanks goes to Professor Jaroslav Koutecký, from whom we have learned about quantum chemistry and whose constant encouragement over the years of work on this book was essential to its completion.

Finally, we are grateful to our editor, Dr. Theodore Hoffman, for his unfailing patience and kind understanding in the face of seemingly interminable delays in the completion of our work.

JOSEF MICHL
VLASTA BONAČIĆ-KOUTECKÝ

Austin, Texas
Berlin, West Germany
April 1990

CONTENTS

Notation — xiii

PART A: BACKGROUND — 1

1 Introduction — 3

1.1 Organic Photochemistry, 3
1.2 Electronic States and Potential Energy Surfaces, 8
    1.2.1 Introducing and Visualizing the Surfaces, 8
    1.2.2 Nuclear Motion on Surfaces, 13
1.3 The Electronic Wave Function, 23
    1.3.1 The Spin Part, 23
    1.3.2 The Space Part, 29
    1.3.3 Classification of Excited States, 43
1.4 Comments and References, 48

2 Photophysical and Photochemical Processes — 50

2.1 An Overview, 50
2.2 Intramolecular Processes, 56
    2.2.1 Initial Excitation, 56
    2.2.2 The Fast Events, 64
    2.2.3 The Slow Events, 67
2.3 Intermolecular Processes, 77
    2.3.1 Initial Excitation, 78
    2.3.2 Molecular Transport Processes, 80
    2.3.3 Electronic Transfer Processes, 81
2.4 Back in the Ground State, 94
2.5 Analysis of Kinetic Data, 95
2.6 A Simple Example: Intermolecular Proton-Transfer Equilibria, 98
2.7 Another Example: 1,4-Dewarnaphthalene, 100
2.8 A Summary: The Need for Theory, 104
2.9 Comments and References, 105

3 Location of Minima, Funnels and Barriers on Surfaces — 109

3.1 General, 109
3.2 Correlation Diagrams, 112
   3.2.1 AO Correlation, 112
   3.2.2 VB Structure and State Correlation, 115
   3.2.3 MO Correlation, 119
   3.2.4 Configuration and State Correlation, 127
3.3 The Use of Interacting Subunits in the Construction of Correlation Diagrams, 135
3.4 Comments and References, 139

PART B: ELEMENTARY PHOTOCHEMICAL STEPS — 141

4 Two-Center Reactions: One Active Orbital per Atom — 143

4.1 General: Introducing the Two-Electron Two-Orbital Model, 143
4.2 Nonpolar Bonds, 147
   4.2.1 Sigma Interactions, 148
   4.2.2 Pi Interactions, 157
4.3 Polar Bonds, 161
   4.3.1 Sigma Interactions, 161
   4.3.2 Pi Interactions, 165
4.4 Biradicals, 169
   4.4.1 Fundamentals, 169
   4.4.2 Perfect Biradicals, 175
4.5 Biradicaloids, 179
   4.5.1 Fundamentals, 179
   4.5.2 Homosymmetric Biradicaloids, 182
   4.5.3 Heterosymmetric Biradicaloids, 184
   4.5.4 Nonsymmetric Biradicaloids, 187
4.6 From Biradicals to Bonds and Lone Pairs: Surface Shapes, 187
   4.6.1 Sigma Interactions, 189
   4.6.2 Pi Interactions, 192
   4.6.3 Limitations of the Simple Model, 196
   4.6.4 Spin–Orbit Coupling in Biradicals and Biradicaloids, 198
4.7 Beyond the Simple Model: Photodissociation of the Single Bond, 200
   4.7.1 Covalent Sigma Bond, 200
   4.7.2 Charged Sigma Bond, 205
   4.7.3 Dative Sigma Bond, 206

4.8 Beyond the Simple Model: Photoisomerization of the Double Bond, 206
    4.8.1 Covalent Pi Bond, 206
    4.8.2 Charged Pi Bond, 217
    4.8.3 Dative Pi Bond, 219
    4.8.4 Pi Bond in the Field of a Charge, 222
4.9 Comments and References, 223

5 Cyclic Multicenter Reactions: One Active Orbital per Atom   225

5.1 Electrons in a Cyclic Array of Orbitals, 226
    5.1.1 Two Nonpolar Bonds, 226
    5.1.2 Two Polar Bonds, 244
    5.1.3 Other Pericyclic Ring Sizes, 246
5.2 Electrocyclic Reactions, 248
    5.2.1 Interconversion of Butadiene and Cyclobutene, 248
    5.2.2 Electrocyclic Processes in Polycyclic Systems, 256
    5.2.3 Electrocyclic Ring Opening in Oxirane, 258
5.3 Bond-Shift Reactions, 260
    5.3.1 Sigmatropic Reactions, 262
    5.3.2 Pseudosigmatropic Reactions, 263
5.4 Cyclic Addition and Reversion Reactions, 263
    5.4.1 Cycloaddition and Cycloreversion, 264
    5.4.2 Excimers, 274
5.5 Comments and References, 285

6 Linear Multicenter Reactions: One Active Orbital per Atom   287

6.1 Electrons in a Linear Array of Orbitals, 287
6.2 Covalent Bonds, 292
    6.2.1 Allylic and Benzylic Bond Cleavage, 292
    6.2.2 Atom and Ion Transfer to Alkenes, 296
    6.2.3 Cis–Trans Isomerization: Butadiene, 302
    6.2.4 Cis–Trans Isomerization: Styrene and Stilbene, 310
    6.2.5 Nonconcerted Ring Opening of Cyclobutene, 314
    6.2.6 Nonconcerted Cycloaddition and Cycloreversion, 314
6.3 Charged Bonds, 317
    6.3.1 Bond Cleavage: Onium Salts, 318
    6.3.2 Atom and Ion Transfer to Charged Double Bonds, 319
    6.3.3 Charge Translocation: Acroleiniminium, 320
    6.3.4 Charge Translocation: Rhodopsin, 323
6.4 Dative Bonds and Other Donor–Acceptor Interactions, 328
    6.4.1 Donor–Acceptor Pairs: CT Complexes, Exciplexes, Ion Pairs, and Free Ions, 328

xii CONTENTS

        6.4.2 Sigma-Bonded Pi-Donor–Pi-Acceptor Pairs: TICT States, 338
  6.5 Comments and References, 345

7 Reactions with More Than One Active Orbital per Atom   **348**

  7.1 Pi-Bond Isomerization, 349
        7.1.1 Formaldimine, 350
        7.1.2 Acroleinimine, 359
        7.1.3 Acrolein, 365
        7.1.4 Diazene, 367
  7.2 Bond Dissociation, 373
        7.2.1 Tritopic. Amines, Alcohols, Norrish I, 374
        7.2.2 Tetratopic. Alkyl Halides, Peroxides, Azo Compounds, 380
        7.2.3 Pentatopic. Vinyl Halides, Ketenes, Diazoalkanes, 384
        7.2.4 Hexatopic. Azides, Bromine, 391
  7.3 Carbonyl Addition to Olefins, 397
  7.4 Atom and Ion Transfer, 398
        7.4.1 Intermolecular Transfer, 400
        7.4.2 Intramolecular Transfer. Phototautomerization, Norrish II, 405
  7.5 Three-Membered Ring Heterocycles: Ring-Opening and Fragmentation, 407
        7.5.1 Oxirane and Aziridine, 408
        7.5.2 Oxaziridine, 411
        7.5.3 Diazirine, 413
  7.6 Comments and References, 415

8 Epilogue   **417**

APPENDIXES

I **The MO and VB Methods Illustrated on the H_2 Molecule**   **421**

II **The Two-Electron Two-Orbital Model of Biradicals**   **433**

III **Computational Methods: SCF-CI and GVB-CI**   **442**

INDEX   **463**

NOTATION

Operators:	
one-electron	$\hat{a}$
many-electron	$\hat{A}$
Vectors:	**a, A**
Matrices:	a, A
Chemical species, structures:	$\mathfrak{A}$
Orbitals:	
general orthogonal	$\mathcal{A}$
most delocalized orthogonal	a
most localized orthogonal	A
most localized nonorthogonal	A
localized on a subunit	a
generalized valence-bond	α
Spinorbitals:	α

ELECTRONIC ASPECTS OF ORGANIC PHOTOCHEMISTRY

PART A
BACKGROUND

CHAPTER 1

Introduction

1.1 ORGANIC PHOTOCHEMISTRY

Vitiligo. The following ode on a plant can be found in the Indian sacred book, Atharva Veda (1400 B.C.):

Born by night art thou, O plant,
Dark, black, sable, do thou,
That art rich in color, Stain
This leprosy and white grey spots.
Even color is the name of thy mother,
Even color is the name of thy father,
Thou O plant producest even color
Render this (spot) to even color.

The disease had been dreaded for millenia before the Roman physician Celsus coined the term *vitiligo* for the cosmetic condition that, to this day, disfigures one out of roughly every hundred humans by forming large chalk-white patches on the skin, similar to the white blotches on a spotted calf (*vitelius* is Latin for calf). Throughout history, it made the lives of innumerable victims miserable by turning them into social outcasts.

The ancients had apparently worked at the problem, with some success. Section 877 of the Ebers papyrus (about 1550 B.C.) is believed to refer to the disease and suggests a remedy whose chief components are fly's dirt, powder of wheat, natron, beans, stibium, and oil. Atharva Veda (Book I, Hymns 23 and 24) goes into considerable detail in describing an herbal treatment. The plant whose curative powers are praised in the above ode most likely was Barachee (*Psoralea corylifolia*). Yet another plant, Aatrillal (*Ammi majus*), has been used for centuries in the Nile valley for the same purpose. According to the thirteenth-century author Ibn al Baytar, the medicinal power of the fruits of Aatrillal was discovered by the Berbers, who sold the drug but managed to keep its nature a secret at first (Figure 1.1).

What makes these herbal remedies interesting to a photochemist? All the sources carefully emphasize that the drugs are only potent when combined with sunshine: It is a photochemical reaction of the active principle in the skin that triggers the desired pigmentation.

Lest the power of ancient herbal medicine be idolized, we note that a yellowish-brown powder named Aatrillal was sold by Egyptian native herbalists as a remedy for vitiligo as recently as the first half of this century; this substance was used in much the same way as recommended by Ibn al Baytar: Daily doses of 4–12 grams,

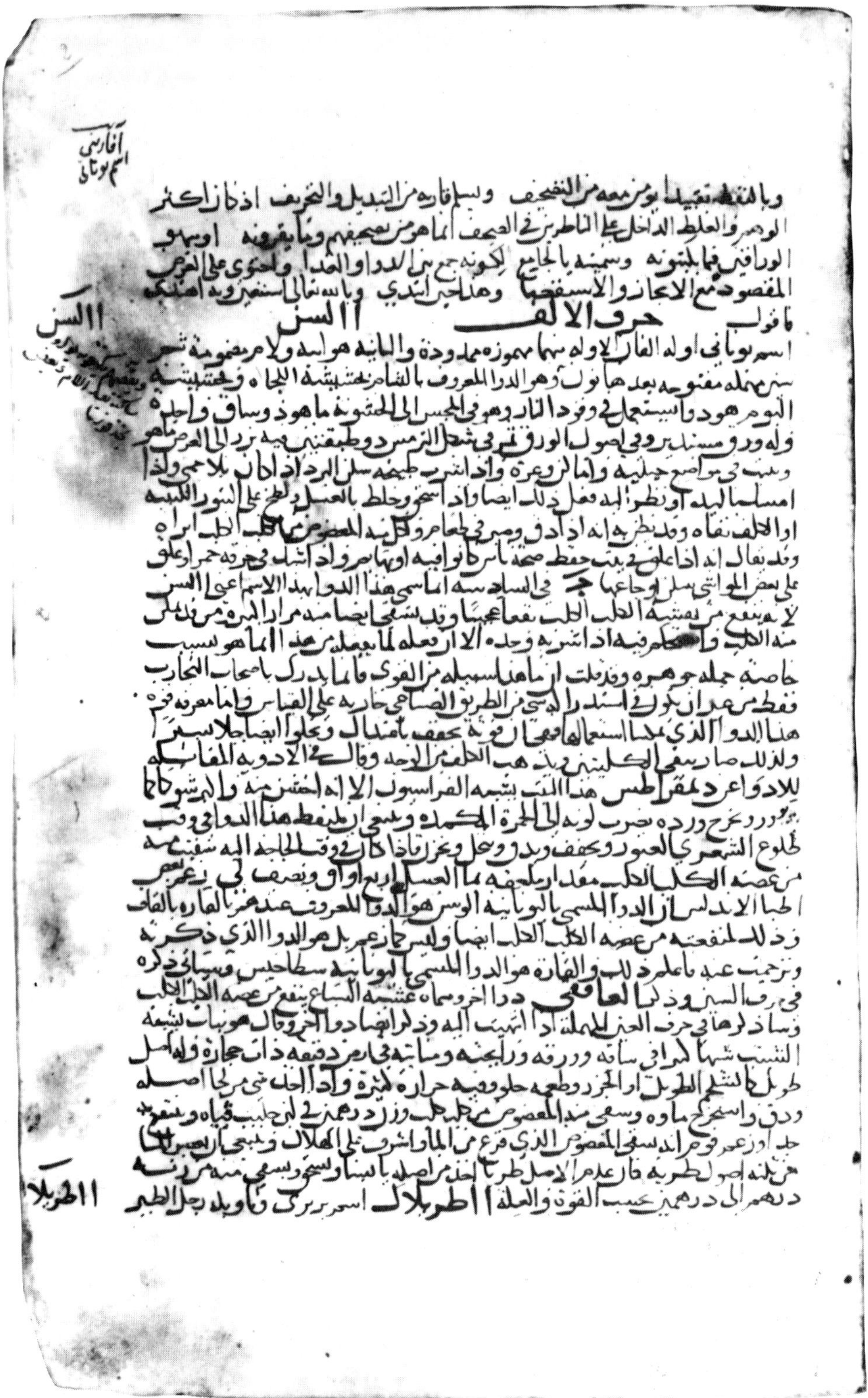

Figure 1.1 A medieval description of the curative powers of Aatrillal. Al-jāmi' li-mufradāt al-adwīya wal-aghdhīya (Materia Medica). By Abdullah Ibn Ahmad al-Andalusi al-Maliqi, known as Ibn al-Baytar [died 646 A.H. (1248 A.D.)]. Photograph purchased from the Chester Beatty Library, Dublin, Ireland.

followed by exposure of the affected patches to the sun until blistering occurred, typically accompanied by nausea and sometimes coma. Only robust patients came back for more treatment, particularly since eventual pigmentation of the white patches of skin by repeated treatment was by no means guaranteed!

Like the ancient Berbers, the dealers who sold Aatrillal would not reveal its nature. In the 1940s a research group at the University of Cairo, led by Professor Fahmy, identified it as powdered seeds of *Ammi majus* and subsequently isolated three crystalline furocoumarins from it. One of these, 5-methoxypsoralen, had been known as a constituent of the oil of bergamot, long used in the perfume industry. It is now suspected to be responsible for the photosensitivity sometimes associated with the use of certain perfumes.

The active principle of *Psoralea corylifolia* is the unsubstituted psoralen (**1**), and a fair number of variously substituted psoralens have now been isolated from many plants, including celery and parsnip. Clinical tests of psoralen for the treatment of vitiligo were initiated by El Mofty at the University of Cairo in 1948. Much success has been reported using both natural and synthetic psoralens with controlled exposure to sunshine or mercury lamp radiation. Under appropriate conditions, the drug has no unpleasant side effects and essentially no toxicity. It cures vitiligo in the majority of cases, but it often needs to be administered for many months or years.

1

The biological mechanism by which pigmentation is triggered by irradiation in the presence of psoralen is not understood. The initial photochemical process appears to involve a 2 + 2 cycloaddition of one or both psoralen C=C double bonds to the double bond of a uracil or a thymine constituent of DNA, into which psoralen intercalates and which thus becomes cross-linked. It has been stated repeatedly that this cycloaddition most likely occurs in the triplet state, but it has not been proven.

The subject of this book does not relate directly to the fascinating biochemical mechanism of induced pigmentation, nor does it relate to the intriguing mechanistic photochemistry of psoralene cycloaddition to DNA constituents. Yet by concentrating on the fundamental nature of elementary reaction steps in organic photochemistry, this text ultimately relates to both of the above.

The many facets of photochemistry. The treatment of vitiligo is only one example of organic photochemistry in action in medicine. Psoralens are also used in the phototherapy for psoriasis. Other examples are well known, such as the treatment of neonatal jaundice by irradiation. These, along with the wider area of photobiology, including vision, photosynthesis, and various kinds of response to light by organisms, represent extraordinarily complicated and immensely exciting subjects for research.

Many other areas of organic photochemistry beckon, from synthetic and preparative use to mechanistic investigations of often deep-seated molecular rearrangements undergone by fleeting reaction intermediates, and investigations of polymer degradation, smog formation, and chemiluminescence. Organic photochemistry ranges from studies of down-to-earth problems such as paint or dyestuff bleaching and photolability, and industrial utilization such as photochlorination, non-silver photography, and other types of optical information storage, photolithography and microelectronics, photopolymerization and polymer curing, to laboratory utilization such as photoaffinity labeling, to photoelectrochemistry and development of photochemical energy storage, to the search for new photoreactions and new structures, and to the purely academic and largely theoretical investigation of the ultimate underlying mechanism and physical nature of all of these processes.

Elementary photochemical mechanisms. By addressing only the last-named subject, we have chosen one which is still only rather poorly understood, but which is of fundamental importance, and whose understanding is essential for in-depth comprehension of any of the others: What are the elementary photochemical reaction steps, and how and why do they take place? Just what happens between the initial electronic excitation of an organic molecule by absorption of light and the appearance of the first ground-state product, which itself perhaps is only a fleeting intermediate?

We attempt to formulate the answers in terms of the basic postulates of physics, primarily of quantum mechanics. Whether the behavior of a living organism can be reduced to these basic postulates is another matter, but there is presently no reason to suspect that the individual elementary reaction steps in organic photochemistry cannot be so reduced. In the following, then, we shall consider a step understood if it is shown to follow from—or, more realistically, to be qualitatively compatible with—the quantum mechanical description of atomic and molecular systems.

Within the usual Born–Oppenheimer approximation, the understanding of an elementary photochemical step can be developed in two stages. The first stage deals with static aspects of molecular electronic structure and consists of estimating the shapes of the potential energy surfaces involved in the transformation—in particular the location of minima, funnels, and barriers. The second stage deals with the dynamics of the transformation. While the first stage relies totally on traditional quantum chemistry, the second stage requires the use of traditional quantum chemistry for the evaluation of nonadiabatic and spin–orbit coupling, as well as the additional tools of chemical dynamics, such as scattering theory, liquid theory, statistical mechanics, and so on.

The present text is dedicated to the static and structural aspects, and only the barest minimum of molecular dynamics is presented in Chapter 2 in order to permit a discussion of photochemical reactivity. We realize that it would be ideal to include both aspects of photochemical theory in a single book, but time limitations and consideration of volume size have made this impractical.

The organization of the material. Part A contains three chapters and provides both the necessary elementary background to those unfamiliar with those aspects of quantum chemical theory which are essential for the understanding of photochemical processes, and a superficial description of these processes in terms of potential energy surfaces. Readers well versed both in quantum theory and photophysics can proceed directly to Part B. In the four chapters of Part B we discuss specific elementary photochemical reaction steps. The coverage is representative but not comprehensive in that the elementary steps involved in many important processes such as the di-π-methane rearrangement and aromatic photosubstitution are not even mentioned. The discussion is largely limited to simpler steps for which *ab initio* computations of relatively good quality had been published or were practical to carry out. With a few exceptions that are noted, at least a double-zeta quality basis set and extensive configuration interaction (CI), or equivalent, were involved.

This does not mean that simple models have been avoided. To the contrary, we feel that results of complex calculations which cannot be rationalized in intuitive terms cannot really be said to have been understood, since not even qualitative predictions for related molecules can be made without additional complex calculations. We attempt to base much of the discussion on such simple models, since we feel that the physical insight they provide is of far longer lasting value than any numerical results obtained by a "state-of-the-art" method, which will be obsolete in a few years. However, we feel really comfortable using a simple model only when we can compare its results with those of a good-quality calculation, preferably "state-of-the-art," so that we understand its shortcomings and limitations—at least to some degree.

Although we have consulted many of the published calculations and have frequently obtained valuable guidance from them, we have ultimately decided to rely primarily on our own computations. This had the advantage of offering immediate access to all needed information on the wave functions, intermediate results, and so on. At the end of each chapter, we give a brief list of references that have been particularly useful to us. We make no attempt to provide an exhaustive coverage of the literature, and we apologize to the many authors of relevant articles that have not been listed.

Chapter 1 summarizes the basic characteristics of potential energy surfaces, electronic wave functions, and light absorption and emission. Chapter 2 provides an overview of the fundamental photophysical and photochemical processes. Chapter 3 deals with the location of the salient features of potential energy surfaces, minima, funnels, and barriers, and it relies heavily on correlation diagrams.

The following chapters cover specific elementary photochemical steps in the order of increasing complexity. In Chapters 4-6 we treat reactions in which each actively involved atom participates through only one of its valence orbitals. In Chapter 4, the relatively simple case of steps affecting one two-electron two-center bond is analyzed in considerable detail. This is the most mathematical chapter and is most likely to be found difficult by a reader unfamiliar with quantum theory. Once mastered, however, it provides a solid background for the remainder of the book. Multicenter bond-breaking and -making are treated in Chapters 5 (cyclic bond arrays, "concerted" reaction paths) and 6 (linear bond arrays, "nonconcerted" reaction paths). Finally, in Chapter 7, the more complicated but very important elementary processes in which at least one atom participates through two or more of its valence orbitals are taken up. Most often, these extra orbitals house "lone-pair electrons" in the ground state. Chapter 8 provides a brief survey of what has and what has not been covered in the book.

The text is provided with an Appendix that describes the basic computational procedures. Bonding in the various electronic states of a simple prototype of a single bond, H_2, is analyzed in Appendix I in terms of both the molecular orbital (MO) and valence bond (VB) methods. This material underlies the traditional qualitative thinking about chemical bonds. It will be familiar to theoreticians but ought to be useful for a beginner with no fear of elementary mathematics. The electronic states of molecules at biradicaloid geometries are analyzed in Appendix II using the simple 3×3 CI model. This provides more mathematical detail to the serious student than does the coverage offered in Chapter 4. Finally, Appendix III provides a reference guide to the common *ab initio* computational procedures. It is quite condensed and requires considerable prior knowledge of quantum chemistry.

We realize that the photochemical elementary steps can be classified in a variety of ways, but we have not actually adopted any one of the previously proposed formalisms for this purpose. We hope that the system used, which has its own shortcomings but which also has the virtue of accommodating nonpolar and polar bond-breaking and bond-making processes, charge translocation, and charge separation processes on equal footing, brings out some interrelations which otherwise might go unnoticed.

Throughout this book, we use units common among U.S. chemists. Their relation to SI units is 1 cal = 4.184 J, 1 Å = 0.1 nm, and 1 eV $\approx 1.602 \times 10^{-19}$ J.

1.2 ELECTRONIC STATES AND POTENTIAL ENERGY SURFACES

At present, and surely for a long time to come, the most useful conceptual framework for the discussion of chemical reactions is the separation of nuclear and electronic motion embodied in the Born–Oppenheimer approximation. This is true even for photochemical reactions, in which this approximation is often inadequate for the description of at least one of the important elementary events. Still, it represents a useful starting point to which corrections can be added.

For this reason, we shall open our discussion of the fundamentals of organic photochemistry by introducing the concept of potential energy surface. The surfaces have no objective existence and are not observable, but they represent very useful crutches for moving about in the inviting, but still only poorly charted, territory of photochemistry.

1.2.1 Introducing and Visualizing the Surfaces

Electronic and nuclear wave functions. In the Born–Oppenheimer approximation, it is assumed that the total wave function Ψ_{TOT} describing the overall state of a molecule can be written as a product of parts describing the state of various separate types of motion. Thus,

$$\Psi_{TOT} = \Psi_{trans} \times \Psi_{rot} \times \Psi_{vib} \times \Psi_{el} \times \Psi_{other} \tag{1.1}$$

where Ψ_{trans} describes the translational motion of the center of mass, Ψ_{rot} describes the rotational motion of the molecule as a whole in the laboratory coordinate system, Ψ_{vib} describes the intramolecular nuclear motion (i.e., vibrations, librations, and internal rotations), Ψ_{el} describes the motion of the electrons, and Ψ_{other} depends on all remaining variables, such as nuclear spins. The total energy can then be written as a sum of contributions from the various types of motion.

In organic photochemistry, Ψ_{vib} and Ψ_{el} are of particular concern. Their separability has its physical roots in the circumstance that nuclei are very much heavier than electrons. As the nuclei slowly move around, the nimble electrons are able to adjust their motion to the instantaneous position of the nuclei without much inertia. As a result, Ψ_{el} then only depends on the instantaneous positions of the nuclei and not on their previous positions (i.e., not on their velocities). For instance, in a diatomic molecule, Ψ_{el} is the same for a given internuclear separation regardless of whether that separation was reached by bringing the nuclei together or pulling them apart.

Since the inertia of the electronic motion is only very small rather than exactly zero, it is not surprising that there are circumstances under which the approximate nature of the Born-Oppenheimer description of molecules comes to light. Such a situation is typically reached at least once during a photochemical process and is thus of obvious importance to the photochemist. In simplest terms, it can be described as the opposite of the above-described behavior of electrons in a diatomic molecule: Because of their finite mass and thus finite inertia, the electrons "remember" what their motion was like in the immediate past so that their behavior will not be fully dominated by the nuclear positions alone and will also be slightly different depending on whether the nuclei are approaching each other or coming apart. Not surprisingly, the memory will be particularly strong if the nuclei move very rapidly and if the rate of change of Ψ_{el} with the internuclear separation is high, and we shall return to this matter in Section 1.2.2. More correctly, the breakdown of the Born-Oppenheimer approximation is described by allowing for the wave mechanical nature of the motion of the nuclei and is due to the perturbing action of the nuclear kinetic energy operator on the zero-order vibronic wave functions.

Nuclear configuration space. Assuming now that the Born-Oppenheimer approximation is valid, Ψ_{el} can be calculated once the relative positions of all the nuclei are given. If N nuclei are present, $3N - 5$ coordinates will be needed in a diatomic and $3N - 6$ in a polyatomic molecule ($3N$ degrees of freedom minus three for the translation of the center of mass, minus three or two for rotation of the molecule as a whole). These coordinates can be collected into a vector **R**. For instance, for the CO_2 molecule the two C—O distances $R(CO_I)$ and $R(CO_{II})$ and the O—C—O angle $\Theta(OCO)$ can be chosen as the coordinates. Each possible molecular geometry, or nuclear configuration, can then be represented by the vector $\mathbf{R} \equiv [R(CO_I), R(CO_{II}), \Theta(OCO)]$.

The set of all possible vectors **R** for a given molecule forms its nuclear configuration space. For CO_2 and all other triatomics, this space is three-dimensional. In general, it is $(3N - 6)$-dimensional. Each point in this space corresponds to one molecular geometry and a one-dimensional path through this space corresponds to a continuous change in molecular geometry, such as a classical vibration.

Each nuclear configuration **R** is associated with a point symmetry group characteristic of the geometry it describes. Diatomics are always of symmetry $C_{\infty v}$ or $D_{\infty h}$, and triatomics are at all geometries of symmetry C_s or higher. For higher polyatomics, the point group for almost all values of **R** is C_1, that is, no elements of symmetry other than the identity operation are present.

Additional symmetry is present if several of the nuclei are identical. Since their interchange has no physical effect, two vectors **R** related by such an interchange must be equivalent in all respects. For instance, in CO_2 the vectors $[r_1,r_2,180]$ and $[r_2,r_1,180]$ represent such a pair. As a result, the nuclear configuration space of each molecule is associated with a permutational symmetry group characteristic of the numbers of identical nuclei in the molecule it describes.

Potential energy surfaces. For any selected geometry **R** the electronic wave function can be determined by solving the Schrödinger equation for the motion of electrons in the field of the stationary nuclei:

$$\hat{H}_{el}(\mathbf{x}, \mathbf{R})\Psi_{el}(\mathbf{x}, \mathbf{R}) = E_{el}(\mathbf{R})\Psi_{el}(\mathbf{x}, \mathbf{R}) \tag{1.2}$$

where the vector **x** collects all the space (**r**) and spin (ζ) coordinates of all electrons. The electronic Hamiltonian operator $\hat{H}_{el}(\mathbf{x},\mathbf{R})$ contains all the factors that govern the motion of the electrons. It is represented by a sum of operators for the kinetic energy of each electron, for the energy of their mutual repulsion, for the energies of electron–nuclear attraction, for the energy of electron spin–orbit coupling, and possibly for other small electronic terms we may wish to include. For convenience, one includes also the nuclear–nuclear repulsion energy, which is a constant for any particular geometry **R** chosen.

In general, equation 1.2 has an infinite number of solutions (states) $\Psi_{el}^{(n)}(\mathbf{x}, \mathbf{R})$; each solution belongs to one irreducible representation of the point group associated with **R**, and each is associated with an energy $E_{el}^{(n)}(\mathbf{R})$. The energy spectrum is usually discrete at lower energies and continuous at higher energies. The discrete solutions are called (electronically) bound states. In such a state the molecule keeps all of its electrons and does not spontaneously ionize. The discrete solutions will be numbered in the order of increasing energy, $n = 1, 2, \ldots$. Solutions in the continuous part of the energy spectrum correspond to ionized states plus a free electron, or to autoionizing states, which eject an electron in a finite time.

Having found and numbered the discrete solutions at a geometry **R**, we can now move an infinitesimal distance away from it in any direction in the nuclear configuration space of the molecule ($3N - 6$ of these are mutually perpendicular) and repeat the calculation. As we continue the process, the point representing the energy of our solution number one, plotted against the position of **R** in its ($3N - 6$)-dimensional space, generates the ground-state surface of dimensionality $3N - 6$ located in a space which also contains energy as an additional dimension, so that its total dimensionality is $3N - 5$. This is sometimes referred to as a ($3N - 6$)-dimensional hypersurface in a ($3N - 5$)-dimensional hyperspace.

The energy of the solution number two will similarly produce a continuous lowest excited surface $E_{el}^{(2)}(\mathbf{R})$, and so on. By the nature of the labeling system, we have

1.2 ELECTRONIC STATES AND POTENTIAL ENERGY SURFACES

$E_{el}^{(1)} \leq E_{el}^{(2)} \leq E_{el}^{(3)} \leq \ldots$, so that the ground-state surface is everywhere below the lowest excited surface, and so on, except at points where they touch. A touching of two or three surfaces at **R** is demanded by symmetry if the corresponding wave functions belong to a doubly or triply degenerate representation of the point group of **R**. A surface touching also occurs frequently at points $\mathbf{R}_t$, where two wave functions belonging to different irreducible representations of the point group of $\mathbf{R}_t$, with energies $E_{el}^{(n)}$ and $E_{el}^{(n+1)}$, change their energetic order as **R** moves along a suitable path. It is also possible for surface touching to occur at points $\mathbf{R}_t$ for which both electronic wave functions belong to the same nondegenerate irreducible representation of the point symmetry group of $\mathbf{R}_t$. There is limited computational evidence for this, and it is usual to assume that in many cases a touching, even if "intended," is at least weakly avoided, so that the surfaces come quite close but do not actually touch. As we shall see later, this fine distinction has little practical impact, if any.

Representation of potential energy surfaces. Diatomics are the only molecules for which potential energy surfaces (in this case, curves) are easily visualized, and a typical plot of a few $E_{el}^{(n)}(\mathbf{R})$ curves is shown in Figure 1.2. Already for triatomics,

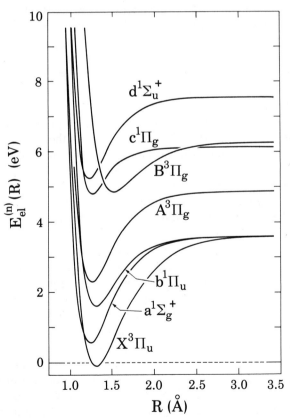

Figure 1.2 Potential curves of the electronic states of C_2.

where one would wish to plot the three-dimensional surfaces $E_{el}^{(n)}(\mathbf{R}) = E_{el}^{(n)}(R_1,R_2,R_3)$ in a four-dimensional space, one needs four-dimensional paper, and the situation gets worse rapidly for larger polyatomics. It is common to plot curves along selected one-dimensional paths in the nuclear configuration space, which then look like the curves for diatomics shown in Figure 1.2. More information can be displayed when surfaces above a selected two-dimensional subspace of the nuclear configuration space are drawn in perspective, as shown in Figure 1.3. Alternatively, one can give up plotting energy against $\mathbf{R}$ and instead show two-dimensional contour drawings (Figure 1.4). In these, two geometric variables, say $R_1 = \phi$ and $R_2 = \theta$, are chosen as axes. The others are held fixed (say at values R_3^0, R_4^0, . . .), and lines joining points of equal energy $E_{el}(R_1,R_2,R_3^0,R_4^0, . . . ,)$ are shown. With some practice, the contour maps are just as easy to read as the plots of energy against geometric coordinates, particularly for those familiar with topographical maps. The concept of a contour map can be generalized to display complete surfaces of triatomic molecules over their three-dimensional nuclear configuration space, or three-dimensional subspaces for larger molecules, as illustrated in Figure 1.5. Here, a perspective view is given of a nested set of two-dimensional equipotential surfaces defined on the coordinate system $R_1 = \phi$, $R_2 = \theta$, $R_3 = \omega$. Paths of low energy correspond to nested tubes, and minima correspond to nested egg shells.

In general, for molecules with four and more atoms, the best graphical display one can produce is to show the surfaces in one-, two-, or three-dimensional sub-

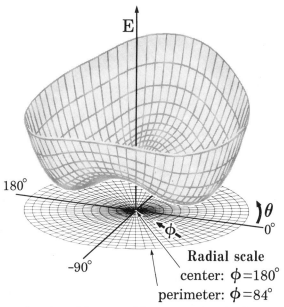

Figure 1.3 A perspective view of the potential energy surface for the ground state of acetone methylimine calculated by modified neglect of differential overlap (MNDO) as a function of the twist angle θ and the valence angle ϕ on nitrogen. The pyramidalization angle ω at the doubly bonded carbon is 40°. All other geometric variables are optimized.

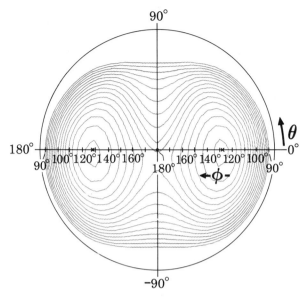

Figure 1.4 A contour plot of the potential energy surface of Figure 1.3. The first contour energy is 1 kcal above the minimum; further energy separation between contours is 3 kcal/mol.

spaces of the nuclear configuration space. The procedure has to be used with extreme caution: What may appear to be a minimum, barrier, or a saddlepoint in one subspace may turn out to be nothing of the kind when viewed in another cross-section.

One final remark is in order: We have been using the word "molecule" in a generalized sense to include the whole photochemical system under consideration, which may consist of one or several molecules (i.e., a "supermolecule"). Thus, the nuclear configuration space of CH_4 not only contains all geometries of what would conventionally be called methane but also contains entities such as $CH_3 + H$, $CH_2 + H_2$, $C + 2H_2$, and so on, which would normally be referred to as several atoms and/or molecules. In general, whenever we need to discuss a set of two molecules, one with N and the other with M atoms, we shall need to work with a $(3N + 3M - 6)$-dimensional nuclear configuration space. Of the dimensions, $3N - 6$ will correspond to the internal coordinates of one molecule, $3M - 6$ will correspond to those of the other, one will describe the distance between the centers of gravity of the two molecules, and five will be needed to describe their mutual orientation.

1.2.2 Nuclear Motion on Surfaces

In this section we provide a brief survey of the processes important for nuclear motion on potential energy surfaces, including jumps between surfaces. To illustrate qualitatively the nature of the quantities that dictate the rates of the processes, we

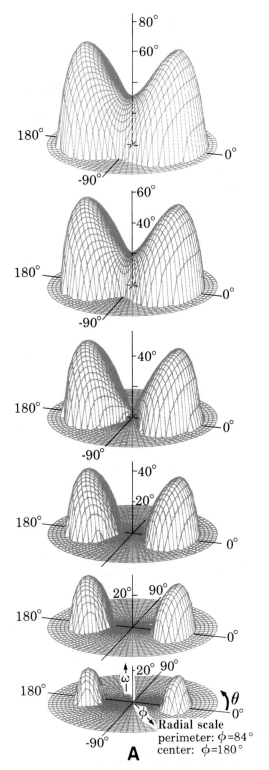

Figure 1.5 A perspective view [(A) exploded; (B) composite] of a nested set of equipotential surfaces for the ground state of acetone methylimine as a function of the twist angle θ, the valence angle ϕ on nitrogen, and the pyramidalization angle ω on the doubly bonded carbon (shown for formaldimine). Energy increase between successive surfaces, from bottom to top: 5 kcal/mol.

1.2 ELECTRONIC STATES AND POTENTIAL ENERGY SURFACES

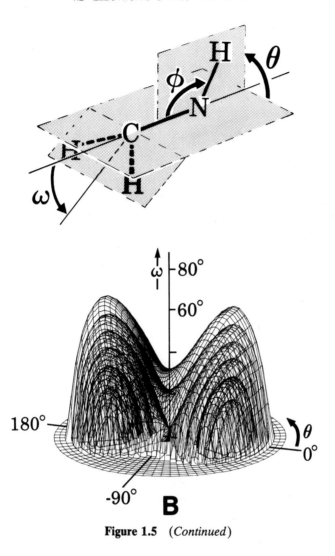

Figure 1.5 (*Continued*)

state the form of the corresponding matrix elements but do not provide any justification or derivation nor any mathematical details. Additional information on some of these is provided in Chapter 2, but they are considered mostly beyond the scope of this book, which concentrates on the electronic aspects of photochemistry and not on the rate processes. The interested reader is referred to Section 1.4 for leading references to further reading material.

Motion on a single surface

The classical viewpoint. With a suitable coordinate choice, the motion of nuclei in a molecule which is in the n-th electronic state, specified by the surface $E_{el}^{(n)}(\mathbf{R})$ and the associated wave function $\Psi_{el}^{(n)}(\mathbf{x},\mathbf{R})$, will be governed by the shape of the surface as in a classical trajectory motion of a marble rolling without friction on the surface. The starting point will be defined by the way in which the initial state was prepared; in a photochemical process, this is frequently a transfer from another surface by electronic excitation. The forces acting on the nuclei at any moment will be given by the negative gradient of the surface at the point $\mathbf{R}'$ describing the molecular geometry at that moment. This is dictated by the matrix element

$$\langle \Psi_{el}^{(n)}(\mathbf{x},\mathbf{R}) | \nabla_\mathbf{R} \hat{H}_{el}(\mathbf{x},\mathbf{R}) | \Psi_{el}^{(n)}(\mathbf{x},\mathbf{R}) \rangle_{\mathbf{R}=\mathbf{R}'}$$

The total energy of the molecule will remain constant during this process unless it interacts with its environment, but the fraction that is present as the kinetic energy of the nuclei will be changing continually. It often happens that the gradient is negligibly small for certain types of nuclear motion, i.e., that the surface is essentially flat and horizontal. This is of little consequence if such motion corresponds to an essentially free rotation of some methyl substituent in an inessential part of the molecule which acts only as a spectator in the photochemical reaction. However, a description of reaction in the direction of a zero gradient may also be essential for a comprehension of the photochemical process. This is the case for the six degrees of freedom of a supermolecule which correspond to changes in the separation and in the mutual orientation of two widely separated component molecules. This type of motion is described by the theories of free translational and rotational diffusion.

The quantum mechanical viewpoint. A more accurate description of the intramolecular nuclear motion requires the recognition of the wave mechanical nature of the nuclei. The nuclear wave function of the initially prepared state, usually fairly localized around some initial geometry (wave packet), will develop in time under the effect of the potential $E_{el}^{(n)}(\mathbf{R})$ in a way dictated by the time-dependent Schrödinger equation. It will generally spread out, perhaps separate in two or more packets, and move against the gradient.

An approximate description of this process can be obtained by performing classical trajectory calculations for the motion of a marble on the surface and averaging over a large number of possible initial positions and velocities. A schematic example of such a trajectory is shown in Figure 1.6A.

The thermal bath. Since organic photochemical reactions are usually carried out in solution, the nuclear motion is complicated tremendously by collisions with the surrounding molecules. These lead to a rapid exchange of the energy of nuclear motion between the reacting molecule and its neighbors, which can be viewed as a part of an infinite thermal bath.

Such vibrational energy exchange with the environment should, in principle, be included in the description of nuclear motion, whether it be wave mechanical or classical. It will generally lead to a rapid thermalization of the vibrational motions within the initially prepared molecular species. Figure 1.6B shows schematically how the trajectory that describes the internal motion of a molecule changes when nuclear kinetic energy is continuously removed by the environment.

1.2 ELECTRONIC STATES AND POTENTIAL ENERGY SURFACES 17

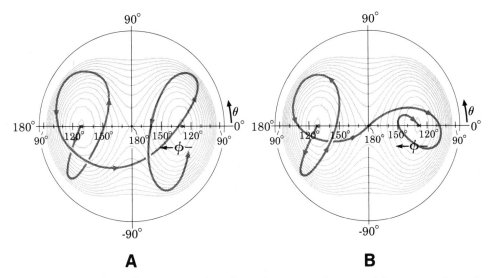

Figure 1.6 A schematic representation of a trajectory on the potential energy surface of Figure 1.3: (A) in an isolated molecule; (B) with continuous loss of vibrational energy to the environment.

Vibrational wave functions. Stationary states for the nuclear motion of an isolated molecule in the n-th electronic state are found by solving the time-independent Schrödinger equation

$$\hat{H}_{nucl}(\mathbf{R})\Psi^{(n)}_{vib}(\mathbf{R}) = E\Psi^{(n)}_{vib}(\mathbf{R}) \tag{1.3}$$

where the nuclear Hamiltonian $\hat{H}_{nucl}$ contains the nuclear kinetic energy operator and the potential energy operator $E^{(n)}_{el}(\mathbf{R})$ for the nuclear motion. The latter contains the kinetic and potential energy of the electrons as well as the nuclear repulsion energy as a function of $\mathbf{R}$ and is nothing but the potential energy surface discussed in Section 1.2.1. If the surface $E^{(n)}_{el}(\mathbf{R})$ contains sufficiently deep minima, there will be one or more (vibrationally) bound levels with nuclear wave functions $\Psi^{(n)}_{vib,v}(\mathbf{R})$ localized in each of these minima, corresponding to the various vibration levels v of a stable molecule in the n-th electronic state. At thermal equilibrium, these levels will be occupied according to Boltzmann's distribution law.

In the limit of small displacements from the minimum of the well $\mathbf{R}_{min}$, the potential energy surface $E^{(n)}_{el}(\mathbf{R})$ can be approximated by a quadratic (harmonic) potential, leading to the usual expressions for normal modes, frequencies, and zero-point energies. The symmetries of the normal modes will be dictated by the point-group symmetry associated with the geometry $\mathbf{R}_{min}$ (in a more general case, by the permutation symmetry associated with the nuclear configuration space).

Tunneling. If several minima are present, the stationary vibrational wave functions $\Psi^{(n)}_{vib,v}$ will generally not be strictly localized in only one of them. Rather, weak or strong tails will reach into the areas of other minima (Figure 1.7). If the molecule is initially prepared in such a fashion that its nuclear wave function Ψ_{vib} is fully

18 INTRODUCTION

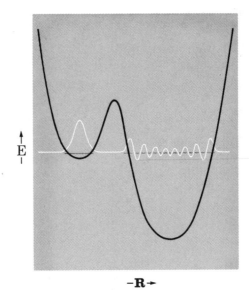

Figure 1.7 A delocalized vibrational wave function in a double-well potential.

localized in one of several minima so that it is not an eigenfunction of the nuclear Hamiltonian (i.e., is nonstationary), tunneling from one minimum to other energetically accessible minima will occur. The matrix element that dictates the rate of tunneling from a state localized in one local minimum and described by Ψ_{vib} to a state localized in another local minimum and described by Ψ'_{vib} is

$$\langle \Psi'_{vib}(\mathbf{R}) | \hat{H}_{nucl}(\mathbf{R}) | \Psi_{vib}(\mathbf{R}) \rangle$$

Often, this process is too slow to have much importance in practice at room temperature. It can be detected occasionally at very low temperatures, since then the otherwise competing thermally activated travel over the barriers separating the minima is suppressed.

Thermal activation. At this point, we again need to recognize the existence of the solvent. Collisions with neighboring molecules will tend to establish thermal equilibrium with the heat bath. Therefore, the nuclei of some molecules will acquire considerable kinetic energy. The molecule will then explore a larger region of the potential surface surrounding $\mathbf{R}_{min}$, and for that fraction of molecules that have been energized sufficiently this will sooner or later provide an opportunity to spill over a saddlepoint and escape from the well surrounding $\mathbf{R}_{min}$, as illustrated in Figure 1.6B. This disturbs the thermal equilibrium in the well, but usually to a negligible degree, and the rate of the process is usually adequately described by Eyring's transition state theory. This thermally activated mechanism for moving from one minimum to another is how chemical reactions on the $E_{el}^{(n)}(\mathbf{R})$ surface usually occur. This description applies to reactions in the general sense, including those which result in no net chemical change, such as automerization, interconversion of conformers, and so on.

Reactions of this type, which occur on a single potential energy surface, are called adiabatic. Although, by their very nature, overall photochemical reactions generally involve motion on more than one potential surface, many of the elementary steps are adiabatic and can be understood in the same way as ordinary ground-state reactions.

Jumps between surfaces. An event that has a profound effect on nuclear motion is a change of the electronic state, referred to as a jump between electronic surfaces. After such a jump, the forces acting on the nuclei are governed by the new surface. The jump can occur either as a result of absorption or emission of light or in a radiationless fashion.

Radiative jumps. The matrix element that dictates the rate of transition from the n-th electronic state to the m-th electronic state by light absorption or emission is the transition dipole moment

$$\langle \Psi_{\text{TOT}}^{(m)}(\mathbf{x},\mathbf{R}) | \hat{\mathbf{M}}(\mathbf{r},\mathbf{R}) | \Psi_{\text{TOT}}^{(n)}(\mathbf{x},\mathbf{R}) \rangle$$

where $\hat{\mathbf{M}}$ is the electric dipole moment operator that approximates the effects of the outside electromagnetic field. Radiative transitions are discussed in more detail in Section 2.2.1.

Radiationless jumps. In order to understand radiationless jumps, we need to recognize the merely approximate nature of the description of a molecule as being in one specific electronic state whose shape governs the motion of its nuclei. Such a description is correct within the limits of the Born–Oppenheimer approximation but is not exact. We have already indicated the qualitative consequences of the nonvanishing inertia of the electronic motion: If $\Psi_{\text{el}}^{(n)}(\mathbf{x},\mathbf{R})$ changes as a function of $\mathbf{R}$ at all, nuclear motion on the surface $E_{\text{el}}^{(n)}(\mathbf{R})$ at any nonzero velocity will cause the real electronic motion at geometry $\mathbf{R}$ to be different from that described by the Born–Oppenheimer wave function $\Psi_{\text{el}}^{(n)}(\mathbf{x},\mathbf{R})$. The reason is that some memory will survive of the somewhat different electronic motion at the immediately preceding geometries, described by $\Psi_{\text{el}}^{(n)}(\mathbf{x},\mathbf{R} - d\mathbf{R})$.

The matrix element that dictates the jump probability from the n-th to the m-th electronic state is

$$\langle \Psi_{\text{vib},v'}^{(m)}(\mathbf{R}) \Psi_{\text{el}}^{(m)}(\mathbf{x},\mathbf{R}) | \nabla_{\mathbf{R}}^2 \Psi_{\text{el}}^{(n)}(\mathbf{x},\mathbf{R}) + 2[\nabla_{\mathbf{R}} \Psi_{\text{el}}^{(n)}(\mathbf{x},\mathbf{R})] \cdot \nabla_{\mathbf{R}} | \Psi_{\text{vib},v}^{(n)}(\mathbf{R}) \rangle$$

where the operator between the vertical bars represents the difference between the full and the Born–Oppenheimer Hamiltonians.

Thus, we see that the nuclear motion and the electronic motion are coupled by the inertia of the electrons. Alternatively, we can say that nuclear motion mixes other electronic states into the dominant one, $\Psi_{\text{el}}^{(n)}$. Such mixing will tend to be strong if other electronic states lie nearby in energy, and this can be seen clearly in an alternative form for the above matrix element, in which the $\nabla_{\mathbf{R}}^2$ term has now been neglected, as is customary unless the $\nabla_{\mathbf{R}}$ term vanishes by symmetry:

$$\langle \Psi_{\text{vib},v'}^{(m)}(\mathbf{R}) | [E_{\text{el}}^{(n)}(\mathbf{R}) - E_{\text{el}}^{(m)}(\mathbf{R})]^{-1} \langle \Psi_{\text{el}}^{(m)}(\mathbf{x},\mathbf{R}) | \nabla_{\mathbf{R}} \hat{H}_{\text{el}}(\mathbf{x},\mathbf{R}) | \Psi_{\text{el}}^{(n)}(\mathbf{x},\mathbf{R}) \rangle \cdot 2\nabla_{\mathbf{R}} | \Psi_{\text{vib},v}^{(n)}(\mathbf{R}) \rangle$$

Since we wish to use the Born–Oppenheimer language of motion on electronic surfaces, we clearly need to allow for the possibility that a molecule may jump from one to another such surface. During such a jump, energy is exchanged between nuclear motion and electronic motion. Processes involving a jump of this kind are called nonadiabatic (a linguistically more correct but less frequently used expression is "diabatic"; sometimes a subtle distinction is made between the meanings of the two).

For the photochemist, the introduction of radiationless jumps between different surfaces is the most important consequence of the coupling of electronic states by nuclear motion. The theory of radiationless transitions is relatively complicated and only the qualitative aspects of the results will be mentioned.

Strong coupling. In the strong coupling case, a molecule whose internal nuclear motion is initially quite well described as being governed by the n-th potential energy surface suffers a brief period of confusion, in which the Born–Oppenheimer approximation is poor. Usually, it then emerges in a state that is again relatively well described by a single potential energy surface, perhaps still the n-th one or perhaps the m-th one. In the latter case, we say that the molecule jumped from the n-th to the m-th surface.

Such a situation typically occurs when the molecular geometry passes a region where two potential energy surfaces touch or nearly touch. At geometries located in the immediate vicinity of the touching, the rate of change of the electronic wave function with a change in **R** is then very large. Under such conditions, the electrons may find it impossible to change their motions fast enough to accommodate the demands of the changing nuclear geometry **R**; in the extreme limit, the molecule will jump between surfaces each time it crosses the area of the touching. Such a jump probability of unity is realized for those paths in the nuclear configuration space which actually lead through a point of surface touching, $\mathbf{R}_t$, at which the lower and the upper electronic wave function mutually exchange their nature (Figure 1.8). The rate of change of the electronic wave function with **R** is then infinite at $\mathbf{R}_t$, and the electrons are given no time to adjust their motion. Invariably, the molecule will then proceed through the touching area by changing surfaces. To specify that the molecule has arrived at geometry $\mathbf{R}_t$ (say, on the first excited-state surface) is quite insufficient for predicting its future behavior: One must also specify from which direction it arrived.

As shown in Figure 1.8, there is a continuous spectrum of possibilities, from the most extreme strong coupling case in which the probability of a jump is unity, through paths along which one encounters a weakly avoided touching, where the probability of a jump is high but not quite unity, to paths that involve a more strongly avoided touching and less probability for a jump. The end of this spectrum of possibilities is known as the case of weak coupling, in which the jump probability is extremely low; this is discussed below.

Even in the extreme case of unavoided touching (curve crossing, Figure 1.8A), the crossing surfaces communicate through perturbations not discussed here, such as rotational motion. Nearly always, suitable geometric distortions will cause the touching (crossing) to be avoided anyway.

Weak coupling. In the weak coupling case, a molecule moves in some vibrational level of an excited surface in a region in which it roughly parallels a lower surface

1.2 ELECTRONIC STATES AND POTENTIAL ENERGY SURFACES

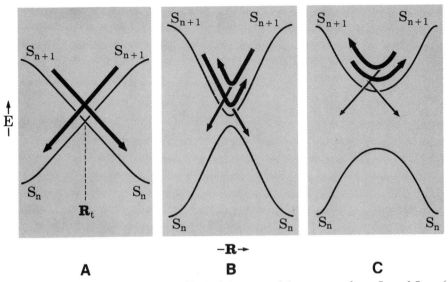

Figure 1.8 Strong coupling—a touching of the potential energy surfaces S_n and S_{n+1}: (A) unavoided; (B) weakly avoided; (C) strongly avoided. Jump probabilities for molecules arriving from the left and from the right are indicated by line thickness.

(Figure 1.9). Each time it undergoes a vibration, there is a tiny but nonvanishing probability that it will jump to an isoenergetic vibrational level of the lower surface. In terms of the above qualitative anthropomorphic discussion, the molecule can be said to be a little confused about its electronic state all the time. After the jump, its motions will be governed by the lower surface. The lost electronic energy is imparted to the nuclei in the form of kinetic energy, preferably in the smallest

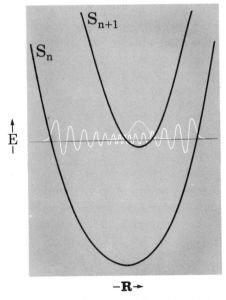

Figure 1.9 Weak coupling—two vibrational wave functions with nearly equal total energies for two almost parallel electronic surfaces.

possible number of vibrational quanta, i.e., in high-frequency vibrational modes. Such conversion of electronic energy into the energy of nuclear motion becomes harder as the amount to be converted increases and as the rate of the process decreases roughly exponentially with the energy gap between the electronic states involved.

Thermal excitation of electronic states. Both in the strong and in the weak coupling case, a jump from one of the levels of the lower electronic surface to an isoenergetic level of an upper surface is also possible but generally has a lower probability. In thermal equilibrium at ordinary temperatures the Boltzmann factor will be quite unfavorable unless the two surfaces have nearly the same energy. Statistics favor the lower electronic state: There are many more ways in which its nuclear motion can accommodate the amount of energy available. In other words, the density of vibronic states at a given total energy generally is many orders of magnitude higher in the lower electronic state than in the upper one. Jumps from the lowest-energy surfaces to higher-energy ones are important in thermal reactions that produce electronically excited products. The phenomenon of light emission from such products is known as chemiluminescence.

Returning now to the more frequent case of jumps from a higher- to a lower-energy surface, we conclude that once the jump to the lower surface occurs, the molecule rarely ever returns to the isoenergetic vibronic level of the higher electronic state. Statistically, the event is improbable, and at ordinary temperatures the vibrationally hot molecules freshly born in the lower electronic state will rapidly come to thermal equilibrium with their solvent environment, making a return to the upper electronic state energetically most unlikely.

Funnels. The regions of touching (or of weakly avoided touching) of two surfaces which provide easy communication between them, with significant jump probability already on the first passage, are referred to as funnels in the upper state, since they effectively suck up molecules from the upper surface and return them to the lower surface. Of course, they promote equally well the travel of sufficiently energetic molecules from the lower to the upper surface. Such a funnel may (Figure 1.10A) or may not (Figure 1.10B) correspond to an actual minimum in the upper surface, depending on the slopes of the touching surfaces. In a more general sense, the expression "funnel" is sometimes used to refer to all minima in the upper surface which provide easy radiationless entry to a lower surface, even if there is time for thermal equilibration in the minimum and concomitant loss of memory of the original direction of arrival.

Special cases of radiationless jumps: Energy transfer and electron transfer. In the case of a supermolecule system composed of two molecules separated well enough to still keep their identity, certain types of jumps between surfaces deserve special mention. The electronic states of such a supermolecule can be described approximately in terms of the states of its two constituents. In its S_0 ground state, both components are in their respective ground states. In its "locally excited" states, one component is in its ground state while the other is in its excited state. In its "charge-transfer excited" states, one component has lost an electron whereas the

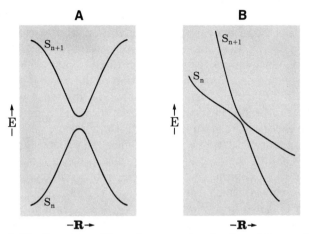

Figure 1.10 Examples of funnels connecting the S_n and S_{n+1} surfaces.

other has acquired it. In "doubly excited" states, both components are electronically excited.

As long as this approximate description is applicable, radiationless jumps between these various kinds of states can be classified as intramolecular and intermolecular. Intramolecular jumps are those in which electronic motion is converted into a light quantum or into nuclear motion within the same component, and vice versa. These are the same processes that can also occur in each component separately when it is isolated. Intermolecular jumps involve an exchange of electronic energy between the two components and are characteristic of the supermolecule.

The matrix element that dictates the rate at which these jumps occur is

$$\langle \hat{\mathscr{A}} \Psi_{\text{TOT}}^{\mathfrak{A},n'} \Psi_{\text{TOT}}^{\mathfrak{B},m'} | \hat{H}_{\text{el}} | \hat{\mathscr{A}} \Psi_{\text{TOT}}^{\mathfrak{A},n} \Psi_{\text{TOT}}^{\mathfrak{B},m} \rangle$$

where $\Psi_{\text{TOT}}^{\mathfrak{A},n}$ and $\Psi_{\text{TOT}}^{\mathfrak{B},m}$ are the localized wave functions describing the initial states of the components $\mathfrak{A}$ and $\mathfrak{B}$, $\Psi_{\text{TOT}}^{\mathfrak{A},n'}$ and $\Psi_{\text{TOT}}^{\mathfrak{B},m'}$ are localized wave functions describing their final states, and $\hat{\mathscr{A}}$ is the antisymmetrizing operator.

If the initial and final states of each component have the same number of electrons, the jump represents an energy transfer between the two components. If they do not, the jump represents an electron transfer.

1.3 THE ELECTRONIC WAVE FUNCTION

1.3.1 The Spin Part

Unlike a nucleus, the electron is so light a particle that its motion within a molecule cannot be handled even approximately by classical means, and a wave mechanical description is imperative from the outset. Not only is it normally unproductive to think in terms of classical trajectories for its motion, but it is also impossible to ignore the fact that it has spin. In contrast, we were able to ignore nuclear spin in our description of nuclear motion with relative impunity.

Electron spin. Each electron in a molecule has four degrees of freedom whose description is collected in the four-dimensional coordinate vector **x**. Three of the coordinates, x, y, and z, describe its position in ordinary space and together represent a three-dimensional vector **r**. The fourth coordinate is known as the *spin coordinate* ξ. Its value describes the orientation of the angular momentum vector associated with the electron—that is, its spin. The existence of electron spin is often considered to be nonclassical in that it is a logical consequence of the application of the theory of relativity, although it can also be derived from classical electromagnetic theory. The spin angular momentum is associated with a magnetic dipole moment. Unlike the coordinates x, y, and z, which can acquire an infinite number of values in a continuous fashion, the spin coordinate of an electron can only acquire two values, which can be chosen as $1/2$ and $-1/2$. If the state of the electron is such that it is known with certainty that its spin coordinate ξ is equal to $1/2$, then the projection of the spin vector into the z axis is $\hbar/2$, where $\hbar = h/2\pi$ and h is Planck's constant (Figure 1.11). The wave function of this state is called $\alpha(\xi)$. If the state of the electron is such that $\xi = -1/2$ with certainty, then the z projection of the spin vector is $-\hbar/2$, and this state is called $\beta(\xi)$. The functions $\alpha(\xi)$ and $\beta(\xi)$ are eigenfunctions of the z component of the spin angular momentum operator $\hat{s}_z$. An infinite number of other states are possible, in which a measurement of the spin coordinate has some nonzero probability to yield either the value $1/2$ or $-1/2$ and which are described by linear combinations of the functions α and β. In each of these states, the projection of the spin angular momentum vector into some particular direction in space (the quantization direction) has the sharp value $\hbar/2$ or $-\hbar/2$. For instance, the spin function $[\alpha(\xi) + \beta(\xi)]/\sqrt{2}$ describes a state in which the angular momentum vector of electron spin has a projection of $\hbar/2$ into the x direction. This, then, is an eigenfunction of the x component of the spin angular momentum operator $\hat{s}_x$. A measurement of the z component of the vector, however, will yield values $\hbar/2$ and $-\hbar/2$ with equal probabilities. The actual length of the spin vector is $\hbar\sqrt{3}/2$, and its direction changes in time about its quantization direction with the Larmor frequency.

It is useful to remember the rules for the action of the three components of the spin angular momentum operator, $\hat{s}_x$, $\hat{s}_y$ and $\hat{s}_z$, on the spin functions $\alpha(\xi)$ and $\beta(\xi)$. This can be summarized in matrix form (Pauli matrices):

$$\begin{array}{ccc} \hat{s}_x & \hat{s}_y & \hat{s}_z \\ \begin{array}{cc} \alpha & \beta \end{array} & \begin{array}{cc} \alpha & \beta \end{array} & \begin{array}{cc} \alpha & \beta \end{array} \\ \begin{array}{c} \alpha: \\ \beta: \end{array} \frac{\hbar}{2}\begin{pmatrix} 0 & 1 \\ 1 & 0 \end{pmatrix} & \frac{\hbar}{2}\begin{pmatrix} 0 & -i \\ i & 0 \end{pmatrix} & \frac{\hbar}{2}\begin{pmatrix} 1 & 0 \\ 0 & -1 \end{pmatrix} \end{array}$$

For instance, $\hat{s}_x\alpha(\xi) = (\hbar/2)\beta(\xi)$, $\hat{s}_y\beta(\xi) = (i\hbar/2)\alpha(\xi)$, etc.

Multiplicity. The spin angular momenta of the many electrons present in an atom or a molecule interact with each other and also with the motion of the electrons in ordinary space (i.e., with the "orbital motion"). In light atoms and in molecules composed of light atoms, it is best to couple the spin angular momenta of all the electrons into a total spin vector and to consider the interaction with the orbit motion (i.e., spin–orbit interaction) as a correction to be introduced subsequently. This is accomplished by adopting the electronic Hamiltonian $\hat{H}_{el}(\mathbf{R})$ which does

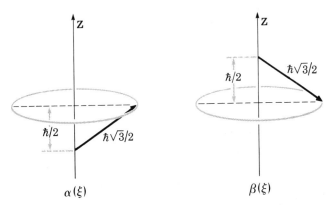

Figure 1.11 The spin angular momentum vector of an electron in spin states α and β.

not reflect any of the magnetic properties associated with electron spin. In an atom, an electron configuration such as p² will then yield several terms of different energies, but in this approximation all levels within each term will be degenerate. In polyatomic molecules, it is customary to refer to terms as states.

The length of the total electron spin vector is dictated by the way in which the spins of the individual electrons are coupled. Its only possible values are $\hbar\sqrt{S(S+1)}$, where S is an integral multiple of 1/2. The quantity 2S + 1 is called the multiplicity: singlet (S = 0), doublet (S = 1/2), triplet (S = 1), etc. The multiplicity reflects the number of possible values that the projection of the total spin vector into the quantization direction can have. If the total number of electrons is odd, the multiplicity must be even. The example shown in Figure 1.11 corresponds not only to the case of a single electron but also to the doublet state of any system with an odd number of electrons. The length of the spin vector for a doublet state is $\hbar\sqrt{3}/2$, and there are two components with projections of $\hbar/2$ and $-\hbar/2$ into the quantization direction. If the total number of electrons is even, the multiplicity must be odd. The length of the spin vector for a singlet state is zero, and there is only one component. Its length for a triplet state is $\hbar\sqrt{2}$, and there are three components with projections $\hbar$, 0, and $-\hbar$ into the quantization direction (Figure 1.12), and so on.

The potential energy surface defined by the eigenvalue $E_{el}^{(n)}(\mathbf{R})$ for a molecule

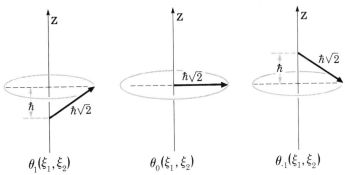

Figure 1.12 The spin angular momentum vector for the three components of a triplet state (Θ_{-1}, Θ_0, and Θ_1).

as a function of its geometry **R** is referred to as a singlet surface, or triplet surface, and so on. It is customary to denote the lowest singlet surface and its electronic wave function $\Psi_{el}^{(n)}(\mathbf{x},\mathbf{R})$ as S_0, the next higher one S_1, then S_2, and so on. The triplet electronic wave functions and surfaces are similarly labeled $T_1, T_2, \ldots$, and so on, in the order of increasing energy. Doublet and quartet states and surfaces can be labeled $D_0, D_1, D_2, \ldots$ and $Q_1, Q_2, \ldots$, respectively; the latter are rarely encountered in organic photochemistry. Surfaces of different multiplicity can cross freely at the present level of approximation (Figure 1.13A).

Although the eigenfunctions of $\hat{H}_{el}(\mathbf{R})$ do not reflect any of the magnetic properties associated with electron spin, they are still affected by the spin motion of the electrons, because of the Pauli principle, and thus have different energies even if they originate in the same configuration. The Pauli principle demands that the probability of finding two electrons with all four coordinates equal must vanish. As a consequence, two electrons of equal spins are less likely to be found in close

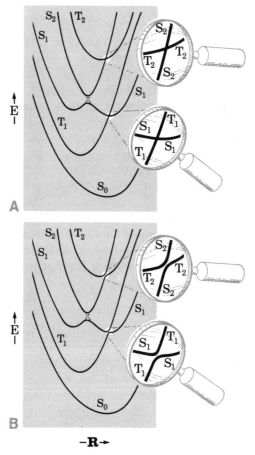

Figure 1.13 The effect of refinements in the Hamiltonian: (A) neglected—singlet and triplet surfaces cross freely; (B) included—singlet–triplet crossings weakly avoided. Only one of the three components of each triplet state is shown.

vicinity of each other than two electrons of opposite spins. If all else is equal, their average electrostatic repulsion is therefore smaller, and this contributes to the differences in $E_{el}(\mathbf{R})$ among states which differ in multiplicity. Hund's first rule states that among the terms resulting from a particular electronic configuration of an atom, the one of highest multiplicity is of lowest energy. The rule is often applied to molecules, particularly organic biradicals. It generally holds for perfect biradicals of the axial kind but not those of the pair kind (see section 4.4.2).

Although one frequently refers to a multiplet, say, a triplet state, as corresponding to a single potential energy surface, it actually corresponds to $2S + 1$ (in this case, three) very closely spaced surfaces. In the approximation adopted so far, spin-free electronic Hamiltonian, these are exactly degenerate in the absence of outside magnetic field. At the next level of approximation, discussed just below, the degeneracy may be lifted, making the multiple nature of the state explicitly apparent. In practice, equilibration between the three molecular states corresponding to the three surfaces of a triplet is usually fast enough that for the purposes of mechanistic photochemistry the surfaces can normally be viewed as one, with a statistical weight of three. At very low temperatures, at very short times, or in very strong magnetic fields, this may be inadequate.

Refinements in the Hamiltonian. Now we are ready to consider the effect of the so far ignored small terms in the Hamiltonian operator $\hat{H}_{el}(\mathbf{R})$ associated with the existence of the spin magnetic moment of the electron.

Electron spin is a clear consequence of a relativistic description of matter. However, it is common not to refer to spin-related phenomena as relativistic effects and to reserve the term "relativistic corrections" for those refinements introduced by the relativistic quantum mechanical description of atoms and molecules that are not associated with the existence of spin alone. These usually need to be considered only when dealing with atoms with a large atomic number Z.

Three of the electron spin effects in the Hamiltonian are the most important. One that we have already alluded to is the spin–orbit coupling operator $\hat{H}_{SO}$, which reflects the interaction between the spin angular momentum of an electron and the orbital motion of this and other electrons. Its importance increases with the fourth power of the atomic number Z of the atoms that constitute the molecule (see the end of Section 1.3.2). Another is the spin–spin dipolar coupling operator $\hat{H}_{SS}$, which reflects the interaction energy of the magnetic moment due to the spin of one electron with that due to the spin of another electron. The presence of these two terms may cause the individual levels of a multiplet to differ slightly in energy even in the absence of an outside magnetic field. In organic molecules, the effect of the dipolar coupling usually dominates the resulting zero-field splitting.

The third small term which often needs to be added to the Hamiltonian is the hyperfine part $\hat{H}_{HF}$, which reflects the interaction between the electron spin magnetic moment and the nuclear spin moments if nuclei with spin are present. The availability of different nuclear spin states increases the overall number of states available; each level of an electronic multiplet then has two or more hyperfine components.

Mixing of pure-multiplicity states and jumps between surfaces. In addition to introducing zero-field splitting, two of the above three spin-related terms in the

Hamiltonian $\hat{H}_{el}(\mathbf{R})$ also have a different effect which we need to discuss; they cause mixing of zero-order states of different multiplicity. In ordinary organic molecules, the spin–orbit coupling term is the most important in this regard. In radical pairs and in biradicals whose radical centers interact only very weakly, the effect of the hyperfine interaction may dominate.

Like the nuclear kinetic energy operator (Section 1.2.2), the spin-related terms can thus also induce jumps between zero-order surfaces defined by means of the spin-free zero-order Hamiltonian. Jumps between singlet and triplet states, which are of particular importance in organic photochemistry, are an example of this. The matrix element that dictates the rate at which these jumps occur is

$$\langle {}^3\Psi_{TOT}|\hat{H}_{SO} + \hat{H}_{HF}|{}^1\Psi_{TOT}\rangle$$

The above description is the usual one in photochemistry: Nuclear motion is considered to be governed by zero-order surfaces corresponding to states of pure multiplicity, and both the nuclear kinetic energy operator and the spin–orbit and hyperfine coupling operators represent perturbations that induce jumps between surfaces (Figure 1.13A).

An alternative description (Figure 1.13B) is often useful as well. Here, the spin-related terms in the Hamiltonian are considered already when constructing the Born–Oppenheimer electronic states. Then, of course, they must not be considered again when it comes to perturbations that induce jumps between surfaces, and the role of the nuclear kinetic energy operator remains unique in this regard.

In this approach, the electronic states are no longer of pure multiplicity but instead, represent mixtures of zero-order contributions of different multiplicities. At almost all geometries, one of these usually predominates by far, so that the states are still called ("impure") singlets, triplets, and so on. These "mixed" wave functions are useful in the calculation of spectroscopic observables such as transition intensities and transition polarizations.

Since the energies associated with the spin-related terms in the Hamiltonian are so very small, one generally uses perturbation theory to describe their effect on the zero-order pure-multiplicity states. Perhaps the most obvious effect on the zero-order potential energy surfaces is that now the crossings of states of different approximate multiplicity are very weakly avoided (Figure 1.13B).

Once the total picture of nuclear motion on surfaces is considered, it does not matter which of the two alternative descriptions is used. In the first, usual picture, the molecule is quite likely to stay on a surface of a given multiplicity even if it crosses a state of another multiplicity, since the spin-related terms that might cause it to jump from one to the other are small. In the second picture, the two surfaces are defined differently. The crossing is avoided, and one surface is below the other at all geometries, being of one approximate multiplicity to the left and of the other approximate multiplicity to the right of the avoided crossing point. The molecule is then nearly certain to jump from one to the other surface as it goes through the avoided crossing region, since the spin-related terms that cause the crossing to be avoided are very small so that the change of the electronic wave function in this region is very abrupt (Figure 1.14). In either description, the most likely ultimate outcome of a passage through the critical region is no change in the approximate multiplicity.

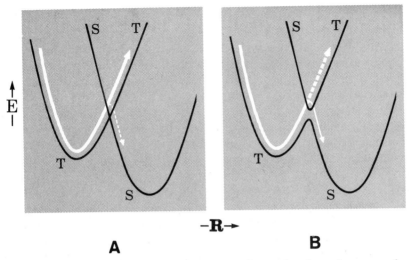

Figure 1.14 Two alternative descriptions of molecular motion through an area of a singlet–triplet crossing using "pure-multiplicity" (A) and "mixed" (B) states. The thickness of the arrows indicates the probability that a particular surface will be followed. Dashed arrows indicate motion after a surface jump.

Weakly interacting spins. A subject that deserves one final note is the effect of the spin-related terms in the Hamiltonian at those geometries at which the zero-order pure multiplicity states are almost exactly degenerate. An example is a pair of radicals linked by a long chain. When should they be thought of as being a part of one system and having a singlet and a triplet state, and when are they better viewed as two free radicals, each in a doublet state? The answer is given by the relative magnitude of the energy splitting provided by our zero-order electronic Hamiltonian, responsible for the singlet–triplet splitting, and of the energy terms provided by the spin-dependent part of the Hamiltonian: the spin–orbit coupling, the spin–spin dipolar coupling, and the hyperfine coupling. As long as the singlet–triplet splitting is larger, it is best to think of the states of the system as being impure singlet and triplet. When the spin-related terms are larger, it is best to think of the states as being those of a pair of impure doublets. Either way, an exact description requires a mixing of the states that we chose for our starting point. This is particularly clear when dealing with the grey area in which the two parts of the total Hamiltonian make comparable contributions. Mathematically, the situation is dealt with by setting up a Hamiltonian matrix that contains terms of both kinds, using either set of spin states as the basis set, and diagonalizing. This procedure automatically yields the final mixed states.

1.3.2 The Space Part

The atomic orbital basis set. Physical intuition suggests that it should be reasonable to build the space part of molecular electronic wave functions using atomic orbitals (AOs) as the starting point. After all, chemical bonding represents only a minor perturbation of the total energy of a molecule, which is nearly exactly given by the sum of the energies of its isolated constituent atoms. This is due primarily to the presence of inner-shell electrons, which are hardly perturbed by bonding at all. Already in CH_4, which has only two inner-shell electrons, the total energy is about 25,000 kcal/mol, to be compared with a single C—H bond energy, on the order of 100 kcal/mol. However, even the valence electrons are perturbed relatively little. Even in H_2, with its exceptionally short bond and no inner-shell electrons, the bonding energy is only 17% of the total energy.

Although it is possible to avoid the use of AOs in the computation of molecular wave functions, they are the starting point for a vast majority of such calculations. Their proper choice is an extremely important part of the calculation. The collection of AOs to be used in the description of a molecule is known as the basis set. It is determined by their location, type, and number. Normally, they are centered at the atomic nuclei. Their type is dictated by the nature of the angular part (s,p,d, . . .) and radial part: $\exp(-ar^2)$ in the "uncontracted" Gaussian-type AOs, $\Sigma c_i \exp(-a_i r^2)$ in the usual "contracted" Gaussian-type AOs, and $\exp(-ar)$ in the Slater-type AOs, where r is the distance from the nucleus, and by the choice of the exponents a or a_i and of the contraction coefficients c_i. The number of AOs to be used in a calculation is normally a compromise between computational feasibility (as few per atom as possible) and desire for accuracy (Appendix III). Computation with Slater-type AOs is much harder than with Gaussian-type AOs, and this is not sufficiently compensated for by the fact that fewer of the former are needed.

The minimum number of AOs that is capable of giving qualitatively meaningful results is dictated by the need to describe, however poorly, the behavior of all the electrons in the molecule. A minimum AO basis set consists of those AOs needed to describe the valence shell for the ground state of each atom present in the molecule: 1s for H and He; 1s, 2s, and 3 × 2p for second-row elements; and so on. Although it is ideal for qualitative discussions and can be parameterized to give very good results in semiempirical calculations, it is woefully inadequate for *ab initio* computations unless the exponent on each AO is optimized separately for each state and each geometry to allow for local expansion or contraction of electron density as a function of atomic charge, and unless each AO is allowed to float off the nucleus to an optimum location to allow for the polarizing effect of the other parts of the molecule. Such nonlinear optimizations would be computationally very expensive and are rarely performed.

A far more efficient way of providing a flexible functional form for the wave function, thus accomplishing the same purpose, is to use a larger number of AOs. As the number of properly chosen AOs increases, so does the quality of the results. When the addition of further AOs no longer has a significant effect, the basis set is said to be saturated. A complete basis set would have to be infinite in size and, in principle, would permit an exact solution. In practice, only quite incomplete basis sets can be used for molecules of interest to organic chemists. In a double-zeta basis set, each valence AO in an atom is represented by two AOs in the calculation, one more and one less compact, permitting charge density to approach the size that is optimal for the particular state and geometry. In a triple-zeta basis set, with three AOs for each valence AO of the atom, this purpose is served even better. Addition of p orbitals on H, d orbitals on second-row atoms, and so on, of a size comparable with that of their s and p orbitals, respectively, permits a description of the polarizing effect of other parts of the molecule. These "polarization functions" are not to be confused with the atomic Rydberg orbitals (2p on

H, 3d on C, etc.), which are far larger in size. The latter play no significant role in the description of the ground states of molecules and of many of the excited states ("valence" excited states). A more mathematical definition of basis sets is given in Appendix III.

The standard models. The two most common ways of constructing approximate many-electron wave functions for electronic states of molecules are known as the valence bond (VB) method and the molecular orbital (MO) method. The procedures are outlined schematically for the case of H_2 in Figure 1.15. The examples given correspond to the construction of the three singlet states of H_2 which can be obtained using a minimum basis set. Both methods proceed by setting up a series of artificial constructs, such as VB ("resonance") structures and MOs, neither of which has a direct exact relation to physical reality. The final many-electron wave function, which is eventually obtained from these intermediate constructs, is normally built from a limited AO basis set and represents only an approximation to the Born–Oppenheimer wave function of the electronic state of the molecule. Of course, even the unobtainable exact relativistic Born–Oppenheimer wave function would be only a simulation of the exact quantum mechanical state function, which, in turn, is nothing more than a simplified model for reality. In this hierarchy of model concepts, we attempt to work at the first level that appears acceptable and still computationally feasible.

The valence bond model. In the VB procedure, the available electrons are assigned to the starting AOs in each of the possible ways that are compatible with the Pauli principle (i.e., up to two electrons per AO). This can be performed in a way which ensures that only pure spin functions result (singlets, triplets, etc.). The resulting many-electron functions are referred to as VB structures and can be

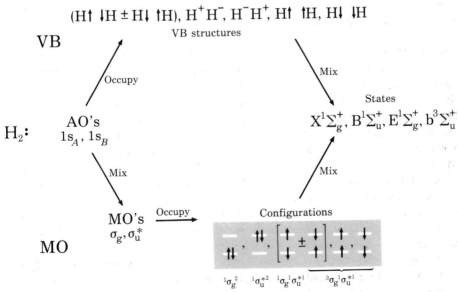

Figure 1.15 The VB and MO routes to full CI wave functions for the states of H_2.

symbolically drawn as resonance structures, usually with some short bonds and some (dashed) long formal bonds between AOs and suitable unpaired electrons or charges on the various AOs. Two types of VB structures that we shall frequently need to refer to are "dot–dot" structures, with two singly occupied AOs (when singlet-coupled, these are usually called "covalent"), and "hole–pair" structures, with an empty and a doubly occupied AO (when the former carries a positive and the latter a negative charge, these are usually called "zwitterionic" or simply "ionic"). Mixing ("resonance") between the VB structures is introduced next and yields the final molecular states. Such states can be characterized roughly by specifying which VB structures contribute most, and expressions such as covalent ("dot–dot") state or zwitterionic ("hole–pair") state are self-explanatory. In principle, of course, all VB structures of appropriate multiplicity and symmetry will contribute, to some degree, to a final state of that symmetry, and their relative contributions will change as the molecular geometry changes. Since the number of possible VB structures is astronomical for any organic molecule, even when only the minimum reasonable set of starting AOs is used, it is necessary to preselect those which can be expected to be of importance on physical grounds, such as the energy of the structure, when an actual calculation is performed. Numerical calculations of this type are difficult to perform on molecules of interest to an organic chemist, since the calculation of the mixing between the VB structures rapidly becomes impractical as the total number of electrons increases, due to the nonorthogonality of the starting AO set. Recently, there have been some indications that this may change in the future.

A very powerful variant of the VB method, known as the generalized valence bond (GVB) method, is particularly useful for the description of biradicals and biradicaloids. Its mathematical form is briefly summarized in Appendix III.

The molecular orbital model. The MO method involves one more step but is computationally much easier and nowadays completely overshadows the VB method in numerical applications.

The self-consistent field (SCF) step. The starting AOs are first combined into one-electron functions that spread over the whole molecule. These one-electron functions are referred to as MOs and are mutually orthogonal. They are normally obtained as eigenfunctions of a suitable one-electron operator (canonical MOs). The eigenvalues ε are known as orbital energies. The matrix diagonalization procedure yields as many eigenfunctions of a one-electron operator as there were AOs. Normally, this number is far larger than what is actually needed to accommodate all available electrons. The unused (empty) MOs are known as virtual MOs. All of the available electrons are assigned to the MOs in various possible ways that satisfy the Pauli principle (i.e., no more than two electrons per MO). The requirement of antisymmetry with respect to interchange of electron coordinates is satisfied by writing the resulting many-electron functions, called configurations, in the form of determinants (Slater determinants). The configurations (more rigorously, configuration state functions) are chosen in a way which guarantees that only pure spin functions result. Usually, one of them, the reference configuration, is selected to represent the state of interest.

A configuration that differs from the reference configuration $|\phi_0\rangle$ by promotion of one electron from MO $|i\rangle$ into a virtual MO $|j\rangle$ is called singly excited and denoted $|i \rightarrow j\rangle$, one in which two electrons are promoted is called doubly excited and denoted $|i,j \rightarrow k,l\rangle$, and so on (Figure 1.16). For pure spin states, multiplicity can be specified by a superscript: $^1|i \rightarrow j\rangle$, $^3|i \rightarrow j\rangle$, and so on. Note that the many-electron wave function resulting from an assignment of electrons into AOs is called a structure, whereas the one resulting from an assignment of electrons into MOs is called a configuration. For each possible spin multiplicity, the total number of possible MO configurations and of possible VB structures is the same, and for a molecule of any reasonable size, both are huge, even when only a minimum AO basis set is used.

The choice of the exact form of the one-electron operator whose eigenfunctions are the MOs (i.e., the AO mixing coefficients in each MO) represents a degree of freedom. It can be selected so as to be extremely simple computationally (e.g., the Hückel Hamiltonian), but usually it is chosen so as to minimize the energy of the reference configuration (the Hartree–Fock operator). This is not simple computationally, since such a one-electron operator must correspond to the energy of an electron in the electrostatic field of the nuclei and the time-averaged field of the other electrons. While the field due to the nuclei is known once their positions have been chosen, the field due to the electrons is not known at the outset, since the electronic wave function is the unknown that is being sought. Thus, in order to define the Hartree–Fock operator for the purpose of finding its eigenfunctions, one must first know the eigenfunctions! The circular nature of the problem is circumvented by using an iterative procedure known as the self-consistent field (SCF) method. To start with, one represents the time-averaged distribution of electron density using an only very approximate but easily obtained set of MOs such as Hückel MOs. One then constructs an approximate Hartree–Fock operator, finds its eigenfunctions, and hopes that these are closer to the correct MOs. They are used to construct the next approximation to the Hartree–Fock operator, whose eigenfunctions are then found and used to construct the next approximation, and so on, until the procedure converges. At this point, the electron density distribution produced by the occupancy of the eigenfunctions of the Hartree–Fock operator, the "canonical" MOs, is identical with the distribution used in the construction of

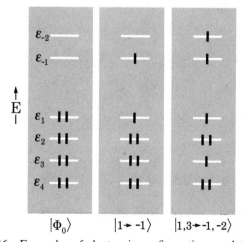

Figure 1.16 Examples of electronic configurations and their labels.

this operator (i.e., the time-averaged field produced by the electrons is self-consistent—hence the name SCF method). Although the SCF solution is the best in the sense of providing the lowest possible energy for the state of interest, it is not the exact solution of the molecular problem, since in reality, electrons do not move in the time-averaged field of other electrons but, instead, move in their instantaneous field, and this permits them to avoid each other better. The resulting reduction in the electron repulsion energy is known as the correlation energy.

Most often, the reference configuration is the ground configuration $|\phi_0\rangle$, which results when all the electrons are assigned to MOs according to the Aufbau principle: The lowest-energy orbital is occupied first, then the next higher one, and so on. (Figure 1.16). The weights of the various AOs in an MO can then be used to approximately describe its nature, and terms like lone-pair orbital (n), sigma orbital (σ), and pi orbital (π) are well known. For those interested in detail, a condensed mathematical description of the SCF procedure is given in Appendix III.

The configuration interaction (CI) step. To obtain the final molecular states from the MO procedure, one now allows all of the configurations to mix, using the variational principle for energy minimization. This procedure is known as configuration interaction (CI). A final state can then be characterized roughly by specifying which configurations contribute most to it, and concepts such as a singly excited or doubly excited state are approximate but self-explanatory. In principle, of course, all configurations of appropriate symmetry will contribute, to some degree, to a final state of that symmetry, and their relative contributions will change as the molecular geometry changes (at the same time, the MO expansion coefficients change as well).

For a given set of AOs, the result of a VB calculation using all structures is identical with that of an MO calculation using all configurations. The wave function is usually referred to as the full configuration interaction (FCI) wave function. Its energy is lower than that of the SCF wave function by the correlation energy. Of course, the true correlation energy would need to be calculated with a complete AO basis set.

As was the case in the VB method, it is usually necessary to truncate the number of terms to be considered in an actual MO calculation as well. The selection of configurations is done using suitable criteria. One of these is the energy of the configuration, since high-energy configurations tend to contribute little individually to the wave functions of low-lying states. However, there are very many of them, and their total contribution is far from negligible. Another criterion is the number of electrons that have been promoted with respect to a reference configuration or configurations, known to be important in the state of interest. Since off-diagonal mixing elements vanish when this number exceeds two, single and double excitations with respect to the reference configuration or configurations are most important to include (single-reference or multi-reference CI).

The results of a truncated VB calculation and of a truncated MO calculation which start from the same AO basis set differ, and both represent an approximation to the exact (FCI) result for this basis set. The calculation of the mixing of configurations is very much easier than the calculation of the mixing of VB structures. The reason is that the MOs are mutually orthogonal, since they all are eigenfunctions of the same Hermitean operator, and this immensely simplifies the evaluation of matrix elements. For additional detail, see Appendix III.

Of course, the whole point of selecting a reference configuration can be to optimize the MOs in such a way as to obtain a large CI coefficient for the reference configuration and small CI coefficients for all the other configurations in the state of interest, usually the ground state. This is usually feasible for molecules containing an even number of electrons at nuclear geometries corresponding to minima in the ground-state surface when the reference configuration is the ground configuration $^1|\phi_0\rangle$. It frequently becomes hard or impossible to do this at other types of molecular geometries, unfortunately including many of those important in photochemical processes. Then, more complex hybrid methods such as multiconfigurational (MC) SCF, which employs several reference configurations simultaneously, are of help in obtaining suitable MOs (Appendix III). We shall return to this point later, in connection with a discussion of electronic states of molecules at biradicaloid geometries.

Frequently, the CI expansion is truncated after the first term; that is, the reference configuration alone is used as an approximation to the final state wave function. When this is done, it is clearly best to use the SCF MOs to construct the configuration $|\phi_0\rangle$. For these MOs, the matrix elements between the closed-shell configuration $^1|\phi_0\rangle$ and all singly excited configurations vanish (Brillouin's theorem). Such SCF calculations are far off in total energy, since the choice of such a single configuration means that the correlation energy is neglected. This becomes pictorially obvious when a wave function of this type is transformed into the VB form and compared with the FCI result: At the SCF level of approximation, zwitterionic structures carry far too large a weight (Section 4.2.1). Nevertheless, SCF wave functions are useful because they produce quite acceptable values for many observable properties at equilibrium geometries and are relatively easy to obtain. The ready availability of this intermediate result represents another advantage of the MO over the VB method. Of course, the limitation of the one-configuration SCF procedure to the neighborhood of equilibrium nuclear geometries is serious in studies of thermal or photochemical reactivity, but most physical measurements are performed at molecular geometries for which the SCF approximation is reasonable. It is therefore still useful to discuss the relation of MO's to spectroscopic observables in some detail.

Orbital energies and configuration energies. The orbital energies ε resulting from a closed-shell single-configuration Hartree–Fock calculation are related to measurable quantities by Koopmans' theorem (Figure 1.17). An energy of an occupied orbital in $^1|\phi_0\rangle$ is equal to minus the ionization potential (IP) for that orbital (i.e., to the amount of energy needed to bring an electron into that orbital from an infinite distance). An energy of a virtual orbital in $|\phi_0\rangle$ is equal to minus the electron affinity (EA) for that orbital (i.e., to minus the amount of energy gained when an electron is brought from infinity and added to the molecule). Unless otherwise specified, the terms ionization potential and electron affinity refer to IP_1 and EA_1—that is, the lowest IP and the highest EA for the molecule. Koopmans' theorem is an *exact* statement about *approximate* wave functions. Its applicability to real molecules is therefore based on some assumptions that are only approximately valid. For most organic molecules the errors seem to cancel (at least for the first few IPs and EAs), and the assignment of photoelectron spectra, which produce IPs, and of electron transmission and electron detachment spectra, which produce EAs, can often be successfully based on calculated SCF orbital energies.

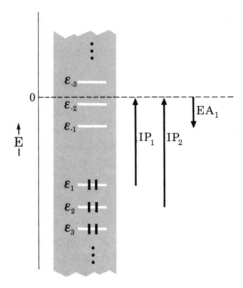

Figure 1.17 Koopmans' theorem.

In photochemistry, we shall be concerned with low-energy excited states. These are frequently, but not always, satisfactorily approximated by a sole singly excited configuration $|i \to j\rangle$ based on the SCF MOs, as can be judged by inspection of large CI wave function. In the cases where such a description applies, one might be tempted to imagine that the excitation energy will be equal to the difference of the energies of the two orbitals i and j between which the electron jump occurs. A glimpse at experimental results quickly dispels that notion: For usual molecules with electron affinities of the order of -1 to 1 eV and ionization potentials of the order of 8–11 eV, the excitation energies are not at all equal to IP − EA but, more typically, are only 3–5 eV. This is true even in the cases where the single configuration approximation should be good when judged by results of more complete CI calculations. Also, the energies of the singlet and the triplet configurations belonging to the same electron occupancy differ and clearly cannot simultaneously both be equal to the orbital energy difference.

In order to understand the relation between the orbital energies ε_i and ε_j and the configuration energies $^{1,3}|i \to j\rangle$, and thus also the relation between singlet and triplet energies for states described adequately by a sole singly excited configuration $|i \to j\rangle$, we need to first briefly review the nature of the singlet and triplet states.

For this type of configuration, electrons can be assigned into orbitals i and j in four ways which differ in the spin functions, so that four wave functions actually result. The assignments can be symbolized by $\alpha(1)\alpha(2)$, $\beta(1)\beta(2)$, $\alpha(1)\beta(2)$, and $\beta(1)\alpha(2)$. The spins of two electrons, one held in orbital $|i\rangle$, the other in $|j\rangle$, can be coupled in a constructive manner to yield the three components of a triplet, as shown on the right in Figure 1.18, or in a destructive manner to yield a singlet, as shown on the left. In the case $\Theta_1(1,2) = \alpha(1)\alpha(2)$, the total projection into the z axis will be $\hbar$; in the case $\Theta_{-1}(1,2) = \beta(1)\beta(2)$, it will be $-\hbar$. In the spin function for the third component of the triplet, one of the electrons has its spin up, the other down, and the projection into the z axis vanishes. Just as for the other two components, the function must respect the fact that either of the two electrons

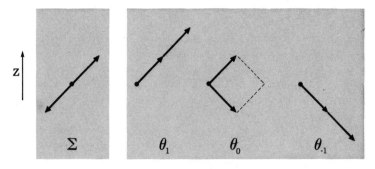

Figure 1.18 The coupling of spin angular momenta of two electrons.

may be in either of the two orbitals, as well as the fact that the length of the spin angular momentum vector is $\hbar\sqrt{2}$. The vector clearly must lie in the xy plane to possess a vanishing projection into the z direction. The proper functional form is $\Theta_0(1,2) = [\alpha(1)\beta(2) + \beta(1)\alpha(2)]/\sqrt{2}$. The destructive addition of the two spin angular moments produces a singlet state, and the spin function for this case is $\Sigma(1,2) = [\alpha(1)\beta(2) - \beta(1)\alpha(2)]/\sqrt{2}$.

In the SCF approximation, one deals only with time-averaged electron repulsions. The time-averaged charge density due to an electron in a normalized orbital $|i\rangle$ is $e\rho_i(\mathbf{r}) = e|i(\mathbf{r})|^2$, and that due to an electron in a normalized orbital $|j\rangle$ is $e\rho_j(\mathbf{r}) = e|j(\mathbf{r})|^2$. If the electron in orbital $|i\rangle$ has spin α and that in orbital $|j\rangle$ has spin β, the expectation value of the energy of their mutual repulsion is given by the classical expression for the repulsion of two clouds of negative charge described by the charge densities $e\rho_i(\mathbf{r})$ and $e\rho_j(\mathbf{r})$, with each of the clouds containing a total charge equal to e (Figure 1.19):

$$\begin{aligned} J_{ij} &= \iint \rho_i(\mathbf{r}_1) \frac{e^2}{r_{12}} \rho_j(\mathbf{r}_2) \, d\mathbf{r}_1 \, d\mathbf{r}_2 \\ &= \langle i(1)j(2) | \frac{e^2}{r_{12}} | i(1)j(2) \rangle \\ &= (ii|jj) \end{aligned} \qquad (1.4)$$

where

$$r_{12} = |\mathbf{r}_1 - \mathbf{r}_2|$$

and two kinds of commonly used notation for two-electron integrals are introduced.

If each of the orbitals is singly occupied, this situation does not represent a stationary spin-adapted state but, instead, contains equal components of a singlet and a triplet state. Since the singlet and triplet wave functions do not interact, the expectation value of the repulsion energy is an average of the eigenvalues pertaining to the two states. We shall return to this subject below.

The positive quantity J_{ij} is referred to as the Coulomb integral between orbitals $|i\rangle$ and $|j\rangle$. The form of the integral follows from the Coulomb law expression for the repulsion energy of charge $e\rho_i(\mathbf{r}_1)d\mathbf{r}_1$ contained in small volume element $d\mathbf{r}_1$ located at $\mathbf{r}_1$ with charge $e\rho_j(\mathbf{r}_2)d\mathbf{r}_2$ contained in the volume $d\mathbf{r}_2$ located at $\mathbf{r}_2$, which

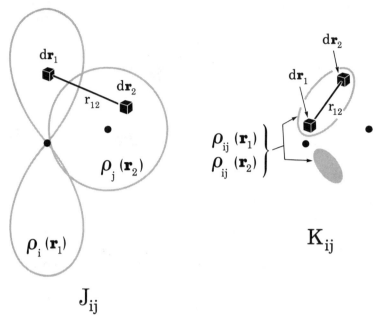

Figure 1.19 Illustration of the physical significance of the electron repulsion integrals J_{ij} and K_{ij} defined by equations 1.4 and 1.5.

is $e\rho_i(\mathbf{r}_1)d\mathbf{r}_1 \times e\rho_j(\mathbf{r}_2)d\mathbf{r}_2/r_{12}$, by integration over all volume elements $d\mathbf{r}_1$ and $d\mathbf{r}_2$ (Figure 1.19). The expression for the repulsion energy contains no allowance for the fact that either electron "feels" the instantaneous position of the other electron, rather than its time-averaged position. This is the origin of the correlation energy error in the SCF method.

As already noted, the SCF method only commits the correlation error fully if the two electrons have opposite spins. Like any other wave function that satisfies the Pauli principle, the SCF wave function does not permit two electrons of the same spin to come arbitrarily close to each other and it thus automatically provides some correlation of their motions. The SCF expression for the expectation value of the repulsion energy of two electrons of the same spin in orbitals $|i\rangle$ and $|j\rangle$ thus is not J_{ij} but is, instead, a smaller quantity written as $J_{ij} - K_{ij}$. The positive number K_{ij} is known as the exchange integral between orbitals $|i\rangle$ and $|j\rangle$ and is a measure of the degree to which an electron in orbital $|i\rangle$ and an electron in orbital $|j\rangle$ come into each other's way. It is defined as the energy of repulsion of the charge density $e\rho_{ij}(\mathbf{r})$ with an identical charge density (Figure 1.19):

$$K_{ij} = \iint \rho_{ij}(\mathbf{r}_1) \frac{e^2}{r_{12}} \rho_{ij}(\mathbf{r}_2)\, d\mathbf{r}_1\, d\mathbf{r}_2$$

$$= \langle i(1)i(2)| \frac{e^2}{r_{12}} |j(1)j(2)\rangle$$

$$= (ij|ij) \qquad (1.5)$$

where

$$e\rho_{ij}(\mathbf{r}) = ei(\mathbf{r})j(\mathbf{r}) \qquad (1.6)$$

is the overlap charge density of orbitals $|i\rangle$ and $|j\rangle$.

If there is only one electron in each of the two orbitals, this situation represents a spin eigenstate, namely, one ($\alpha\alpha$) or another ($\beta\beta$) of the three components of the triplet state. Thus, there is no doubt that a measurement on a system described by such an uncorrelated wave function would yield the value $J_{ij} - K_{ij}$ for the repulsion energy of the two electrons.

The third component of the triplet state, $\alpha\beta + \beta\alpha$, must have the same energy, $J_{ij} - K_{ij}$. We now see that the repulsion of two singlet-coupled electrons in orbitals $|i\rangle$ and $|j\rangle$ must have the value $J_{ij} + K_{ij}$ because otherwise the average of the singlet and the triplet values would not be J_{ij}, as it must, based on the prior consideration of the mixed spin state $\alpha\beta$.

The probability density $\rho_i(\mathbf{r})$ integrates to unity over all space for a normalized orbital $|i\rangle$ since the particle may be delocalized but the probability that it will be found somewhere is one. On the contrary, the integration of the overlap probability density $\rho_{ij}(\mathbf{r})$, which has no such straightforward physical interpretation, can yield any number between -1 and 1. This integral is known as the overlap between orbitals $|i\rangle$ and $|j\rangle$:

$$S_{ij} = \int \rho_{ij}(\mathbf{r}) \, d\mathbf{r} \qquad (1.7)$$

The overlap probability density $\rho_{ij}(\mathbf{r})$ is usually positive in some parts of space, where orbitals $|i\rangle$ and $|j\rangle$ have the same sign, and negative in others, where the two orbitals have different signs (Figure 1.19). It is small in all regions of space except those in which both orbitals $|i\rangle$ and $|j\rangle$ have a large amplitude simultaneously (i.e., where they overlap), and this explains its name. It is also sometimes referred to as the transition density, and the magnitude and direction of its dipole moment determine the intensity and polarization of the transition from $^1|\phi_0\rangle$ to $^1|i \rightarrow j\rangle$. Even if the two orbitals $|i\rangle$ and $|j\rangle$ overlap considerably so that there is a region of space where $\rho_{ij}(\mathbf{r})$ is large, the overlap integral S_{ij} can still be zero if negative and positive contributions cancel exactly in the integration (e.g., a p and an s orbital on the same atom).

In order to understand the relation between orbital energies ε_i and ε_j and excitation energies, let us consider a cycle that produces the configurations $^{1,3}|i \rightarrow j\rangle$ from the ground configuration $^1|\phi_0\rangle$ in two ways (Figure 1.20). First consider the triplet excitation energy for the $i \rightarrow j$ promotion, $E(^3|i \rightarrow j\rangle)$. Starting with the ground configuration $^1|\phi_0\rangle$, remove an electron of spin β from orbital $|i\rangle$ to infinity, leaving its partner of spin α in place. By Koopmans' theorem, this costs an amount of energy equal to IP. Next, bring an electron of spin α from infinity and place it into orbital $|j\rangle$. If the β electron in orbital $|i\rangle$ had not been removed, this would have given us back energy equal to EA by Koopmans' theorem, and the total expended in both steps would be IP $-$ EA. In reality, the β electron in orbital $|i\rangle$ had been removed, so that the α electron will enter orbital $|j\rangle$ more easily. The

40 INTRODUCTION

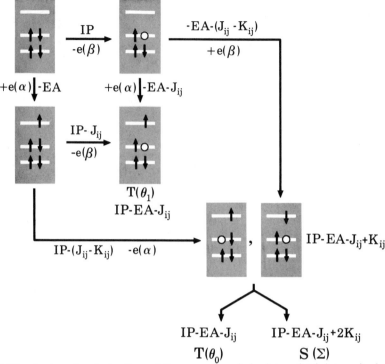

Figure 1.20 A stepwise derivation of the energies of singlet and triplet excited configurations.

missing repulsion will be that with the previously removed β electron in orbital $|i\rangle$, and it will be J_{ij} since their spins differ. The total energy needed to convert $^1|\phi_0\rangle$ into the $\alpha(1)\alpha(2)$ component of the triplet configuration $^3|i \to j\rangle$ will therefore be

$$E(^3|i \to j\rangle) = IP - EA - J_{ij} \tag{1.8}$$

An analogous process producing the $\beta(1)\beta(2)$ component by removal of an electron of spin α from orbital $|i\rangle$ and addition of an electron of spin β to orbital $|j\rangle$ is readily seen to require the same amount of energy. The third spin component of $^3|i \to j\rangle$, namely $\alpha(1)\beta(2) + \beta(1)\alpha(2)$, also must have the same excitation energy.

To derive an expression for the singlet excitation energy $E(^1|i \to j\rangle)$, let us consider a process which starts with $^1|\phi_0\rangle$, removes a β electron from orbital $|i\rangle$, and places a β electron in orbital $|j\rangle$, producing the configuration $|i(\alpha)j(\beta)\rangle$, which is a 50:50 mixture of a pure singlet and a pure triplet state and therefore has an energy equal to the average of the two, namely, $[E(^3|i \to j\rangle) + E(^1|i \to j\rangle)]/2$. The first step will cost an amount of energy equal to IP. The second would have produced an amount of energy equal to EA had the β electron not been removed from orbital $|i\rangle$, but in reality it will produce more since one electron repulsion is missing, namely that between a β electron in orbital $|j\rangle$ and a β electron in orbital $|i\rangle$. Since their spins are equal, the missing repulsion energy amounts to $J_{ij} - K_{ij}$, so that we obtain a value of $IP - EA - (J_{ij} - K_{ij})$ for the energy of excitation of a 50:50 singlet–triplet mixture.

Since the excitation energy of $^3|i \rightarrow j\rangle$ is IP − EA − J_{ij} and the average of the excitation energies of $^3|i \rightarrow j\rangle$ and $^1|i \rightarrow j\rangle$ is IP − EA − J_{ij} + K_{ij}, obviously the singlet excitation energy must be

$$E(^1|i \rightarrow j\rangle) = \text{IP} - \text{EA} - J_{ij} + 2K_{ij} \quad (1.9)$$

The singlet–triplet splitting for the two states arising from the i → j promotion therefore amounts to $2K_{ij}$. Given approximate shapes of the orbitals $|i\rangle$ and $|j\rangle$, we can roughly estimate this quantity: The less the two orbitals overlap in space, the less the total amount of overlap charge density will be, the less its self-repulsion will be, and the less the singlet–triplet splitting will be. For instance, charge transfer and n → π* transitions, in which orbitals $|i\rangle$ and $|j\rangle$ are predominantly located in different parts of space, will have small singlet–triplet splittings. Since their overlap charge density will be small, its dipole moment will also tend to be small, and such transitions will usually have low intensity.

Density matrices and natural orbitals. An important concept in MO theory is the charge–bond-order matrix, which provides information about the distribution of electron density described by a configuration. Its diagonal elements, i.e., the charge densities q_μ, and its off-diagonal elements, i.e., the bond orders $p_{\mu\nu}$, are defined as sums of contributions from all occupied MOs. The contribution from the i-th orbital is weighted by the occupation number of that orbital, n_i, which has the value 2 for a doubly occupied MO, 1 for a singly occupied MO, and 0 for a virtual (empty) MO. In the simplest case, well known from Hückel theory, $q_\mu = \Sigma_i n_i c_{\mu i}^2$, and $p_{\mu\nu} = \Sigma_i n_i c_{\mu i} c_{\nu i}$, where $c_{\mu i}$ is the coefficient of the μ-th AO in the i-th MO. In a closed-shell configuration, such as the ground configuration $|\phi_0\rangle$, all occupation numbers n_i are either zero or two. A generalization of the concept of a charge–bond-order matrix to all types of wave functions is known as the density matrix.

The orbitals that diagonalize the density matrix of an electronic state are known as the natural orbitals (NOs) of that state; the off-diagonal "bond orders" between them vanish and the diagonal "charge densities" are the NO occupation numbers n. Like the q_μ's, they are non-integral but are restricted to the interval 0–2. All observable one-electron properties of an electronic state can be expressed as sums of contributions from the individual NOs, each weighted by its occupation number. An important use of the NOs is in CI expansions: When used in place of the usual SCF MOs, they greatly accelerate the convergence of the CI process. Of course, in order to find them, one needs to diagonalize the density matrix, whose determination, in turn, requires the knowledge of the unknown final wave functions. It is therefore possible to proceed in an iterative fashion reminiscent of the SCF procedure of finding MOs; this is known as the iterative natural orbital (INO) method. It is more common to approximate the NOs in other ways.

For an SCF ground configuration $|\phi_0\rangle$, the NOs are identical with the SCF MOs, and their occupation numbers n are 2 or 0. For correlated wave functions such as CI, one finds that for ordinary "closed-shell" molecules most of the n's are still very close to 2 or to 0, but for "biradicaloid" molecules this is no longer so: Two of the n's now have values close to 1. This property of the NOs can actually be used to define biradicals and biradicaloids (Chapter 4).

Spin-orbit coupling. We are now in position to return to the spin–orbit coupling part of the Hamiltonian operator $\hat{H}_{SO}$. For a photochemist, this is the most important of the three spin-dependent small terms mentioned in Section 1.3.1.

The usual approximation for $\hat{H}_{SO}$ is

$$\hat{H}_{SO} = \frac{e^2}{2m^2c^2} \sum_{j=1}^{N} \sum_{\alpha=1}^{M} \frac{Z_\alpha}{|\mathbf{r}_{\alpha j}|^3} \hat{\mathbf{l}}_j \cdot \hat{\mathbf{s}}_j \quad (1.10)$$

where N is the number of electrons, M is the number of nuclei, $|e|Z_\alpha$ is the nuclear charge of the α-th nucleus, $\mathbf{r}_{\alpha j}$ is the position vector of the j-th electron relative to the α-th nucleus, $\hat{\mathbf{s}}_j$ is its spin angular momentum operator, and $\mathbf{l}_j$ is its orbital angular momentum operator $\hat{\mathbf{l}}_j = \hat{\mathbf{r}}_j \times \hat{\mathbf{p}}_j$, where $\hat{\mathbf{p}}_j$ is the linear momentum operator.

The operator $\hat{H}_{SO}$ thus is a sum of one-electron operators, each of which is a sum of contributions from each atom α in the molecule. Contributions from the heavier atoms normally dominate, not only because Z_α is larger but also because, on the average, $|\mathbf{r}_{\alpha j}|$ is smaller as the electron comes closer to the nucleus ("heavy-atom effect"). The scalar product $\hat{\mathbf{l}}_j \cdot \hat{\mathbf{s}}_j$ can be imagined to measure the interaction of the intrinsic magnetic moment of the j-th electron due to its spin with the magnetic moment due to the current caused by the "motion of the nucleus around the electron," viewed from the vantage point of the electron.

The $\hat{\mathbf{s}}_j$ operator acts on the spin part of the total wave function, and the $\hat{\mathbf{l}}_j$ operator acts on the space part. Since the scalar product contains three terms, $\hat{l}_{xj}\hat{s}_{xj} + \hat{l}_{yj}\hat{s}_{yj} + \hat{l}_{zj}\hat{s}_{zj}$, its application to an electronic wave function is capable of causing a change in multiplicity.

Consider the evaluation of matrix elements of $\hat{\mathbf{l}}_j \cdot \hat{\mathbf{s}}_j$ over one-electron wave functions, concentrating on contributions from a single nucleus α. The $|\mathbf{r}_{\alpha j}|^{-3}$ factor in equation 1.10 effectively eliminates contributions from AOs located on other nuclei. The action of $\hat{\mathbf{l}}_j \cdot \hat{\mathbf{s}}_j$ converts the one-electron wave function into a sum of three terms; in two of these the electron spin has been flipped, and all three have been acted on by the angular momentum operator $\hat{\mathbf{l}}_j$. Now, this is the operator that differentiates s,p,d, ... orbitals. Its action on an electron in an s orbital annihilates it because the orbital is invariant to rotation. Application of $\hat{l}_z$ to a p_z orbital annihilates it, too, because p_z is invariant to rotation about the z axis, but application of $\hat{l}_y$ and $\hat{l}_x$ converts it to a multiple of a p_x orbital and of a p_y orbital, respectively (i.e., effectively rotates it by 90° around the y or the x axis). Since any contribution to the wave function from s orbitals on nucleus α will be annihilated by the $\hat{\mathbf{l}}_j$ part of the operator, we consider only the action on p orbitals, assuming that the electron spin is α:

$$(\hat{l}_{xj}\hat{s}_{xj} + \hat{l}_{yj}\hat{s}_{yj} + \hat{l}_{zj}\hat{s}_{zj})|p_z(j)\alpha(j)\rangle$$
$$= |\hat{l}_{xj}p_z(j)\hat{s}_{xj}\alpha(j)\rangle + |\hat{l}_{yj}p_z(j)\hat{s}_{yj}\alpha(j)\rangle + |\hat{l}_{zj}p_{zj}\hat{s}_{zj}\alpha(j)\rangle$$
$$= -\frac{i\hbar^2}{2}|p_y(j)\beta(j)\rangle - \frac{\hbar^2}{2}|p_x(j)\beta(j)\rangle + 0 \quad (1.11)$$

We see that the operator produced a sum of two terms. In one of these, the z projection of the electron spin was flipped from the +z direction (α) into the −z direction (β) by rotation about the x axis, while the orbital axis was flipped from the z direction (p_z) to the y direction by rotation about the same axis in the opposite direction, conserving angular momentum. In the other term, similar events occurred, but now by rotation about the y axis.

In summary, the $\hat{l}_j \cdot \hat{s}_j$ operator moves an electron from an initial p orbital into a p orbital rotated by 90° while simultaneously flipping its spin so as to preserve angular momentum. It is then understandable that the $\hat{H}_{SO}$ operator, which contains a sum of these terms, can change the multiplicity of a many-electron wave function to which it is applied, thus permitting the mixing of triplet with singlet functions, and so on. It is also intuitively clear that each of the three components of the triplet state, which differ in their spin functions, will interact with any one singlet function to different degrees. Finally, it is understandable why the removal of spin-forbiddenness by spin–orbit coupling is particularly effective in the presence of heavy atoms (large Z, small $|r_{\alpha j}|^{-3}$) and for combining states that differ in promotion of an electron from one to another p orbital on the same atom.

1.3.3 Classification of Excited States

The wave function of many of the low-energy excited states of organic molecules can be built to a satisfactory degree of accuracy from atomic orbitals all of which belong to the valence shells of the participating atoms. Although such wave functions can, in principle, always be improved by the inclusion of more diffuse orbitals of higher quantum numbers in the basis set, such improvement has no qualitative effect. The states of this kind are called valence excited states.

There are also molecular excited states at energies that are frequently not much higher, whose wave functions cannot be described even qualitatively in such a manner, while the inclusion of atomic orbitals of a higher principal quantum number in the basis set leads to a satisfactory description. These are known as molecular Rydberg excited states.

In the present section, we shall describe briefly the nature of those electronically excited states that most often play an important role in organic photochemistry, and we shall also mention their experimental spectroscopic characteristics.

Valence states. These are the most common initial excited states in organic photochemical processes. Typically, the photochemical reactions of organic compounds are performed in solvents that are transparent down to perhaps 200 nm (up to about 6 eV in photon energy), and this is also the region in which most convenient photochemical light sources are available. Although work at shorter wavelengths in the vacuum UV and work in the gas phase have been becoming more popular recently, only limited understanding of the spectroscopy of organic molecules at these higher energies is available; we shall focus our attention on the more common case.

Saturated hydrocarbons do not have singlet–singlet absorptions in the readily accessible region and are frequently used as solvents. Photochemical substrates ordinarily contain a functional group responsible for the absorption of light in the region above 200 nm; this group is called a chromophore.

It is convenient to classify the excited states of organic molecules using the local symmetry properties of the chromophores even if these are not symmetry elements of the whole molecule. The most important such symmetry element is a plane of symmetry in chromophores containing multiple bonds. Orbitals that are symmetric with respect to reflection in such a plane are referred to as σ orbitals, and those that are antisymmetric are called π orbitals. It is common to indicate antibonding orbitals by an asterisk and to indicate bonding orbitals by a lack thereof. Also, lone-pair orbitals that are symmetric with respect to the plane are usually called n. Lone pairs contained in orbitals that are antisymmetric are usually more strongly delocalized and are considered to form a part of the π system.

Thus, the molecular orbitals of the chromophore are generally classified into the σ, σ^*, π, π^*, and n categories. Unless the chromophore by itself represents the whole molecule, these orbitals are not truly the canonical MOs of the molecule; however, they are often considered as such in that their interaction with the rest of the molecule is neglected to the first approximation. In this way, attention is focused on that part of the overall molecule which is most intimately involved in the act of light absorption. It should be remembered, however, that this is only an approximation and that in reality the electronic excitation of the chromophore is felt by the remainder of the molecule as well; and it should also be remembered that the canonical orbitals are really delocalized.

Figure 1.21 is a schematic presentation of the orbital energy levels in a typical chromophore. Single electron promotions from one of the occupied into one of the unoccupied MO levels of the ground configuration will yield singly excited singlet and triplet configurations that can be used to describe the excited states. Promotions of the $\pi \rightarrow \pi^*$, $\sigma \rightarrow \sigma^*$, and n $\rightarrow \sigma^*$ types will produce total wave functions that are symmetric with respect to mirroring in the symmetry plane; in this respect, these wave functions resemble the ground-state wave function. In general, many such configurations will have to be mixed to produce a good description of any one excited state, but since the promotions of the $\pi \rightarrow \pi^*$ type generally are at the lowest energies, one or more of them typically predominate in this kind of an excited state; such a state is referred to as a "$\pi\pi^*$ state" for brevity. Promotions of the n $\rightarrow \pi^*$, $\sigma \rightarrow \pi^*$, and $\pi \rightarrow \sigma^*$ types similarly produce configurations that are antisymmetric with respect to the plane, and these mix among themselves. Transitions from the ground state into these antisymmetric states are typically of low intensity and are only detected readily in absorption spectroscopy if they occur at lower energies than all of the normally much more intense $\pi \rightarrow \pi^*$ transitions. This usually happens only if one or more nπ^* configurations are present. These states are referred to as "nπ^*." States of purely $\pi\sigma^*$ and $\sigma\pi^*$ character are rarely observed in organic compounds.

As soon as the symmetry of the chromophore is broken (say, by a geometric distortion or by the presence of an unsymmetric remainder of the molecule), the labels σ and π become only approximate but are usually still considered useful. Perturbations due to geometric distortions or due to other groups in the molecule are then said to mix states of different symmetries, such as nπ^* and $\pi\pi^*$. When the deviation from the symmetric geometry becomes excessive, as is often the case along photochemical reaction paths, this mixing may become so strong that the labels eventually lose their utility.

In the singly excited configuration pair generated by the promotion of an elec-

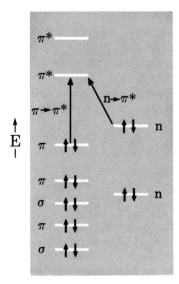

Figure 1.21 Electronic $\pi \rightarrow \pi^*$ and $n \rightarrow \pi^*$ transitions.

tron from one of the occupied orbitals, $|i\rangle$, into one of the unoccupied orbitals, $|j\rangle$, the singlet lies above the triplet by an amount equal to $2K_{ij}$. This quantity tends to be much larger for $\pi\pi^*$ configurations than for $n\pi^*$ configurations, since the n and the π^* orbitals tend to avoid each other efficiently in space. Even after configuration mixing, this situation remains largely unchanged and it is a good rule of thumb to assume that the energy separation between the lowest $\pi\pi^*$ triplet and the lowest $\pi\pi^*$ singlet is considerably larger than a similar separation between an $n\pi^*$ triplet and an $n\pi^*$ singlet. Among $\pi\pi^*$ states, those with the smallest singlet–triplet splitting involve promotions between orbitals $|i\rangle$ and $|j\rangle$ that are located in different regions in space, so that the transition density $e\rho_{ij}$ is very small, that is, they involve charge-transfer transitions. Such transitions are also characterized by relatively low intensity, since the transition moment is given by the dipole moment of the transition density (Section 1.2). The first transitions in azulene and acenaphthylene serve as examples.

Intermolecular charge-transfer transitions are also known. Their excited states are dominated by configurations in which an electron is taken from an orbital of one molecule, typically the highest occupied π orbital, and promoted to an orbital of another molecule, typically the lowest unoccupied π^* orbital. In order for a transition from the ground state into such an intermolecular charge-transfer state to have any observable intensity, the two molecules have to be in close vicinity so that the two orbitals involved have significant overlap and the transition density does not vanish. This condition is fulfilled in molecular complexes (charge-transfer complexes) and during intermolecular collisions (collision-induced charge-transfer absorption).

A transition can have low intensity for other reasons as well, particularly in molecules of high symmetry, but we shall refer the reader to books on electronic spectroscopy for this type of information. These sources also contain additional information on the characteristic UV spectral properties of systems of double bonds in linear or cyclic conjugation. The spectral properties of many chromophores of specific interest in organic photochemistry will be discussed in some detail in Part B of this book.

Another type of chromophore that leads to absorption above 200 nm is due to the presence of lone pairs on atoms that form relatively weak bonds, such as carbon–halogen bonds or sulfur–sulfur bonds. These are the $n\sigma^*$ excited states, in which the leading configuration is obtained by promoting an electron from a lone pair on an atom to a σ^* bond orbital located between that atom and its partner. Like $n\pi^*$ states and $\pi\sigma^*$ states, the $n\sigma^*$ states have relatively low intensity and a small singlet–triplet splitting.

Rydberg states. Rydberg states occur in series. Their energies E_n obey the relation

$$E_n = IP - R/(n - \delta)^2 \tag{1.12}$$

where IP is the ionization potential towards which the series converges as the integer n grows beyond all bounds, R is the Rydberg constant (13.60 eV, or 109,737.1 cm^{-1}), and δ is known as the quantum defect. Typical ionization potentials of organic molecules are in the range of 8–11 eV. The lowest line of a Rydberg series usually lies about 2 eV below the ionization limit, and this indicates that Rydberg excited configurations are likely to lie in the energy region around 200 nm or perhaps even a little lower in energy, so that they can mix with valence excited configurations that occur in this region. This is of limited importance in the photochemistry of those substrates whose chromophores have excited states at much lower energies. However, the existence of Rydberg excited configurations in this region may be essential for the understanding of the photochemistry of those chromophores that do not have low-energy valence states, such as an isolated double bond.

The spatial extent of a Rydberg orbital (i.e., an orbital with a high principal quantum number) can be huge (Figure 1.22). This has several important consequences.

First, interaction matrix elements between Rydberg configurations and valence excited configurations tend to be small, so that the Rydberg configurations often tend to remain aloof from the general mixing of configurations in this high energy region and form an identifiable series which converges to the ionization limit. It is generally admissible to talk about a Rydberg excited state, represented well by a single Rydberg configuration. This is not to say that no mixing occurs at all, and these series frequently show signs of perturbation.

Second, the energy difference between the triplet and singlet members of a Rydberg configuration pair tends to be small, since the exchange integral K_{ij} is small.

Third, Rydberg transitions tend to be easily identified only in the gas-phase spectra as relatively weak and often very sharp bands. In solution, the space required by the electron in a large Rydberg orbital is already pre-empted by the electron density belonging to neighboring solvent molecules, and this pushes the energy of the Rydberg transition up and causes the transition to be very diffuse and almost unobservable.

Fourth, because the electron in a large Rydberg orbital spends the bulk of its time quite far away from the center of the molecule, the excited species can be viewed very crudely as an ion pair composed of the radical cation of the molecule

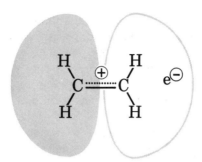

Figure 1.22 A p-symmetry Rydberg orbital (schematic).

and of a strongly delocalized electron playing the role of a normally completely localized anion. To a degree, then, the photochemical behavior of the Rydberg excited states can be analyzed in terms of the behavior of molecular radical cations.

Rydberg excited states tend to remain constant in energy or even to increase in energy upon distortion from equilibrium geometries of the ground state along photochemical reaction paths. This is most fortunate for those attempting to interpret organic photochemical reactions in simple ways. Examples of such behavior, sometimes referred to as de-Rydbergization of the S_1 or T_1 state, are given in Part B. At those geometries which are of particular interest in the determination of the outcome of photochemical reactions (i.e., biradicaloid geometries), Rydberg states rarely represent the S_1 or T_1 surface of an organic molecule, particularly in solution.

Assignment criteria. Although we do not wish to go into the details of electronic spectroscopy to any depth, it is perhaps useful to summarize some of the experimental criteria which can be used to assign a nature of electronically excited states. Each electronic transition in absorption or emission is characterized by its energy, intensity (in emission, radiative lifetime), polarization, and fine structure. Additional important properties are a response to minor perturbations such as introduction of substituent or change of solvent. Also, further spectral characteristics such as behavior in circular dichroism (CD), magnetic circular dichroism (MCD), or two-photon absorption spectroscopy are useful. The magnetic properties of the excited state and its response to quenching agents such as oxygen provide important clues.

The assignment of Rydberg states is often, but not always, straightforward. They are normally only observed in the gas phase, and they form a series of sharp lines converging on the ionization potential. Difficulties arise in the case of Rydberg-valence mixing.

The two assignment problems that an organic photochemist faces most often are a distinction between a singlet and a triplet excited state and the distinction between a $\pi\pi^*$ and an $n\pi^*$ excited state, using the standard notation introduced by Kasha.

Singlet–triplet absorption is normally so weak that it can hardly be mistaken for singlet–singlet absorption. Its intensity is often sensitive to external perturbations by the presence of heavy atoms (such as argon, mercury, or xenon) or by the presence of paramagnetic molecules (such as oxygen) in the solvent. The low absorption intensities are associated with a long radiative lifetime (and often also a long observed phosphoresence lifetime) and sensitivity to quenchers. The usual absence of detectable absorption causes the phosphorescent emission to appear in a spectral region separated from the absorption region by a considerable gap. Furthermore, it is often possible to detect the paramagnetism of the triplet state by electron spin resonance (ESR) measurements on solid solutions, usually at low temperatures.

The distinction between $n \to \pi^*$ and $\pi \to \pi^*$ transitions is usually based on the difference in their intensities (with the former normally being much weaker) and in their polarizations (in-plane for $\pi \to \pi^*$ versus out-of-plane for $n \to \pi^*$), and on the effects of perturbations. Electron-donating substituents that raise the energy of the π^* orbital tend to cause a hypsochromic shift for $n \to \pi^*$ transitions. This they rarely do for $\pi \to \pi^*$ transitions because they typically increase the energy of the π orbital even more than they increase the energy of the π^* orbital. In hydrogen-bonding solvents, $n \to \pi^*$ transitions tend to be shifted to higher energies as well, since the n orbital is stabilized.

It is important to note that in Kasha's notation, lone pairs that are not of σ symmetry but are exactly or approximately of π symmetry (such as the nitrogen lone pair in aniline or the arsenic lone pair in triphenylarsine) should not be labeled "n" but, instead, should be labeled "l". They normally enter into π and, to a smaller degree, π^* MOs.

1.4 COMMENTS AND REFERENCES

Readers interested in the history of the treatment of vitiligo with **psoralens** and ultraviolet light should consult the following references: A. M. El Mofty, *Vitiligo and Psoralens*, Pergamon, Oxford, 1968; T. B. Fitzpatrick and M. A. Pathak, *J. Invest. Dermatol.* **32**, 229 (1959); and M. A. Pathak, D. M. Kramer, and T. B. Fitzpatrick, in *Sunlight and Man* (T. B. Fitzpatrick, Ed.), University of Tokyo Press, 1973. The photochemistry and photobiology of psoralens are summarized in E. Ben-Hur and P.-S. Song, *Adv. Radiat. Biol.* **11**, 131 (1984).

A simplified and much shorter version of the present treatment of the theoretical aspects of **organic photochemistry,** with many examples and applications, has recently appeared in German: M. Klessinger and J. Michl, *Lichtabsorption und Photochemie Organischer Moleküle*, Verlag Chemie, Weinheim, Germany, 1989. Many general books on photochemistry are available. Examples of introductory texts are J. D. Coyle, *Introduction to Organic Photochemistry*, Wiley, New York, 1986 and the more physically oriented R. P. Wayne, *Principles and Applications of Photochemistry*, Oxford Science Publications, Oxford, 1988. More complete coverage of organic photochemical reactions is available in W. M. Horspool, Ed., *Synthetic Organic Photochemistry*, Plenum, New York, 1984 and in the older but still very useful book, N. J. Turro, *Modern Molecular Photochemistry*, Benjamin/Cummings, Menlo Park, New Jersey, 1978. Detailed information on progress in photochemistry is summarized yearly in the specialist periodic reports *Photochemistry*, published by The Royal Society of Chemistry.

A lucid, rigorous, yet brief discussion of the **Born–Oppenheimer approximation, potential energy surfaces,** and related subjects is found in J. Simons, *Energetic Principles of Chemical Reactions*, Jones and Bartlett Publishers, Boston, 1983. A book dedicated to potential energy surfaces has appeared recently: P. G. Mezey, *Potential Energy Hypersurfaces*,

Elsevier, Amsterdam, 1987. A discussion of the vibrational wave function of the initial electronically excited species is given in J. Simons, *J. Phys. Chem.* **86**, 3615 (1982).

The aspects of **molecular symmetry** that deal with permutations of nuclei are treated well in P. R. Bunker, *Molecular Symmetry and Spectroscopy,* Academic Press, New York, 1979.

Electronic wave functions are treated in numerous standard textbooks. A classic that we have found useful is R. McWeeny and B. T. Sutcliffe, *Methods of Molecular Quantum Mechanics,* Academic Press, London, 1969. Much information on spin is condensed in a short text, R. McWeeny, *Spins in Chemistry,* Academic Press, London, 1970. Relations to photochemistry are emphasized in L. Salem, *Electrons in Chemical Reactions: First Principles,* Wiley, New York, 1982. The physical origin of Hund's rule is analyzed in J. Katriel and R. Pauncz, *Adv. Quantum Chem.* **10**, 143 (1977).

The mechanics of the **computation of electronically excited wave functions** (including spin–orbit coupling terms) at the semiempirical level are also described in R. L. Ellis and H. H. Jaffé, in *Modern Theoretical Chemistry,* Vol. 8 (G. A. Segal, Ed.), Plenum, 1977. See Appendix III for references to *ab initio* computational methods for excited states.

The discussion of the relation between orbital energy differences and singlet versus triplet excitation energies is taken from J. Michl and E. W. Thulstrup, *Tetrahedron* **32**, 205 (1976).

Information on **electronic excited states of organic molecules** is available in texts on UV-visible spectroscopy: J. N. Murrell, *The Theory of Electronic Spectra of Organic Molecules,* Methuen, London, 1963; H. H. Jaffé and M. Orchin, *Theory and Applications of Ultraviolet Spectroscopy,* Wiley, New York, 1962; H. Suzuki, *Electronic Absorption Spectra and Geometry of Organic Molecules,* Adademic Press, New York, 1967; N. Mataga and T. Kubota, *Molecular Interactions and Electronic Spectra,* Marcel Dekker, New York, 1970; J. Fabian and H. Hartmann, *Light Absorption of Organic Colorants,* Springer Verlag, Berlin, 1980. Spectra of linear polyenes are treated in B. S. Hudson, B. E. Kohler, and K. Schulten, in *Excited States* (E. C. Lim, Ed.), Vol. 6, Academic Press, New York, 1982. For a discussion of $\sigma\sigma^*$, $\pi\pi^*$, $\sigma\pi^*$, $l\pi^*$, $d\pi^*$, $n\pi^*$ and $n\sigma^*$ labels for excited states, see M. Kasha and H. R. Rawls, *Photochem. Photobiol.* **7**, 561 (1968).

A review of **molecular geometries in excited states** is available in K. K. Innes, in *Excited States,* Vol. 2 (E. C. Lim, Ed.), Academic Press, New York, 1975. For a symmetry-based procedure for estimating the preferred directions for molecular distortion upon excitation, see V. Bachler and O. E. Polansky, *J. Am. Chem. Soc.* **110**, 5972 (1988) and V. Bachler, *J. Am. Chem. Soc.* **110**, 5977 (1988).

Finally, **Rydberg states,** largely ignored in the present text, which concentrates on solution photochemistry, are treated in detail in M. B. Robin, *Higher Excited States of Polyatomic Molecules,* Academic Press, New York, Vol. I, II (1975), III (1985).

CHAPTER 2
Photophysical and Photochemical Processes

This chapter provides an introductory treatment of the fundamentals of photophysical and photochemical dynamics. Although strictly speaking, this lies outside the scope of the intended coverage, a review of the basic aspects may be useful for full appreciation of the main subject.

2.1 AN OVERVIEW

Photophysical processes. Photochemical reactions of interest in this book are initiated by excitation of a reactant molecule from its ground state, typically S_0, into an electronic excited state. Absorption of a photon usually produces an excited singlet state, say S_1. Triplet energy transfer is commonly used to excite the molecule into the lowest excited triplet state T_1. The singlet excited molecule can jump from the S_1 surface to the triplet surface T_1, which usually lies somewhat lower in energy than S_1. This intramolecular process of multiplicity change is known as intersystem crossing, and its rate is generally enhanced by the presence of heavy atoms (the "heavy atom effect").

The molecule can also jump from the S_1 surface back to the ground S_0 surface, either with emission of a quantum of light or with conversion of its electronic energy into vibrational energy. The former, emissive intramolecular process is known as fluorescence; the latter, radiationless one is known as internal conversion. There also are two intramolecular processes for returning from the T_1 state to the ground state S_0. A return with emission of a quantum of light is referred to as phosphorescence, and a radiationless return is again called intersystem crossing.

Most of the intramolecular events described so far are photophysical, since they produce no net chemical change. They are summarized in the state diagram shown in Figure 2.1, usually called the Jablonski diagram. We use the standard convention in which radiative processes are shown by straight lines and where nonradiative processes are shown by wavy lines.

Photochemical processes. There are two ways in which the original electronic excitation can produce a chemical change. First, the conversion of the energy of the electronic excitation into the energy of the vibrational motion in one of the photophysical processes just discussed may generate a molecule that carries a very large amount of vibrational energy. This molecule may undergo a ground-state chemical transformation before it can cool off (hot ground-state photochemical

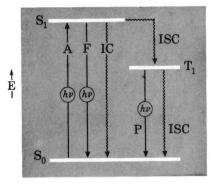

Figure 2.1 The Jablonski diagram. A, absorption; F, fluorescence; P, phosphorescence; IC, internal conversion; ISC, intersystem crossing.

reaction). Second, the excited molecule may undergo a chemical change while still in the excited state. This may correspond to motion on a single potential energy surface to some new geometry, but jumps between various excited-state surfaces may also be involved. Radiative or radiationless return to the ground-state surface at the new geometry will then produce a ground-state species that may be different from the starting molecule (excited-state photochemical reaction).

The fundamental photophysical and photochemical processes are summarized in a hypothetical photochemical system in Figures 2.2 and 2.3. The special script letters $\mathfrak{A}$–$\mathfrak{F}$ represent a series of geometries at which one or more of the states involved (S_0, S_1, and T_1) have a local minimum. The state of the molecule is shown in parentheses; thus $\mathfrak{A}(S_0)$ represents the starting molecule $\mathfrak{A}$ in its singlet ground state. In our hypothetical example, some fraction of the singlet excited molecules $\mathfrak{A}(S_1)$ reacts to yield a new excited species $\mathfrak{B}(S_1)$, which has no ground-state counterpart. Therefore, a subsequent jump to the S_0 state results in further chemical change, either to product $\mathfrak{C}$ in a diabatic photochemical reaction (i.e., with a nonradiative surface jump) or back to the starting material $\mathfrak{A}$. In competition with return to the S_0 state, the excited molecules $\mathfrak{B}(S_1)$ may also react further in the excited state to yield the excited state of the product, $\mathfrak{C}(S_1)$, in an adiabatic photochemical reaction (i.e., without a nonradiative surface jump). In the hypothetical example chosen, this state is depopulated only by fluorescence, assumed to be much faster than other potentially competitive processes such as internal conversion or intersystem crossing. It yields a stable ground-state product $\mathfrak{C}(S_0)$.

Another fraction of the singlet excited molecules $\mathfrak{A}(S_1)$ follows a course that leads to a point where the S_1 and S_0 surfaces touch, which is referred to as a conical intersection or a funnel. These molecules have no time for vibrational equilibration with the environment in the excited state and proceed directly to the ground state of a different product, $\mathfrak{E}(S_0)$, in a "direct" photochemical reaction. They possess a very large amount of vibrational energy and since they do not cool off fast enough, some of them can overcome a relatively high barrier and proceed to the final product, $\mathfrak{F}(S_0)$, in a hot ground-state photochemical reaction.

Figure 2.2 indicates that yet another fraction of the excited singlet molecules $\mathfrak{A}(S_1)$ returns to the ground state $\mathfrak{A}(S_0)$ by fluoresence. In this case, it is assumed that vertical internal conversion plays a negligible role.

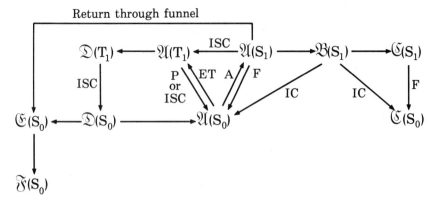

Figure 2.2 A summary of photochemical processes resulting from the excitation of a hypothetical chemical species $\mathfrak{A}$. The electronic state is indicated in parentheses. For labels on the arrows, see caption to Figure 2.1, ET = energy transfer; "return through funnel" corresponds to very fast IC. Unlabeled arrows represent chemical reactions producing species $\mathfrak{B}$, $\mathfrak{C}$, $\mathfrak{D}$, $\mathfrak{E}$, and $\mathfrak{F}$.

In the triplet state, which can be reached either from $\mathfrak{A}(S_0)$ by energy transfer or from $\mathfrak{A}(S_1)$ by intersystem crossing, $\mathfrak{A}(T_1)$ reacts to yield a different species $\mathfrak{D}(T_1)$, which is bound also in the ground state $\mathfrak{D}(S_0)$. However, in the example chosen, it is very reactive and reacts in the ground state with only small activation energies to yield either the product $\mathfrak{E}$ or the starting molecule $\mathfrak{A}$. Figure 2.2 indicates that some of the triplet molecules $\mathfrak{A}(T_1)$ also return to the ground state $\mathfrak{A}(S_0)$ by phosphorescence or intersystem crossing.

These processes are conveniently thought of as occurring on potential energy surfaces, as shown in Figure 2.3. It is perhaps advisable to point out that the exact shapes and energies of the potential energy surfaces are most simply thought of to be effectively a function of the solvent, although this is only a crude approximation. For instance, those geometries at which a molecule has a large dipole moment, or perhaps actually consists of a pair of ions, are stabilized in polar solvents, $n\pi^*$ states are destabilized in hydrogen-bonding solvents, and so on. In many cases, it is essential to represent the effective structure of the solvent around the solute molecule explicitly as at least one additional geometric variable which then adds at least one more dimension to the $3N - 6$ already under consideration.

A closer look. The above listing and description of the processes involved in our hypothetical photochemical system is greatly oversimplifed. We shall now briefly consider what needs to be added in preparation for the following sections in which the sequence of elementary events is taken up in some detail.

There are three main factors that bring us from the basic steps outlined in Figure 2.2 to the more complex reaction scheme given by Figure 2.4.

First, the light used in the initial excitation is usually not monochromatic and populates many vibrational states of several electronic states. Energy transfer, which represents an alternative means of initial excitation, is usually exothermic and may also produce a vibrationally hot species. The various simultaneously occurring excitation processes are shown by thick arrows in Figure 2.4. The arrows

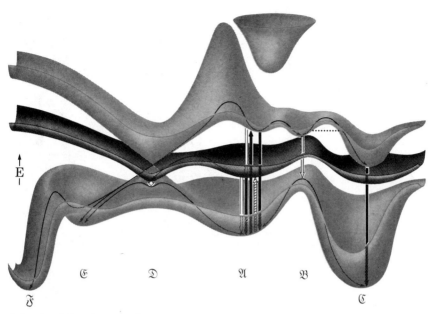

Figure 2.3 A schematic potential energy surface representation of the photochemical processes depicted in Figure 2.2. Light surfaces: Singlets S_0, S_1, and S_2. Dark surface: Triplet T_1. Vertical arrows represent jumps between surfaces: radiative transitions (black), energy transfer (hatched), internal conversion (grey), intersystem crossing (white). Adiabatic motion on a single surface is shown by thin curved arrows: full arrows denote "singlet photochemistry," and a dashed arrow denotes "triplet photochemistry." Tunneling is indicated by a dotted arrow.

are black for excitations to singlet states and are hatched for excitations to triplet states (only one triplet state is shown).

Second, all of the elementary steps are, in principle, reversible. Only in the case of some of them will the reverse step play a detectable role in practice. We have shown this explicitly for a few steps in Figure 2.4.

Third, the role of vibrational excitation needs to be considered more carefully. For the present purposes, we define as a vibrationally cool species one whose vibrational energy is not sufficiently high for a chemical reaction. In terms of potential energy surfaces, it is represented by a point that moves about in a local minimum but cannot escape from it except by tunneling. Hot species are labeled with an asterisk: $\mathfrak{A}(S_0)^*$, $\mathfrak{B}(S_1)^*$, and so on. If a molecule stays hot long enough, it will undergo a chemical reaction (the point that represents its geometry will escape over a barrier from the local minimum on a surface).

At any temperature above absolute zero, there will be a finite probability that random thermal fluctuations in the environment will impart enough vibrational energy to a molecule to make it hot. If the molecule does not cool off fast enough thereafter, it will undergo a reaction. In addition to this ordinary thermal activation mechanism of permitting molecules to acquire sufficient vibrational energy to react, we now need to consider two other mechanisms that produce vibrationally excited molecules. One of these is the initial excitation, as already mentioned. The other

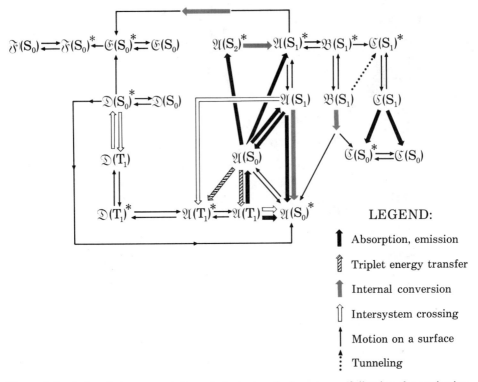

Figure 2.4 A detailed representation of photochemical processes following the excitation of the chemical species 𝔄. Asterisk denotes vibrational excitation. The thick arrows symbolize jumps between surfaces, using the conventions introduced in Figure 2.3. The thin full arrows symbolize adiabatic motion and a thin dotted arrow tunneling on a single surface.

is the conversion of electronic into vibrational energy associated with nonradiative surface jumps during internal conversion and intersystem crossing, and radiative surface jumps during fluorescence and phosphorescence into vibrationally excited levels of the ground state. In a nonradiative jump, all of the lost electronic energy is converted into energy of nuclear motion (vibration, internal rotation, libration, etc.). In a radiative jump, only a fraction is so converted and the remainder appears as the energy of a photon (note that emission most often produces a vibrationally excited level of the lower electronic state). Only in the limiting case of zero–zero electronic transition, all of the lost electronic energy is converted into radiation.

If a reaction occurs as a result of vibrationally hot molecules being produced by the initial excitation or by surface jumps, and reacting faster than they can cool off, it is referred to as a hot reaction. If it occurs on the S_0 surface, it is a hot ground-state reaction, which we have already mentioned. If it occurs on an excited surface, it is a hot excited-state reaction.

In Figure 2.4, we have used thin arrows to indicate processes in which molecular geometry changes adiabatically (i.e., without a change in the electronic state); this corresponds to the motion of a point on a surface. Radiative and radiationless jumps between surfaces are indicated by fat arrows, grey for internal conversion

TABLE 2.1 Photophysical[a] and Photochemical[b] Processes

	Intramolecular	
Section 2.2.1: *Initial Excitation*	Section 2.2.2: *Fast Events* (~ps)	Section 2.2.3: *Slow Events* (~ns to ~s)
$S_0 \to S_n$ absorption $S_0 \to T_n$ absorption Chemiexcitation	$S_n \to S_1$ internal conversion $T_n \to T_1$ internal conversion Vibrational relaxation Hot state reactions "Direct" reactions	$S_1 \to S_0$ internal conversion $T_1 \to S_0$ intersystem crossing $S_1 \to T_n$ intersystem crossing Adiabatic reactions Fluorescence, phosphorescence
	Intermolecular[c]	
Section 2.3.1: *Initial Excitation*	Section 2.3.2: *Molecular Transport Processes*	Section 2.3.3: *Electronic Transfer Processes*

Radiative electron
  transfer—absorption
Energy transfer
Ion recombination
Radical recombination
	Rotational diffusion
	Translational diffusion and
	  solute–solute encounters
		Energy transfer
		Radiative electron
		  transfer—emission
		Nonradiative electron
		  transfer

[a]The ground state of the initial molecule is recovered at the end.
[b]The ground state of a different molecule is recovered at the end.
[c]Typically followed by one or more processes listed as "intramolecular."

and white for intersystem crossing. Some steps involve both a jump and a substantial geometry change, and this is indicated by composite arrows.

Finally, tunneling, which permits even a vibrationally cool molecule to undergo a chemical reaction, is indicated by a dotted arrow. In the surface picture, the reaction occurs by passage under the barrier. Only one example is shown in Figure 2.4, although, in principle, tunneling contributes to some degree to all thermally activated chemical reactions.

The complications introduced in Figure 2.4 relative to Figure 2.2 can also be introduced explicitly into the surface picture shown in Figure 2.3. This is a particularly useful way of viewing the situation, and we shall use it in the following sections.

Photophysical and photochemical processes are summarized in Table 2.1.

The Fermi golden rule. The rates of most elementary steps encountered in this chapter can be satisfactorily approximated by an expression known as the Fermi golden rule:

$$k = (2\pi/\hbar)|\hat{H}_{pert}|^2 \rho \qquad (2.1)$$

Here, $\hat{H}_{pert}$ is the matrix element of the perturbation that induces the elementary step, taken between the initial and the final state, and ρ is the density of the final states at the energy of the initial state.

In most cases, $\hat{H}_{pert}$ can be factored into a purely electronic matrix element and a Franck–Condon overlap integral for each final state. Then, the above expression can be taken to involve a perturbation factor $\hat{H}_{pert}$ that acts only on the electronic wave function and a Franck–Condon weighted density of final states ρ. The expressions given in the following for the rates of the various elementary processes can be derived from this general result.

2.2 INTRAMOLECULAR PROCESSES

2.2.1 Initial Excitation

The most common intramolecular process that leads to initial electronic excitation in organic photochemistry is light absorption. At times, chemiexcitation needs to be considered as well.

Light absorption

General remarks. The exchange of energy between electromagnetic waves and molecules can be described in a semiclassical fashion as being due to an interaction of the electric field of the wave with an oscillating electric dipole moment set up in a molecule; this dipole moment is the result of the perturbation by the outside field. The oscillation frequency ν is given by the difference ΔE of the energies of the lower and upper state, $\nu = \Delta E/h$. When the frequency of the oscillating dipole moment and the frequency of the field agree, a resonance occurs and energy can flow from the field into the molecule (light absorption) or from the molecule into the field (stimulated light emission).

The oscillating electric dipole moment produced in the molecule has an amplitude and direction determined by a vector $\mathbf{M}$, known as the electric dipole transition moment. The oscillations do not distinguish between the positive and the negative end of the vector $\mathbf{M}$, and the choice of the polarity is arbitrary.

The electronic transition moment vector $\mathbf{M}_{0n}$ is related to the wave functions of the initial state $\Psi_{el}^{(0)}$ and the final state $\Psi_{el}^{(n)}$ by

$$\mathbf{M}_{0n} = \langle \Psi_{el}^{(n)} | \hat{\mathbf{M}} | \Psi_{el}^{(0)} \rangle$$
$$\hat{\mathbf{M}} = -|e| \sum_i \mathbf{r}_i + |e| \sum_\alpha Z_\alpha \mathbf{R}_\alpha \qquad (2.2)$$

where $\mathbf{r}_i$ is the position vector of the i-th electron, $\mathbf{R}_\alpha$ is the position vector of the αth nucleus, e is the electron charge, Z_α is the charge of the αth nucleus; the sum over i runs over all electrons, and the sum over α runs over all nuclei in the molecule. A multiplication of $\Psi_{el}^{(0)}$ or $\Psi_{el}^{(n)}$ by -1 or another complex unity causes a change in $\mathbf{M}_{0n}$ by the same multiplicative factor but has no effect on any observable quantities.

The integrated absorption probability is proportional to $|\mathbf{M}_{0n}|^2$, and the polarization direction of the transition is given by the direction of the vector $\mathbf{M}_{0n}$.

The Franck–Condon principle. Electronic excitation of molecules by light absorption is usually said to be a "vertical" process, that is, one in which the molecular geometry is preserved. On an energy surface diagram, this means that the position of the point which represents the molecular geometry on the horizontal axis remains the same before and after a jump between surfaces, so that an arrow connecting the starting point with the end point of the jump is vertical. The statement is only approximately true, since the wave nature of the nuclei cannot really be ignored, so that a molecule cannot be characterized by a single sharply defined geometry in any of its stationary electronic states.

In the Born–Oppenheimer approximation, the initial state of the absorbing molecule is described sufficiently for our purposes by the product of the electronic wave function of the S_0 state and an appropriate vibrational wave function, say, that of the lowest vibrational state: $\Psi_{el}^{(0)}(\mathbf{x},\mathbf{R})\,\Psi_{vib,0}^{(0)}(\mathbf{R})$ (cf. equation 1.1). Here, the coordinates of the electrons are collected in the vector $\mathbf{x}$, and the coordinates of the nuclei are collected in the vector $\mathbf{R}$. The superscripts on the wave functions label the various electronic and vibrational levels, and the subscript on the vibrational wave function indicates the electronic level to which the function belongs. Transitions into various vibrational levels of the excited electronic state S_1 are possible. Each of these is characterized by the same electronic wave function $\Psi_{el}^{(1)}(\mathbf{x},\mathbf{R})$; they differ in the vibrational wave functions $\Psi_{vib,v}^{(1)}(\mathbf{R})$, where $v = 0, 1, \ldots$ (Figure 2.5).

The delocalized nature of the nuclei is reflected in the fact that all of the vibrational wave functions have nonzero values over a whole range of geometries $\mathbf{R}$ centered around the equilibrium geometry. If the assumption is made that the

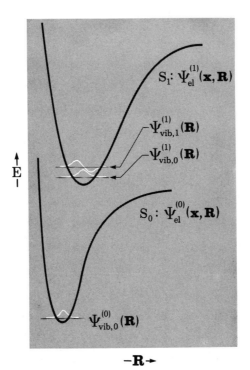

Figure 2.5 Vibronic levels in S_0 and S_1.

electronic transition moment does not depend on the geometry **R** within this restricted range, as is usually reasonable for allowed transitions, we arrive at the Franck–Condon principle. This states that the probability of a radiative transition (absorption or emission) from the v-th vibrational level of the n-th electronic state to the v'-th vibrational level of the n'-th electronic state, W(n,v → n',v'), is

$$W(n,v \rightarrow n',v') = \text{const} \times \left(\int \Psi^{(n')}_{\text{vib},v'} \Psi^{(n)}_{\text{vib},v} \, d\mathbf{R} \right)^2$$
$$= \text{const} \times |\langle \Psi^{(n')}_{\text{vib},v'} | \Psi^{(n)}_{\text{vib},v} \rangle|^2 \quad (2.3)$$

The integral $\langle \Psi^{(n')}_{\text{vib},v'} | \Psi^{(n)}_{\text{vib},v} \rangle$ is known as the Franck–Condon overlap integral, and the square of its absolute value is known as the Franck–Condon factor. The proportionality constant in front of the expression contains the square of the electronic transition moment, $|\mathbf{M}_{nn'}|^2$. In our example of $S_0 \rightarrow S_1$ absorption (Figure 2.5), the Franck–Condon factors are $|\langle \Psi^{(1)}_{\text{vib},v} | \Psi^{(0)}_{\text{vib},0} \rangle|^2$. Figure 2.6 shows three cases of such transitions that differ in the relative disposition of the potential energy surfaces S_0 and S_1. When the equilibrium geometries in the two states are similar (small lateral displacement) and the force constants are similar as well (little difference in curve shapes), the overlap of $\Psi^{(0)}_{\text{vib},0}$ with $\Psi^{(1)}_{\text{vib},0}$ is large and its overlaps with all of the other vibrational wave functions of the excited electronic state ($\Psi^{(1)}_{\text{vib},1}$, $\Psi^{(1)}_{\text{vib},2}$, etc.) are small. Such a transition is referred to as Franck–Condon-allowed. In the absorption spectrum the 0 → 0 peak will appear with high intensity, whereas the 0 → 1, 0 → 2, . . . peaks at higher energies will appear with rapidly decreasing intensity (Figure 2.6). If the S_0 and S_1 surfaces were to have identical equilibrium geometries and identical shapes, only the 0–0 absorption peak (the "origin" of the band) would appear and the others would carry zero intensity. This limit is normally not reached in practice.

If the S_0 and S_1 surfaces differ in shape but not in the equilibrium geometry, additional peaks begin to appear in the vibrational fine structure. For instance, for antisymmetric vibrational modes in symmetric molecules, 0–0, 0–2, 0–4, . . . peaks become allowed in the harmonic approximation.

As the equilibrium geometries begin to differ, spectral intensity shifts from the 0–0 peak in the spectrum to 0–1, 0–2, and so on, and the transition gradually becomes Franck–Condon-forbidden. When the difference in the equilibrium geometries is large, it may be extremely difficult to detect the lower-energy peaks at all, and an absorption maximum may appear, say, at the 0 → 10 transition (Figure 2.6). Roughly speaking, the most intense peaks in the spectra will be those whose vibrational wave functions have classical turning points located near the equilibrium geometry of the S_0 state; that is, the "vertical" transitions.

If monochromatic light is used for the excitation, a particular vibrational level of S_1 can be populated exclusively. In this level, the nuclei may be significantly delocalized over a region of geometries that is far larger than that in the vertical state, or they may be mostly localized in a region of geometries that differs from the initial one. If the initial and final vibrational wave functions are such that the transition corresponds to a large relocation of nuclear probability density (i.e., to a large change in geometry), the transition will be highly improbable but still possible. In this sense the transition may appear not to be vertical. In a sense, however, it really is vertical, since the weak tails of the vibrational wave function

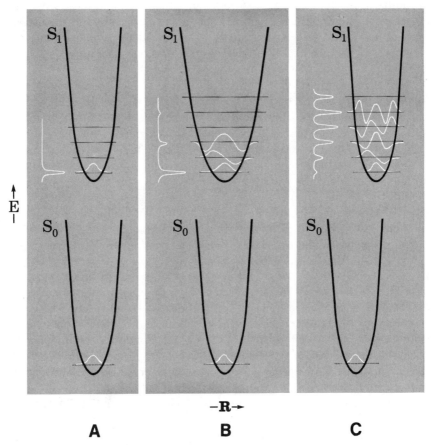

Figure 2.6 Electronic transitions: (A) Fully Franck–Condon "allowed"; (B) Partially Franck–Condon "allowed"; (C) Franck–Condon "forbidden." The spectral intensities of transitions from the lowest vibrational level of the ground state to the vibrational levels of the electronically excited state, as well as the vibrational wavefunctions of these levels, are shown schematically.

in the S_0 and in the S_1 states represent a weak delocalization of the nuclei into all space in both states, and it is this feature that then permits the integral $\langle \Psi_{vib,v}^{(1)} | \Psi_{vib,0}^{(0)} \rangle$ to be nonzero. Thus, although improbable, the transition can still occur even though the two equilibrium geometries are different and do not lie above each other in Figure 2.6, but it is mostly nuclear amplitude located away from the equilibrium geometry that is being used to induce the jump.

The physical reason for the requirement of vertical excitation can be found in the much larger mass of the nuclei relative to electrons. At the light frequencies needed for electronic excitation, electrons can follow the changes in the direction of the rapidly oscillating electric field, but the inertia of the nuclei is far too high for resonant excitation. This is true regardless of the length of time the molecule spends in the electromagnetic field. Thus, the popular statement that the Franck–Condon principle is a consequence of the short duration of the electronic excitation event is not really correct.

Broad-band (rather than monochromatic) irradiation is often used in photochemical experiments. Since the usual sources of such light are incoherent, its action can be viewed as a large number of independent excitations into various vibrational levels proceeding in parallel to each other, with probabilities determined by the Franck–Condon factors.

A clear demonstration of the "vertical" nature of electronic excitation would be obtained in the so far hypothetical case of coherent white light, such as an ultrashort light pulse. Its interaction with a molecule would produce a coherent superposition of vibrational states: It would transform the initial wave function $\Psi_{el}^{(0)}\Psi_{vib,0}^{(0)}$ into a final wave function that differs only in the electronic part, $\Psi_{el}^{(1)}\Psi_{vib,0}^{(0)}$. Now it would be highly unlikely for $\Psi_{vib,0}^{(0)}$ to be equal to one of the eigenfunctions $\Psi_{vib,v}^{(1)}$ of the vibrational Hamiltonian for the S_1 state, whose surface has a different shape. It can, however, be thought of as a superposition of a large number of vibrational eigenfunctions of S_1. A measurement of energy can only yield as an answer the energy of one of these eigenfunctions, and it will do so with probabilities given by the squares of the projections of $\Psi_{vib,0}^{(0)}$ into the individual $\Psi_{vib,v}^{(1)}$'s (i.e., with the Franck–Condon factors).

In experiments with common light sources, either monochromatic or incoherent polychromatic, the change in the wave function upon excitation is more complicated. The vibrational wave function of the molecule immediately after excitation is then not only limited to those regions of the nuclear configuration space where the initial vibrational wave function $\Psi_{vib,0}^{(0)}$ has a large amplitude, but it is also further restricted to regions where the difference between the starting and the final electronic potential energy surface is very close to the energy of the photons absorbed.

In many molecules, it is not appropriate to think of the geometry of the ground state as represented by points **R** localized in a small neighborhood. If low-frequency vibrational modes are available and, in particular, if the molecule can exist in several minima separated by low barriers (e.g., as a mixture of conformers), it is necessary to consider a whole range of points when referring to "the ground-state geometry." This is also true for "supermolecules" composed of two independent molecules, since the degree of freedom that represents the separation and the five that represent the mutual orientation of the two components are described by parameters whose values are distributed randomly if the initial solution is isotropic.

After excitation, the excited-state geometry will be "blurred" in a similar manner as the ground-state geometry. Different sections of this set of points may have different excitation probabilities. If their post-excitation fates differ, this will introduce wavelength dependence into the photochemical results. Temperature is also a consideration, since it will affect the degree to which different geometries of flexible molecules are populated.

Singlet-singlet excitation. This spin-allowed process represents the usual means of initiating a photochemical reaction. Under the usual experimental conditions employed in organic photochemistry, initial light absorption produces vibrationally hot molecules. If the light used is polychromatic, a whole variety of vibrational levels of several low-lying electronic states are excited. Even if monochromatic light is used, its wavelength is usually dictated by convenience and thus by the nature of the light source used, and it need not produce vibrationally thermalized molecules. In order to accomplish this, nearly monochromatic light of a wavelength corresponding to the energy difference of the lowest vibrational levels of the S_0 and S_1 states would have to be used.

In Figure 2.7, we have chosen to show simultaneous absorption from the S_0 into the S_1 and S_2 states, producing some vibrationally cool $\mathfrak{A}(S_1)$ molecules, some vibrationally hot $\mathfrak{A}(S_1)^*$ molecules, and some vibrationally hot molecules in the S_2 state, $\mathfrak{A}(S_2)^*$. We also show excitation into the T state by energy transfer, to be discussed in Section 2.3.

With few exceptions, the rate of internal conversion from S_2 and higher excited states of organic molecules to their lowest singlet state S_1 is at least comparable to the rate of vibrational cooling and may even be higher. This situation is schematically depicted in Figure 2.8 by showing vibrationally excited $\mathfrak{A}$ molecules in the S_2 state, $\mathfrak{A}(S_2)^*$, jumping to the S_1 state during the vibrational cooling process. In the jump, additional vibrational energy is released.

Singlet–triplet excitation. In principle, absorption of a photon of light by the S_0 state of a molecule can produce not only an excited singlet but also an excited triplet state. However, the probability of the latter event is very low because of its "spin-forbidden" nature, with the corresponding extinction coefficient being many orders of magnitude smaller than usual in singlet–singlet absorption processes. This makes it difficult to initiate photochemical reactions directly by photon absorption into a triplet state. In the few cases in which this was done, the substrate contained one or more heavy atoms, whose presence reduces the spin-forbiddenness of the singlet–triplet absorption process. It is also possible to incorporate heavy atoms into the solvent by using for instance, liquid xenon. Another possibility is the incorporation of paramagnetic molecules, such as oxygen, into the environment,

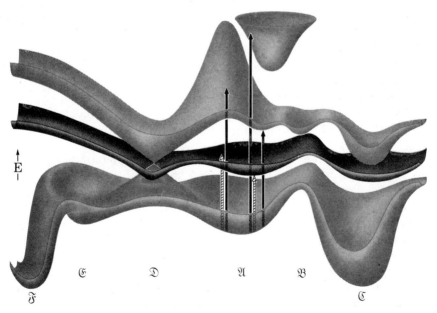

Figure 2.7 The initial excitation processes in the photochemistry of chemical species $\mathfrak{A}$. See caption to Figure 2.3.

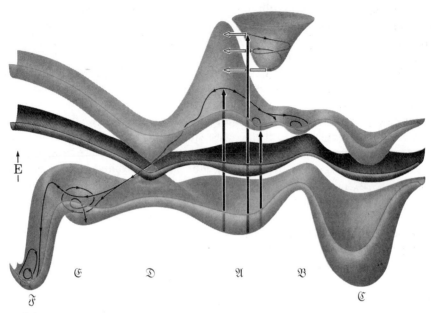

Figure 2.8 The fast processes in the singlet photochemistry of chemical species 𝔄. See caption to Figure 2.3.

but they are often not inert chemically and their presence may cause other complications. Since the singlet–singlet absorption is so much stronger than the singlet–triplet absorption, effective excitation into a triplet state is only possible at wavelengths longer than that which corresponds to the onset of singlet–singlet absorption. Thus, the fact that T_1 lies below S_1 is essential. In radicals, the lowest quartet state, Q_1, generally lies above the lowest excited doublet state D_1, and a selective excitation $D_0 \rightarrow Q_1$ is then impossible.

In practice, excitation into the triplet state is most commonly accomplished by triplet energy transfer (Section 2.3.2). Molecules can also be excited with particles other than photons. For some of these modes of excitation, singlet–triplet transitions are just as allowed as singlet–singlet transitions. This is true, for instance, for excitation by collision with relatively slow electrons.

The effect of excitation energy and excitation density. The total amount of energy delivered to a molecule by one of the methods just discussed may have an important effect on the photochemical processes that follow, since it determines which electronic excited state surface is reached and how much vibrational energy is delivered to the nuclei. Later on, we shall illustrate the importance of variables such as the excitation wavelength.

However, the density of the excitation, i.e., the concentration of the excited species, may also play an important role. This variable can also be controlled, e.g., by varying the intensity of the light used for the excitation. If a molecule that has received two quanta of excitation, e.g., absorbed two photons in sequence, reacts differently from a less energetic molecule that has only received one quantum, the photochemical outcome will again depend on the excitation density. This is particularly important for reactions in rigid solutions at low temperatures. Under these conditions, significant populations of T_1 molecules often build up and these can then absorb another photon. Reactions of the type $S_0 \rightarrow S_1 \rightarrow T_1 \rightarrow T_n \rightarrow$ product can then occur.

Simultaneous absorption of two photons is also possible and occurs when very high light intensities are used.

Chemiexcitation. The term chemiexcitation is somewhat vague and is usually used to denote production of electronically excited states from ground-state species at the expense of chemical energy. Although it is rarely used in photochemical practice, it is closely enough related to photochemical processes that we have decided to include it here.

As mentioned in Section 1.2.2, a vibrationally excited ground-state molecule can jump from the ground-state surface to an isoenergetic vibrational level of an excited surface. If the energy separation of the two surfaces is small, the vibrational energy that needs to be delivered to the starting molecule may be readily available already at room temperature. The probability of the event will be particularly high if the two surfaces touch at the geometries accessible to the molecule during its vibrational excursions on the ground-state energy surface, but this is not necessary.

The thermally induced jump to the excited state will have no chemical consequences if it is followed only by return to the ground state. Such photophysical type of thermal equilibration between the ground electronic state and excited electronic state is common in triplet ground-state molecules with a low-lying singlet state, such as certain carbenes (Figure 2.9A). Since most organic molecules have very limited thermal stability, only low-energy excited states can be populated in this way.

However, if the molecule in the excited state can escape over a small barrier to a deeper well from which it will not return, a chemical transformation will take place. If in the process the separation of the ground-state and first excited-state surfaces has increased (Figure 2.9B), the resulting excited molecule may reveal its presence by emitting light (chemiluminescence) and may undergo further chemical

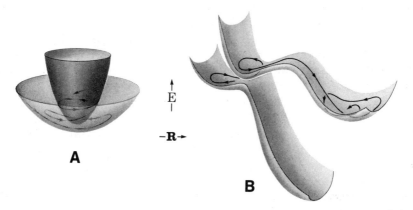

Figure 2.9 Thermally activated chemiexcitation from the S_0 state to a T_1 state (A) or an S_1 state (B). Note the S_0–S_1 "avoided touching."

transformations ("photochemistry without light"). The large drop in the energy of the ground state along the reaction coordinate, which is necessary if an excited state of substantial excitation energy is to be produced by this mechanism, corresponds to a large decrease in the chemical energy content (hence, chemiexcitation).

2.2.2 The Fast Events

The term fast events will be used to describe those processes that occur on a time scale of much less than a nanosecond. The most common among them are intramolecular and intermolecular vibrational relaxation and equilibration with the surrounding solution, as well as interconversion among the many low-lying excited states of equal multiplicity ("internal conversion"). Under special circumstances, other fast events can occur: intersystem crossing, hot excited-state reactions, direct reactions, and hot ground-state reactions. The hot and direct reactions can be viewed as special cases of intramolecular vibrational relaxation, in the latter case combined with a simultaneous surface jump.

The discussion of the Franck–Condon principle in Section 2.2.1 made it clear that the vibrational wave function of the initially produced excited state will depend in a fairly complicated way on the exact circumstances surrounding the excitation. For a qualitative discussion, it is best to adopt a classical picture, in which the point representing the molecule on the excited surface will start rolling downhill immediately after its birth (Figure 2.8).

The time scale of this motion can be subpicosecond and depends on the steepness of the slope of the surface on which the point representing the excited molecule landed. Even quite large geometry changes can occur very fast if this slope is steep.

Internal conversion. If the initial excitation was into one of the higher singlets ($S_0 \rightarrow S_n$) or triplets ($S_0 \rightarrow T_n$), we need to consider an additional subpicosecond type of event, namely, internal conversion ($S_n \rightarrow S_1$ or $T_n \rightarrow T_1$). With few exceptions, such as the $S_2 \rightarrow S_1$ conversion of azulene, this process is even faster than vibrational energy equilibration; we have indicated this schematically in Figure 2.8. Typical rates that have been measured for aromatics correspond to lifetimes of a few dozen femtoseconds to a few hundred femtoseconds. Expressions for the rate of internal conversion are discussed in Section 2.2.3.

Two special kinds of internal conversion are discussed in Section 2.3.3. They occur in "bichromophoric" molecules, which can be approximated as an assembly of two molecular species with relatively low excitation energies (chromophores), coupled by a relatively inert saturated framework. In one case, internal conversion is between states one of which is largely localized on one chromophore, while the other is largely localized on the other chromophore (intramolecular energy transfer). In the second case, internal conversion is between states which differ in the number of electrons allocated to each chromophore (intramolecular charge transfer or charge translocation).

Intersystem crossing. In some organic molecules, intersystem crossing may be very fast (aromatic ketones, molecules with heavy atoms), although it apparently hardly ever occurs in the subpicosecond domain. When it is unusually fast, it can compete with vibrational cooling; in such a case the first thermally equilibrated species formed after an initial S_0–S_1 excitation may find itself in a minimum in the T_1 surface. Expressions for the rate of intersystem crossing are discussed in Section 2.2.3.

Vibrational relaxation and barrier-free adiabatic reactions. Intramolecular vibrational relaxation (IVR)—that is, redistribution of vibrational energy among the various vibrational modes of a molecule—appears to occur on the 100-fs–100-ps scale in fluid solutions and is normally the fastest process after the $S_n \rightarrow S_1$ internal conversion. It can be considerably slower in isolated small molecules.

The initial adjustment of the solvent environment to the electronic excitation of the solute can be extremely fast (50–200 fs in hexane at room temperature), but it takes a little longer before the energetic vibrations of the excited species subside as a result of energy losses upon collisions with surrounding solvent molecules, typically picoseconds or tens of picoseconds (intermolecular vibrational relaxation). Then, the molecule will settle into a minimum in the excited S_1 surface if the initial excitation was of the S_0–S_1 type or in the T_1 surface if it was of the S_0–T_1 type, producing a thermally equilibrated S_1 or T_1 excited molecule of the starting material.

If the first minimum into which the point representing the molecular geometry settles after the initial excitation is quite distant from the initially vertically excited geometry, it may be useful to think of the electronically adiabatic vibrational relaxation process as a "barrier-free" chemical reaction. This is particularly useful for solution reactions involving a large change in the electronic structure and in the molecular charge distribution, whose rate is then controlled by solvent properties such as the dielectric relaxation time. N-Arylaminonaphthalenesulfonic acids and their derivatives, as well as certain triphenylmethane dyes in polar solvents, are good examples of this type of behavior.

Hot excited-state reactions. In many molecules, the energetic vibrational motions that rapidly follow the initial excitation, even if the excitation is to one of the first few higher excited states, result in no new chemical reactions. This means that the motions do not take the point representing the molecular geometry irretrievably over any barriers which it would not overcome anyway as a simple result of subsequent thermal activation. In such a case, the photochemical result is wavelength-independent. However, as one goes to even more highly excited states in the vacuum ultraviolet region, the picture often changes and new reaction channels open. Now, travel over barriers competes with vibrational cooling, and the thermally equilibrated molecule that is being produced in the S_1 and T_1 state is a chemical species other than the starting molecule—a hot reaction has occurred.

For some organic molecules, it is already known (and for many others, it can be suspected) that such competition between chemical reactions of a vibrationally hot molecule and its cooling begins upon excitation into low-lying electronic states. Indeed, it can occur upon excitation in the higher vibrational levels of the lowest states S_1 or T_1. These processes are referred to as hot excited-state reactions. Some of the motion over otherwise prohibitive barriers in S_1 or T_1 which can occur upon excitation into the higher states S_n or T_n, respectively, may be due to nuclear motion on these higher surfaces before internal conversion to S_1 or T_1 takes place.

For large organic molecules, the experimental distinction is difficult and it may be best to talk in such a case of "hot and/or higher excited-state" reactions.

Thus, it would seem that after some picoseconds, or tens of picoseconds, the point representing the excited molecule will have settled in one or another of the various minima of the S_1 or T_1 surface. Depending on the nature of the minimum reached, this could then correspond to a photophysical or photochemical outcome. Which particular minimum will be reached with what probability is very hard to predict. Here, not only the energy of the initial excitation but also the nature of the solvent will play a role. Both the viscosity of the solvent and the rate at which it removes excess vibrational energy from the molecule may be important.

Figure 2.8 shows an example of the hot excited species $\mathfrak{A}(S_1)^*$ cooling off to $\mathfrak{A}(S_1)$ in the photophysical way and, in competition, cooling off with a concomitant passage over a barrier to yield $\mathfrak{B}(S_1)$. For simplicity, the competition between vibrational cooling and intersystem crossing is not shown.

"Direct" reactions (passage through a funnel). There is an alternative fate that an excited molecule can meet on the time scale of picoseconds, shown on the left in Figure 2.8. In the course of the energetic vibrational motions provoked by the initial excitation, the molecule can arrive in the vicinity of a funnel in the S_1 state, i.e., an area where an $S_1 \rightarrow S_0$ jump is highly probable. A conical intersection shown in Figure 2.8 is an extreme example of such a situation. Then, the molecule will never have a chance of thermally equilibrating in the excited state at all. Rather, it will find itself back on the S_0 surface within picoseconds or tens of picoseconds after the initial event, and it will thermally equilibrate into one of the minima in S_0, $\mathfrak{C}(S_0)$ in Figure 2.8. Such a process is referred to as a direct photochemical reaction.

The simplest formulation of the jump probability P in an area of an avoided crossing is due to Landau and Zener, who write

$$P = 1 - \exp(-\pi^2 \Delta E^2 / hv\Delta s) \qquad (2.4)$$

where ΔE is the energy separation of the two surfaces at the geometry of close approach, v is the velocity of nuclear motion through the area of the avoided crossing, and the positive quantity Δs is the difference of slopes the two surfaces would have at the crossing point if their crossing were not avoided. This formulation is not applicable in the case of weak coupling (large ΔE).

More accurately, the rate of a transition from a vibronic level of the (n + 1)st electronic state $\Psi_{el}^{(n+1)}\Psi_{vib,v}^{(n+1)}$ to any of the isoenergetic vibronic levels of the ground state $\Psi_{el}^{(n)}\Psi_{vib,v'}^{(n)}$ is given by

$$W = \frac{2\pi}{\hbar} \sum_{v'} |\langle\Psi_{el}^{(n)}|\langle\Psi_{vib,v'}^{(n)}|\nabla_R^2\Psi_{el}^{(n+1)}\rangle + 2\nabla_R\Psi_{el}^{(n+1)}\cdot\nabla_R|\Psi_{vib,v}^{(n+1)}\rangle|^2 \delta(E_{v'}^{(0)} - E_v^{(n)}) \qquad (2.5)$$

The presence of the $\delta(E_{v'}^{(0)} - E_v^{(n)})$ term, which vanishes unless $E_{v'}^{(0)} = E_v^{(n)}$, guarantees energy conservation. The term in ∇_R^2 is usually small and is normally neglected. The matrix element can then be written as

$$2\langle\Psi_{vib,v'}^{(n)}|\langle\Psi_{el}^{(n)}|\nabla_R|\Psi_{el}^{(n+1)}\rangle\cdot\nabla_R|\Psi_{vib,v}^{(n+1)}\rangle$$

The quantity $\nabla_R|\Psi_{el}^{(n+1)}\rangle$ is a vector in the nuclear configuration space and points in the direction in which the electronic wave function $\Psi_{el}^{(n+1)}$ changes the fastest with the change in geometry. Its length is a measure of the rate of this change. The change in $\Psi_{el}^{(n+1)}$ can be thought of as the admixing of other electronic states to $\Psi_{el}^{(n+1)}$. The matrix element $\langle\Psi_{el}^{(n)}|\nabla_R|\Psi_{el}^{(n+1)}\rangle$ provides a measure of how much of this admixture is contributed by the lower state $\Psi_{el}^{(n)}$.

In a region of an avoided crossing, where the two states gradually exchange the nature of their electronic wave functions, $\langle\Psi_{el}^{(n)}|\nabla_R|\Psi_{el}^{(n+1)}\rangle$ can be expected to be particularly large. This is illustrated in Figure 2.10 for the avoided crossing between the lowest two excited states of singly pyramidalized ethylene upon twisting. Although the pure twisting motion, which changes only the twist angle Θ and leaves all other geometrical parameters of the ethylene molecule intact, undoubtedly does not coincide exactly with the direction of the gradient $\nabla_R|S_2\rangle$, it is likely to represent a reasonable approximation and we use $\partial/\partial\Theta$ in place of ∇_R. The magnitude of the electronic matrix element $\langle S_1|\partial/\partial\Theta|S_2\rangle$ is particularly large if the S_1–S_2 state crossing is only weakly avoided, and it decreases as the change in the nature of the wave functions becomes more gradual. This is illustrated by a comparison of parts A and B of Figure 2.10. It is further documented by comparison of Figure 2.10 with Figure 2.11, which shows $\langle S_0|\partial/\partial\Theta|S_1\rangle$ for the strongly avoided crossing of the S_1 and S_0 states of pyramidalized ethylene.

Another factor that favors surface jumps at weakly avoided crossings becomes clear upon consideration of the isoenergetic vibrational wave functions. If the two surfaces are far apart, these two wave functions will differ widely in the rate of oscillation as a function of geometry, since v' then must be much larger than v (Figure 2.11). If the two surfaces are close in energy, this difference will be minimized and the integral over nuclear coordinates will be more likely to have a larger value. The highly localized nature of the peak in $\langle\Psi_{el}^{(n)}|\nabla_R|\Psi_{el}^{(n+1)}\rangle$ as a function of geometry, typical of a weakly avoided crossing, will further help minimize the cancellation due to the oscillatory nature of the integrand.

Hot ground-state reactions. We have used Figure 2.8 to illustrate yet another event that can also occur on the time scale presently discussed: a hot ground-state reaction. Some of the molecules of the product 𝕰 may take so long to cool to their equilibrium geometry that they pass the relatively high barrier in the S_0 surface separating 𝕰 from 𝔉 in the meantime. Then, they yield 𝔉 as the final product even though 𝕰 is essentially indefinitely stable at the temperature of the experiment once it is in thermal equilibrium with its surroundings.

As we now proceed to the discussion of slower events in the next two sections, we shall repeatedly reencounter the fast processes discussed presently, since even the slow processes can generate vibrationally hot molecules.

2.2.3 The Slow Events

Some of the slower processes that we shall now consider occur on a single surface. These can be viewed as ordinary thermally activated chemical reactions and correspond to motion from one minimum to another. Each such process also has a tunneling component that may but need not be significant in the total picture.

Others among the slower processes represent jumps between surfaces. If they

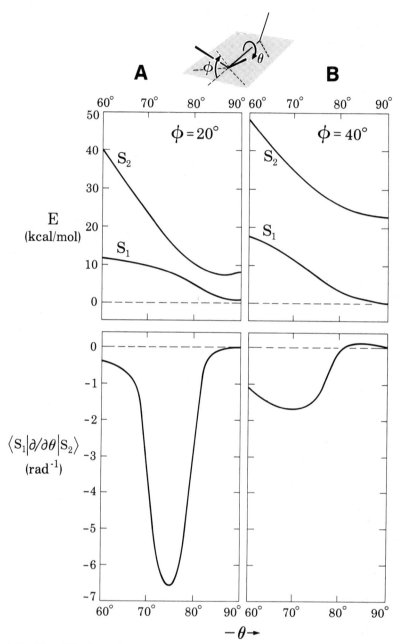

Figure 2.10 Top: The S_1 and S_2 surfaces of singly pyramidalized ethylene (A, by 20°; B, by 40°) as a function of the twist angle θ from 60° to 90°. Bottom: The non-Born–Oppenheimer coupling element $\langle S_1|\partial/\partial\theta|S_2\rangle$ as a function of θ. Adapted by permission from M. Persico and V. Bonačić-Koutecký, *J. Chem. Phys.* **76**, 6018 (1982).

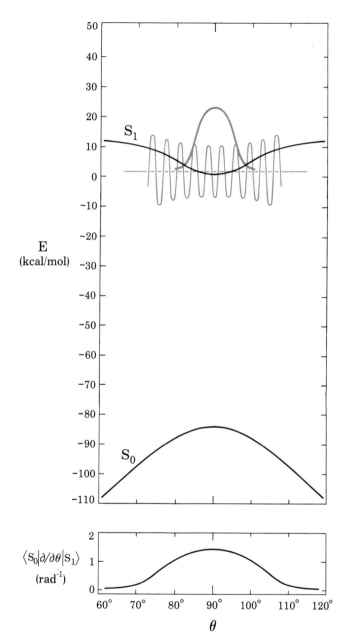

Figure 2.11 Top: The S_0 and S_1 surfaces of ethylene as a function of the twist angle θ from 60° to 120° (solid lines). The form of the vibrational wave functions of the lowest vibrational level in S_1 and of the isoenergetic level of S_0 is shown schematically (light lines). Bottom: The non-Born–Oppenheimer coupling element $\langle S_0|\partial/\partial\theta|S_1\rangle$ as a function of θ. Adapted by permission from M. Persico and V. Bonačić-Koutecký, *J. Chem. Phys.* **76,** 6018 (1982).

occur with light emission, they correspond to fluorescence ($S_1 \rightarrow S_0$) or phosphorescence ($T_1 \rightarrow S_0$). If they occur in a radiationless fashion, they correspond to internal conversion ($S_1 \rightarrow S_0$) or to intersystem crossing ($S_1 \rightarrow T_n$ or $T_1 \rightarrow S_0$).

All of these processes start immediately after the initial excitation. In most cases, they occur so slowly that their contribution is negligible until the nanosecond regime is reached. Phosphorescence is usually insignificant well into the microsecond or even millisecond regime.

This does not mean that the processes cannot be detected at shorter time scales, but it is often difficult. For instance, very weak fluorescence can often be observed immediately after the initial excitation. With some effort, fluorescence has actually been observed from upper excited states such as S_n of several aromatics, whose lifetime is as short as 10 fs. What the slowness of these processes does mean is that they do not dominate the course of events on the picosecond time scale discussed in the previous section but, instead, represent only a minor perturbation.

Another point to keep in mind is the existence of exceptions to almost all the generalizations we have been making. To a good approximation, the rates of motion from one minimum on a surface to the next have an inverse exponential dependence on the barrier height. Thus, given a small enough barrier, such motion can be important on the picosecond time scale. As already mentioned, the rate of internal conversion between two surfaces increases roughly exponentially with the decreasing gap between them. Thus, if the $S_1 \rightarrow S_0$ separation is small, the internal conversion $S_1 \rightarrow S_0$ can also be fast on the picosecond time scale, as $S_n \rightarrow S_1$ usually is. Even intersystem crossing can be competitive on the picosecond time scale if the molecular structure favors it, as already pointed out.

Presently, however, we consider as representative the case of molecules in which all of these processes come to play an important role only in the nanosecond or even slower time regime. Thus, we shall now consider processes starting with molecules thermally equilibrated in one of the minima in the S_1 surface or in the T_1 surface. All processes starting with the initial excitation up to and including the slow events are shown in Figure 2.12 for our hypothetical example. The slow singlet processes are then shown once again separately in Figure 2.13A. The triplet processes starting with the formation of the triplet up to and including the slow events are shown separately in Figure 2.13B.

Adiabatic reactions with a barrier. The rate constants of these processes are inherently temperature-dependent. The motion from one local minimum to another, either in the S_1 surface or in the T_1 surface, is quite well described by Eyring's transition state theory of chemical reactions with corrections for tunneling contributions as necessary. In the simple form without tunneling corrections, this theory yields the following expression for the rate constant k of the adiabatic reaction:

$$k = \kappa \frac{kT}{h} e^{\Delta S^\ddagger/k} e^{-\Delta H^\ddagger/kT} \qquad (2.6)$$

where κ is the transmission coefficient, k is Boltzmann's constant, T is the absolute temperature, h is Planck's constant, $\Delta S^\ddagger$ is the entropy of activation, and $\Delta H^\ddagger$ the enthalpy of activation. Often, the empirical Arrhenius equation is used:

$$k = A 10^{-E_a/kT} \qquad (2.7)$$

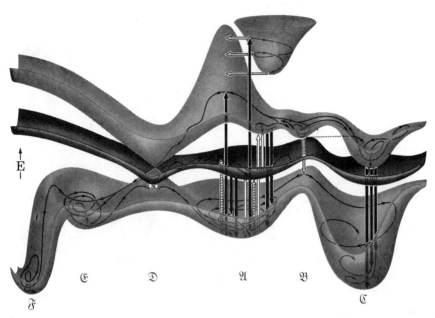

Figure 2.12 A schematic potential energy surface representation of the photochemical processes depicted in Figure 2.4. Light surfaces: Singlets S_0, S_1, and S_2. Dark surface: Triplet T_1. Vertical arrows represent jumps between surfaces: radiative transitions (black), energy transfer (hatched), internal conversion (grey), intersystem crossing (white). Adiabatic motion on a single surface is shown by thin curved arrows: full arrows denote "singlet photochemistry," and a dashed arrow denotes "triplet photochemistry." Tunneling through a barrier is indicated by a dotted arrow.

where A is the frequency factor and E_a is the activation energy. If they are to play a significant role, the adiabatic reactions must occur at rates competitive with the radiative and the radiationless jumps to lower surfaces, which also depopulate the minima in the S_1 and T_1 surfaces, and this is only possible if the activation barriers are quite small.

For instance, at room temperature, a unimolecular reaction on the S_1 surface with an Arrhenius activation energy of 6 kcal/mol and a frequency factor of 10^{13} s^{-1} will compete evenly with $S_1 \rightarrow S_0$ fluoresence if the radiative lifetime of the latter is 2 ns.

We shall see below that the processes competing with adiabatic chemical reactions are usually faster for molecules in the S_1 surface than for those in the T_1 surface, giving the triplet molecules a longer period of time and thus a better chance to overcome larger barriers. In Figures 2.12 and 2.13A we show the reversible adiabatic thermal transformation between the species $\mathfrak{A}(S_1)$ and $\mathfrak{B}(S_1)$ and an effectively irreversible adiabatic reaction of $\mathfrak{B}(S_1)$ to $\mathfrak{C}(S_1)$, for which we also explicitly show a tunneling contribution. Figures 2.12 and 2.13B show the reversible adiabatic reaction between $\mathfrak{A}(T_1)$ and $\mathfrak{D}(T_1)$. These steps correspond to the processes shown in the reaction scheme in Figure 2.4.

Reactions such as $\mathfrak{B}(S_1) \rightarrow \mathfrak{C}(S_1)$, in which a product molecule in its vertical excited state is formed, are uncommon in organic photochemistry. Best known

72 PHOTOPHYSICAL AND PHOTOCHEMICAL PROCESSES

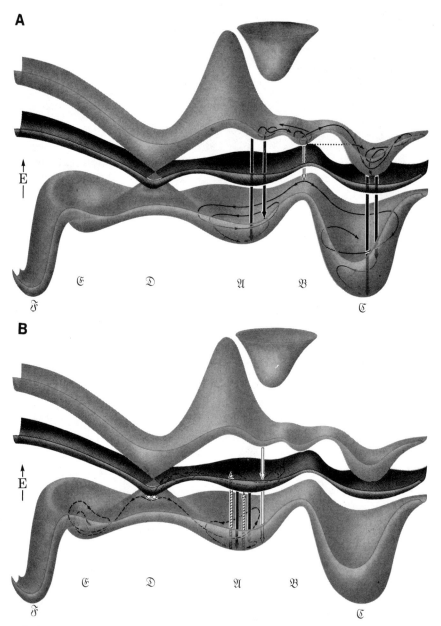

Figure 2.13 The slow processes in the singlet (A) and triplet (B) photochemistry of chemical species 𝔄. See caption to Figure 2.12.

examples are proton transfer reactions, exciplex and excimer formation, and electrocyclic rearrangements of dewar aromatics.

Solvent viscosity effects. The transmission coefficient κ in equation 2.6 can be smaller than unity for several reasons, e.g., if the activation barrier results from a very weakly avoided surface crossing, so that a jump to a higher surface, rather than travel over the barrier on the lower surface, occurs a significant fraction of the time. A commonly encountered factor that reduces the transmission coefficient below unity is recrossing of the top of the potential barrier after a successful initial passage as a result of collisions with solvent molecules.

The simplest model that describes the effects of solvent viscosity on rates of reactions with an activation barrier is due to Kramers. This author treated the problem as the motion of a Brownian particle escaping over a one-dimensional potential energy barrier and arrived at an approximate solution for the rate constant k:

$$k = \frac{\omega}{2\pi} \frac{\beta}{2\omega'} \left\{ \left[1 + \left(\frac{2\omega'}{\beta}\right)^2 \right]^{1/2} - 1 \right\} e^{-E_a/kT} \qquad (2.8)$$

The rate constant depends on four parameters: ω and ω' are the frequencies corresponding to the bottom of the initial potential energy well and to the top of the barrier, respectively. Both are taken to be parabolic. E_a is the intrinsic barrier height, and β is the reduced friction coefficient (i.e., the friction coefficient divided by the effective mass of the reacting molecule). Of these, β is usually the most contested contributor. In the hydrodynamic approximation, it is taken as proportional to the solvent macroscopic shear viscosity, η_s, but it can also be derived phenomenologically from measured microscopic properties of the solvent such as rotational reorientation times, translational diffusion coefficients, or solvent self-diffusion coefficients.

The Kramers model is applicable only if the motions of the solvent are fast compared to motion along the reaction coordinate. The frictional coefficient then reflects the net frictional drag experienced by the reacting molecule as it changes its shape. If the time for barrier crossing is comparable with times for solute–solvent collisions, the model breaks down since the motion over the barrier is then correlated with individual solute–solvent collisions. This does not happen for reactions with lifetimes longer than ~10 ps, whose rate constant falls off regularly with increasing solvent viscosity.

Radiative jumps between surfaces

Fluorescence and phosphorescence. Although the rate constants of these processes may show a temperature dependence, it is frequently negligible. The rate of light emission by a molecule that finds itself in a minimum in an excited S_1 or T_1 surface is related to the magnitude of the electronic transition moment $\mathbf{M}_{0n}$ between the two states connected by the transition. The larger the transition moment, the faster the rate at which light is spontaneously emitted. The same quantity governs the probability of absorption of light by the lower state. If the $S_0 \rightarrow S_1$ transition is intense, the radiative rate constant k_F for the fluorescence $S_1 \rightarrow S_0$ will be large. For a transition between nondegenerate levels, the quantitative relation between the two is given by the Strickler–Berg equation:

$$k_F = 2.88 \times 10^{-9} n_0^2 \langle \tilde{\nu}_e^{-3} \rangle^{-1} \int [\varepsilon(\tilde{\nu})/\tilde{\nu}] \, d\tilde{\nu} \quad (2.9)$$

where n_0 is the average refractive index of the solution in the region of the emitting transition, $\langle \tilde{\nu}_e^{-3} \rangle$ is the mean value of $\tilde{\nu}^{-3}$ over the emission spectrum, and $\tilde{\nu}$ is wavenumber; the decadic molar extinction coefficient ε is in its usual units of liter mol^{-1} cm^{-1}.

Since the multiplicity-forbidden $S_0 \to T_1$ transition has a very small transition moment and is extremely weak, the radiative rate constant k_P for the phosphorescence $T_1 \to S_0$ will be small. Typical values are 10^6–10^9 s^{-1} for fluorescence and 10^2–10^{-2} s^{-1} for phosphorescence. In Figures 2.12 and 2.13 the fluorescence from $\mathfrak{A}(S_1)$ and the phosphorescence from $\mathfrak{A}(T_1)$ into several vibrational levels of $\mathfrak{A}(S_0)$, as well as the fluorescence from $\mathfrak{C}(S_1)$ into several vibrational levels of $\mathfrak{C}(S_0)$, are shown in accordance with the scheme in Figure 2.4. The $\mathfrak{A}(S_1) \to \mathfrak{A}(S_0)$ and $\mathfrak{A}(T_1) \to \mathfrak{A}(S_0)$ processes return the molecules back to the starting minimum in the S_0 surface. Although they may leave them with an excess of vibrational energy which might then be used for a hot ground-state reaction, they are usually unproductive in the photochemical sense and represent energy wastage. On the other hand, the $\mathfrak{C}(S_1) \to \mathfrak{C}(S_0)$ process represents the final step of a sequence of photochemical events, in one of which an excited starting material reacted to produce an excited product.

If spontaneous light emission were the only process depopulating the excited state, the number $N(t)$ of molecules that remain excited at time τ after an initial population $N(0)$ was established would be given by the relation

$$N(t) = N(0)e^{-k_R t} \quad (2.10)$$

where k_R stands for the radiative rate constant k_F or k_P. The radiative or "natural" lifetime τ_R of the excited state would then be equal to the observed lifetime and would be given by $\tau_R = 1/k_R$.

The three components of a triplet state have different radiative behavior in that each has its own transition moments connecting it to other states. At the temperature of liquid nitrogen and higher, this normally makes no difference because the populations of the three nearly degenerate components equilibrate very rapidly relative to the lifetime of the triplet state. However, at very low temperatures, this is not so, and the radiative decay of each of the three populations proceeds independently. Under the usual conditions of spectral resolution, all three emissions are observed jointly. Then, the time dependence of emission intensity derived after the initial excitation event consists of three superimposed exponentials, whose weights are determined by the relative size of the initial three populations. The results can be affected strongly by an external magnetic field that can cause a mixing of the three zero-field spin eigenfunctions.

The degree to which phosphorescent emission and singlet–triplet absorption are forbidden is a fairly sensitive function of molecular structure and of the nature of the states involved. These spin-forbidden processes are generally enhanced by the introduction of heavy atoms, and they are also faster when the transition involves a motion of an electron between two p orbitals located on the same center. Thus, spin–orbit mixing of a closed-shell ground state (or a $\pi\pi^*$ excited state) with an $n\pi^*$ state is more efficient than their mixing with a $\pi\pi^*$ state.

Emission from higher excited states. Radiative decay from higher excited states into lower excited states or into the ground state occurs as well. It is rarely observable because these states are too short-lived. To appreciate this, consider an S_n state that undergoes internal conversion into S_1 with a rate constant of 10^{14} s^{-1}. If the $S_n \rightarrow S_0$ fluorescence is the only competing process and if its rate constant is 10^9 s^{-1} (near the upper limit of the typical values), only one molecule will fluoresce for every 10^5 molecules that undergo the internal conversion to S_1.

Stimulated light emission. In addition to spontaneous light emission, which we have considered so far, stimulated light emission needs to be mentioned. If the population of the minimum in the excited surface exceeds that in the corresponding area of the lower surface, the intensity of coherent light of wavelengths that are appropriate for the energy difference between the surfaces will not be reduced as it passes through the sample. Rather, it will be amplified because it will stimulate radiative jumps from the upper to the lower surface. This action is known as lasing, and the probe then represents a laser. The enhanced rate of depopulation of the excited state will shorten its lifetime below τ_R even if no other processes compete, in a way which depends on the intensity of the incident light.

A particularly favorable situation for achieving the population inversion required for laser action is obtained in systems in which an adiabatic excited-state reaction produces an excited chemical species that is not present in the ground state. In Figures 2.12 and 2.13 the fluorescence $\mathfrak{C}(S_1) \rightarrow \mathfrak{C}(S_0)$ is such a case, since no ground-state $\mathfrak{C}(S_0)$ molecules are present initially and a population inversion results as soon as the excited molecules $\mathfrak{C}(S_1)$ are formed. In order to produce an operational laser system on this principle, the ground-state $\mathfrak{C}(S_0)$ molecules need to convert spontaneously to the starting ground-state species $\mathfrak{A}(S_0)$, and this is accomplished in the commercial "excimer laser." This is not the case in Figures 2.12 and 2.13 shown here.

Environmental effects. So far, we have given no consideration to the possibility that the radiative rates depend on the nature of the solvent and on temperature. The perturbation by the solvent has an effect on the potential energy surfaces and on the electronic wave functions and therefore also on the transition moments. This effect is particularly noticeable if the transition moment of the isolated molecule is small or zero, corresponding to a long lifetime τ_R. Consequently, the introduction of a small additional solvent-induced transition moment can have a dramatic effect on τ_R. One example is provided by the fluoresence lifetimes of molecules whose $S_1 \rightarrow S_0$ emission is symmetry-forbidden (benzene) or, for other reasons, only weakly allowed (naphthalene, pyrene). These lifetimes are a very sensitive function of solvent polarity and polarizability (this phenomenon is known as the Ham effect). Another example is offered by the phosphorescence lifetimes of $\pi\pi^*$ triplets of aromatics, which are sensitive to the presence of heavy atoms in the solvent.

Temperature effects. Temperature is important inasmuch as it changes the properties of the solvent, and inasmuch as it controls the population of higher vibrational levels of the excited state which may have different transition probabilities to the ground state than the lowest level does. Moreover, it controls the radiationless transition of molecules from lower to higher electronic states, important in chemiluminescence, E-type delayed fluorescence ($T_1 \rightarrow S_1 \rightarrow S_0 + h\nu$), and elsewhere.

Radiationless jumps between surfaces

Internal conversion and intersystem crossing. The multiplicity-allowed process of internal conversion ($S_1 \rightarrow S_0$) and the multiplicity-forbidden processes of intersystem crossing ($S_1 \rightarrow T_1$ and $T_1 \rightarrow S_0$) convert electronic energy into the energy of vibrational motion. Although their rates depend on the particular vibrational level of the excited state in which they start, this dependence is usually sufficiently weak, at least for the thermally accessible levels, that they can normally be regarded as temperature-independent. If the minimum in S_1 or T_1 occurs at a geometry where the excited and the ground surfaces touch or nearly touch, the probability of the $S_1 \rightarrow S_0$ jump upon each passage is high and is given approximately by the Landau–Zener formula given in equation 2.4.

If the minimum in S_1 or T_1 occurs at a geometry where the excited and the ground surface S_0 are well separated and run more or less parallel to each other, the rate of the radiationless transition is much lower and decreases exponentially with the increasing $S_0 \rightarrow S_1$ energy difference. This "energy-gap law" takes the following specific form for aromatic hydrocarbons:

$$k_{IC} \cong 10^{13} \exp\{-4.5[E(S_1) - E(S_0)]\} \qquad (2.11)$$

where the energies of the S_1 and S_0 states are in units of eV and k_{IC} is in s^{-1}. This means that, in practice, internal conversion from S_1 to S_0 in these molecules is negligible when the two states are separated by more than about 2.5 eV.

Figures 2.12 and 2.13A show internal conversion from the S_1 to the S_0 state starting at the geometry of species $\mathfrak{B}$, namely, $\mathfrak{B}(S_1) \rightarrow \mathfrak{B}(S_0)$. Figures 2.12 and 2.13B show intersystem crossing $\mathfrak{A}(S_1) \rightarrow \mathfrak{A}(T_1)$, $\mathfrak{A}(T_1) \rightarrow \mathfrak{A}(S_0)$, and $\mathfrak{D}(T_1) \rightleftarrows \mathfrak{D}(S_0)$. The nonradiative processes $\mathfrak{A}(S_1) \rightarrow \mathfrak{A}(S_0)$ and $\mathfrak{C}(S_1) \rightarrow \mathfrak{C}(S_0)$ are not shown, since they would occur over large energy gaps in regions where equation 2.11 is applicable, and most likely would be very slow. In aromatics, such processes occur with rates of 10^6 s^{-1} and less unless the energy gap becomes quite small. On the other hand, internal conversion $S_1 \rightarrow S_0$ in the regions of avoided crossing is generally much faster and probably occurs in the picosecond-to-nanosecond regime.

The multiplicity-forbidden intersystem crossing processes $S_1 \rightarrow T_1$ and $T_1 \rightarrow S_0$ are favored by their generally smaller energy gaps. This is particularly true of the $S_1 \rightarrow T_1$ and $S_1 \rightarrow T_n$ processes for which the energy gap may be quite close to zero. This largely compensates for the spin-forbiddenness, so that these processes are generally faster (typically 10^8 s^{-1}) than internal conversion in aromatics. The strong dependence of the intersystem crossing rate on the energy difference makes it likely that the jump actually is not between the S_1 and T_1 surfaces if there are other triplet states T_n in the gap between S_1 and T_1. Even if a T_2 state is located slightly above the S_1 state so that its population requires some thermal activation, the activated process $S_1 \to T_2 \to T_1$ may be much faster than the more direct process $S_1 \to T_1$. In such a case, the intersystem crossing rate will exhibit a thermal dependence of the Arrhenius type from which the S_1-T_2 gap may be deduced.

The rate of intersystem crossing follows a similar energy gap law as the rate of internal conversion. Thus, for the $T_1 \to S_0$ radiationless transition in aromatic hydrocarbons, the following expression is approximately valid (in eV and s^{-1}):

$$k_{ISC} \cong 10^{4.3} \exp\{-4.5[E(T_1) - E(S_0)]\} \qquad (2.12)$$

Note the much smaller value of the factor in front of the exponential for k_{ISC} relative to that for k_{IC}, due to the spin-forbidden nature of intersystem crossing (equations 2.11 and 2.12). Like the rate of spin-forbidden radiative processes, the rate of intersystem crossing is enhanced (i) when heavy atoms are present in the molecule and, to a lesser degree, in the solvent and (ii) when the two states involved differ by promotion of an electron between two p orbitals on the same atom: Going between the ground state or a $\pi\pi^*$ state on the one hand and an $n\pi^*$ state on the other is much more facile than going between two states from the same group (El Sayed's rule).

The rates of crossing into the three separate sublevels of the triplet state are different. At very low temperatures, where the populations of the three levels are not in equilibrium, the preferential population of one or two of them in the $S_1 \to T_1$ process can be detected.

2.3 INTERMOLECULAR PROCESSES

The formal representation of electronic state energies in terms of hypersurfaces in nuclear configuration space is identical whether the system under consideration consists of only one molecule or several, and we do not need to repeat the figures shown in Section 2.2.

However, the presence of two at most only weakly interacting subunits in the case of intermolecular processes has significant consequences that make it desirable to handle these processes separately. These are due, first, to the small degree of overlap between the subunits and, second, to the relatively slow nature of rotational and translational diffusion, which corresponds, respectively, to five degrees and one degree of freedom on the hypersurface.

Sometimes it is advantageous to view two parts of a single molecule that are only weakly coupled to each other, perhaps by a saturated hydrocarbon chain, as two separate molecules. Then, events such as energy or charge transfer from one chromophore to another within the same large molecule are labeled and treated as if they were intermolecular, although, strictly speaking, they are intramolecular.

In the present section, we concentrate on those aspects that are characteristic for the bimolecular processes, and we do not repeat what has already been said in Section 2.2 about events occurring within each subunit. It should be noted that in photochemistry, those nonradiative processes in which an excited state of a photochemical substrate is converted into the ground state of the same species due to encounter with another molecule are referred to as "quenching" ("self-quenching" if the second molecule is of the same kind). We shall not use this term much,

78 PHOTOPHYSICAL AND PHOTOCHEMICAL PROCESSES

although we shall discuss briefly the main mechanisms by which quenching occurs.

2.3.1 Initial Excitation

The most common bimolecular events involved in initial electronic excitation of a photochemical substrate in organic chemistry are light absorption by complexes and energy transfer. Ion recombination and radical recombination and disproportionation can also produce excited states.

Light absorption by complexes. In "supermolecule" systems composed of two separate molecules or of two molecules only weakly coupled through an "inert" framework, only those excited states in which the excitation is localized in one or the other component can be reached by absorption with large probability. Transitions of this type have already been covered in Section 2.2.1. Transitions into charge-transfer states, in which an electron has been transferred from one component to the other, will have negligible extinction coefficients if the overlap of the two components is negligible, since both the transition density and its dipole moment are negligible. If the overlap of the two components is small but not negligible, weak absorption intensity is observable. This is referred to as the charge-transfer absorption of a complex of the two components ("charge-transfer" complex). Since the transition density is small, the singlet–triplet splitting in charge-transfer complexes is small.

Supermolecules composed of two identical components represent a limiting case of considerable interest. In this case, the mixing of various locally excited states and possibly of charge-transfer states is particularly likely if the components interact. This again leads to absorption bands that are absent in the isolated monomer component (exciton bands). If the two components as well as their environments are truly identical, there will be no net electron transfer from one to the other upon excitation, but in practice this is rare.

Strictly speaking, cases in which singlet–triplet absorption by a molecule is enhanced by a heavy atom (such as Xe), a heavy-atom-containing molecule (such as CH_3I or CS_2), or a paramagnetic species (such as oxygen) in the solvent belong to the category of bimolecular processes, but for practical reasons they have already been mentioned in Section 2.2.1.

The effect of excitation density. As was the case with unimolecular processes, light intensity can play a significant role as a photochemical variable in that it controls the concentration of the excited species and of the reactive intermediates. This is important if an excited molecule can react with another excited molecule or with a reactive intermediate.

Energy transfer. Standard nomenclature for energy-transfer processes indicates the nature of excitation in the energy donor before the transfer as well as the nature of excitation in the energy acceptor after the transfer. The most common types are triplet–triplet energy transfer and singlet–singlet energy transfer, both allowed by the spin conservation rule.

Triplet–triplet transfer. Energy transfer is the most important practical method of selective $S_0 \rightarrow T_1$ excitation and is symbolized in Figures 2.4 and 2.7. In this process, a donor molecule in its triplet state is allowed to approach the ground-state substrate molecule in its ground singlet state (the acceptor) closely. If the S_0–T_1 gap of the acceptor is comparable with or smaller than the S_0–T_1 gap of the donor, an energy transfer will take place. In this process, the triplet excitation is transferred from the donor to the acceptor molecule. Subsequently, the former emerges in its ground S_0 state, and the latter emerges in its excited T_1 state.

This method of bringing a substrate molecule to its T_1 state obviously requires a good supply of triplet donor molecules of suitable energies. To be useful, these so-called triplet sensitizers must be chemically inert and must satisfy two conditions. First, they must absorb light of wavelengths that are not at all (or only weakly) absorbed by the actual photochemical substrate, so that they can be excited selectively. Second, such absorption must produce their triplets with high efficiency. This has two practical implications. First, the S_1–T_1 gap of the sensitizer should be small so as to permit the absorption of light of long wavelengths, not absorbed by the substrate, and yet to offer a high-energy triplet state capable of transfer to many acceptors. Second, the intersystem crossing efficiency in the sensitizer must be high.

Singlet–singlet transfer. Electronic excitation from the S_0 into the S_1 state can also be accomplished by energy transfer—this time from another molecule in its singlet excited state.

Unlike triplet–triplet energy transfer, singlet–singlet energy transfer can take place rapidly over quite large separations, of the order of dozens of angstrom units. It, too, will in general initially produce vibrationally hot $\mathfrak{A}(S_1)^*$ molecules. Singlet sensitization is, however, rarely intentionally used in photochemical work.

Additional information on energy-transfer processes and their rates is provided in Section 2.3.3.

Ion and radical recombination. As already noted, charge-transfer states of a "supermolecule" consisting of two well-separated components cannot be reached directly from the ground state by light absorption. In these states, one of the components is positively charged and the other is negatively charged. For instance, the pair of ground-state ions derived from 9,10-diphenylanthracene (**1**)—namely, the radical cation (**1**$^+$) and radical anion (**1**$^-$)—represent an electronically excited state of the pair of neutral molecules **1**–**1**, if we choose to define these together as a "supermolecule."

This may appear artificial at the first sight, but the usefulness of the definition becomes clear when the two ions are allowed to approach closely, since internal conversion then produces lower-energy electronically excited states of the system, such as those that could also result from the approach of an excited neutral molecule

(**1***) and a neutral one (**1**). If both partners are in their singlet states, the supermolecule is in a singlet state. If they both are in doublet states, the supermolecule has the statistical probability of one to be in the singlet state and three to be in the triplet state. This dictates the multiplicity of the lower-energy excited state into which a charge-transfer state is usually transformed by internal conversion when the two component parts approach. Deviations from statistical behavior can appear in the presence of an external magnetic field and hyperfine interactions within the components. The internal conversion process from a charge-transfer state to an ordinary (locally excited) state corresponds to an electron transfer or translocation. Information on these processes is found in Section 2.3.3.

Charge-transfer excited states of "supermolecules" can be produced readily by mixing the solutions of the two ground-state ions if the latter are stable (chemiexcitation). The preparation of the component ions is usually accomplished by chemical or electrochemical oxidation and reduction of the neutral precursor. For instance, electrochemiluminescence results when radical cations and radical anions of a fluorescent molecule such as **1** are generated alternately on an electrode and allowed to diffuse together.

Exothermic recombination and disproportionation of neutral radicals in an overall triplet state can yield triplet products [e.g., $2\text{PhCO}\cdot \rightarrow$ triplet $(\text{PhCO})_2^*$ and $2\,\text{R}_2\text{CHO}\cdot \rightarrow \text{R}_2\text{CHOH} + $ triplet R_2CO^*; chemiexcitation].

2.3.2 Molecular Transport Processes

Bimolecular processes depend on the ability of the reacting partners to move into close vicinity or actual contact, and they often require their proper mutual orientation as well. The principles of the dynamics involved in the transport of molecules towards each other will be briefly discussed in this section. A key role is played by the shear viscosity of the solvent η. For many solvents, its temperature dependence is given by the Arrhenius equation, which expresses η at temperature T as the product of a frequency factor A_η and a temperature-dependent factor $\exp(-E_\eta/kT)$, where E_η is the activation energy for viscous flow.

As already noted, surfaces with essentially zero slope in six directions occur for a supermolecule composed of two molecules in those regions of geometries where these are far apart and essentially independent. Five of the directions are related to rotational diffusion and represent the relative orientation of the components, and the sixth describes the separation of the centers of gravity of the two partners.

Rotational diffusion. The rate of rotational diffusion depends on the properties of the solute and the solvent as well as on temperature. For large molecules, the so-called stick hydrodynamics are appropriate. In these, the solvent layer adjacent to the solute molecule is assumed to move with the molecule. An approximate relation due to Debye, Einstein, and Stokes relates the rotational diffusion coefficient D_{rot} (whose inverse is the rotational correlation time τ_{rot}) to solvent viscosity η, to the volume V_{rot} of the molecule (which is assumed to be spherical), and to the absolute temperature T:

$$D_{rot} = \tau_{rot}^{-1} = kT/6V_{rot}\eta \tag{2.13}$$

where k is Boltzmann's constant. More complicated expressions are available for molecules whose shape is a symmetric ellipsoid. For the rotation of very small molecules, slip hydrodynamics may apply, in which the surface layer of solvent does not follow the motion of the molecule.

In the usual solvents, the rotational motion of ordinary aromatics occurs on the time scale of picoseconds or dozens of picoseconds at room temperature. In an encounter complex, it is normally much faster than translation.

Translational diffusion. The frequency with which motion along the coordinate representing translational diffusion brings the two components of a supermolecule together in a solvent of given viscosity is proportional to the concentrations of both solutes. For uncharged solutes, the encounter rate constant is given very approximately by

$$k_{enc} = kT/1500\eta \qquad (2.14)$$

This extremely simple expression is useful but can be in error by as much as a factor of 5. More sophisticated expressions for the rate of diffusion-controlled processes are available (Section 2.9). For oppositely charged species, the rate is increased by the factor $|Z_1 Z_2 e^2/\varepsilon_D kT|$, where $Z_1 e$ and $Z_2 e$ are the charges, and ε_D is the dielectric constant.

In ordinary solvents such as hexane, k_{enc} is of the order of 10^{10} s^{-1} at room temperature. Generally, it is safe to assume that motion along all other degrees of freedom has already thermally equilibrated in both components of a supermolecule before they actually encounter each other, unless they were in contact to start with, as in a complex or when the reactant is a neat liquid or solid.

2.3.3 Electronic Transfer Processes

Two kinds of transfer processes that are essential in the creation as well as destruction of electronically excited states as a result of close or distant encounters of molecular partners are electronic energy transfer and electron transfer. In practice, their rates are often limited by the rate of diffusion, which was addressed in Section 2.3.2. Here, we consider briefly the intrinsic nature of these two processes, ignoring the possibly rate-limiting role of diffusion. Thus, we assume that the two partners are separated by distance R, leaving aside the sticky question as to whether it is better thought of as a center-to-center or an edge-to-edge separation. More realistically, we need to assume that we are treating a static statistical distribution of partners such as would be found in a random glassy solution. In order to remove the complication introduced by the statistical distribution of the distances R, studies have been performed on "bichromophoric" systems in which the two partners (chromophores) are held at a fixed distance by means of a presumably inert spacer such as a saturated hydrocarbon framework. In practice, of course, no spacer is truly inert, and the energy or electron transfer may be strongly affected by its presence. To a lesser degree, this is true even of the "inert"solvent. The investigation of the "conductivity" of various types of rigid spacers with respect to the transfer of energy or charge is a lively current area of research, as is the examination of the dynamics introduced into the transfer of energy or charge by the use of flexible chains.

Energy transfer. The simplest way in which electronic excitation energy can be transferred from one molecule to another is for the excited molecule to emit a photon of electromagnetic radiation and for the other to absorb it. This process occurs over macroscopic distances and is known as "radiative" or "trivial" electronic energy transfer. It is responsible for the apparent increase in emission lifetimes by "radiation trapping." Here, we concentrate on energy transfer over microscopic distances, involving specific interactions between a pair of molecules at a time.

The interaction between two molecules (chromophores) is usually written as a sum of a direct (Coulomb) term and an exchange term. The existence of these two terms is responsible for the existence of two mechanisms of nonradiative electronic energy transfer between two weakly interacting molecules.

The Coulomb mechanism can be thought of as being due to the electrostatic interaction of the transition charge density corresponding to electronic deexcitation in the initially excited partner (the donor) with that corresponding to electronic excitation in the initially unexcited partner (the acceptor). In the transfer process, the two transitions occur simultaneously, and energy is lost by the donor molecule and acquired by the acceptor molecule in a resonant fashion. This process corresponds to the coupling of two oscillators of electromagnetic field by a "virtual" rather than a "real" photon. The electrostatic interaction is the same as that responsible for exciton (Davydov) splitting in crystals and dimers.

If the interaction between the two transition charge densities is expanded in a multipole series and only the dipole–dipole interaction is kept, i.e., only the transition dipole moments $\mathbf{M}_{0n}$ (Section 2.2.1) are considered, the Förster approximation to the Coulomb transfer mechanism results. This is an excellent approximation as long as neither of the transition charge densities has a vanishing or nearly vanishing dipole moment $\mathbf{M}_{0n}$ and as long as the two partners, and therefore the two densities, are far apart. Since the energy of the interaction of two point dipoles falls off with their distance R as R^{-3}, and since the transfer rate is porportional to the square of the Hamiltonian matrix element (equation 2.1), the transfer rate falls off as R^{-6}. In spite of this rapid fall-off, Förster energy transfer can be efficient over a distance of many dozens of angstroms. However, since only multiplicity-conserving transitions have large transition dipoles, the Förster mechanism is effective only for singlet–singlet but not for triplet–triplet energy transfer.

The exchange mechanism of energy transfer can be thought of as being due to the overlap of the wave functions of the two partners in space. In the transfer process, an electron can be said to tunnel from one partner to the other, while another electron tunnels in the opposite direction. This leaves the overall spin multiplicity intact, but it works equally well regardless of the initial spin multiplicity of each of the partners. Thus, the exchange mechanism of transfer, also known as the Dexter mechanism, works both for singlet–singlet and for triplet–triplet energy transfer. Its rate falls off exponentially with the distance between the partners, since it relies on the overlap between their wave functions. It therefore operates only over a much shorter range than does the Förster mechanism.

Energy transfer by either mechanism conserves energy. It occurs from a particular vibrational level of a particular electronic state of the donor to an isoenergetic vibrational level of a particular electronic state of the acceptor (a small mismatch can be taken up by rotational or translational degrees of freedom of the partners or by motion of the solvent). Ordinarily, the electronic excitation energy of the acceptor is smaller than that of the donor, and some fraction of the transferred energy appears as vibrational (rather than electronic) excitation and is rapidly dissipated. If the electronic excitation energy of the acceptor is higher ("endothermic" energy transfer), enough thermal (vibrational) energy needs to be provided to the system to make up for the difference. Under such circumstances, the rate of energy transfer will exhibit thermal activation, with the energy defect serving as the activation energy. Because of the typically quite short lifetimes of the excited states of donors, only small activation energies can be tolerated before the transfer is shut off (a few kilocalories per mole at room temperature).

Energy transfer has Franck–Condon character. If there is only a weak interaction between the partners, so that the potential energy surfaces of their various electronic states remain unperturbed, the transfer is "vertical" in both partners in the same sense as absorption or emission. Recognizing the somewhat delocalized nature of the nuclear wave functions, this means that the rate of energy transfer falls off rapidly if nuclear positions in either partner have to change significantly during the transfer in order to make it energetically feasible. Energy transfer that requires a substantial geometric distortion in one or both partners and/or relies strongly on the overlap of vibrational wave functions localized in fairly different regions of the nuclear configuration space in the initial and the final state is often referred to as "nonvertical."

Förster energy transfer. The dipole–dipole transfer rate is given by

$$k = (3/2)\kappa^2 \tau_D^{-1}(R_0/R)^6 \qquad (2.15)$$

where τ_D is the lifetime of the donor in the absence of the acceptor, and κ reflects the mutual orientation of the transition dipole moments:

$$\kappa^2 = \sin\theta_D \sin\theta_A \cos\phi - 2\cos\theta_D \cos\theta_A \qquad (2.16)$$

Here, θ_D and θ_A are the angles that the transition dipoles of the donor and the acceptor, respectively, make with the line joining A and D, and ϕ is their azimuthal angle. The "critical transfer distance" R_0 is related to the spectral overlap of the emission of an isolated donor and the absorption of an isolated acceptor (both occurring between those states that are involved when the transfer occurs in the donor–acceptor pair):

$$R_0^6 = \frac{9000 \ln 10\ \phi_D}{128\pi^5 n^4 N} \frac{2}{3} \int_0^\infty f_D(\tilde{v}) \varepsilon_A(\tilde{v}) \tilde{v}^{-4}\, d\tilde{v} \qquad (2.17)$$

In this expression, ϕ_D is the quantum yield of donor emission, n is the refractive index of the solvent, N is Avogadro's number, $f_D(\tilde{v})$ is donor fluorescence intensity at wavenumber $\tilde{v}$ (normalized to unit area on a wavenumber scale), and $\varepsilon_A(\tilde{v})$ is the decadic molar extinction coefficient of the acceptor at wavenumber $\tilde{v}$ in the usual units (liter mol^{-1} cm^{-1}). The reason why the emission shape needs to be normalized to unity, while the absorption shape does not, is the presence of the term τ_D^{-1} in the expression for k, which already contains information on the absolute magnitude of the emitting transition moment.

Thus, fast energy transfer requires strongly allowed transitions in the donor and the acceptor, as well as a good spectral overlap between the emission of the donor and the absorption of the acceptor. The expression makes it clear why triplet–triplet energy transfer by this mechanism is virtually impossible: For singlet–triplet absorption, ε_A, and τ_D^{-1}, are very small and the integration gives almost exactly zero even if the spectral regions of the absorption and the emission overlap.

Averaging over a random distribution of distances and orientations yields the fluorescence decay law for the donor:

$$F(t) = \exp[-t/\tau_D - (4/3)g\pi^{3/2}n_A R_0^3 (t/\tau_D)^{1/2}] \qquad (2.18)$$
$$g = \sqrt{(3/2)\langle \kappa^2 \rangle}$$

where n_A is the concentration (number density) of acceptor molecules, and $\langle \kappa^2 \rangle$ stands for the average value of the orientation factor κ^2. If all molecules reorient randomly and infinitely fast, we have $g = 1$. If the molecular orientations are random but do not change in time, we obtain $g = 0.845$.

The result needs to be modified if donor–donor energy transfer is important and also if translational diffusion (Section 2.3.2) is introduced. In practice, the energy transfer rate becomes diffusion controlled when the transfer radius R_0 is less than about 10 Å.

Dexter energy transfer. The exchange transfer rate from a donor D to an acceptor A is given by

$$k = (2\pi/\hbar) KJ \exp(-2R/L) \qquad (2.19)$$

where the constants K and L reflect the ease of electron tunneling between the donor and the acceptor and are not simply related to measurable quantities. L is known as the "average orbital radius" and reflects the rate of attenuation of the wave functions with distance. The spectral overlap integral J is now defined between the emission shape $f_D(\tilde{\nu})$ and absorption shape $a(\tilde{\nu})$, both normalized to unit area on the wavenumber scale:

$$J = \int_0^\infty f_D(\tilde{\nu}) a(\tilde{\nu}) \, d\tilde{\nu} \qquad (2.20)$$

This clearly reflects the fact that now the transition dipole moments $\mathbf{M}_{0n}$ connecting the excited state of D to its final (ground) state and connecting the ground state of A to its final (excited) state are immaterial. The exchange mechanism therefore works well for both triplet–triplet energy transfer and singlet–singlet energy transfer, provided that the emission of the donor and the absorption of the acceptor overlap, and provided that the partners are very close together. If they are not connected by an "inert" spacer but actually are two distinct molecules, they typically need to be separated by no more than about 10 Å. In fluid solution, triplet–triplet energy transfer is therefore virtually always limited by diffusion (Section 2.3.2).

Even when it is exothermic, exchange energy transfer does not need to occur on every encounter, and the triplet–triplet energy transfer rate can be somewhat lower than diffusion-controlled. This is due to the finite transfer rate even at contact distances, and it can be related qualitatively to factors such as overlap of the donor and acceptor orbitals involved in the transfer. A few of these rates have been measured; for example, benzophenone triplet transfers its energy to piperylene or 1-methylnaphthalene solvent in about 10 ps.

The rate constant of endothermic triplet–triplet "vertical" energy transfer k_{TT} follows the Sandros equation:

$$k_{TT} = k_{lim}/\{1 + \exp[-(E_T^D - E_T^A)/kT]\} \qquad (2.21)$$

where E_T^D and E_T^A are the triplet energies of the donor and the acceptor, respectively, and k_{lim} is close to the diffusion-controlled rate constant. A reference to a more general and more complex expression describing not only "vertical" but also "non-vertical" triplet sensitization is given in Section 2.9.

Molecules with triplet ground state represent a special case. By far the most common among them is molecular oxygen, which quenches excited singlets at a diffusion-controlled rate and converts them into triplets. Usually, it remains in its triplet state itself so that, in effect, it only catalyzes intersystem crossing in the collision partner. However, if the singlet–triplet energy gap in the partner exceeds 22.5 kcal/mol, O_2 can be excited to its $^1\Delta_g$ state in an energy-transfer process. Since ground state O_2 also quenches triplets by a triplet–triplet annihilation process (see below) in viscous or rigid solutions, it can quench an excited singlet partner all the way to the ground state.

Intramolecular energy transfer. In the few rigidly connected donor–acceptor pairs that have been investigated, a Dexter-type exponential decrease of the rate with distance is obeyed by triplet–triplet energy transfer as well as by singlet–singlet energy transfer, at least when the transition dipoles as well as the chromophore separation are relatively small, so that the Förster mechanism is suppressed. It is not easy to tell whether the exponential dependence follows the through-space distance or the through-bond distance, which would be proportional to the number of bonds. In one series of bichromophoric molecules, a comparison was possible with electron-transfer rates, which also reflect electron tunneling. In what will probably become known as the "Closs rule," the triplet–triplet energy-transfer rate was found to be proportional to the product of the electron-transfer and hole-transfer rates. This is a qualitatively sensible result, considering that an electron and a hole are tunneling simultaneously in the Dexter energy-transfer process.

Triplet–triplet annihilation. A special case of Dexter energy transfer is of considerable importance. In this, both partners are in their triplet states to start with, and after the energy transfer the donor is in its singlet ground state and the acceptor is in a highly excited state, having collected the excitation energies of both partners. Because it proceeds by the exchange mechanism, it requires close contact between the two participating molecules. Since excited states normally form quite densely packed manifolds, large energy gaps rarely inhibit this exothermic radiationless energy-transfer process. When two triplet molecules meet, the statistical probability of a singlet encounter is 1/9, that of a triplet encounter is 1/3, and that of a quintet encounter is 5/9. These probabilities can be perturbed by the presence of an external magnetic field and by hyperfine interactions (Section 2.9).

If the two components come together in their triplet states in such a way that the overall state of the doubly excited supermolecule is a singlet, the acceptor will be excited into a high-energy singlet state, and a rapid internal conversion to the lowest singly excited state S_1 normally follows. This is known as triplet–triplet annihilation. If the components subsequently separate, only one of them is excited, and emission from this singlet excited state can often be detected as fluorescence (P-type delayed fluorescence, $T_1 + T_1 \to S_1 \to S_0 + h\nu$).

If the overall state of the supermolecule upon encounter is a doubly excited triplet, a rapid internal conversion to a lower singly excited triplet state T_1 follows. This could be referred to as triplet–triplet quenching, since subsequent separation of the components yields one of them in the ground state and the other in the triplet state.

If the overall multiplicity of the doubly excited supermolecule is a quintet, no internal conversion is usually possible because this already is the lowest-energy quintet.

If the total state of the supermolecule is a triplet, an intersystem crossing is needed for conversion to a singlet state, and if it is a quintet, two intersystem crossings would be needed. Such intersystem crossing may be promoted by hyperfine interactions and would be affected by the presence of an outside magnetic field (Section 2.9).

A common case of triplet–triplet annihilation involves molecular oxygen, whose ground state is a triplet ($^3\Sigma_g^-$). Upon encounter with a triplet-energy donor, this can be excited to the $^1\Delta_g$ singlet which lies 22.5 kcal/mol higher while the donor returns to its singlet ground state ("quenching of triplets by oxygen"). When the triplet energy of the donor is less than 22.5 kcal/mol, equation 2.21 applies except that the statistical factor of 1/9 must be introduced. Sensitized production of O_2 ($^1\Delta_g$) is a standard procedure for the generation of this reactive intermediate in the laboratory and in biological objects and is also responsible for the fading of certain dyes in air.

Electron transfer. A trivial mechanism for electron transfer combines ejection of an electron from one partner into the solvent conduction band followed by an uncorrelated capture by another partner, typically located far away. The initial state of the donor needs to be an unbound or an autoionizing state. This mechanism is involved in the photoejection of electrons from solutes into the solvent, well known both as a one-photon process and as a two-photon process, usually with an intermediate triplet state, and can lead to photochemical events if the resulting "solvated electrons," the photoreduced partner, or the photooxidized partner are reactive. The relation to Rydberg excitations is clear. Such solution photoionization of neutrals or photodetachment from anions, although a photochemical event, leads to chemical reactions that can be considered as ground-state chemistry of a species richer or poorer by an electron, and we shall not discuss them further.

Here, we shall be interested in electron transfer from a molecule to a specific nearby partner. This can be either spontaneous (each partner initially in its electronic ground state) or photoinduced (donor or acceptor initially in an excited electronic state). The spontaneous transfer may be nonradiative (no energy lost as radiation during the transfer in the form of a photon) or radiative (photon emitted during the transfer). Like other photophysical processes, electron transfer reactions represented as occurring on a single potential energy surface are called adiabatic whereas those represented as occurring by a jump from one to another potential energy surface are called nonadiabatic (or diabatic).

In principle, the final state may be the ground electronic state of the total system (each partner in the ground state) or an excited electronic state of the total system (with neither partner, one partner, or possibly both partners in their respective electronic ground states).

The initial and the final state of the total system may have the same multiplicity (spin-allowed transfer, where the transferred electron can be said to have kept its spin) or different multiplicities (spin-forbidden transfer, where the transferred electron has inverted its spin). Spin-forbidden transfer is promoted by the mixing of states of different multiplicity through the spin–orbit coupling and the hyperfine parts of the Hamiltonian.

Radiationless electron transfer from a bound state of one partner to a bound state of another is normally referred to as electron tunneling because in the part of space separating the two molecules the potential energy for electron sojourn is higher than the initial total energy of the electron. As a result of its quantum mechanical (as opposed to classical) nature, the electron can tunnel through such a barrier into an isoenergetic level of the new partner. Examples are a spin-allowed electron transfer from an aromatic radical anion to a neutral aromatic, forming a new neutral aromatic and a new aromatic radical anion, and a spin-forbidden electron transfer from a radical anion to a radical cation, the two forming a contact ion pair in its triplet state, to form a pair of neutral molecules in their ground states.

Radiative electron transfer, or charge-transfer emission, corresponds to fluorescence or phosphorescence, depending in the usual way on spin-forbiddenness, that is, on the multiplicity of the "supermolecule" before and after the event. An example is provided by emission from an exciplex.

Photoinduced electron transfer, or charge-transfer absorption, is the opposite event, in which a "supermolecule" absorbs a photon and uses the energy to transfer an electron from one partner to another. Absorption by charge-transfer complexes, yielding charge-transfer excited states, has already been mentioned in Section 2.3.1.

In all three cases of interest here, some degree of overlap of the wave functions of the partners between which the electron transfer is to take place is necessary. The maximum possible separation will be clearly dictated by the length of time for which one is willing to wait before the event takes place; most of our discussion will pertain to molecules that are in contact with each other or that are connected by an "inert" spacer.

The principal factors of interest in the investigations of electron transfer are the effect of the nature, separation and relative orientation of the donor and acceptor orbitals, the free energy change, and the effect of the solvent and temperature.

Classification of electron-transfer processes. In addition to the distinction between spontaneous radiationless, spontaneous radiative, and photoinduced electron-transfer processes, the distinction between adiabatic and nonadiabatic transfer processes, as well as the distinction between spin-allowed and spin-forbidden electron-transfer processes, another type of classification is useful, based on the charges present on the electron donor and the electron acceptor in their initial state.

Charge recombination has already been mentioned in Section 2.3.1. The donor is initially negatively charged, the acceptor is positively charged, and both molecules are neutral after the transfer event.

TABLE 2.2 Electron-Transfer Processes

Charge transfer

$\mathfrak{A}^+ + \mathfrak{B}^- \rightarrow \mathfrak{A} + \mathfrak{B}$	Charge recombination
$\mathfrak{A} + \mathfrak{B} \rightarrow \mathfrak{A}^- + \mathfrak{B}^+$	Charge separation

Charge translocation

$\mathfrak{A} + \mathfrak{B}^- \rightarrow \mathfrak{A}^- + \mathfrak{B}$	Electron translocation
$\mathfrak{A}^+ + \mathfrak{B} \rightarrow \mathfrak{A} + \mathfrak{B}^+$	Hole translocation

Charge translocation (*charge shift*) corresponds to one of two cases. Either the donor is initially negatively charged and the acceptor neutral, so that after the transfer the donor is neutral and the acceptor negative (negative charge translocation, electron translocation), or the donor is initially neutral and the acceptor positively charged, such that after the transfer the donor is positively charged and the acceptor neutral (positive charge translocation, hole translocation).

Charge separation is the opposite of charge recombination: Both the donor and the acceptor are neutral initially. After the transfer, the donor is positively and the acceptor negatively charged.

We shall refer to charge recombination and charge separation jointly as *charge transfer*, as opposed to charge translocation. The nomenclature we shall adopt is summarized in Table 2.2. Other cases are possible but of lesser importance.

Like energy transfer, electron transfer conserves energy and occurs between isoenergetic levels of two different electronic states of the total system if it is nonradiative, and between levels differing by the energy of the photon if it is radiative.

Photoinduced and spontaneous radiative electron transfer. The probability of electron transfer induced by absorption of a photon or associated with emission of a photon is related to the size of the transition dipole moment in the usual way (Sections 2.2.1 and 2.2.3). Since this is the dipole moment of the transition density, it vanishes unless the wave functions of the donor and the acceptor overlap, and it is small even if the partners are in close contact. The extinction coefficients characterizing the charge-transfer bands of donor–acceptor complexes are small, and the radiative lifetimes are very long. This is true even for the spin-allowed case and more so for the spin-forbidden case.

Nonradiative electron transfer. Like energy transfer, electron transfer is of Franck–Condon character, and its rate falls off very rapidly if nuclear positions in either partner have to change significantly during the transfer process in order to make it energetically feasible. This requirement frequently imposes an activation energy, since thermal excitation may be needed to bring the geometries of both partners, as well as the surrounding solvent, to a point at which energies of the initial and the final electronic states would be the same if their crossing were not avoided.

Constant partner separation. Since the electron-transfer process depends on electron tunneling, it is expected to exhibit an exponential dependence on the separation

R of the partners. From studies on glassy solutions and on bichromophoric molecules with a rigid spacer inserted between the donor and the acceptor in order to ensure a constant separation, the rate constant for electron transfer has indeed been found to follow a relation similar to the Dexter expression for energy transfer:

$$k(R) = (2\pi/\hbar)V_0^2 \, \rho \, \exp[-\alpha(R - R_0)] \qquad (2.22)$$

Here, V_0^2 is a parameter characterizing the donor–acceptor pair as well as the intervening medium in terms of their aptitude towards electron tunneling, $R - R_0$ is the edge-to-edge distance between the donor and the acceptor, and ρ is the Franck–Condon weighted density of states, which depends on the standard free energy of the reaction, solvent polarity, temperature, and the difference in equilibrium geometries of the initial and the final electronic state. It can be thought of as the probability of bringing the system by zero-point motion and by thermal fluctuations to a geometry where both electronic arrangements, with the electron on the donor and with the electron on the acceptor, have identical energies. At this geometry, the crossing of the two potential energy surfaces is avoided by an amount equal to $2V_0 \exp[-\alpha(R - R_0)/2]$.

The attenuation parameter α typically equals 1.2–1.3 Å^{-1}. In intramolecular transfer through a saturated hydrocarbon spacer, it seems to be the number of intervening bonds that counts, with the rate decreasing by an order of magnitude each time that two more bonds are added.

Variable partner separation. The exponential dependence of the electron-transfer rate on the intermolecular distance R implies that in fluid solution, where the distance varies with time, virtually all transfer occurs only upon collisions of the two partners. In this respect, electron transfer resembles triplet–triplet energy transfer. The exponential factor in the expression for the rate constant now becomes equal to unity, and the process can be viewed as an ordinary bimolecular reaction characterized by a free energy of activation $\Delta G^\ddagger$ and a transmission factor, which reflect ρ and V_0^2, respectively.

Marcus equation. A simplified treatment of the situation has produced a relation between $\Delta G^\ddagger$ and the standard free energy $\Delta G°$ of the electron-transfer reaction. This relation is known as the Marcus equation. In order to appreciate its origin, we return to Table 2.2. Either the assignment of electrons and charges shown on the left in Table 2.2, or the one shown on the right, represents an excited electronic state of the overall system. For simplicity, we assume that the other arrangement shown represents the ground electronic state. Even in the case of charge translocation between two identical species in solution, $\mathfrak{A} + \mathfrak{A}^+ \leftrightarrows \mathfrak{A}^+ + \mathfrak{A}$, the two electronic states differ in energy when the system is at its equilibrium geometry, since the solvent molecules are normally arranged very differently around the charged partner and around the uncharged one and the molecular nuclear geometries are different (Figure 2.14). In the ground state, the solvent stabilizes the charged partner strongly and the uncharged one weakly. A vertical electron transfer, i.e., one in which nuclear geometries of each partner and its solvent surrounding remain unchanged, generates an electronically excited state. This is higher in energy by an amount known as the reorganization energy λ. As discussed further below,

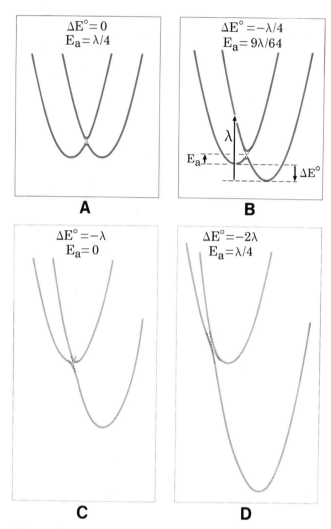

Figure 2.14 Adiabatic (solid lines) and diabatic (dotted lines) potential energy curves for an electron-transfer reaction (schematic). (A, B) Normal region ($|\Delta E°| < \lambda$). (C) The barrier-free case ($|\Delta E°| = \lambda$). (D) Inverted region ($|\Delta E°| > \lambda$).

when the charge translocation process involves different partners 𝔄 and 𝔅 in the charge-transfer process, the magnitude of the energy difference between the two electronic states may well be much larger still, but it can also be smaller.

The rate of the radiationless electron-transfer conversion between the two states is very high at nuclear geometries at which their energies are very close and drops off rapidly as the energy separation increases. Therefore, as already pointed out, the way in which the electron transfer takes place at normal temperatures is for the touching partners to wait until random thermal fluctuations in the solvent and in the nuclear geometries of the partners make the energies of the two states equal or almost equal, at which point the electron jump occurs nearly instantaneously. If they separate first, they have to wait for the next encounter. The charge translocation $𝔄^+ + 𝔄 \rightarrow 𝔄 + 𝔄^+$ therefore shows an activation energy whose magnitude is dictated by the reorganization energies of the solvent and of 𝔄 itself.

Going now to the exothermic case $\mathfrak{A}^+ + \mathfrak{B} \to \mathfrak{A} + \mathfrak{B}^+$ or the usually even more exothermic case $\mathfrak{A}^+ + \mathfrak{B}^- \to \mathfrak{A} + \mathfrak{B}$, the activation energy decreases and the radiationless process becomes faster. In highly viscous solvents, the rate of solvent reorganization could actually be rate-limiting if the partners are next to each other to start with. As the exothermicity increases further, it again becomes harder to reach a nuclear configuration at which the two electronic states are degenerate, and the rate begins to drop (Figure 2.14D).

A consideration of a plot similar to that in Figure 2.14, but for free energy instead of potential energy, assuming that the potential energy curves are simple parabolas, leads to the Marcus relation between the free energy of activation $\Delta G^\ddagger$, the standard free energy of the electron-transfer reaction ΔG°, and the reorganization energy λ:

$$\Delta G^\ddagger = (\lambda/4)(1 + \Delta G^\circ/\lambda)^2 \tag{2.23}$$

It shows that the rate increases as ΔG° becomes more negative, but only until it reaches the value $-\lambda$. At this point, the free energy of activation vanishes. The rate constant is at its maximum and, unless it is diffusion-limited, provides a direct measure of the electronic coupling elements V_0. If ΔG° becomes even more strongly negative, the rate constant begins to drop.

This region of large negative values of ΔG° is known as the Marcus inverted region; an increase in the equilibrium constant actually decreases the rate constant. As ΔG° becomes even more negative, sooner or later an excited state of the system becomes energetically accessible. At that point, the electron transfer generating the ground state of the system will have to compete with an electron transfer generating the electronically excited state. Since ΔG° for the latter process will be increased by an amount equal to the excitation energy and will thus be less negative, the process that generates the excited state will be in an excellent position to compete with the process generating the ground state. Since additional higher-energy excited states are always available, the electron-transfer rate from the donor to the acceptor will never drop to zero even for very large negative values of ΔG°.

In the neighborhood of zero activation energies, $\Delta G^\circ = -\lambda$, the transfer rate tends to be much faster (10^{11}–10^{12} s^{-1}) than the rate at which the partners collide in solution under ordinary circumstances, so that the electron-transfer reaction becomes diffusion-controlled and therefore independent of ΔG°.

Both of the above factors combined make it difficult to operate in the Marcus inverted region in practice, but it has been observed under special conditions. However, as we shall see below, it is relatively commonly observed in photochemical processes, where it becomes just another embodiment of the energy-gap law (see Section 2.2.3).

Charge transfer in photochemistry. In organic photochemistry, charge-transfer processes, both charge separation and charge recombination, play a particularly important role, while bimolecular charge translocation has been investigated much less. The species $\mathfrak{A}$ and $\mathfrak{B}$ are normally of closed-shell and the ions $\mathfrak{A}^{\ddot+}$ and $\mathfrak{B}^{\ddot-}$ of open-shell nature, so that upon dissociation of the tight encounter complex, the latter represent a radical pair as well as a loose ion pair. The singlet $^1(\mathfrak{A}^{\ddot+}\mathfrak{B}^{\ddot-})$ and triplet $^3(\mathfrak{A}^{\ddot+}\mathfrak{B}^{\ddot-})$ states of the tight ion pair, and particularly those of the loose pair, are nearly degenerate since there is almost no overlap between the partners. In addition, the locally excited singlet states $^1(\mathfrak{A}^*\mathfrak{B})$ and $^1(\mathfrak{A}\mathfrak{B}^*)$ and triplet states $^3(\mathfrak{A}^*\mathfrak{B})$ and $^3(\mathfrak{A}\mathfrak{B}^*)$ need to be considered, typically only the lower energy one in each pair.

The relative ordering of the energies of these states, which is affected very strongly by the solvent, determines the direction in which the electron-transfer process will proceed:

(i) The locally excited states are higher in energy than the ion-pair states of the complex. An approach of $^1\mathfrak{A}^*$ to $\mathfrak{B}$ or $\mathfrak{A}$ to $^1\mathfrak{B}^*$ may lead to an electron transfer at some small distance, producing a loose ion pair $^1(\mathfrak{A}^{\ddagger} + \mathfrak{B}^{\bar{}})$, or only upon close contact, producing a tight ion pair $^1(\mathfrak{A}^{\ddagger}\mathfrak{B}^{\bar{}})$. The two can subsequently interconvert, and the former can dissociate into a pair of free ions $\mathfrak{A}^{\ddagger}$ and $\mathfrak{B}^{\bar{}}$. Loose ion pairs are susceptible to intersystem crossing by the hyperfine mechanism which interconverts the singlet and the triplet ion pair (Section 2.9). The singlet tight ion pair, known as the singlet exciplex, may decay to the ground state $^1(\mathfrak{A}\mathfrak{B})$ by "back electron-transfer" charge recombination, either radiative (exciplex fluorescence) or radiationless. If the energy gap is large (Marcus inverted region), the latter is a relatively slow process. Exciplex fluorescence, as well as intersystem crossing, may be competitive. This intersystem crossing, as well as the reformation of a tight ion pair from a loose singlet ion pair that underwent intersystem crossing to the triplet state by the hyperfine mechanism, produces triplet exciplexes. These result also from an initial approach of $^3\mathfrak{A}^*$ to $\mathfrak{B}$ or $\mathfrak{A}$ to $^3\mathfrak{B}^*$, either directly or from a loose ion pair. Their dissociation into a loose ion pair or into free ions is again possible. However, electron transfer to yield the ground state, whether radiative (exciplex phosphorescence) or radiationless, is now slowed down by spin-forbiddenness, so that a return to singlet exciplex via transformation into a loose ion pair, intersystem crossing to singlet by the hyperfine mechanism, and transformation back to a tight ion pair represent a good possibility. Clearly, the outcome of the competition among the various possibilities can be quite different for the singlets and the triplets.

Finally, and most important for photochemists, the radical ions can undergo chemical reactions, such as proton transfer and attachment of a nucleophile, with other species present or with each other.

A general scheme for the photophysical processes involved is quite complex:

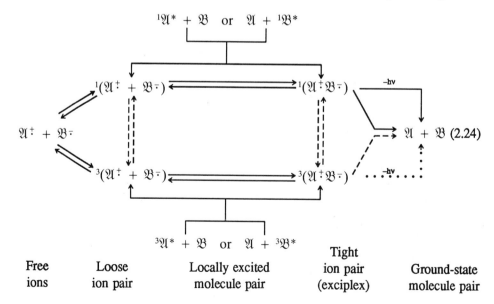

(2.24)

| Free ions | Loose ion pair | Locally excited molecule pair | Tight ion pair (exciplex) | Ground-state molecule pair |

Spin-forbidden processes are indicated by lines that are dotted (unlikely) or dashed (more probable). Chemical reactions remain to be added to complete the scheme.

(ii) The locally excited states, both singlet and triplet, are lower in energy than the ion-pair states of the complex. Free ions form loose and then tight ion pairs $^1(\mathfrak{A}^{\ddagger}\mathfrak{B}^{\bar{}})$ or $^3(\mathfrak{A}^{\ddagger}\mathfrak{B}^{\bar{}})$, with a statistical probability of 1:3 in the absence of magnetic perturbations (Section 2.9). These rapidly undergo radiationless conversion by charge recombination to form the locally excited states $^1(\mathfrak{A}^*\mathfrak{B})$ or $^3(\mathfrak{A}^*\mathfrak{B})$. It is very difficult to observe the Marcus inverted region in this case even when the energy of the ion pair is much higher than that of the locally excited states, since typically higher excited states are spaced closely together and thus additional electronic states will be available and will offer alternative final excited states for the charge recombination step, and for these the energy gap will be small.

The facile nature of electron transfer to or from a locally excited state can be understood by reference to the fact that an electronically excited state is both a better electron donor and a better electron acceptor than the ground state of the same molecule—after all, it has an electron in an antibonding orbital and a hole in a bonding orbital.

Now, a general scheme for the photophysical processes will be inverted relative to the one given previously:

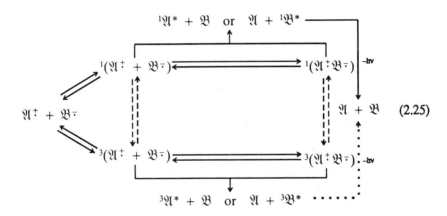

(2.25)

Even more complicated schemes result when the radical ion pair energies lie below that of the lowest locally excited singlet state but above that of the lowest locally excited triplet. The latter, but not the former, can then be populated selectively by ion recombination, and a new path for intersystem crossing opens for the initial partners:

$$^1\mathfrak{A}^* + \mathfrak{B} \rightarrow {}^1(\mathfrak{A}^{\ddagger}\mathfrak{B}^{\bar{}}) \rightarrow {}^3(\mathfrak{A}^{\ddagger}\mathfrak{B}^{\bar{}}) \rightarrow {}^3\mathfrak{A}^* + \mathfrak{B} \qquad (2.26)$$

The chemical consequences of photoinduced charge separation depend on the nature of the constituents of the ion pair. For example, a radical cation can often act as an acid or a nucleophile, a radical anion can act as a base or an electrophile, and a weak one-electron bond in a radical cation can dissociate. However, all of these processes represent ground-state chemistry and are not considered further in this book.

2.4 BACK IN THE GROUND STATE

Once the molecule has returned to the ground state after a more or less complicated set of excursions on various excited-state surfaces as exemplified in Figure 2.12 (p. 71), its further fate will be governed by the same S_0 surface that also governs its ordinary thermal chemistry. It might therefore appear that the knowledge of thermal reactivity will be sufficient to predict all remaining events that will occur before the molecule attains thermal equilibrium in one or another minimum in the ground-state surface and thus terminates its photochemical reaction proper. A whole series of further thermally activated processes on the ground state may follow, but we shall declare them to be ordinary ground-state chemistry and need not consider them further in this text.

Unfortunately, two circumstances complicate matters. First, the molecule will be reborn in the S_0 state at the geometry dictated by the position of the minimum in the excited surface in which it settled before it returned, or by the position of the funnel through which it returned plus the direction in which the funnel was reached (Figure 2.12). With the exception of a return from a minimum such as $\mathfrak{C}(S_1)$ in Figure 2.12, which is located above a minimum in the S_0 surface, it will therefore be born at a geometry that is highly unusual for a ground-state species, and perhaps quite energetic. Since our knowledge of ground-state chemistry is generally limited to the low-energy regions of the S_0 surface, it will then not be of much help in deciding in which direction the molecular geometry will develop and with which probabilities the various nearby minima in the S_0 surface will be reached. This means that it will be hard to predict yields and perhaps even the nature of products.

Second, if the return to S_0 was nonradiative, as is generally the case, and even if it was radiative but did not lead to the lowest vibrational level, the molecule will possess considerable vibrational energy after the jump to the S_0 surface. In general, it will be a hot species, whose further fate will depend not only on the shape of the S_0 surface but also on the rate at which the surrounding solvent takes up the excess vibrational energy. We have already pointed out that at this point the molecule can overcome ground-state barriers that otherwise would be prohibitive at the temperature of the experiment (hot ground-state reactions).

In Figures 2.12 and 2.13A, we illustrate this on the internal conversion of the species $\mathfrak{B}$ from S_1 to S_0. The landing is on a ridge between two minima, and the vibrational energy generated in the jump is most likely to propel the point representing the molecule in one of the two directions shown, which lie along the line in which the S_0–S_1 touching is avoided. Depending on the direction taken, the molecule ends up as the starting material $\mathfrak{A}(S_0)$ or as the product $\mathfrak{C}(S_0)$.

In Figures 2.8 and 2.12, we show molecules returning from the S_1 surface through a funnel located at geometry $\mathfrak{D}$. Some of the initially vibrationally hot molecules settle in the minimum $\mathfrak{C}(S_0)$, whereas others utilize their extra energy to move over the barrier separating $\mathfrak{C}$ and $\mathfrak{D}$ and yield the product $\mathfrak{D}(S_0)$.

So far, we have always referred to the energy of the nuclear motion as vibrational. This is correct as long as the system lives as a single molecule. If it separates into two or more molecules, or if it started as two or more molecules from the outset, we need to think of it as a supermolecule; in that case, some of the nuclear kinetic energy will actually be of translational rather than vibrational nature. For instance, the irradiation of HI produces rapidly moving (translationally "hot") H atoms.

2.5 ANALYSIS OF KINETIC DATA

So far, this chapter has been devoted to the listing and systematization of all the processes involved in organic photochemical reactions, but nothing has been said about the ways in which the relative importance of these processes and their rates can be disentangled. Although this is properly the subject of experimental photochemistry while the present text is limited to photochemical theory, it is perhaps appropriate to indicate how the elementary steps and their rates can be extracted from observable quantities.

The photochemical observables. The quantities commonly measured in quantitative studies of photochemical mechanisms are of two kinds, namely, static and dynamic.

Steady-state illumination experiments provide the quantum yields Φ of the appearance of all the products, of the disappearance of the starting material, and of all the emissions that appear upon excitation of the starting material or the products. These are defined as the number of moles of the product formed, starting material removed, or photons produced upon initial excitation of one mole of a substrate.

Pulsed illumination experiments permit the determination of the time dependence of the concentration of the starting material, various intermediate species such as excited states, and the final products under nonequilibrium conditions. The concentrations are usually measured by time-resolved spectrophotometry in absorption or, in the case of emitting excited species, in emission.

Often, the measurements are performed repeatedly as a function of freely adjustable variables, such as (i) the concentration of the substrate, (ii) the solvent, to which various additives may be added at known concentrations as quenchers or reaction partners, (iii) the temperature, (iv) the energy of the initial excitation, (v) the intensity of the initial excitation, and possibly (vi) the presence and strength of an outside magnetic field.

In the following, we give a few simple examples demonstrating how the rates of the elementary steps contributing to a photochemical reaction mechanism can be extracted from the observed data. As usual in kinetic studies, it is common to accept the simplest mechanism among all those compatible with all the observations as the "correct" one, with the understanding that future data may demand the adoption of a more complicated mechanism.

Relations between quantum yields, efficiencies, and rates. A photochemical reaction mechanism can generally be represented as a temporal sequence of steps characterized by rate constants. The rates at which the steps proceed are proportional to the rate constants and to the concentrations of the species involved. Thus, if the excited singlet state of $\mathfrak{A}$ [namely, $\mathfrak{A}(S_1)$] can fluoresce with the rate constant k_F, undergo intersystem crossing with the rate constant k_{ISC}, adiabatically react with the ground state of a reagent $\mathfrak{H}$ to yield an exciplex with rate constant k_E, be quenched with the rate constant k_Q to the ground state $\mathfrak{A}(S_0)$ by a ground-state quencher $\mathfrak{Q}$, and isomerize diabatically with rate constant k_I to yield an isomer $\mathfrak{C}(S_0)$, the total rate of its disappearance will be given by

$$-\frac{d[\mathfrak{A}(S_1)]}{dt} = k_F[\mathfrak{A}] + k_{ISC}[\mathfrak{A}] + k_E[\mathfrak{A}][\mathfrak{H}] + k_Q[\mathfrak{A}][\mathfrak{Q}] + k_I[\mathfrak{A}] \quad (2.27)$$

The efficiency of a step is equal to its rate divided by the sum total of the rates of all steps that originate in the same intermediate. Thus, the total efficiency of the destruction of $\mathfrak{A}(S_1)$ is unity by definition. The efficiency of fluorescence is

$$e_F = k_F/(k_F + k_{ISC} + k_E[\mathfrak{H}] + k_\mathfrak{Q}[\mathfrak{Q}] + k_I) \qquad (2.28)$$

The efficiency of intersystem crossing (e_{ISC}), of excimer formation (e_E), of quenching ($e_\mathfrak{Q}$), and of isomerization (e_I) are given similarly. If the concentrations of the reagents $\mathfrak{H}$ and $\mathfrak{Q}$ remain constant, and if no continued supply of $\mathfrak{A}(S_1)$ exists, as would be the case in a pulsed excitation experiment, integration of equation 2.27 yields an exponential decay for the starting species $\mathfrak{A}$:

$$[\mathfrak{A}(S_1)] = [\mathfrak{A}(S_1)]_0 \, e^{-t/\tau_{\mathfrak{A}(S_1)}} \qquad (2.29)$$

where $[\mathfrak{A}(S_1)]_0$ is the initial concentration and $\tau_{\mathfrak{A}(S_1)}$ is the lifetime of $\mathfrak{A}(S_1)$, related to the various rate constants by

$$1/\tau_{\mathfrak{A}(S_1)} = k_F + k_{ISC} + k_E[\mathfrak{H}] + k_\mathfrak{Q}[\mathfrak{Q}] + k_I \qquad (2.30)$$

Thus, the efficiencies of the individual steps originating in $\mathfrak{A}(S_1)$ can be written as

$$e_{ISC} = k_{ISC} \tau_{\mathfrak{A}(S_1)} \qquad (2.31)$$

$$e_E = k_E[\mathfrak{H}] \tau_{\mathfrak{A}(S_1)} \qquad (2.32)$$

$$e_\mathfrak{Q} = k_\mathfrak{Q}[\mathfrak{Q}] \tau_{\mathfrak{A}(S_1)} \qquad (2.33)$$

$$e_I = k_I \tau_{\mathfrak{A}(S_1)} \qquad (2.34)$$

It follows that the sum of the efficiencies of all steps originating in the same intermediate is equal to unity.

If two steps follow each other, and their efficiencies are e_1 and e_2, respectively, one can define the overall two-step efficiency even if the second product cannot be obtained from the starting molecule directly in a single elementary step. This efficiency is given by the product $e_1 e_2$. This definition is useful since frequently the intermediacy of a short-lived species may be suspected but has not been proven.

This leads us to the concept of the quantum yield Φ for the production of a species, which is equal to the efficiency of its formation from an originally excited starting material, usually through a series of steps. For instance, in Figure 2.12, the quantum yield $\Phi_{\mathfrak{A}(S_1)}$ of the formation of thermally equilibrated $\mathfrak{A}(S_1)$ upon initial excitation of $\mathfrak{A}(S_2)^*$ is less than unity as drawn, since a fraction of the excited molecules escapes via the funnel in $\mathfrak{D}$ to $\mathfrak{E}(S_0)$ and $\mathfrak{F}(S_0)$ and a fraction reaches $\mathfrak{B}(S_1)$ and $\mathfrak{C}(S_1)$ without ever settling in $\mathfrak{A}(S_1)$. The quantum yield Φ_F of the $S_1 \rightarrow S_0$ fluorescence from $\mathfrak{A}(S_1)$ is equal to $e_{\mathfrak{A}(S_1)} e_F$, where e_F is the efficiency of the fluorescence. In general, the quantum yield of a step starting in a species is given by its efficiency multiplied by the quantum yield of that species. This means that the ratio of the efficiencies of two steps originating in the same species is equal to the ratio of their quantum yields, which can be measured. Now, the ratio of efficiencies is also related to the ratio of rate constants so that the measurement of quantum yields allows the determination of ratios of rate constants.

Rate constant determination. This can be illustrated on our first example:

$$\frac{\Phi_F}{\Phi_I} = \frac{e_F}{e_I} = \frac{k_F}{k_I} \qquad (2.35)$$

$$\frac{\Phi_F}{\Phi_E} = \frac{e_F}{e_E} = \frac{k_F}{k_E[\mathfrak{H}]} \qquad (2.36)$$

where Φ_F, Φ_I, and Φ_E are the quantum yields of fluorescence, isomerization, and exciplex formation. In order to determine the absolute values of the rates rather than their ratios alone, the lifetime of the starting species [in our example, $\mathfrak{A}(S_1)$], or the risetime of at least one of the stable products, needs to be known. For example, if the left-hand side in equation 2.30 is known and all terms on the right-hand side are expressed through ratios of quantum yields and a single rate constant (say k_F), an equation for this constant results:

$$1/\tau_{\mathfrak{A}(S_1)} = k_F + \frac{\Phi_{ISC}}{\Phi_F} + \frac{\Phi_E}{\Phi_F}k_F + \frac{\Phi_{\mathfrak{Q}}}{\Phi_F}k_F + \frac{\Phi_I}{\Phi_F}k_F \qquad (2.37)$$

Once one of the rate constants is known absolutely, either from equation 2.37 or from risetime measurements, the others can be obtained from relations of type 2.35 and 2.36.

If the lifetimes or risetimes cannot be measured directly, a Stern–Volmer analysis of bimolecular kinetics may be helpful. Assume, for simplicity, that $\mathfrak{A}(S_1)$ has been produced with unit efficiency by photon absorption directly into the lowest vibrational level of the S_1 state (often, it can be produced with unit efficiency even using shorter absorption wavelengths). Then, the efficiencies in the equations 2.28 and 2.31–2.34 are equal to quantum yields. If the quantum yield of one of the processes is measured as a function of the concentration of either the quencher, [$\mathfrak{Q}$], or the reaction partner, [$\mathfrak{H}$], information about the lifetime of the reactive excited state will result. For instance, let us assume that the fluorescence quantum yield has been measured as a function of the concentration of the quencher. In the absence of the quencher, the quantum yield of the fluorescence is Φ_F^0 and the lifetime τ^0 is unknown. We have

$$\Phi_F^0 = k_F/(k_F + k_{ISC} + k_E[\mathfrak{H}] + k_I) = k_F\tau^0 \qquad (2.38)$$

In the presence of a known concentration of the quencher, the fluorescence quantum yield is

$$\Phi_F = k_F/(k_F + k_{ISC} + k_E[\mathfrak{H}] + k_I + k_{\mathfrak{Q}}[\mathfrak{Q}]) = k_F\tau \qquad (2.39)$$

The Stern–Volmer equation is obtained by combining equations 2.38 and 2.39:

$$\frac{\Phi_F^0}{\Phi_F} = 1 + k_Q \tau^0 [Q] \qquad (2.40)$$

If Φ_F^0/Φ_F is plotted against $[Q]$ for a series of quencher concentrations, a straight line of slope $k_Q \tau^0$ will result. If it can be assumed that the rate of quenching is dictated by the rate of diffusion and thus has an approximately known rate constant k_Q, the lifetime τ^0 in the presence of a known concentration $[H]$ of the reaction partner can then be obtained.

A completely analogous analysis could be made of the determination of the quantum yield of the isomerization reaction Φ_I as a function of the quencher concentration. Also, instead of varying the concentration of the quencher, one could vary the concentration of the reaction partner, H. If the rate constant for exciplex formation can be estimated, a value for τ^0 will again result.

Ordinarily, the experimenter has complete control over the composition of the system, and the determination of the lifetime τ^0 would be performed in the presence of either Q or H but not both, so that it would yield the value of $1/(k_F + k_{ISC} + k_I)$.

Nowadays, direct measurements of fluorescence lifetimes by the single-photon counting technique are facile and, in the absence of Q and H, yield the value of $1/(k_F + k_{ISC} + k_I)$ directly. This could then be used to separate the product $k_Q \tau^0$ or $k_E \tau^0$ into its two parts without assumptions concerning the rate of diffusion. Since $\Phi_F^0/\Phi_F = \tau^0/\tau$, Stern–Volmer plots can also be obtained from measurements of the changes in the observed fluorescence lifetime as a function of the concentration of the quencher or a reaction partner.

These very simple examples are only meant to indicate the type of procedures that are available for the analysis of photochemical kinetics. Many real systems are more complicated in that they contain reversible steps, multiple paths to the same products, and so on.

2.6 A SIMPLE EXAMPLE: INTERMOLECULAR PROTON-TRANSFER EQUILIBRIA

Our first example of an actual photochemical reaction is so simple that it might be considered a part of photophysics. It involves the reversible transfer of a proton from the X—H bond of one molecule to a lone pair on atom Y: of another molecule. Such acid–base equilibria are very well known in the S_0 ground state, and in aqueous solution the strengths of acids are characterized by the familiar pK_a values. If electronic excitation does not involve the electrons of the X—H bond nor those of the Y: lone pair, the same reversible proton-transfer reaction is possible on the electronically excited surface, and can occur without loss of electronic excitation, which plays fundamentally only the role of a spectator. However, since such excitation affects the charge distribution within the molecule, it has the potential of affecting the rates and equilibrium constants of the proton-transfer reaction, so that each electronic state is characterized by its own pK_a value. Only the S_1 and T_1 excited states normally have a long enough lifetime to permit the acid–base equilibrium to be approached or established, and three different pK_a values can be measured: $pK_a(S_0)$, $pK_a(S_1)$, and $pK_a(T_1)$. In aromatics, the situation is usually complicated by competing quenching of the S_1 state and, less commonly, the T_1 state, caused by protonation of the aromatic ring.

If the electronic excitation is localized in a chromophore that is quite distant from the X—H or the Y: group, the three pK_a values will not differ significantly. The interesting cases are those in which other electrons on atoms X or Y are involved in the excitation, since the three pK_a values can then differ dramatically.

For instance, in phenols, the acidic O—H bond is a part of the σ system, but one of the lone pairs on the oxygen atom is located in a $2p_z$ orbital and participates in the π system. Since the S_1 and T_1 states normally are of the π,π^* type, this provides an opportunity for a large decrease of electron density in the $2p_z$ orbital on oxygen upon excitation. The pK_a value is dictated by the degree to which the charge-transfer configuration Ar^-—O^+H participates in the electronic state wave function, and this can be quite different in the S_0, S_1, and T_1 states. Typically, the T_1 state of a phenol is a little more acidic than the S_0 state, and the S_1 state is orders of magnitude more acidic.

$$Ar-OH + H_2O \xrightarrow{h\nu} Ar-O^- + H_3O^+ \qquad (2.41)$$

An experimental determination of the $pK_a(S_1)$ and $pK_a(T_1)$ values can be made directly, by measuring the concentrations of the X—H and X^- or Y: and YH^+ species using absorption or emission methods. If one assumes that the entropy change associated with the proton-transfer process is the same in the ground state and the excited state, it is possible to determine $pK_a(S_1)$ and $pK_a(T_1)$ from a simple thermodynamic cycle (the "Förster cycle") that requires only the knowledge of excitation energies of the two forms, X—H and X^- (or Y: and YH^+) as shown in Figure 2.15. Since the energies of thermally relaxed species are needed, they are usually obtained as the average of excitation energies measured in absorption and in emission.

As Figure 2.15 shows, of the two members of a conjugate acid–base pair, the one that absorbs at longer wavelengths is stabilized in the excited state relative to the other. The increased acidity of phenols in the S_1 state can then be viewed as

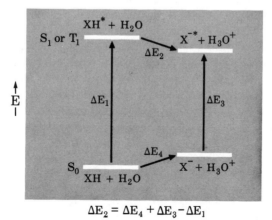

$\Delta E_2 = \Delta E_4 + \Delta E_3 - \Delta E_1$

Figure 2.15 The Förster cycle.

100 PHOTOPHYSICAL AND PHOTOCHEMICAL PROCESSES

a simple consequence of the red shifts in the absorption and emission spectra of phenolate anions relative to unionized phenols.

The direction of the spectral shift is, of course, not independent of the changes in electron density upon excitation. This is readily seen at the level of first-order perturbation theory when it is noted that protonation inevitably increases the electronegativity of an atom.

Simple proton-transfer equilibria of the kind described here are thus particularly easily handled by theory, since it is possible to obtain useful information on the charge distributions from quite simple semiempirical calculations of the π-electron variety. Unfortunately (or perhaps fortunately for the employment of theoreticians), this is a highly atypical situation (see Chapters 4–7). Even photochemical proton-transfer reactions can become much more complicated when the transfer is to a double bond or an aromatic ring rather than to a lone pair. These processes normally involve a rapid return to the S_0 surface (sections 6.2.2, 6.3.2, and 7.4).

2.7 ANOTHER EXAMPLE: 1,4-DEWARNAPHTHALENE

Our second example is still very simple chemically but relatively complex mechanistically. The chemical transformation is the valence isomerization of a Dewar isomer of naphthalene (**2**) into naphthalene (**3**), a process well known in the ground state. Although it is of the "forbidden" kind in the sense of orbital symmetry rules because it is constrained to the disrotatory path, it is quite facile in the ground state (E_a = 23.7 kcal/mol). We have chosen this example because it shows how much more complicated the mechanism of a simple thermal transformation can become when it is carried out photochemically, since it illustrates the way in which photochemical kinetic data are analyzed, and since it displays three features that are of current interest to organic photochemists: Some of the transformation proceeds by tunneling, some produces an excited state product, and some can proceed as a hot excited-state reaction even in a cold solid solution if the initial excitation is into a higher vibrational level of the S_1 state. It is also noteworthy that in spite of the huge excess of energy stored in the relaxed S_1 state of the strained Dewar isomer (~150 kcal/mol), its transformation to ground-state naphthalene competes only about equally with fluorescence and with intersystem crossing.

The photochemical and photophysical processes triggered by the excitation of 1,4-dewarnaphthalene (**2**) into its S_1 state are indicated in Figure 2.16 by arrows. Full arrows denote radiative processes, and wavy arrows represent nonradiative processes. The dark dots show the initial state after light absorption, as defined by the photon energy. The percentages shown at the arrows indicate the fraction of molecules that participate in the process symbolized by the arrow when the excitation is into the lowest vibrational level of the S_1 state (0–0 excitation). The partitioning of the 44% that return to S_0 in the center of the diagram between the

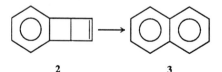

2   3

ground states of **2** and **3** is not known. The rate constants are given where known. The percentages in parentheses indicate the fraction of the vibrationally hot S_1 molecules that escape over the barrier to the right after the initial excitation to the vibrational level stated on the left; the remainder relax into the lowest vibrational level of S_1.

The results shown in Figure 2.16 were obtained from the following primary data collected in a matrix of solid nitrogen at 10 K. Upon 0–0 excitation of **2**, one observes fluorescence from **2** [$\Phi_F(2) = 0.13 \pm 0.02$, $\tau_F(2) = 14.3 \pm 1$ ns] and from **3** [$\Phi_F(3) = 0.020 \pm 0.003$, $\tau_F(3) = 177 \pm 10$ ns] and phosphorescence from **3** [$\Phi_P(3) = 0.019 \pm 0.003$, $\tau_P(3) = 2.3 \pm 0.1$ s].

When **3** is excited in an N_2 matrix at 10 K, it fluoresces with a quantum yield of $\Phi_F^0(3) = 0.14 \pm 0.02$ and phosphoresces with a yield of $\Phi_P^0(3) = 0.040 \pm 0.006$.

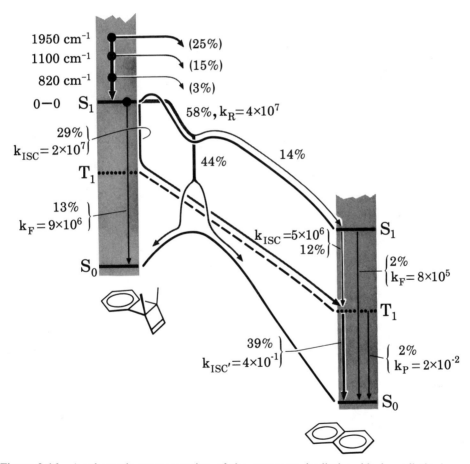

Figure 2.16 A schematic representation of the processes (radiative, black; radiationless, grey) triggered by the excitation of 1,4-dewarnaphthalene (**2**) into the S_1 state, with rate constants. Percentages of molecules that follow the various paths are indicated, as are the potential energy surfaces proposed to account for the results. For details see text. Adapted by permission from S. L. Wallace and J. Michl, in *Photochemistry and Photobiology* (A. H. Zewail, Ed.), Vol. II, Harwood, Chur, Switzerland, 1983. p. 1191.

Both emissions have the same lifetimes and spectral shapes as those observed upon excitation of **2**.

The analysis of the data is based on two assumptions: (i) The rate of vertical internal conversion from the S_1 state of **2** to its S_0 state is negligible next to the sum of the rates of S_1 fluorescence and intersystem crossing. This is already known to be true of **3**. (ii) Triplet **2** is converted to triplet **3** with 100% efficiency. Supporting evidence for these assumptions was obtained by comparison with related aromatics and in triplet sensitization experiments.

The processes to consider, along with their efficiencies (e) and rate constants (k), can now be listed, and the values of some can be specified immediately:

$S_1(\mathbf{2})$:

Fluorescence: $\quad e_F(\mathbf{2}) = \Phi_F(\mathbf{2}) = 0.13$

$k_F(\mathbf{2}) = e_F(\mathbf{2})/\tau_F(\mathbf{2}) = 9 \times 10^6 \text{ s}^{-1}$

Intersystem crossing S_1–T_1: $\quad e_{ISC}(\mathbf{2})$: unknown

$k_{ISC}(\mathbf{2})$: unknown

Nonvertical return to $S_0(\mathbf{2})$: $\quad e_{RET}(\mathbf{2})$: unknown (2.42)

$k_{RET}(\mathbf{2})$: unknown

Adiabatic production of $S_1(\mathbf{3})$: $\quad e_A(\mathbf{2}) = \Phi_F(\mathbf{3})/\Phi_F^0(\mathbf{3}) = 0.14$

$k_A(\mathbf{2}) = e_A(\mathbf{2})/\tau_F(\mathbf{2}) = 1 \times 10^7 \text{ s}^{-1}$

Diabatic production of $S_0(\mathbf{3})$: $\quad e_D(\mathbf{2})$: unknown

$k_D(\mathbf{2})$: unknown

$T_1(\mathbf{2})$:

Adiabatic production of $T_1(\mathbf{3})$: $\quad e_{TA}(\mathbf{2}) = 1.0$ (assumed) (2.43)

$S_1(\mathbf{3})$:

Fluorescence: $\quad e_F(\mathbf{3}) = \Phi_F^0(\mathbf{3}) = 0.14$

$k_F(\mathbf{3}) = e_F(\mathbf{3})/\tau_F(\mathbf{3}) = 8 \times 10^5 \text{ s}^{-1}$

Intersystem crossing: $\quad e_{ISC}(\mathbf{3}) = 1 - e_F(\mathbf{3}) = 0.86$

$k_{ISC}(\mathbf{3}) = e_{ISC}(\mathbf{3})/\tau_F(\mathbf{3}) = 5 \times 10^6 \text{ s}^{-1}$ (2.44)

$T_1(\mathbf{3})$:

Phosphorescence: $\quad e_P(\mathbf{3}) = \Phi_P^0(\mathbf{3})/e_{ISC}(\mathbf{3}) = 0.047$

$k_P(\mathbf{3}) = e_P(\mathbf{3})/\tau_P(\mathbf{3}) = 2 \times 10^{-2} \text{ s}^{-1}$

Intersystem crossing to $S_0(\mathbf{3})$: $\quad e_{ISC'}(\mathbf{3}) = 1 - e_P(\mathbf{3}) = 0.95$

$k_{ISC'}(\mathbf{3}) = e_{ISC'}(\mathbf{3})/\tau_P(\mathbf{3}) = 4 \times 10^{-1} \text{ s}^{-1}$ (2.45)

To obtain the values of the remaining unknowns, we note that the efficiency with which the triplet of **3** is formed is the sum of contributions from the singlet adiabatic path $e_A(2) \, e_{ISC}(3) = 0.12$ and from the triplet adiabatic path $e_{ISC}(2) \, e_{TA}(2) = e_{ISC}(2)$. However, this efficiency can also be obtained from $\Phi_P(3)/e_P(3) = 0.41$. It follows that

$$e_{ISC}(2) = 0.29 \tag{2.46}$$

and we see that the triplet adiabatic reaction is responsible for about 70% of the total triplet of **3** formed. From this value of $e_{ISC}(2)$, we obtain

$$k_{ISC}(2) = e_{ISC}(2)/\tau_F(2) = 2 \times 10^7 \, s^{-1} \tag{2.47}$$

Finally, we note that the efficiency with which the initially excited **2** adiabatically passes the initial barrier in the S_1 surface is given by $1 - e_F(2) - e_{ISC}(2) = 0.58$. Since the efficiency with which excited singlet **3** is formed is only $e_A(2) = 0.14$, it follows that

$$e_{RET}(2) + e_D(2) = 0.44 \tag{2.48}$$

In order to obtain separately the efficiency of unproductive return to S_0, $e_{RET}(2)$, and the efficiency of the diabatic photochemical process, $e_D(2)$, the total quantum yield of formation of **3** would have to be known. This has not been measured under the conditions of these experiments, but room-temperature results indicate that neither efficiency is negligible.

The specific mechanism proposed in Figure 2.16 for the radiationless loss of three-quarters of the 58% of all singlet excited molecules of **2** that pass over or through the barrier in the S_1 surface involves brief sojourn in a minimum characteristic of ground-state "forbidden" electrocyclic reactions (see Chapter 5). A fraction of the molecules escape from this minimum on the S_1 surface in an adiabatic fashion, but most return to S_0 in a diabatic process and partition between the ground-state product **3** and the starting material **2**. In less exothermic electrocyclic processes the barrier for adiabatic escape is higher, and only a diabatic ring opening is observed.

The rate of passing the initial barrier in the S_1 surface at 10 K is given by

$$k_R = [e_A(2) + e_{RET}(2) + e_D(2)]/\tau_F(2) = 4 \times 10^7 \, s^{-1} \tag{2.49}$$

This is both relatively slow and quite fast. It is slow enough to justify the assumption that it represents the rate-determining step in the processes leading to S_1 or S_0 of **3** and to S_0 of **2**. Then, the minimum postulated in the central region of the S_1 surface may still provide a favorable point for return to the S_0 surface, but its existence will have no kinetic consequences. However, k_R is quite fast for passage over a barrier of almost any size at 10 K. Interestingly, it remains constant when the temperature is more than doubled to 24 K, implying that the passage occurs solely through (rather than over) the barrier, which is presumably quite thin and primarily requires motion of the bridgehead hydrogens in **2**. At higher temperatures in other media, k_R increases and thermally activated passage over the barrier dominates.

Excitation into higher vibrational levels of the S_1 state of **2** changes the ratios of the quantum yields of the three observed emissions. For excitation at the four points shown in Figure 2.16, $\Phi_F(\mathbf{2})/\Phi_F(\mathbf{3})$ gradually decreases from 6.5 to 2.3 and $\Phi_F(\mathbf{3})/\Phi_P(\mathbf{3})$ increases from 1 to 2.2. The direction of these changes is sensible when it is recalled that only 30% of the triplet **3** results from the excited singlet adiabatic process whereas 70% is formed via intersystem crossing in the S_1 of **2**. If the higher energy of excitation promotes the passage of the excited singlet molecules of **2** over the initial barrier in S_1, thus producing more excited singlet **3**, it will be not only at the expense of the fluorescence from the excited singlet **2** but also at the expense of the intersystem crossing from excited singlet **2**, and therefore at the expense of the population of triplet **3** and its subsequent phosphorescence. An analysis of the two emission intensity ratios as a function of the exciting wavelength permits a test of the self-consistency of the reaction scheme and leads to the fractions of molecules undergoing hot excited-state reaction, as listed in Figure 2.16.

2.8 A SUMMARY: THE NEED FOR THEORY

At first encounter, the long list of photochemical processes outlined in this chapter may be somewhat discouraging. How is one to understand why certain of the many competing processes occur preferably, given a molecular structure and reaction conditions? Worse still, how is one to predict what will be the outcome of experiments on new substrates?

At present, true *a priori* predictions for the photochemistry of novel types of substrates are hardly possible, and experimental experience with related molecules is still the best guide. The role of theory lies more in the rationalization and systematization of observations already made. This can be done at two levels of sophistication.

First, theory can be asked for information on the paths taken by the molecule between the points at which it has been observed. Very often, no experimental observations exist between the thermally relaxed S_1 or T_1 state of the starting material and the thermally relaxed S_0 state of the product; sometimes, even the S_1 and T_1 states are so short-lived that they have not been observed. Theory can be asked to rationalize why certain paths seem to be followed, whereas others, equally reasonable at first sight, seem to be avoided. Why do some reactions proceed with retention of stereochemistry, whereas others do not? Why does singlet excitation often yield different products from triplet excitation? Qualitative answers to these questions should be forthcoming from a theoretical analysis of the positions of minima and funnels in the S_1 and T_1 states, which determine the geometries at which molecules return to the ground state, and of the location and heights of the barriers in the S_1 and T_1 states, which may prevent some of the minima or funnels from being accessible. Finally, a knowledge of the shape of the S_0 surface in certain critical regions will be necessary for the understanding of the structures of the products finally formed.

Second, for systems where the static picture of the potential energy surfaces is sufficiently well developed, one might wish to proceed to a full dynamic description and to compute the rate of the individual elementary steps. This is generally even harder than the tasks outlined above, since only one kind of rate constants, namely those of radiative processes, are relatively easy to calculate from the observed spectra or, generally less accurately, from computed transition moments.

As mentioned in Chapter 1, only the static and structural aspects of the theoretical description of photochemical processes are addressed in this book in any depth.

2.9 COMMENTS AND REFERENCES

A glossary of **photochemical terminology** was published by the IUPAC commission on photochemistry: S. E. Braslavsky and K. N. Houk, *Pure Appl. Chem.* **60**, 1055 (1988).

Much more is known about the **photophysics** of aromatics than about that of any other class of organic molecules. Somewhat outdated but still useful and very thorough surveys of all aspects of photophysics are: J. B. Birks, *Photophysics of Aromatic Molecules,* Wiley, New York, 1970; J. B. Birks, Ed., *Organic Molecular Photophysics,* Wiley, New York, Vol. 1 (1973), Vol. 2 (1975).

More recent information is found in two series of volumes on excited states: One starts with E. C. Lim, Ed., *Excited States,* Vol. 1, Academic Press, New York, 1974; the other starts with A. A. Lamola, Ed., *Creation and Detection of the Excited States,* Vol. 1A, B, Marcel Dekker, New York, 1971.

A text on fluorescence is J. R. Lakowicz, *Principles of Fluorescence Spectroscopy,* Plenum, New York, 1983; an older but still useful book is C. A. Parker, *Photoluminescence in Solution,* Elsevier, Amsterdam, 1968. Excellent recent books on time-resolved laser spectroscopy are: G. R. Fleming, *Chemical Applications of Ultrafast Spectroscopy,* Oxford University Press, Oxford, 1986; K. B. Eisenthal, Ed., *Applications of Picosecond Spectroscopy to Chemistry,* NATO ASI series, Reidel, Dordrecht, 1983.

The relation of photophysical to photochemical processes has been discussed in: J. Michl, *Mol. Photochem.* **4**, 243 (1972); J. Michl, *Topics Curr. Chem.* **46**, 1 (1974); J. Michl, in *Excited States in Quantum Chemistry* (C. A. Nicolaides and D. R. Beck, Eds.), D. Reidel, Dordrecht, Holland, 1978, p. 417.

Most of the subjects that we have only lightly touched upon here are elaborated in detail in specialized reviews and monographs.

A recent introduction to the principles of **light absorption,** Franck–Condon principle, and transition intensities and polarizations and their relation to transition moments is found in J. Michl and E. W. Thulstrup, *Spectroscopy with Polarized Light,* VCH Publishers, Deerfield Beach, Florida, 1986, and, in a more elementary form, in E. W. Thulstrup and J. Michl, *Elementary Polarization Spectroscopy,* VCH Publishers, New York, 1989. A more advanced discussion of the nature of molecule–light interaction during the absorption event is provided in J. D. Macomber, *The Dynamics of Spectroscopic Transitions,* Wiley, New York, 1976. For the Strickler–Berg equation (equation 2.9), see S. J. Strickler and R. A. Berg, *J. Chem. Phys.* **37**, 814 (1962). A similar result was obtained by J. B. Birks and D. J. Dyson [*Proc. Roy. Soc.* **A275**, 135 (1963)], who used n_F^3/n_A in place of n^2, but since the index of refraction at the wavelength of fluorescence (n_F) and at the wavelength of absorption (n_A) are nearly the same, this difference has no practical consequence.

The theory of **radiationless surface jumps** is outlined in a brief but lucid manner in the book by J. Simons quoted in Section 1.4. A very detailed survey of methods for very small molecules is available in M. Baer, *Theory of Chemical Reaction Dynamics,* Vol. 2, CRC Press, Boca Raton, Florida, 1985, p. 219. Internal conversion and intersystem crossing dynamics are discussed in: G. W. Robinson and R. P. Frosch, *J. Chem. Phys.* **37**, 1962 (1962); G. W. Robinson and R. P. Frosch, *J. Chem. Phys.* **38**, 1187 (1963); J. Jortner,

S. A. Rice, and R. M. Hochstrasser, *Adv. Photochem.* **7,** 149 (1969); K. F. Freed, *Topics Curr. Chem.* **31,** 105 (1972); K. F. Freed, in *Radiationless Processes in Molecules in Condensed Phase* (F. K. Fong, Ed.), Springer, Berlin, 1976, p. 23; K. F. Freed, *Accounts Chem. Res.* **11,** 74 (1978). For the energy-gap law, see W. Siebrand, *J. Chem. Phys.* **46,** 440, 2411 (1967). For references to funnels (conical intersections), see F. Bernardi; S. De, M. Olivucci, and M. A. Robb, *J. Am. Chem. Soc.* **112,** 1737 (1990).

An early example of an actual calculation of the lifetime of an excited molecule (twisted stilbene) is found in G. Orlandi and G. Marconi, *Nuovo Cimento* **63B,** 332 (1981). The computation of the electronic matrix elements needed for such calculations is described, for instance, in M. Persico and V. Bonačić-Koutecký, *J. Chem. Phys.* **76,** 6018 (1982) (*ab initio*) and in G. J. M. Dormans, G. C. Groenenboom, W. C. A. van Dorst, and H. M. Buck, *J. Am. Chem. Soc.* **110,** 1406 (1988) (semiempirical) for internal conversion, and in S. D. McGlynn, T. Azumi, and M. Kinoshita, *Molecular Spectroscopy of the Triplet State,* Prentice-Hall, Englewood Cliffs, New Jersey, 1969, and S. P. McGlynn, L. C. Vanquickenborne, M. Kinoshita, and D. G. Carroll, *Introduction to Applied Quantum Chemistry,* Holt, Rinehart and Winston, New York, 1972, for intersystem crossing. A short book dedicated to the theory of spin-orbit coupling is W. G. Richards, H. P. Trivedi, and D. L. Cooper, *Spin-Orbit Coupling in Molecules,* Clarendon Press, Oxford, 1981.

The **RRKM theory** of reaction rates is described, for instance, in P. J. Robinson and K. A. Holbrook, *Unimolecular Reactions,* Wiley-Interscience, New York, 1972.

Adiabatic reactions have been reviewed in N. J. Turro, J. McVey, V. Ramamurthy, and P. Lechtken, *Agnew Chem. Int. Ed. Eng.* **91,** 597 (1979). Additional examples are described in H.-D. Becker, *Pure Appl. Chem.* **54,** 1589 (1982) and in J. Wirz, G. Persy, E. Rommel, I. Murata, and K. Nakasuji, *Helv. Chim. Acta* **67,** 305 (1984). Individual adiabatic steps can be treated in terms of **transition-state theory;** see, for example, K. J. Laidler, *Theories of Chemical Reaction Rates,* McGraw-Hill, Newport, 1969; for a more recent review of its status, see: D. G. Truhlar, W. L. Kase, and J. T. Hynes, *J. Phys. Chem.* **87,** 2664 (1983); M. M. Kreevoy and D. G. Truhlar, in *Investigation of Rates and Mechanisms of Reactions,* 4th ed. (C. F. Bernasconi, Ed.), Wiley, 1986, Chapter 1. See also: E. Pollack, in *Theory of Chemical Reaction Dynamics* (M. Baer, Ed.), Vol. 3, CRC Press, Boca Raton, Florida, 1985, p. 123; D. G. Truhlar, A. D. Isaacson, and B. C. Garrett, Vol. 4, p. 65 of the same series. For the theory of **reactions in solution,** see J. T. Hynes, in Vol. 4, p. 171 of the same series. Dynamic solvent effects [see, e.g., P. F. Barbara and W. Jarzeba, *Acc. Chem. Res.* **21,** 195 (1988); B. Bagchi, G. R. Fleming, and D. W. Oxtoby, *J. Chem. Phys.* **78,** 7375 (1983)], solvent control of fast intramolecular charge transfer [see, e.g., D. Huppert, V. Ittah and E. M. Kosower, *Chem. Phys. Lett.* **144,** 15 (1988)] and the **Kramers equation** (equation 2.8) [see H. A. Kramers, *Physica* **7,** 284 (1940)] have received considerable attention lately. For the use of translational diffusion coefficients to define microviscosity, see Y.-P. Sun and J. Saltiel, *J. Phys. Chem.,* **93,** 8310 (1989).

Early examples of **photochemical theoretical studies with trajectory computations** are R. R. Birge and L. M. Hubbard, *J. Am. Chem. Soc.* **102,** 2195 (1980) (using a semiempirical surface) and I. Ohmine and K. Morokuma, *J. Chem. Phys.* **74,** 564 (1981) (using an *ab initio* surface). For trajectory calculations that explicitly model the motion of the solvent, see I. Ohmine, *J. Chem. Phys.* **85,** 3342 (1986). The methods of trajectory calculations are discussed, for instance, by L. M. Raff and D. L. Thompson, in *Theory of Chemical Reaction Dynamics* (M. Baer, Ed.), Vol. 3, CRC Press, Boca Raton, Florida, 1985, p. 1. For leading references to Gaussian wave packet dynamics, see D. Huber and E. J. Heller, *J. Chem. Phys.* **87,** 5302 (1987).

The **tunneling** of protons through barriers has been reviewed many times. For an early summary, see E. F. Caldin, *Chem. Rev.* **69,** 135 (1969); more recent reviews are found in R. P. Bell, *The Tunnel Effect in Chemistry,* Chapman and Hall, London, 1980, and in J. Jortner and B. Pullman, Eds., *Tunneling,* The Jerusalem Symposia on Quantum Chemistry and Biochemistry, Vol. 19, Reidel, Dordrecht, 1986. For a very nice example, see K.-H. Grellmann, A. Schmidt, and H. Weller, *Chem. Phys. Lett.* **88,** 40 (1982). Heavy atom tunneling is discussed much less frequently; see, for instance, S. L. Buchwalter and G. L. Closs, *J. Am. Chem. Soc.* **101,** 4688 (1979) and M. B. Sponsler, R. Jain, F. D. Coms, and D. A. Dougherty, *J. Am. Chem. Soc.* **111,** 2240 (1989).

Reactions in hot and/or upper excited states have been reviewed by N. J. Turro, V. Ramamurthy, W. Cherry, and W. Farneth, *Chem. Rev.* **78,** 125 (1978). For a recent direct observation of a hot-ground-state reaction of a triene, see P. J. Reid, S. J. Doig, and R. A. Mathies, *Chem. Phys. Lett.* **156,** 163 (1989).

For spectra of **charge-transfer complexes,** see R. Foster, *Organic Charge-Transfer Complexes,* Academic Press, New York, 1969; the factors contributing to their stability are discussed in M. J. S. Dewar and C. C. Thompson, Jr., *Tetrahedron (Suppl.)* **7,** 97 (1966). References to **excimers** and **exciplexes** are discussed in Section 5.5.

The classical articles on **energy transfer** are: Th. Förster, *Naturwissenschaften* **33,** 166 (1946); Th. Förster, *Disc. Faraday Soc.* **27,** 7 (1959); D. L. Dexter, *J. Chem. Phys.* **21,** 836 (1953). For a book, see J. T. Yardley, *Introduction to Molecular Energy Transfer,* Academic Press, New York, 1980. The Sandros equation (equation 2.21) was obtained by K. Sandros, *Acta Chem. Scand.* **18,** 2355 (1964) and B. Stevens and M. S. Walker, *Proc. Roy. Soc.* **26,** 27, 109 (1964). A more general equation suitable for both "vertical" and "nonvertical" triplet energy transfer has been proposed; see V. Balzani, P. Bolletta, and F. Scandola, *J. Am. Chem. Soc.* **102,** 2152 (1980). For a measurement of the rate of triplet–triplet transfer between molecules in contact, see R. W. Anderson, R. M. Hochstrasser, H. Lutz, and G. W. Scott, *J. Chem. Phys.* **61,** 2500 (1974). For studies of bichromophoric molecules, see, for instance, S. Hassoon, H. Lustig, M. B. Rubin, and S. Speiser, *J. Phys. Chem.* **88,** 6367 (1984) (singlet–singlet) and S. Speiser, S. Hassoon, and M. Rubin, *J. Phys. Chem.* **90,** 5085 (1986) and G. L. Closs, M. D. Johnson, J. R. Miller, and P. Piotrowiak, *J. Am. Chem. Soc.* **111,** 3751 (1989) (triplet–triplet). Additional information is found in N. J. Turro, *Pure Appl. Chem.* **49,** 405 (1977).

Triplet–triplet annihilation has been used for the production of highly excited singlet states and has permitted the observation of emission from the latter; see, for instance, B. Nickel, *Helv. Chim. Acta* **61,** 198 (1978), which also provided evidence that, in pyrene, vibrational relaxation in S_2 is slow relative to the $S_2 \rightarrow S_1$ internal conversion.

General information on **electron transfer** is found in R. A. Marcus and N. Sutin, *Biochim. Biophys. Acta* **811,** 265 (1985). The mechanism and spin dynamics of photoinduced electron-transfer reactions in solution are discussed in A. Weller, *Z. Phys. Chem. (NF)* **130,** 129 (1982). A review of photoinduced electron transfer and its role in organic photochemistry can be found in G. J. Kavarnos and N. J. Turro, *Chem. Rev.* **86,** 401 (1986). Electron transfer by tunneling in solids is reviewed in K. V. Mikkelsen and M. A. Ratner, *Chem. Rev.* **87,** 113 (1987). References to studies of the Marcus inverted region can be found in I. R. Gould and S. Farid, *J. Am. Chem. Soc.* **110,** 7883 (1988) and in G. L. Closs, L. T. Calcaterra, N. J. Green, K. W. Penfield, and J. R. Miller, *J. Phys. Chem.* **90,** 3673 (1986), who give an extensive list of references to long-distance electron transfer. An *ab initio* calculation of stereoelectronic effects on intramolecular electron transfer is described in K. Ohta, G. L. Closs, K. Morokuma, and N. J. Green, *J. Am. Chem. Soc.* **108,** 1319 (1986).

A four-volume compendium on all aspects of photoinduced electron transfer is available: M. A. Fox and M. Chanon, Eds., *Photoinduced Electron Transfer,* Elsevier, Amsterdam, 1988.

Diffusion is treated in: R. E. Weston, Jr., and H. Schwarz, *Chemical Kinetics,* Prentice-Hall, Englewood Cliffs, New Jersey, 1972; J. Saltiel and B. W. Atwater, *Adv. Photochem.*

14, 1 (1988); H. Fischer and H. Paul, *Acc. Chem. Res.* **20,** 200 (1987). A classical review of diffusion-controlled reactions is R. M. Noyes, *Prog. React. Kinet.* **1,** 129 (1961), and diffusion effects on rapid bimolecular reactions are surveyed in J. Keizer, *Chem. Rev.* **87,** 167 (1987). For references to chemiluminescent radical recombination, see E. Lissi and J. de la Fuente, *J. Chem. Soc. Perkin Trans. II,* 819 (1986).

The **effects of magnetic field and hyperfine interactions** on intersystem crossing, triplet–triplet annihilation, and electron-transfer processes are analyzed in detail in U. E. Steiner and T. Ulrich, *Chem. Rev.* **89,** 51 (1989); effects on reaction products are surveyed in I. R. Gould, N. J. Turro, and M. B. Zimmt, *Adv. Phys. Org. Chem.* **20,** 1 (1984).

Standard photochemical texts provide additional information on the **analysis of kinetic data;** see, for example, N. J. Turro, *Modern Molecular Photochemistry,* Benjamin/Cummings, Menlo Park, New Jersey, 1978.

Proton-transfer equilibria in excited states are reviewed in J. F. Ireland and P. A. H. Wyatt, *Adv. Phys. Org. Chem.* **12,** 131 (1976), H. Shizuka, *Accounts Chem. Res.* **18,** 141 (1985), and in E. M. Kosower, *Annu. Rev. Phys. Chem.* **37,** 127 (1986).

By restricting our attention to medium-size molecules in solutions we omit from consideration much interesting **photochemistry in Rydberg states.** This has received considerable theoretical attention. For a review and leading references, see: E. M. Evleth, H. Z. Cao, and E. Kassal, in *Photophysics and Photochemistry Above 6 eV* (F. Lohman Ed.), Elsevier, Amsterdam, 1985, p. 479; J.-P. Malrieu, *Theor. Chim. Acta* **59,** 251 (1981). Vacuum-UV photochemistry in solution is beginning to attract attention as well: M. G. Steinmetz, in *Organic Photochemistry,* (A. Padwa, Ed.), Marcel Dekker, New York, 1987, Vol. 8, p. 67.

For the photochemistry of 1,4-dewarnapthalene, see S. L. Wallace and J. Michl, *Photochemistry and Photobiology: Proceedings of the International Conference,* January 10, 1983, University of Alexandria, Egypt, Vol. II (A. H. Zewail, Ed.), Harwood, Chur, Switzerland, 1983, p. 1191; also see references therein.

CHAPTER 3

Location of Minima, Funnels and Barriers on Surfaces

3.1 GENERAL

As we leave Chapter 2, we see that the course of photochemical reactions of organic molecules, at least those occurring in solution, is determined primarily by the shapes of the S_1 and T_1 surfaces, along with that of the S_0 surface. The most important characteristics of these surfaces are the locations of minima, funnels (touching points of S_0 and S_1), barriers, and singlet–triplet intersection points.

The need for the understanding of the changes in electronic wave functions during large-amplitude geometry variations provides a special flavor to theoretical aspects of photochemistry relative to those of ordinary electronic spectroscopy. In the present chapter, we focus our attention on the principles used in qualitative arguments for the examination of geometries at which minima and barriers occur in the S_1 and T_1 states, relying primarily on the use of correlation diagrams. Many specific applications of the correlation diagram technique to individual molecules can be found in Part B of this book.

The shape of the potential energy surfaces is intimately tied to the changes in the electronic wave functions as a function of molecular geometry. In spite of the general statement in the introductory paragraph in this chapter, a mere knowledge of the surface shapes and energies is not really adequate. Rather, an understanding of the nature of the electronic wave functions associated with the individual molecular geometries in the various electronic states is needed as well.

There are several reasons for this. First, if a quantum mechanical program is used as a black box to produce the energies of states at a series of geometries and no thought is given to the nature of the electronic wave function, one runs the same danger as with all black boxes, in theory or in experiment: The answer may well be wrong. After all, even the most sophisticated *ab initio* calculations for molecules of interest to the organic photochemist, and in particular for their excited states, are only rather approximate. An appreciation of the nature of the wave function will provide some guidance as to the degree of reliability of the energies produced by a computer program.

Second, for molecules of interest here, $3N - 6$ usually is a very large number and it is unrealistic to expect that all regions of the nuclear configuration space could be explored. For a molecule with only four atoms, and with a sparse grid of only 10 points per geometrical variable, energies of all states of interest would already have to be computed at a million different geometries for a complete mapping. Clearly, judgment has to be exercised in the choice of pathways to be explored and of regions in which a search for minima, saddlepoints and surface touchings should be performed. A qualitative understanding of the wave function will help a great deal. In particular, it will permit estimates of the effect of small variations in the geometry using the principles of perturbation theory.

Third, a knowledge of the wave function will permit the use of perturbation theory to derive results for photochemical substrates related to the one under study. After all, photochemical reactions occur in familes of related transformations on similar molecules. Ordinarily, one seeks the abstract concepts that allow a rationalization of a general type of transformation rather than only the rationalization of the reaction of a particular substrate. Here again, qualitative appreciation of the nature of the electronic wave function is essential.

When faced with the problem of rationalizing the existence of a minimum in the ground-state (S_0) surface of an organic molecule, or predicting its existence and location, a chemist will most often begin by applying the normal rules of valence, probably quite subconsciously. Structures with unusual bond lengths, bond angles, or coordination numbers will be more or less automatically discarded unless some unusual circumstance calls for their inclusion. At times, molecular models, which represent an embodiment of this experience accumulated over the decades, will be needed. Possibly, their quantitative offspring, molecular mechanics, will be called for. Quantum mechanical calculations may be needed when exact geometries are to be predicted, particularly in less usual bonding situations, but only relatively rarely will they produce totally unexpected results.

At present, no comparable body of experience and intuition exists for the excited states. In this respect, locating minima in the S_1 and T_1 surfaces is a game reminiscent of the early days of ground-state chemistry. In this regard, quantum theory, both in its qualitative form and in the form of numerical computations, is of invaluable assistance. The excited-state surfaces appear to have many more minima than do the ground-state surfaces, frequently at geometries that are quite unfavorable in the ground state.

There appear to be at least three classes of geometries at which the S_1 and T_1 surfaces of organic molecules tend to have minima.

First, for many molecules, the S_1 and/or T_1 surface has a minimum at a geometry that lies in the vicinity of the ground-state minimum. The excitation from the S_0 state into S_1 or T_1 is then of the "vertical" or nearly vertical type, and relatively easy to observe spectroscopically. These minima in S_1 and T_1 are known as the spectroscopic minima. Return through these minima is of limited interest in condensed-phase photochemistry in that it usually leads right back to the starting minimum in the S_0 surface. Yet, spectroscopic minima are of pivotal importance in photochemistry in that they provide the means of initiating the photochemical transformation. If excitation into a spectroscopic minimum in S_1 or T_1 is followed by escape from the minimum to other areas on the S_1 or T_1 surface, the photochemical chain of events has been triggered. In the case of bimolecular photochemical reactions, the spectroscopic minima serve the crucial role of reservoirs that hold one of the excited partners until diffusion brings the other one close.

Second, in many cases, molecules in their S_1 and T_1 states are "stickier" and more prone to complexation than they are in their ground states. At relatively low temperatures, most pairs of molecules will hold together weakly even in the ground state as a result of van der Waals and other intermolecular forces, at least in the gas phase if not in a solution, but the binary complexes are often not stable enough in solution at room temperature for spectroscopic observation. In effect, then, there is no significant minimum in the ground state at geometries corresponding to an intermolecular complex. Yet, in a T_1 and particularly in the S_1 state, such a minimum is often present. These minima are referred to as excimer minima if the two partners are identical and exciplex minima if they are different. They can only rarely be observed in absorption since the ground-state complex normally cannot be prepared and kept for a sufficient length of time, and in that sense they are "nonspectroscopic." Often, however, they are observed in emission. Intramolecular analogs of these minima also exist and can be reached by approach of two parts of a single molecule which are not directly bonded in the ground state.

The return from excimer and exciplex minima to the S_0 state is normally also unproductive in that no new ground-state product is formed: The partners, now nonbonded, diffuse apart. Again, however, these minima are of crucial importance in photochemistry, most of all in bimolecular photochemistry: They hold the reaction partners together for an extended time, frequently in a particular orientation, and thus permit motion to geometries representing other regions of the S_1 or T_1 surface which become available when the two partners are next to each other. Their relatively long lifetimes may also provide an opportunity for an adjustment of molecular geometry and/or rearrangement in the solvation shell which causes electron transfer from one partner to the other, leading to the eventual production of an ion pair.

Third, there is a class of geometries that are favorable in the S_1 and T_1 states but at which the S_0 state hardly ever has a minimum, or, if it does, it is so shallow and of such high energy that it does not correspond to a species isolable at room temperature. These are "biradicaloid geometries," defined as those at which the molecule in its S_0 state has two electrons in a pair of nearly degenerate approximately nonbonding orbitals. The reason why these geometries are energetically so unfavorable in the ground state is obvious: If the geometry is distorted so as to allow the two nonbonding orbitals to overlap and to produce a bonding and an antibonding combination, both electrons can be placed in the former and the molecule has acquired an additional bond. Quite generally, biradicaloid geometries can be viewed as geometries at which molecules in their S_0 state have one fewer bond than they ought to. In the S_1 and T_1 states, this reason for relative instability is absent. The same geometrical distortion that brought relief in the ground state because it offered the possibility of placing both electrons into a bonding orbital now brings none: In the excited state, one of the electrons goes into the bonding, the other into the antibonding combination. Often, the antibonding combination is destabilized more than the bonding one is stabilized, and then this geometric distortion is actually energetically unfavorable, so that a local minimum in S_1 or T_1 results.

Return from biradicaloid minima is generally photochemically productive. A very high energy region of the S_0 surface is reached, and further motion on this surface frequently leads to new nooks and crannies and thus to new products. It is perhaps fair to say that of all three classes of minima in S_1 and T_1 states, the biradicaloid minima are the most important and also the hardest to understand. They are almost never observable by direct absorption from the S_0 state, nor do they produce light emissions that would be readily observable, because of their usually very long wavelength and their typically very short lifetimes. In this case, quantum theory is quite essential for rudiments of understanding.

An important class of biradicaloid minima is comprised of those that occur halfway along the paths of ground-state "forbidden" pericyclic reactions, and their existence is most readily rationalized by the use of correlation diagrams. The presence of significant barriers along various reaction paths also is often readily discerned using the technique of correlation diagrams. In the following, we discuss the principles of this technique in considerable detail.

Finding the exact location of a biradicaloid minimum in S_1 typically calls for a lowering of symmetry and is difficult. It generally requires a search for the best "critically heterosymmetric biradicaloid" geometry, at which a funnel in S_1 is expected to occur (Chapter 4). However, this degree of detail is not required for the approximate understanding sought in the present chapter.

3.2 CORRELATION DIAGRAMS

In an effort to obtain some intuitive understanding of the results of elaborate calculations, and also to obtain at least approximate surface shapes for cases in which such calculations are not practical, it is extremely handy to use a simple tool known as state correlation diagrams. In these diagrams, a particular reaction path is chosen with the starting geometry on the left and the final geometry on the right. State energies in these two limits are plotted vertically and are connected from the left to the right by suitably constructed lines that indicate the approximate course of the state energy along the reaction path. The presence of barriers and minima in the resulting curves is then assigned a physical significance.

It is frequently not simple to construct a state correlation diagram in one step. The usual procedures are based on sequences that approach the state correlation diagram from one of two starting points, either in the valence-bond (VB) framework, or in the molecular orbital (MO) framework.

In the present section, we shall describe the ways in which simple state correlation diagrams can be constructed. In Section 3.3, we shall describe a technique that provides additional physical insight and permits an extension to more complicated molecules as well.

3.2.1 AO Correlation

The reaction center. The first thing that needs to be done in either the VB or MO approach is to separate the reacting molecule into the reaction region and the more or less inactive remainder. This is done by considering which bonds are transformed along the reaction path and which ones play the role of a passive spectator. For instance, in the cis–trans isomerization of ethylene, the C—H bonds and the σ component of the C—C bond can be viewed as inactive, whereas the π bond is clearly active. The somewhat arbitrary nature of this division is perhaps best illustrated by the fact that the understanding of the exact shape of the energy hypersurfaces for the twisting of ethylene is made easier if the electrons of the C—C σ bond are treated as being active, that is, as a part of the reaction region (see Section 4.8.1).

The active AOs. Once the reacting part of the molecule has been isolated, it is necessary to establish and to define a minimum set of atomic orbitals that will provide an adequate description of the electronic wave function of the reacting part of the molecule along the reaction path. It is desirable to keep the number of these orbitals to a minimum. Usually, some of these will be hybrid AOs, whereas others will be unhybridized AOs. Frequently, the hybridization will change somewhat along a reaction path. The exact nature of the hybrids to use is usually established fairly readily by consideration of the atoms bonded to each reacting center. It is helpful that usually only one or two bonds at any reacting atom participate in the reaction process.

AO correlation. Once the minimum basis of AOs has been established for each end of the reaction path, it is important to check that their number is the same on both sides. The next task is to establish a correlation that assigns to each orbital on the left-hand side an AO on the right-hand side of the diagram. This is straightforward if only one participating AO is located on each atom in the reacting part of the molecule. Then, the changes between the start and the end of the reaction path involve primarily translation and rotation of the direction of this orbital and possibly some change in its hybridization. An example is the conversion of butadiene to cyclobutene along the concerted disrotatory path (Figure 3.1). Two of the four active orbitals of butadiene, 2 and 3, keep their hybridization as well as their orientation. Two others, 1 and 4, rotate by 90° and change their hybridization away from pure p by acquiring partial s character.

The correlation of atomic orbitals between the two sides of the correlation

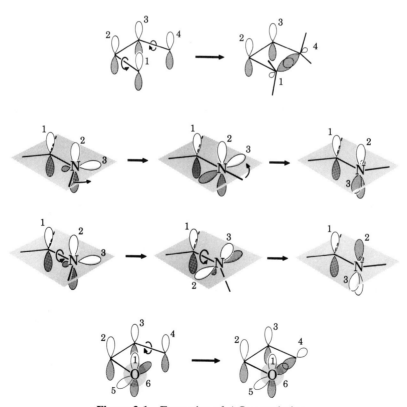

Figure 3.1 Examples of AO correlation.

diagram is less straightforward if more than one orbital has to be considered on at least one of the reacting atoms. In many cases, however, the assignment of the correlation is quite unambiguous. For example, for the two commonly considered limiting paths of syn–anti isomerization of the C=N double bond, either by a pure in-plane inversion or by a pure twisting motion (Figure 3.1), the AO correlation is quite clear. For the inversion motion, one of the AOs on the nitrogen involved in the process keeps its identity and remains parallel to the p orbital on carbon which participates in the double bond, while the other changes its hybridization from approximately sp^2 through p to again approximately sp^2; along the twisting path, one of the rotating orbitals on nitrogen remains an essentially pure p, while the other remains approximately sp^2-hybridized.

In other cases, the degree of arbitrariness in the way in which the AOs are correlated is higher. In the case of the ring-opening of oxetene (Figure 3.1), AOs describing the π component of the double bond change neither their direction nor their hybridization. The participating orbital at the CH_2 group changes its hybridization and rotates 90°. The two participating orbitals on the oxygen are most simply viewed as one keeping its direction parallel to the orbitals of the C—C π bond, and the other moving away from the carbon of the CH_2 group in the CCO plane. If one wished, however, one could rotate the oxygen orbital that participates in the C—O bond in a way that could be either disrotatory or conrotatory relative to the sense of the rotation of the CH_2 group, while at the same time rotating the "lone-pair" oxygen orbital so as to keep the two orthogonal.

AO labels. It is frequently necessary to refer to the atomic orbitals on either side of the diagram, or possibly even in the middle, by suitable labels. If only one participating orbital is present on every participating atom, the label can be simply the number or name of the atom. The situation becomes more complicated if two participating orbitals are present on the same atom. Then they can be characterized by their approximate state of hybridization or by their orientation relative to other parts of the molecule.

An important classification tool for the labeling of participating AOs is their local symmetry. It is very helpful to be able to refer to these AOs as being a part of a local π system if they are antisymmetric with respect to a local plane of symmetry or as being a part of a local σ system if they are symmetric with respect to such a plane. More or less localized orbitals of σ symmetry which contain two electrons in the ground state and do not interact strongly with other participating orbitals are usually referred to as being n ("lone pair") orbitals.

AO phase. While the overall phase of an atomic orbital is of no interest in itself, the orientation of orbitals that have both positive and negative lobes needs to be carefully kept track of. Thus, a rotation of a p orbital that accompanies a disrotatory motion of the attached atoms does not change the sign of either of its lobes. Similarly, a change in the hybridization of an orbital does not change the absolute sign of either of its lobes.

Once the AO correlation is established, one can proceed toward the next step in the construction of state correlation diagrams, which are our ultimate goal. One can proceed either in the VB framework discussed in Section 3.2.2 or in the MO framework discussed in Sections 3.2.3 and 3.2.4.

3.2.2 VB Structure and State Correlation

Notation. In many situations, the VB description (Appendix I) is the most natural to the organic photochemist. To provide such a description for the relevant fragment of the reacting molecule, one indicates occupancies of the participating AOs by an appropriate number of dots that represent electrons and, if necessary, by arrows indicating the spin state. We shall indicate local singlet coupling between two electrons in two atomic orbitals by attaching the letter s to a line adjoining the two orbitals, and we shall indicate local triplet coupling by labeling a similar but dotted line with the letter t. In ordinary chemical structures, the singlet coupling is normally indicated by a full valence line if the two atoms are bonded and by a dotted line (long bond) if they are not. In the case of double occupancy in a single AO, we do not indicate singlet coupling explicitly, since the Pauli principle permits no other possibility.

In many cases, a single VB structure of this type does not adequately represent the electronic state of the molecule, but the other contributing structures are obvious and need not be indicated explicitly. For example, in an $n\pi^*$ excited state of a ketone, a single dot indicates one-electron occupancy in an oxygen lone-pair orbital, and the three electrons available for the π system can be distributed in two ways: (i) two on the oxygen and one on the carbon p orbital or (ii) two on the carbon and one on the oxygen p orbital. Both of these contribute, but we shall generally only indicate one as representative of both.

Construction of the VB structure correlation diagram. Once the AO correlation has been defined, it is an easy matter to construct the VB correlation diagram for any starting electron assignment: The occupancies of the AOs and the spin coupling are simply kept constant to produce the corresponding VB structure at the other end of the diagram. It is then necessary to estimate the energies of the individual structures and place them at their proper relative positions on the vertical energy scale. This is, by far, the most difficult part of the task unless it is possible to estimate the structure energies from the knowledge of state energies as can be done if there is an approximate one-to-one correspondence between these. Otherwise, general guiding rules are that an increase in the number of singlet pairings between interacting atomic or hybrid orbitals is favorable for the energy and that an increase in either the number of triplet pairings or in charge separation is unfavorable. A few examples will illustrate the procedure.

Single-bond dissociation. The simplest case is the VB correlation diagram for the dissociation of a single bond (Figure 3.2). The equilibrium bond-length geometry is indicated on the left-hand side of the diagram, and the broken-bond situation is illustrated on the right-hand side. On the left-hand side, the lowest-energy VB structure is the covalent ("dot–dot") singlet, and this is followed by the dot–dot triplet. The charge-separated structures are degenerate by symmetry if the two groups connected by the bond are equal, and this has been chosen to be so in our example. On the right-hand side, the dot–dot triplet and singlet structures are of equal energy because the two AOs no longer interact. Their energy lies between the two energies of the dot–dot structures on the left-hand side. The ionic ("hole–pair") structures are now at higher energy because the degree of charge separation is larger. It is a simple matter to connect the left-hand side

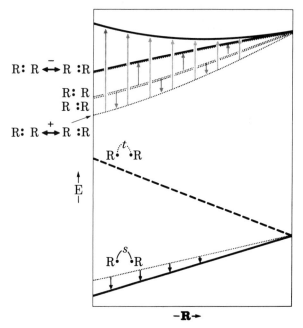

Figure 3.2 Dissociation of a nonpolar sigma bond. VB structure and state correlation diagram. Solid lines denote singlets; dashed line denotes triplet. Intermediate results (VB structure energies) are represented by dotted lines.

with the right-hand side by keeping the electron occupancy in each atomic orbital constant.

In order to describe the electronic states of molecules, one generally needs to mix VB structures. The outcome of the mixing ("resonance") is frequently predictable in a qualitative fashion, and in such a case the VB structure correlation diagram is easily converted into a state correlation diagram. Perhaps the most important rule to notice is that VB structures of different symmetries cannot mix whereas those of equal symmetries will. As a result, if the VB structure correlation leads to the crossing of lines, this crossing will remain if the two lines correspond to structures of different symmetries, and it will be avoided if they do not. A mixing is particularly important if the structures have the same energy, and it generally decreases in importance as the energy difference of the two structures increases. Also, structures that differ in the assignment of more than two electrons to AOs do not mix in the first approximation, and those that differ in the assignment of two electrons mix significantly only if the orbitals with different occupancies have a significant overlap.

Given these basic rules, we can now proceed from the VB structure correlation diagram in Figure 3.2 to the desired state correlation diagram, in two steps. In the first step, we consider only the interaction of the two degenerate hole–pair ("ionic") configurations at high energy. Mixing of these two configurations is promoted by their degeneracy, but it is hindered by the fact that they differ in the assignment of two electrons. On the left-hand side of the diagram where the two orbitals overlap significantly, the resulting splitting of the two degenerate energies is relatively large, as indicated by vertical arrows, and as one proceeds to the right-hand side where the two orbitals no longer overlap, it reduces to zero. The resulting energy curves are shown in Figure 3.2 by dotted lines.

Having taken care of the interaction of those VB structures that were degenerate, we can now consider the rest. The triplet is not capable of mixing with a singlet so that the original VB structure correlation line also represents the state correlation line. Since we have chosen the molecule to be symmetric, the two VB structure mixtures that resulted from the mixing of the hole–pair configurations are symmetry-adapted, one being symmetric and the other antisymmetric with respect to reflection in the middle of the bond. We shall see in Chapter 4 that the in-phase combination indicated by the plus sign in our diagram is the one that is more stable, whereas the out-of-phase combination indicated by a minus sign is the less stable one. Since the covalent singlet configuration is also symmetric, it can only interact with the symmetric combination of the hole–pair configurations. The effect of this interaction is indicated by arrows that show the origin of the fat lines in the final state correlation diagram. The resulting state correlation diagram can be compared with the outcome of the calculation for H_2 shown in Figure 4.3.

Double-bond twisting. Another simple case is that of the cis–trans isomerization of a double bond indicated in Figure 3.3. Here the path leads from a planar molecule to a 90° twisted molecule and back to a planar 180° twisted molecule in a symmetric fashion. We shall assume for simplicity that both termini of the double bond are equivalent.

In the planar molecule on the left in Figure 3.3, the dot–dot π-bonded VB structures are again present as a low-energy singlet and a high-energy triplet. Two hole–pair VB structures follow at higher energies and are degenerate in our case. The reasons for this ordering are the same as in the case of the single bond. At the orthogonally twisted geometry, the two dot–dot structures have nearly the same energies, lying between the two shown on the left-hand side. The hole–pair structures are again degenerate. The connection between the left-hand-side and right-hand-side limits is again established easily by connecting configurations with identical orbital occupancies.

To proceed towards the state correlation diagram we consider first the mixing of those structures that are degenerate. This produces the now familiar in-phase and out-of-phase combinations of the hole–pair structures as indicated in Figure 3.3 with dotted lines. Once again the in-phase combination lies lower. Finally, permitting the interaction of all VB structures that are not degenerate, we shall have similar spin and symmetry limitations as we did in Figure 3.2, and the final state diagram will have the form indicated by the thick lines in Figure 3.3. The triplet structure has no partners to mix with, the covalent singlet structure is of the same symmetry as the in-phase combination of the hole–pair structures, and the out-of-phase combination has no further partners with which to mix. Once again, the resulting correlation diagram is to be compared with the state energy curves calculated in Chapter 4 (Figure 4.10).

Dissociation of a lone-pair-carrying single bond. Our next example deals with a case of single-bond dissociation in which an excited singlet state correlates with the lowest dissociation limit. The molecules in question are methyl halides, and

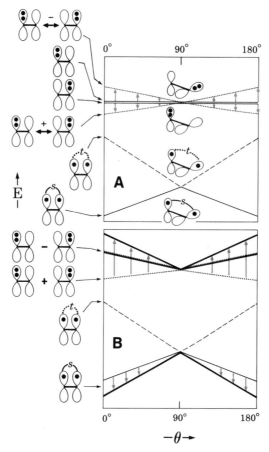

Figure 3.3 Twist of a double bond. VB structure (A) and state (B) correlation diagram. See caption to Figure 3.2.

the reaction is the dissociation of the carbon–halogen bond. The left-hand side of Figure 3.4 shows the three relevant singlet VB structures. The one that is of the lowest energy is the covalent structure, and at higher energies the degenerate pair of structures describing $n\sigma^*$ excitation is shown. Each of these in reality has a counterpart in which two of the three electrons in the σ bond are located on the carbon and only one on the halogen, and these also contribute but are not shown. On the right-hand side we have a combination of a methyl radical and an infinitely distant halogen atom. Since three possible occupancies of the p orbitals in the halogen atom are possible, the level is triply degenerate. Triplet configurations are not shown in order to keep the diagram simple.

The correlation between the structures on the left and those on the right is established easily, and the next step is the introduction of interactions between the VB structures. It is seen that the three lines each correspond to a different symmetry, the bottom one being symmetric with respect to any plane containing the carbon–halogen bond axis and the upper two being chosen so that one is antisymmetric with respect to one and the other antisymmetric with respect to another

3.2 CORRELATION DIAGRAMS

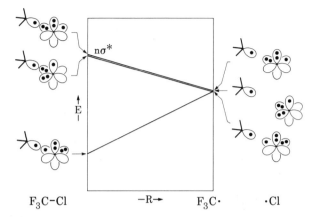

Figure 3.4 Dissociation of a sigma bond carrying lone pairs. VB structure and state correlation diagram for singlet states.

plane of symmetry containing the carbon–halogen axis, with the two planes perpendicular to each other. The lowest configuration belongs to an A representation, and the upper two belong to the degenerate E representation of the C_{3v} group. As a result, there is no VB structure mixing in this case, and the lines shown represent not only a VB structure correlation but also simultaneously a state correlation diagram.

More complicated situations. The reader will have noticed that the increasing complexity of the active part of the molecule rapidly complicates the VB description by increasing considerably the number of structures that need to be taken into account. As the number of participating AOs increases, the procedure becomes impracticable. What then is still occasionally possible is not to deal with individual VB structures but to represent known states by mixtures of configurations at both sides of the diagram, identifying the dominant one in each case, and then to correlate the dominant structure from one side to the other, finally introducing their VB structure mixing and thus arriving at a state correlation diagram. As the situation becomes increasingly complex, one is ultimately reduced to using only the symmetry of the individual states on both sides of the diagram to estimate how the correlation lines are to be drawn ("Salem diagrams"). Examples of these procedures can be found in Part B of this book.

3.2.3 MO Correlation

Although the VB method is intuitive and thus very useful for a description of chemical phenomena, as the complexity of the system to be described increases, it rapidly becomes impractical to use. In spite of its other disadvantages, the MO description (Appendix I) does not suffer from this difficulty to nearly the same degree. It is therefore essential to be familiar with the MO method of construction of state diagrams if one wishes to deal with large systems, such as those typically encountered in organic photochemistry.

MOs of reactants and products. In order to set up an MO correlation diagram for a reaction path, one first establishes the shape and energy sequence of the MOs at both ends of the path. For many simple systems, this construction is self-evident. The nodal properties and energy arrangement of MOs of conjugated linear and cyclic chains are well known, and MOs for many other systems can be derived from them by using the principles of perturbation theory. We shall see in the end that it is primarily the nodal properties of the orbitals, rather than the exact magnitudes of their coefficients that are of primary interest, and these can usually be estimated fairly easily. It is also well known how the number and location of nodal planes are related to the energy of an orbital: A node between two atomic orbitals interacting through a large resonance integral increases the energy, whereas an antinode decreases it.

In more complicated cases, it will be advisable to rely on explicit numerical calculations of the MO coefficients. Tables and books with such information are available. Alternatively, it is easy to perform simple calculations by methods such as the extended Hückel or the intermediate neglect of differential overlap (INDO) to obtain the required information in an approximate fashion.

MO following. The next task is to follow the development of orbital energies during the course of the reaction path. This can be done relying on the AO correlation already discussed in Section 3.2.1, combining it with knowledge of the changing coefficients on the individual contributing AOs. For this purpose, a calculation is ordinarily necessary. The procedure has been referred to as MO following. The reason why calculations are usually needed for this process is that quite a number of orbitals can mix as one proceeds from one end of the reaction path to the other.

In many instances, useful information can be obtained without any calculations. We shall discuss two such cases. In the first case sufficient symmetry is present to limit the number of possible interactions to a manageable level. In the second case sufficient formal symmetry is absent but enough is known about nodal properties of the orbitals involved to act as if formal symmetry actually were present. Finally, in Section 3.3 we shall consider a procedure that introduces the total complication in a step-by-step manner and thus, at times, permits construction of the correlation diagram even in fairly complicated instances.

MO correlation by formal symmetry. Those symmetry operations that remain preserved along a particular reaction path can be used to classify orbitals at all points of the path. Each orbital preserves its symmetry along the whole path, and orbitals of like symmetry generally do not cross. Orbitals of different symmetry can cross freely.

Single-bond dissociation. Our first example is the already familiar case of dissociation of a single bond between two equivalent saturated groups R. Here, we shall seek guidance from the H_2 calculation of Chapter 4 (Figure 4.4). Figure 3.5 shows the equilibrium distance between the two residues on the left and an infinite distance between them on the right. The energies of the bonding σ and antibonding σ^* orbitals are widely different on the left, and they are equal on the right. This is

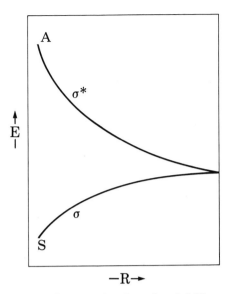

Figure 3.5 Dissociation of a nonpolar sigma bond. MO correlation diagram.

because on the left the two atomic orbitals that combine to form the MOs shown do overlap and have a large resonance integral, while on the right they do not overlap and their resonance integral vanishes. With respect to reflection in a plane perpendicular to the bond and located at its midpoint, the bonding combination is symmetric (S) whereas the antibonding combination is antisymmetric (A).

Double-bond twisting. Our next example is the cis–trans isomerization of a double bond. Two substituents that are on the same side of the double bond on the left-hand side of Figure 3.6 are on opposite sides of the double bond on the right-hand side. Halfway across, the bond is orthogonally twisted. Four MOs are shown at each end of the reaction path. These are the σ and σ^* orbitals and the π and π^* orbitals characteristic of the double bond. The symmetry elements preserved during the twisting path depend on the nature of the substitution on the double bond. In the simplest case, all four substituents are equal and three mutually perpendicular twofold rotation axes are preserved throughout (Figure 3.6). At special points along the path, additional elements of symmetry may be present. In this case they are symmetry planes at 90° twist. Such elements are generally less useful for the construction of correlation diagrams, although they can help us decide whether a particular crossing is avoided or not.

Label the three axes x, y, and z, with z being parallel to the C–C bond axis. Symmetry or antisymmetry relative to twofold rotation around these axes will be labeled by S and A in the usual manner, and the symbols will be arranged in the order x, y, and z. Inspection of the form of the MOs shown in Figure 3.6 shows that the σ orbital is symmetric with respect to all three rotations (SSS), whereas the σ^* orbital is symmetric with respect to rotation around z but antisymmetric with respect to rotation around the x and y axes (AAS). The π orbital on the left-hand side, where the substituent atoms are located in the xz plane, is symmetric

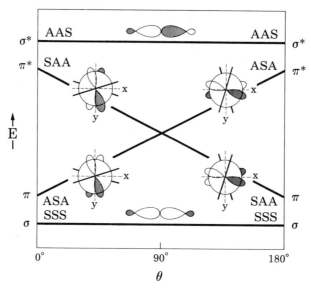

Figure 3.6 Twist of a double bond. MO correlation diagram. Symmetries with respect to the three C_2 axes are shown.

with respect to rotation around the y axis and antisymmetric with respect to the other two rotations (ASA). This is also the symmetry of the antibonding π^* orbital on the right-hand side, where the substituent atoms are located in the yz plane after the rotation has been completed. The symmetry of the π^* orbital on the left-hand side is similarly seen to be SAA, the same as the symmetry of the bonding π orbital on the right-hand side. If we now connect the orbitals carrying equal symmetry labels, we obtain the diagram shown in Figure 3.6. The π orbital of the starting material correlates with the π^* orbital of the product, and vice versa. At orthogonal twist, the orbitals are degenerate.

We can now consider the effects of symmetry lowering. Making the two substituents on atom 1 equal to R_1 and R_2 and the two on atom 2 also equal to R_1 and R_2 leads to two possibilities. In the cis arrangement the twofold symmetry axis x is preserved, whereas in the trans arrangement the symmetry axis y is preserved. In neither case does z remain as a symmetry axis along the reaction path. Thus, only one of the three symbols jointly designating the symmetry properties of the orbitals along the reaction path in Figure 3.6 remains. However, this is sufficient to keep the correlation intact when orbitals of like symmetry on both sides are connected. Next, we consider a substitution pattern in which both substituents on the first carbon are R_1 and both on the second carbon atom are R_2. Now only the rotation axis z is preserved along the reaction path. In this case both the π and the π^* orbital on the left and on the right all carry the same symbol A, and the crossing present in the diagram in Figure 3.6 will be avoided. The situation would be the same if one substituent were different from the other three, and it would also be the same in all other cases in which the two carbon atoms are not symmetry equivalent at orthogonal twist. Then, no symmetry symbols will be left, and all crossings will be avoided.

3.2 CORRELATION DIAGRAMS

Purely formally, then, one might expect that a lowering of symmetry of the molecule or, equivalently, a lowering of the symmetry of the reaction path by a suitable geometrical distortion, will change the correlation diagram dramatically. However, the changes are only gradual as we shall see below during the discussion of the use of nodal properties of orbitals for the construction of correlation diagrams.

Note that the slopes of the energy curves for the individual MOs in Figure 3.6 can be easily understood in terms of perturbation theory: As overlap between orbital lobes of like signs decreases or overlap between orbital lobes of opposite signs increases, the energy of the orbital goes up. This is because resonance integrals are, to a good approximation, proportional to overlap.

2 + 4 Cycloaddition. The two examples given so far are too simple to illustrate the full power of the MO correlation diagram procedure. Several more complicated examples of such diagrams for actual photochemical paths are given in Part B of this book. There is a third example in the present section. We take a reaction path along which photochemical reactivity is actually not predicted, namely, a concerted 2 + 4 cycloaddition of the Diels–Alder type.

Figure 3.7 shows schematically the geometry of approach of olefin to a diene molecule selected for the reaction path, in which a plane of symmetry perpendicular to the individual planes of both molecules is assumed. The MO energies for the starting material and product cyclohexene are also shown in Figure 3.7 and are labeled by their symmetry with respect to reflection in the plane S or A. Note

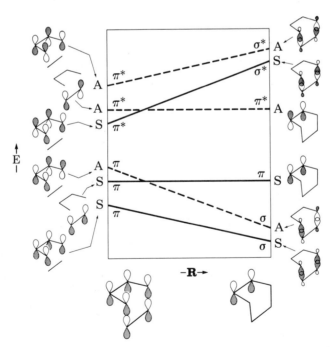

Figure 3.7 Diels–Alder addition of ethylene and 1,3-butadiene. MO correlation diagram.

that symmetry-adapted combinations of equivalent orbitals of equal energy have been used.

Invoking the noncrossing rule and connecting levels of like symmetry leads to the correlation diagram shown in Figure 3.7. Once again, it can be noted that the slopes of the energy curves for the individual MOs can be easily rationalized in terms of perturbation theory. Overlap between orbital lobes of like signs is favorable energetically, and overlap between orbital lobes of opposite signs is unfavorable. The correlation diagram for the Diels–Alder reaction (Figure 3.7) differs decisively from that for the twisting of a double bond (Figure 3.6) in that the former does not and the latter does contain crossing of a bonding with an antibonding orbital between the reactants and the products. The reaction paths characterized by the former kind of diagram are often called ground-state "allowed," and those characterized by the latter kind are often called ground-state "forbidden."

The HOMO (highest occupied MO) of ethylene is of the same symmetry as the LUMO (lowest unoccupied MO) of butadiene, and the interaction that sets in as the molecules proceed along the reaction path therefore stabilizes the more stable of the two orbitals and destabilizes the less stable one, as is clear in Figure 3.7. At the same time, the HOMO of butadiene is of the same symmetry (A) as the LUMO of ethylene, and these two orbitals interact, leading again to a stabilization of the former and a destabilization of the latter. This type of argument is known as the Frontier Molecular Orbital Theory and is usually justified by reference to the minimal energy gaps involved in HOMO–LUMO interactions. The situation is a bit more complicated than has just been indicated, since there are not two, but three, MOs of symmetry S and not two, but three, MOs of symmetry A. There are avoided crossings between the lowest two orbitals of symmetry S and between the highest two orbitals of symmetry A, so that, in effect, the energies of the orbitals that start as ethylene HOMO and LUMO do not change very much along the reaction path. If the correlation diagram is constructed by simply drawing straight lines between the levels on the left and on the right, these two orbitals will appear not to change in energy at all.

MO correlation by nodal properties. The argument of the type just described as Frontier Molecular Orbital Theory can be applied even in the absence of formal symmetry. While the presence of such symmetry permits clear-cut statements about which orbital pairs can interact and which ones cannot, the actual reason for vanishing interactions in the case of an A–S combination is that the outcome of integration in the half-space located on one side of the symmetry plane will be exactly canceled by the result of integration in the half-space located on the other side of the plane. However, planes and other surfaces across which an orbital changes sign (nodal planes) are common even in the absence of formal symmetry. Clearly, as long as we know where such nodal planes or surfaces are located, we can estimate when such cancellation will occur even in the absence of perfect symmetry, at least approximately. The cancellations may not be complete, but they can be clearly expected to be significant relative to interactions of the type in which the functions to be integrated contribute with the same sign in both halves of the total space. This type of argument is commonly used to explain the lack of significant effect by substituents on the course of correlation diagrams. In effect, the nodal properties of the orbitals, and not their formal symmetry, are of ultimate consequence.

1,3 Bond shift. As an example of this type of argument, we can consider the 1,3 shift in propene and in allyltrimethylsilane. In Figure 3.8 we show the bonding and antibonding orbitals of the two bonds participating in the reaction of propene, assuming for simplicity that there is no hyperconjugative interaction in the starting materials and products so that the two types of orbitals can be clearly separated. The orbitals of the π part of the double bond, π and π^*, are closer to the nonbonding level, and the σ and σ^* orbitals of the C—H bond are drawn farther from the nonbonding level. The reaction path is not characterized by any element of symmetry except at the midpoint, so that the formal symmetry procedures just outlined cannot be applied. The interaction introduced by moving along the reaction path consists of an overlap of the orbital at the hydrogen atom with the π orbital of the terminal carbon and of an overlap of the p orbital of the central carbon with the sp^3 hybrid of the carbon of the C—H bond. At the same time, the original π-bonding and σ-bonding interactions are weakened. Applying again the criterion of the simple perturbation theory, namely, that positive overlap is favorable and negative overlap is unfavorable for orbital energy, we see that the interaction of the two bonding orbitals, σ and π, pushes them apart, making π rise in energy, and that a similar interaction of the two antibonding orbitals causes π^* to decrease

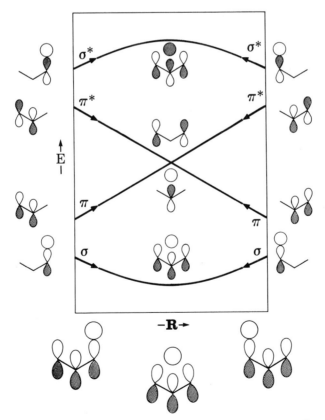

Figure 3.8 Hydrogen 1,3-sigmatropic shift. MO correlation diagram.

126 LOCATION OF MINIMA, FUNNELS AND BARRIERS ON SURFACES

in energy. From the direction of change of the HOMO and LUMO, we suspect that these two orbitals are heading for crossing so that the reaction path will be of the "forbidden" type. Note that the interactions between the bonding and the antibonding orbitals have been neglected because of the large energetic separation. This is no longer justified when one reaches the center of the correlation diagram where the four orbitals mix quite equally, but this is beyond the frame of a frontier molecular orbital description.

The situation can be contrasted with the case of allylsilane (Figure 3.9), in which the migrating silicon atom uses two opposite lobes of its p orbital for binding to one and the other terminal along the reaction path with inversion. Arguments similar to the above show that the net effect of the phase change is to make the reaction path "allowed" in the ground state.

A simple device for the recognition of "allowed" and "forbidden" pericyclic reaction paths exists and is based on the recognition that in the former the electronic

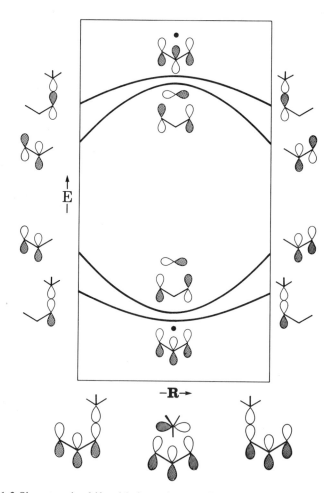

Figure 3.9 1,3-Sigmatropic shift with inversion on the migrating group. MO correlation diagram.

states halfway along the correlation diagram resemble those of an aromatic cyclic conjugated system, while for the latter they resemble those of an antiaromatic cyclic conjugated system. This is very useful for recognizing "allowed" and "forbidden" reactions in ground-state processes, but by itself it does not permit a construction of a correlation diagram and is therefore of rather limited use for photochemical reaction paths. It does, however, permit a prediction as to whether a biradical geometry is reached halfway through the reaction, and this in itself is significant.

3.2.4 Configuration and State Correlation

Once the MO correlation diagram for a reaction path is available, one can proceed with the construction of the configuration correlation diagram. In this diagram, the energies of the various electronic configurations are plotted as a function of molecular geometry. The estimation of configuration energies is based on the notion that the total energy of a many-electron system roughly reflects a sum of the contributing one-electron energies. This is certainly not strictly valid, but it is frequently possible to relate at least the changes in configuration energies to changes in the energies of occupied orbitals. It is essential to rely only on the use of those orbital changes which are very large; for instance, doubly excited configurations, and to a lesser degree, singly excited configurations, tend to be considerably higher in energy then the ground configuration.

Once the configuration correlation diagram is set up, a consideration of possible configuration mixing, as permitted by symmetry, will produce the final state correlation diagram. It is frequently possible to anchor the end points of lines in a state correlation diagram using experimental values for excitation energies and for the heats of formation. At times, high-quality calculations may be available for such end points and can serve as a substitute for the experimental values.

We have seen in Section 3.2.3 that there are two fundamental kinds of MO correlation diagrams, namely "allowed" and "forbidden." In the MO correlation diagrams of the allowed type, there is no orbital crossing between the bonding orbital group and the antibonding orbital group (see Figure 3.7 for an example). In the MO correlation diagrams of the forbidden type, there is such a crossing for at least one pair of orbitals; that is, at least one bonding orbital of the reactant becomes antibonding in the product, and at least one of the antibonding orbitals of the reactant becomes bonding in the product (see Figure 3.6 for an example).

In the consideration of ground-state reactivity, this distinction is all that is needed, and the terms "allowed" and "forbidden" refer to the relative ease with which two thermal processes that are otherwise equivalent with respect to factors such as steric hindrance or steric strain will proceed. In photochemical processes, additional detail is useful.

Ground-state "allowed" paths. In Figure 3.10, we show the MO configuration correlation diagram (part A, cf. Figure 3.7) and the state correlation diagram (part C) for a typical ground-state "allowed" reaction, such as the Diels–Alder reaction of Section 3.2.3. The configuration correlation diagram (part B) was constructed by considering various low-energy ways of assigning electrons to the orbitals on the left-hand side of the MO correlation diagram, and following their fate as one moves to the right in the diagram. In this process, the degree of occupation of

128 LOCATION OF MINIMA, FUNNELS AND BARRIERS ON SURFACES

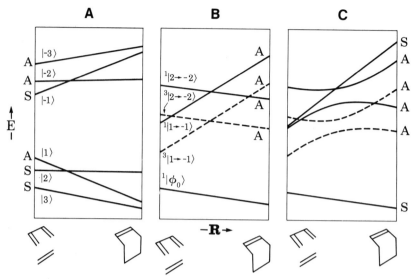

Figure 3.10 Diels–Alder addition of ethylene and 1,3-butadiene. MO (A), configuration (B), and state (C) correlation diagram. Singlets are denoted by solid lines, and triplets are denoted by dashed lines. To minimize congestion, configurations of symmetry S are not shown in part B. Three of them ($^1|1 \to -2\rangle$, $^1|2 \to -1\rangle$, $^1|1,1 \to -1,1\rangle$) combine to yield the S state shown in part C.

each MO is kept constant. Since there is no crossover between bonding and antibonding orbitals, it is clear that the ground configuration on the left correlates with the ground configuration on the right, that all singly excited configurations on the left correlate with singly excited configurations on the right, and so on. The relative energies of the various singly excited configurations may change in the process, and these changes may be more or less exactly retained when configuration interaction is introduced to produce the final state correlation diagram (Fig. 3.10C). But it is clear that there is no particular reason to expect minima in the S_1 or T_1 surface halfway, or anywhere else along the reaction path, except for either the starting or the ending point. For the particular choice of behavior of MO energies made in Figure 3.10, barriers actually result in the S_1 and T_1 states, starting either on the left or on the right. In such a case, an excited molecule would not tend to follow the reaction path considered in the construction of the correlation diagram, unless it were provided with a very large amount of vibrational energy. If the barriers were lower or absent altogether, perhaps because of a very large exothermicity of the ground-state process, one might expect an adiabatic photochemical reaction in which the S_1 surface is followed from one side of the diagram to the other so that an excited product is formed either in the S_1 or the T_1 state.

Ground-state "forbidden" paths. The ground-state "forbidden" paths are of more intrinsic interest in photochemistry. A casual inspection of the MO diagrams shows that a biradicaloid geometry is reached halfway along the reaction path. Since we have come to expect high energies in the ground state, but low energies in the S_1 and T_1 states for such geometries, they will often be accessible in the excited state and will provide a potential point of return to the ground state in a photochemical process.

This expectation is obviously fulfilled both in Figure 3.11 and Figure 3.12, in which we show the MO correlation diagram (A), the configuration diagram (B), and the state correlation diagram (C) for two different ground-state "forbidden" reaction paths. However, the mere presence of a minimum in the S_1 or T_1 state along the reaction path is not yet sufficient to provide a point of return to the ground state for excited molecules. An important consideration is: Is this minimum accessible from the starting geometry, or is it separated from it by forbidding barriers? In the latter case, only those molecules that initially have excess vibrational energy may be able to proceed to the minimum and on to the photochemical product.

The two examples of forbidden MO correlation diagrams shown in Figures 3.11 and 3.12 differ in the nature of the orbital crossing. The example chosen for Figure 3.11 is one in which it is the HOMO ($|1\rangle$) and the LUMO ($|-1\rangle$) of the reactant which correlate with the LUMO and the HOMO of the product, respectively (a "normal" MO crossing). This type of MO correlation diagram offers the best opportunity for photochemically active reaction paths, since it is least likely to lead to a barrier in the excited state between the starting geometry and the biradicaloid minimum through which return to the ground state is to occur. It is clear from Figure 3.12 that there is an opportunity for the presence of a barrier in the S_1 surface when the MO crossing is of the "abnormal" type. The same argument holds for the T_1 surface.

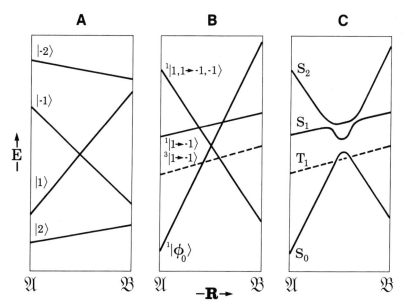

Figure 3.11 A ground-state "forbidden" reaction with "normal" MO crossing. MO (A), configuration (B), and state (C) correlation diagram. Singlets are denoted by solid lines, and triplets are denoted by dashed lines.

130 LOCATION OF MINIMA, FUNNELS AND BARRIERS ON SURFACES

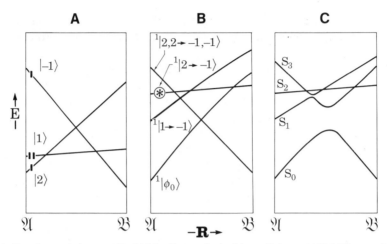

Figure 3.12 A ground-state "forbidden" reaction with an "abnormal" MO crossing. MO (A), configuration (B), and state (C) correlation diagrams. Singlets only. The configuration $|2 \to -1\rangle$ whose MO occupancy is shown in A and which is labeled with an asterisk in B is the "characteristic" configuration defined later in the text.

Configuration correlation. As we proceed to construct the configuration correlation diagram (Figures 3.11B and 3.12B) from the MO correlation diagram (Figures 3.11A and 3.12A), we note that all configurations in which the HOMO is doubly occupied will rise steeply in energy as one proceeds towards the product, since this orbital rapidly becomes antibonding. One of these is the ground configuration that clearly correlates with a doubly excited configuration of the product, but there are many others. Obviously, in Figure 3.11B, only those configurations that have the same occupancy in HOMO and LUMO will not change their energy much along the correlation path. The most important among these is the singly excited $^1|1 \to -1\rangle$ configuration. Those configurations that have a higher occupancy in LUMO than in HOMO will descend rapidly as one proceeds from the reactant to the product. The most important among these is the doubly excited configuration $^1|1,1 \to -1,-1\rangle$ in which HOMO is empty and LUMO is occupied twice. This becomes the ground configuration of the product. In Figure 3.12B, the above still applies except that the second HOMO, $|2\rangle$, replaces the HOMO, $|1\rangle$, with obvious consequences.

State correlation. Since all configurations that contain only doubly occupied orbitals are totally symmetric, the crossing of the ground and the doubly excited $^1|1,1 \to -1,-1\rangle$ configurations of the starting material, present in the configuration correlation diagram in Figure 3.11B, will be absent when configuration interaction is considered (Figure 3.11C). This still provides a high barrier for the thermal ground-state process, since the crossing is normally not avoided very strongly (note that two electrons differ in their orbital occupancy in the two configurations that mix), and this is the justification for the label ground-state "forbidden" path. Configuration interaction also yields the anticipated minimum in the S_1 surface, provided that the crossing is not avoided so strongly that the upper of the two resulting states is pushed above the $^1|1 \to -1\rangle$ singly excited configuration.

If the nature of the system under consideration is such that configuration interaction in the reactant does not change the order of energies anticipated in the first approximation, that is, if the lowest excited state is dominated by the HOMO → LUMO singly excited configuration, no correlation-imposed barrier is expected for this reaction path. This does not yet mean that no barrier in the S_1 surface will be present, since there may be other causes for the presence of such barriers not apparent from correlation diagrams, and since there may be an overall endothermicity to the ground-state process which tilts the correlation lines in an unfavorable direction. Similar constraints on the use of the correlation diagrams for the estimation of organic reactivity are already well known from ground-state processes.

Frequently, particularly in aromatic molecules, configuration interaction in the reactant may be so strong that the lowest excited state of the molecule does not correspond to the HOMO → LUMO singly excited configuration but, rather, to some more or less complicated mixture of other configurations, which then usually correlate with even more highly excited configurations of the product. This type of situation then often results in the presence of a barrier in the state correlation diagram, although none was apparent in the configuration correlation diagram for the singlet state. It is uncommon in the case of the triplet T_1 state, even in aromatic compounds. A physical reason for the difference is the fact that the lower triplet state of an alternant hydrocarbon is a well-correlated "dot-dot" state even when described by a single configuration wave function, while for the lowest excited singlet state a high degree of electron correlation, which removes ionic terms from the wave function and lowers the state energy, is only introduced into the wave function when configuration interaction is considered.

In the "abnormal" MO crossing case, the state correlation diagram (Figure 3.12C) contains a barrier in S_1 even if configuration interaction does not dominate the state order in the reactant.

Relation of ground-state and excited-state "allowedness." The results described in Figure 3.11 reflect the standard symmetry rules for chemical reactivity in which ground-state "allowed" processes are photochemically "forbidden" and ground-state "forbidden" reaction paths are photochemically "allowed" except for the added complication caused by configuration interaction changing the order of reactant states which may cause barriers in the excited states S_1 or T_1 to appear even if the reaction is formally allowed.

These rules are applicable under the assumption that the two orbitals which are responsible for the crossing are the HOMO and LUMO of the reactant, that is, when the orbital crossing is of the "normal" type. When this is not the case, that is, for the case of "abnormal" orbital crossing, more or less significant barriers have to be expected in the S_1 and T_1 surfaces between the starting geometry and that of the biradicaloid minimum already at the stage of the configuration correlation diagram. Although they may be lowered when configuration interaction is introduced, it is unlikely that they will disappear altogether. Reaction paths of this type are then both "allowed" by the correlation since a minimum in S_1 for the return to the ground state is provided, and "forbidden" by the correlation since barriers are imposed on the way to the minimum.

A specific illustration of an abnormal MO crossing is provided in Figure 3.13, which shows the MO correlation diagram for the disrotatory ring-opening of a polycyclic derivative of cyclobutene. The HOMO and the LUMO are both localized on the naphthalene chromophore in the reactant. Although the latter descends in

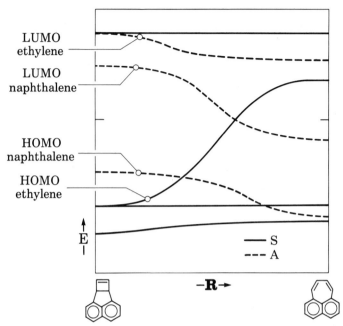

Figure 3.13 An MO correlation diagram for a disrotatory electrocyclic reaction with an "abnormal" MO crossing. Adapted by permission from J. Michl, *Mol. Photochem.* **4**, 257 (1972).

energy along the reaction path to become the HOMO of the pleiadiene product, so does the former, and the two do not cross. Instead, a quite different bonding orbital of the reactant rises in energy and becomes antibonding in the product, crossing the LUMO on the way. This reactant orbital is best described as the HOMO of the ethylene subunit.

The reactant configuration that correlates with the lowest excited state of the product therefore is of ethylene-to-naphthalene charge-transfer type and lies high in energy. The low-lying excited states of the reactant, localized in the naphthalene chromophore, correlate with very highly excited states of the product. This is just the situation shown in Figure 3.12 to produce large barriers in the S_1 surface; the T_1 surface behaves similarly.

The "characteristic configuration." In summary, we see that simple photochemical processes such as the breaking of a σ bond or a π bond, or pericyclic processes characterized by a cyclic array of overlapping atomic orbitals halfway through the reaction path which are of the ground-state "forbidden" nature, can be characterized by a "characteristic configuration." This is the singly excited configuration of the reactant in which an electron has been removed from that bonding orbital which is becoming antibonding along the reaction path and promoted into that antibonding orbital which is becoming bonding along that path. This is the lowest energy configuration likely to remain at approximately constant energy along the path. If it is heavily represented in the lowest excited state of the starting material, S_1 or T_1, the correlation will not impose a barrier, and the path on the excited surface to the biradicaloid minimum for the return to the ground state is clear. If instead it is heavily represented in one of the higher excited states of the starting material, other configurations necessarily represent the lowest excited state, S_1 or T_1, which then rises in energy along the reaction path until it meets the state represented primarily by the characteristic configuration so that barriers in the excited surface are imposed (compare Figures 3.11 and 3.12). The unfavorable order of states may be either due to effects of configuration interaction in the reactant (in the case of a normal orbital crossing) or due to the fact that the HOMO → LUMO configuration is not the characteristic configuration (abnormal orbital crossing).

Our last example deals with the dissociation of the C–N bond in the anilinium cation (Figure 3.14). First, we note the effect of symmetry lowering on the MO correlation diagram. Along nonlinear dissociation paths in which the departing NH_3 group moves out of the aromatic symmetry plane, the sharp distinction between σ and π orbitals is lost and new interactions appear. If the nitrogen remains in the plane of symmetry containing the C–N bond and perpendicular to the ring,

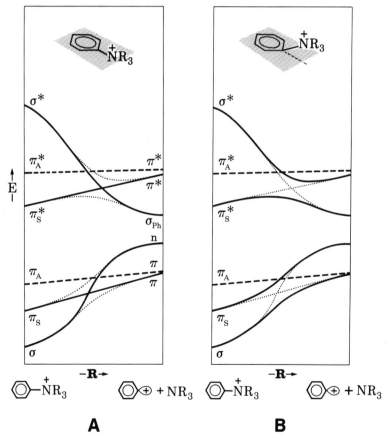

Figure 3.14 In-plane (A) and out-of-plane (B) sigma-bond dissociation path for an anilinium ion. MO correlation diagrams. Solid lines denote symmetric orbitals, and dashed lines denote antisymmetric orbitals. Dotted lines mark the avoided crossing.

134 LOCATION OF MINIMA, FUNNELS AND BARRIERS ON SURFACES

those π orbitals that are antisymmetric with respect to this plane still cannot interact with the σ and σ^* orbitals, but those that are symmetric can, as indicated in the Figure 3.14. If the dissociation path violates both symmetry elements, all of the orbitals will interact freely.

Direct construction of state correlation diagrams. In favorable cases it is possible to construct state correlation diagrams without recourse to MO and configuration correlation diagrams, and this is particularly advantageous if the effects of configuration interaction are strong, as is typically the case in aromatic chromophores. Good examples of situations in which such analysis is possible are the dissociation of σ bond or twisting of a π bond effectively isolated by symmetry from the remainder of the molecule. The dissociation of the C—N bond in the anilinium ion is such a case.

In Figure 3.15 we show the singlet-state correlation diagram for the dissociation of the anilinium cation, whose MO correlation diagram was shown in Figure 3.14. We first consider the linear dissociation path that preserves both symmetry planes: the one given by the plane of the benzene ring, and the one perpendicular to it and containing the C—N bond. We draw first the known course of the curves representing the ground and $\sigma\sigma^*$ singly excited states of the C—N bond. If there were no interaction between the benzene ring and the dissociating C—N bond, adding the locally excited states of the perturbed benzene ring would amount to

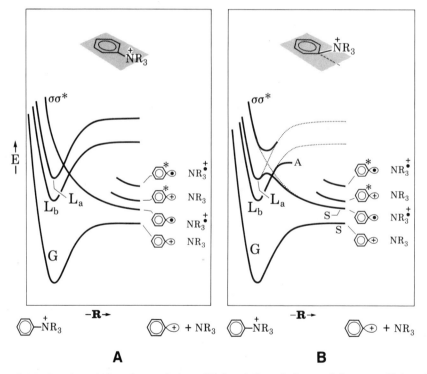

Figure 3.15 In-plane (A) and out-of-plane (B) bond dissociation path for an anilinium ion. Singlet-state correlation diagrams. Dotted lines mark the avoided crossing.

adding a series of parallel curves separated by distances equal to the benzene-ring excitation energies. For the lowest excited singlet state of the benzene chromophore, L_b, which is antisymmetric with respect to the plane containing the C—N bond and perpendicular to the aromatic ring, this interaction indeed vanishes by symmetry and the crossing is not avoided, as indicated in Figure 3.15. For the higher-energy singlet L_a state, which is symmetric with respect to this plane, the interaction is no longer avoided by symmetry.

For reaction paths in which the nitrogen atom swings out of one or both planes of symmetry initially present in the molecule, the crossing of the singlet L_b with the $\sigma\sigma^*$ state will no longer be exactly avoided, but some barrier will most likely remain.

3.3 THE USE OF INTERACTING SUBUNITS IN THE CONSTRUCTION OF CORRELATION DIAGRAMS

The discussion of the dissociation of the C—N bond in the anilinium cation in Section 3.2.4 represents an example of a general approach to the correlation diagram analysis of reaction paths of relatively complicated systems which can be formally decomposed into a series of subunits. The approach consists of two steps. One first assumes that the individual subunits do not interact at all and constructs a correlation diagram. This is usually quite easy because most of the subunits can be chosen so as not to be affected by motion along the reaction path, so that their energies remain constant across the whole diagram. In a second step, interaction between the subunits is added, usually relying on simple principles from first-order perturbation theory and using symmetry or nodal properties where these are known.

Correlation diagrams of this type are sometimes referred to as natural correlation diagrams. The transition from the first step to the second can take place at the level of MO correlation diagrams, at the level of configuration and state correlation diagrams, or at the level of VB correlation diagrams:

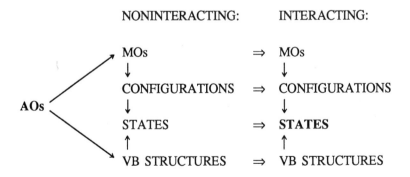

MO correlation. We shall first illustrate the procedure at the MO correlation diagram level by considering two examples. The first is the effect of push–pull substitution on ethylene on the MO correlation diagram for a twisting motion. This case will also demonstrate the difficulties that can be encountered in an attempt to go from an MO correlation diagram to a configuration correlation diagram

136 LOCATION OF MINIMA, FUNNELS AND BARRIERS ON SURFACES

without a detailed consideration of electron repulsion terms. The second example is the dissociation of the C—N bond in the benzylammonium cation.

Twisting a push–pull substituted double bond. Let us assume that position 1 of ethylene carries a π-electron-donating substituent D such as the dimethylamino group, and that carbon 2 carries a π-electron-withdrawing substituent A, such as the nitrile group. We shall assume further that during the twisting motion along the cis–trans isomerization coordinate, both substituents remain aligned with their respective carbon atom neighbors such that the donating and withdrawing interactions are carried through the whole reaction coordinate. Figure 3.16A indicates the energies of the orbitals along the reaction coordinate as they would look if there were no π interactions across the C═C bond. Then the energies of both the bonding and the antibonding combination made within both pairs, each consisting of a carbon p orbital and a substituent p or π orbital, would be independent of the angle of rotation. Approximate shapes and nodal properties of the four orbitals are indicated in the figure. When the interaction between the two π-symmetry

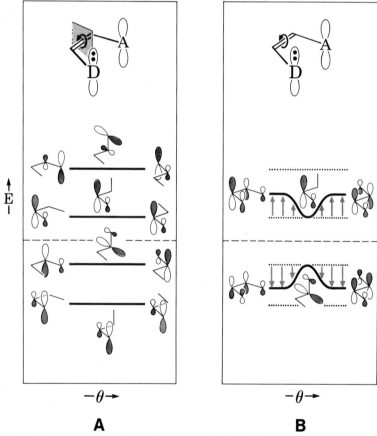

Figure 3.16 Twist of a push–pull substituted double bond (D = π-donor, A = π-acceptor). MO correlation diagrams before (A) and after (B) subunit interaction. The twist angle θ varies from 0° to 180°.

orbitals on the carbon atoms is introduced, its magnitude will vary as the cosine of the twist angle θ, and its effect on the energies of HOMO and LUMO will be that shown in Figure 3.16B. The resulting MO correlation diagram is similar to that obtained previously for ethylene (Figure 3.6), but the crossing at orthogonal twist is now avoided because of the perturbation by the substituents.

If one now wished to proceed to a configuration and state correlation diagram, an estimate of the relation between orbital energies and configuration energies as a function of twist angle would be necessary. This is a difficult objective to attain in the absence of additional information because at orthogonal twist the configuration in which the lower orbital HOMO is occupied twice, and the singly excited configuration in which HOMO and LUMO are each occupied once, can have approximately equal energies when the substituents are chosen correctly. This fact is not obvious from a simple consideration of MO energies, and its derivation requires a more detailed knowledge of the nature of electronic repulsion contributions to the energy of the system (see Chapter 4).

Dissociation of a benzylic bond. Another example of decomposition into subsystems is shown in the correlation diagram for the dissociation of the C—N bond in the benzylammonium ion. Consider at first the case in which the π-interactions between the benzene ring and the rest of the molecule do not exist (Figure 3.17A). Then, the MO correlation diagram consists simply of lines showing the dissociation of the C—N bond, with the σ and σ^* orbitals on the left and the nitrogen lone pair and methyl cation orbital energies on the right, and lines for the orbitals of the benzene ring. The energies of the latter are constant as a function of the reaction coordinate, since no π-interactions across the exocyclic bond are permitted.

When the interactions are introduced, the C—N bond can hyperconjugate with the π system of the ring with the results shown on the right in Figure 3.17B. The system now dissociates into benzyl cation and ammonia in its ground state. It should be noted that Figure 3.17A is virtually identical to Figure 3.14A, since the subunits of the benzene ring and the C—N σ bond are noninteracting in both cases.

State correlations. The use of subunits in the construction of correlation diagrams is particularly useful for the construction of state correlation diagrams. When the fundamental bond-breaking process can be isolated in a single subunit, often as a σ bond dissociation or π bond twisting, the characteristic configuration for the reaction is easily defined and the course of its energy along the reaction path is identified. Locally excited states of other subunits are then independent of progress along the reaction path and are easily added to the diagram from the knowledge of the spectroscopy of the subunits. If necessary, charge-transfer excited states, in which an electron has been moved from one subunit to another, and doubly excited states, particularly those in which each subunit is excited to its lowest triplet state, can be added as well. In the second step of the procedure the interactions between states of the individual subunits and possibly charge-transfer and doubly excited states are considered by use of symmetry and perturbation theory. Viewed in this light, photochemical bond-breaking is seen not to require vertical $\sigma \rightarrow \sigma^*$ excitations in single bonds and $\pi \rightarrow \pi^*$ excitations in isolated double bonds, which would correspond to the characteristic configurations. The reason is the availability of

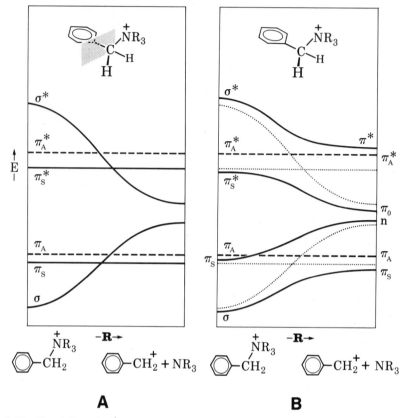

Figure 3.17 Bond dissociation in a benzylammonium ion. MO correlation diagram before (A) and after (B) subunit interaction. Solid lines denote symmetric orbitals, and dashed lines denote antisymmetric orbitals. Dotted lines mark the avoided crossing.

cooperation from those chromophore subunits that are located in sufficient proximity to conjugate with the orbitals that form the bond to be broken.

Dissociation of a benzylic bond. An example of the procedure is shown in Figure 3.18 for the case of the dissociation of the C—N bond in the benzylammonium cation. Part A shows the states of the two noninteracting subunits, namely, the aromatic ring and the C—N σ bond. The ground-state curve rises from the left to the right by an amount corresponding to the C—N bond strength, and the S_1 curve descends to the limit represented by the lowest excited singlet state of the fully dissociated system, a combination of the benzyl radical and the ammonia radical cation. Two locally excited states of the benzene chromophore are also shown, displaced above the ground state by amounts corresponding to the singlet L_b and singlet L_a excitation energies of benzene. Charge-transfer states corresponding to electron transfer out of the σ orbital or into the σ^* orbital are even higher in energy and are not shown.

Figure 3.18B shows the changes that occur when interaction across the exocyclic bond is permitted. The S_1 state of the dissociating σ bond is symmetric with respect

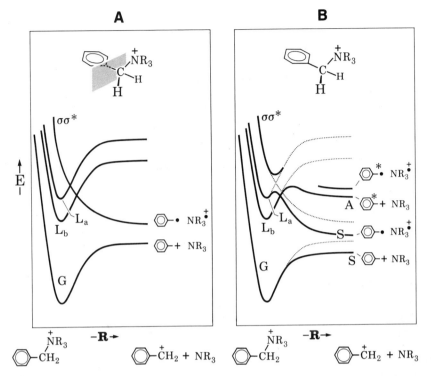

Figure 3.18 Bond dissociation in a benzylammonium ion. Singlet-state correlation diagram before (A) and after (B) subunit interaction. Dotted lines mark the avoided crossing.

to a plane that contains the C—N bond and is perpendicular to the aromatic ring, whereas the singlet L_b state is antisymmetric with respect to this plane. The crossing of these two is therefore not avoided; however, the locally excited singlet L_a state is symmetric with respect to this plane, and its crossing with the S_1 state of the dissociating C—N bond is avoided as shown in Figure 3.18B. The conclusion from the diagram is that along the reaction paths that keep the high symmetry there ought to be a barrier in the S_1 surface. Further lowering of symmetry may cause the unavoided crossing indicated in Figure 3.18B to be avoided, but a sizable barrier is likely to remain in the S_1 surface.

3.4 COMMENTS AND REFERENCES

General qualitative discussions of the factors determining the **location of minima, funnels, and barriers on surfaces** relevant in photochemistry can be found in: J. Michl, *Mol. Photochem.* **4**, 257 (1972); J. Michl, *Topics Curr. Chem.* **46**, 1 (1974); J. Michl, in *Chemical Reactivity and Reaction Paths* (G. Klopman, Ed.), Wiley, New York, 1974, Chapter 8; J. Michl, *Photochem. Photobiol.* **25**, 141 (1977); R. C. Dougherty, *J. Am. Chem. Soc.* **93**, 7187 (1971); E. M. Evleth, in *Photochemistry of Macromolecules* (R. F. Reinisch, Ed.), Plenum Press, New York, 1970, p. 167; T. Förster, *Pure Appl. Chem.* **24**, 443 (1970); L. Salem, *Science* **191**, 822 (1976); L. Salem, *Israel J. Chem.* **14**, 89 (1975); H. E. Zimmerman, *Science* **191**, 523 (1976); H. E. Zimmerman, *Acc. Chem. Res.* **15**, 312

(1982); N. D. Epiotis and S. S. Shaik, *J. Am. Chem. Soc.* **99**, 4936 (1977); N. D. Epiotis and S. S. Shaik, *J. Am. Chem. Soc.* **100**, 1, 8, 29 (1978); V. Bonačić-Koutecký, J. Koutecký, and J. Michl, *Angew. Chem. Int. Ed. Engl.* **26**, 170 (1987); F. Bernardi, S. De, M. Olivucci, and M. A. Robb, *J. Am. Chem. Soc.* **112**, 1737 (1990).

The **"method of interacting subunits"** for the construction of correlation diagrams has been introduced in J. Michl, *Mol. Photochem.* **4**, 287 (1972); a more recent embodiment of the same principle is the **"natural correlation"** method: See B. Bigot, A. Devaquet, and N. J. Turro, *J. Am. Chem. Soc.* **103**, 6 (1981); see also V. G. Plotnikov, *Radiat. Phys. Chem.* **26**, 519 (1985). The **"MO following"** procedure is outlined in H. E. Zimmerman, *Acc. Chem. Res.* **5**, 393 (1972). J. Katriel and E. A. Halevi, *Theoret. Chim. Acta* **40**, 1 (1975), summarize the **"orbital correspondence analysis in maximum symmetry"** method, and E. A. Halevi and C. Trindle, *Israel J. Chem.* **16**, 283 (1977) apply it to spin-forbidden processes.

An example of a **conical intersection** (S_0–S_1 touching) is found in V. Bonačić-Koutecký and J. Michl, *Theoret. Chim. Acta* **68**, 45 (1985). Avoided surface crossings and touchings are discussed in: L. Salem, *J. Am. Chem. Soc.* **96**, 3486 (1974); A. Devaquet, *Pure Appl. Chem.* **41**, 455 (1975); A. Devaquet, A. Sevin, and B. Bigot, *J. Am. Chem. Soc.* **100**, 479 (1978); L. Salem, C. Leforestier, G. A. Segal, and R. Wetmore, *J. Am. Chem. Soc.* **97**, 479 (1975); V. Bonačić-Koutecký, *Pure Appl. Chem.* **55**, 213 (1983).

Books dealing with the use of **correlation diagrams** for organic photochemical reactions are, for instance: R. B. Woodward and R. Hoffmann, *The Conservation of Orbital Symmetry*, Verlag Chemie, Weinheim, 1970; R. G. Pearson, *Symmetry Rules for Chemical Reactions*, Wiley, New York, 1976; L. Salem, *Electrons in Chemical Reactions: First Principles*, Wiley, New York, 1982; N. D. Epiotis, *Theory of Organic Reactions*, Springer Verlag, Heidelberg, 1978.

For higher energy surfaces and the concept of **de-Rydbergization**, see: E. M. Evleth, H. Z. Cao, and E. Kassab, in *Photophysics and Photochemistry Above 6 eV* (F. Lahmani, Ed.), Elsevier, Amsterdam, 1985, p. 479; J.-P. Malrieu, *Theoret. Chim. Acta* **59**, 251 (1981); E. M. Evleth and E. Kassab, *Theor. Chim. Acta* **60**, 385 (1982); C. Sandorfy, in *Progress in Theoretical Organic Chemistry* (I. G. Csizmadia, Ed.), Vol. 2, Elsevier, Amsterdam, 1977, p. 384. We do not attempt to cover the photochemical processes initiated by light of wavelengths below 200 nm. For a survey of the experimental observations, see M. G. Steinmetz, in *Organic Photochemistry* (A. Padwa, Ed.), Marcel Dekker, New York, 1987, Vol. 8, p. 67.

PART B

ELEMENTARY PHOTOCHEMICAL STEPS

In Part A we have described the basic theoretical tools necessary for developing an understanding of organic photochemical processes. In Part B we shall apply these tools to a series of simple photochemical processes that frequently represent the elementary steps in photochemical reaction sequences.

In Chapter 4 we consider the making and breaking of σ and π bonds with the primary involvement of only two orbitals, one on each of the participating atoms (bitopic reactions). In the next two chapters, we consider the simultaneous making and breaking of several bonds, arranged either in a linear array (Chapter 5) or in a cyclic array (Chapter 6), again involving no more than a single valence orbital per participating atom.

Finally, in Chapter 7 we consider the processes of making and breaking of σ and π bonds involving more than one orbital per atom (tritopic and polytopic).

We have attempted to provide at least one specific example for each kind of elementary process discussed. These are presented in the form of potential energy diagrams for the reactions of prototype molecules composed of relatively few atoms, computed at a sufficient level of theory to make them qualitatively reliable. Whenever possible, we attempt to compare these results of *ab initio* calculations with simple correlation diagrams.

Even for the quite small organic molecules, no full searches of the relevant parts of potential energy hypersurfaces and reaction path optimizations have been performed. This situation is far less satisfactory than is the case in ground-state reactions, where a fair number of good-quality calculations have been used to locate transition states and minima in the ground-state surface. In view of the frankly quite dismal state of detailed knowledge of the excited-state surfaces, it is presently not possible to make even semiquantitative statements about reaction rates, about partitioning among several reaction paths, about relative rates of intersystem crossing or internal conversion at different geometries, and so on. Thus, although it is now reasonably clear what needs to be done in order to make such predictions, at least in a semiquantitative fashion, the actual execution of such a program depends critically on advances in quantum chemical methodology and computer development.

As a result, the cuts through potential energy hypersurfaces which are displayed for more or less arbitrarily selected reaction paths through the body of this chapter represent hardly more than glorified correlation diagrams. In the absence of full geometric optimizations, they cannot be used to deduce the presence or absence of small energy barriers in reaction paths, although very large barriers, mostly those induced by the nature of correlation diagrams, are to be taken seriously, at least in a semiquantitative sense. The exact location of minima in the excited-state surfaces, and the exact location of crossing points or touching points between surfaces, also cannot be read off the plots shown.

Having pointed out the pitfalls that await the unwary as they attempt to interpret the diagrams shown, it is perhaps also important to point out their usefulness. Semiquantitative potential energy diagrams of the kind shown have significant advantages over the simple correlation diagrams that we have discussed in previous chapters. First of all, the excitation energies at the initial geometry of the starting material can be calculated with a fair degree of accuracy, thus tying the end points of the correlation lines onto their proper places on the vertical scale and placing them in the correct order except for nearly degenerate states. Basing the curves drawn for the more or less arbitrarily selected paths on a calculation of fair quality is certainly more reliable than simply drawing straight lines with a ruler as we have done so far. In particular, it becomes possible to estimate roughly the degree to which intended crossings are actually avoided.

The primary role of the energy diagrams shown in this chapter is thus the same as that of the correlation diagrams that we have discussed so far, namely, to provide *a priori* rationalization for observed photochemical processes. They are somewhat more reliable than the simple correlation diagrams and thus increase our confidence that these rationalizations are actually meaningful. It is hoped that the sorry state of affairs in this part of theoretical photochemistry will serve as an impetus for further development.

CHAPTER 4

Two-Center Reactions: One Active Orbital per Atom

This chapter deals with the properties of σ and π bonds between two atoms, each of which participates through only one of its valence orbitals in the bond dissociation process in any of the low-lying electronic states. Therefore, the total number of atomic (or hybrid) orbitals available to the two electrons of the bond as it is being broken or made is two, and such processes are called bitopic. Processes in which two or more orbitals on an atom participate actively are called polytopic and will be discussed in Chapter 7.

4.1 GENERAL: INTRODUCING THE TWO-ELECTRON TWO-ORBITAL MODEL

Wave functions describing two electrons delocalized over two interacting atomic orbitals on two different atoms are the simplest among those important for photochemical problems. In the ground state, such a wave function represents a chemical bond. In a sense, then, in this section we shall consider the various states of excitation of a chemical bond. Strictly speaking, this nomenclature is not appropriate since in some of these states the two electrons do not bond the two centers together and may even push them apart, in effect constituting an "antibond."

The MO–VB transformation. We shall consider the electronic states of a chemical bond from both viewpoints outlined in Section 1.3.2 and described in detail in Appendix I, namely, molecular orbital (MO) and valence bond (VB). Each offers certain advantages, and familiarity with both is very useful to a photochemist. A transformation between the two descriptions of a two-electron system is simple and shall be described first.

The MO description. Let a and b be two orthogonal MOs. Occupancy of each by one electron leads to the four already familiar configurations with spin functions $\Theta_1 = \alpha(1)\alpha(2)$, $\Theta_0 = (1/\sqrt{2})[\alpha(1)\beta(2) + \beta(1)\alpha(2)]$, $\Theta_{-1} = \beta(1)\beta(2)$ (triplet), and $\Sigma = (1/\sqrt{2})[\alpha(1)\beta(2) - \beta(1)\alpha(2)]$ (singlet). In symbolic notation for the three components of a triplet wave function, we shall use the superscripts 3(1), 3(0), and 3(−1) to indicate the triplet spin functions Θ_1, Θ_0, and Θ_{-1}, respectively. Occupancy of either orbital by both electrons provides us with two additional configurations. We adopt the following notation for the six configurations of the two-electron system:

$$^1|a^2\rangle = \frac{1}{\sqrt{2}} |a\bar{a}| \tag{4.1}$$

$$^1|ab\rangle = \frac{1}{2} (|a\bar{b}| - |\bar{a}b|) \tag{4.2}$$

$$^1|b^2\rangle = \frac{1}{\sqrt{2}} |b\bar{b}| \tag{4.3}$$

$$^{3(1)}|ab\rangle = \frac{1}{\sqrt{2}} |ab| \tag{4.4}$$

$$^{3(0)}|ab\rangle = \frac{1}{2} (|a\bar{b}| + |\bar{a}b|) \tag{4.5}$$

$$^{3(-1)}|ab\rangle = \frac{1}{\sqrt{2}} |\bar{a}\bar{b}| \tag{4.6}$$

where an occupancy of an orbital with an electron of spin β is indicated by a bar, and occupancy with an electron of spin α is denoted by the absence of a bar. The symbols within the vertical bars stand for the diagonal of a determinant. In full detail, we then have

$$^1|a^2\rangle = \frac{1}{\sqrt{2}} [a(1)\alpha(1)a(2)\beta(2) - a(2)\alpha(2)a(1)\beta(1)] \tag{4.7}$$

and similar expressions can be written for the other configurations. The determinantal form guarantees that the wave functions satisfy the Pauli principle, since the interchange of the coordinates of two electrons amounts to the exchange of two rows of the determinant, that is, to multiplication by -1. The usual graphical representation of the six configurations is shown in Figure 4.1.

The triplet spin functions Θ_1, Θ_0, and Θ_{-1} have been chosen in a way that is convenient but not unique. Since the three components of the triplet state are degenerate in the present approximation (no spin-related terms in the Hamiltonian), other linear combinations are equally acceptable. The three functions used presently are eigenfunctions of the $\hat{S}_z$ operator and would be appropriate even for the full Hamiltonian, which removes the triple degeneracy, in the limit of a strong magnetic field directed along Z. In Appendix II, we give another choice of three spin functions, Θ_x, Θ_y, and Θ_z, appropriate for the full Hamiltonian in the limit of no outside magnetic field. All choices are equivalent for the purposes of this chapter until we discuss spin–orbit coupling in Section 4.6.4.

The VB description. Let us further consider a pair of AOs $|A\rangle$ and $|B\rangle$, normalized but not necessarily orthogonal, located on atoms A and B, and related to the MOs

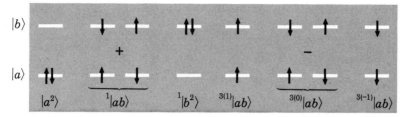

Figure 4.1 Spin-adapted MO configurations for two electrons in two delocalized orbitals.

by

$$|a\rangle = \lambda_1|A\rangle - \lambda_2|B\rangle \tag{4.8}$$

$$|b\rangle = \lambda_3|A\rangle + \lambda_4|B\rangle \tag{4.9}$$

Now, the AOs $|A\rangle$ and $|B\rangle$ are used to construct the corresponding six VB structures, $|A^2\rangle$, $^1|AB\rangle$, $|B^2\rangle$, $^{3(1)}|AB\rangle$, $^{3(0)}|AB\rangle$, and $^{3(-1)}|AB\rangle$, not necessarily orthogonal, in a way quite analogous to equations 4.1–4.6 except that $|A\rangle$ and $|B\rangle$ are now used in place of $|a\rangle$ and $|b\rangle$:

$$^1|A^2\rangle = \frac{1}{\sqrt{2}}|A\overline{A}| \qquad (A^-B^+) \tag{4.10}$$

$$^1|AB\rangle = \frac{1}{2\sqrt{1+S_{AB}^2}}(|A\overline{B}| - |\overline{A}B|) \qquad (A{-}B) \tag{4.11}$$

$$^1|B^2\rangle = \frac{1}{\sqrt{2}}|B\overline{B}| \qquad (A^+B^-) \tag{4.12}$$

$$^{3(1)}|AB\rangle = \frac{1}{\sqrt{2(1-S_{AB}^2)}}|AB| \qquad (A{\cdot\cdot}B) \tag{4.13}$$

$$^{3(0)}|AB\rangle = \frac{1}{2\sqrt{1-S_{AB}^2}}(|A\overline{B}| + |\overline{A}B|) \qquad (A{\cdot\cdot}B) \tag{4.14}$$

$$^{3(-1)}|AB\rangle = \frac{1}{\sqrt{2(1-S_{AB}^2)}}|\overline{AB}| \qquad (A{\cdot\cdot}B) \tag{4.15}$$

The usual graphical representation of the six VB structures is shown in Figure 4.2. The structures $^1|A^2\rangle$ and $^1|B^2\rangle$ are of the hole–pair type, and the others are of the dot–dot type.

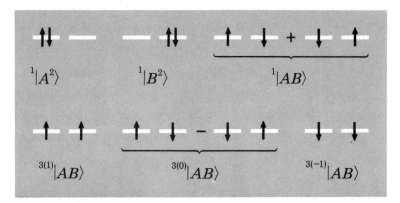

Figure 4.2 Spin-adapted VB-like structures for two electrons in two localized orbitals.

The transformation. It can be verified by direct substitution that the relation between the MO configurations and the VB structures is

$$^1|a^2\rangle = \lambda_1^2\,^1|A^2\rangle + \lambda_2^2\,^1|B^2\rangle - \sqrt{2}\,\lambda_1\lambda_2\,^1|AB\rangle \quad (4.16)$$

$$^1|ab\rangle = \sqrt{2}\,\lambda_1\lambda_3\,^1|A^2\rangle - \sqrt{2}\,\lambda_2\lambda_4\,^1|B^2\rangle - (\lambda_2\lambda_3 - \lambda_1\lambda_4)\,^1|AB\rangle \quad (4.17)$$

$$^1|b^2\rangle = \lambda_3^2\,^1|A^2\rangle + \lambda_4^2\,^1|B^2\rangle + \sqrt{2}\,\lambda_3\lambda_4\,^1|AB\rangle \quad (4.18)$$

$$^{3(1)}|ab\rangle = {}^{3(1)}|AB\rangle \quad (4.19)$$

$$^{3(0)}|ab\rangle = {}^{3(0)}|AB\rangle \quad (4.20)$$

$$^{3(-1)}|ab\rangle = {}^{3(-1)}|AB\rangle \quad (4.21)$$

Note that the triplet MO configurations are equal to the triplet VB structures. The inverse expressions for the singlet VB structures in terms of the singlet MO configurations are

$$^1|A^2\rangle = \frac{1}{(\lambda_1\lambda_4 + \lambda_2\lambda_3)^2}\{\lambda_4^2\,^1|a^2\rangle + \lambda_2^2\,^1|b^2\rangle + \sqrt{2}\,\lambda_2\lambda_4\,^1|ab\rangle\} \quad (4.22)$$

$$^1|AB\rangle = \frac{-1}{(\lambda_1\lambda_4 + \lambda_2\lambda_3)^2}\{\sqrt{2}\,\lambda_3\lambda_4\,^1|a^2\rangle - \sqrt{2}\,\lambda_1\lambda_2\,^1|b^2\rangle + (\lambda_2\lambda_3 + \lambda_1\lambda_4)\,^1|ab\rangle\} \quad (4.23)$$

$$^1|B^2\rangle = \frac{1}{(\lambda_1\lambda_4 + \lambda_2\lambda_3)^2}\{\lambda_3^2\,^1|a^2\rangle + \lambda_1^2\,^1|b^2\rangle - \sqrt{2}\,\lambda_1\lambda_3\,^1|ab\rangle\} \quad (4.24)$$

A common situation in which such transformations are practically useful is encountered when an SCF MO calculation is performed for a molecule at a biradicaloid geometry, that is, one which is characterized by the presence of only two electrons in approximately nonbonding mutually orthogonal MOs $|a\rangle$ and $|b\rangle$. The configurations built for the two nonbonding electrons are those listed in equations

4.1–4.6, and in a minimum configuration interaction (CI) treatment the resulting three singlet states are expressed by

$$\Psi_G = C_{a^2,G} {}^1|a^2\rangle + C_{ab,G} {}^1|ab\rangle + C_{b^2,G} {}^1|b^2\rangle \quad (4.25)$$

$$\Psi_S = C_{a^2,S} {}^1|a^2\rangle + C_{ab,S} {}^1|ab\rangle + C_{b^2,S} {}^1|b^2\rangle \quad (4.26)$$

$$\Psi_D = C_{a^2,D} {}^1|a^2\rangle + C_{ab,D} {}^1|ab\rangle + C_{b^2,D} {}^1|b^2\rangle \quad (4.27)$$

In general, the MOs $|a\rangle$ and $|b\rangle$ are delocalized. To obtain a localized picture, which is generally more likely to provide a simple view of the physical nature of the electronic states, one needs to transform the one-electron basis (MOs) to a localized form $|A\rangle$, $|B\rangle$. In the simplest case, the two localized orbitals will simply be two atomic orbitals, as we shall see below in the minimum basis set description of H_2 and ethylene. More generally, they will be localized group orbitals, each of which extends over more than one AO. For instance, in the description of 90° twisted ethylene, it may be recognized that each 2p orbital in the π system interacts hyperconjugatively with the opposed CH_2 group.

As stated above, the localized orbitals $|A\rangle$ and $|B\rangle$ do not need to be orthogonal. If they are related to the orthogonal delocalized orbitals $|a\rangle$ and $|b\rangle$ by equations 4.8 and 4.9, the three singlet states defined in equations 4.25–4.27 can be written as

$$\Psi_G = C_{A^2,G} {}^1|A^2\rangle + C_{AB,G} {}^1|AB\rangle + C_{B^2,G} {}^1|B^2\rangle \quad (4.28)$$

$$\Psi_S = C_{A^2,S} {}^1|A^2\rangle + C_{AB,S} {}^1|AB\rangle + C_{B^2,S} {}^1|B^2\rangle \quad (4.29)$$

$$\Psi_D = C_{A^2,D} {}^1|A^2\rangle + C_{AB,D} {}^1|AB\rangle + C_{B^2,D} {}^1|B^2\rangle \quad (4.30)$$

where

$$C_{A^2,K} = C_{a^2,K} \lambda_1^2 + C_{b^2,K} \lambda_3^2 + \sqrt{2}\, C_{ab,K} \lambda_1 \lambda_3 \quad (4.31)$$

$$C_{AB,K} = -[\sqrt{2}\, C_{a^2,K} \lambda_1 \lambda_2 - \sqrt{2}\, C_{b^2,K} \lambda_3 \lambda_4 + C_{ab,K} (\lambda_2 \lambda_3 - \lambda_1 \lambda_4)] \quad (4.32)$$

$$C_{B^2,K} = C_{a^2,K} \lambda_2^2 + C_{b^2,K} \lambda_4^2 - \sqrt{2}\, C_{ab,K} \lambda_2 \lambda_4 \quad (4.33)$$

with $K = G, S, D$.

4.2 NONPOLAR BONDS

We shall now consider two nonpolar two-electron two-center model cases. The first is an example of a σ interaction, namely, the minimum basis set model of the hydrogen molecule. The second is an example of a π interaction, namely, a simple π model of ethylene. These will serve as prototypes of nonpolar σ and π bonds, respectively. In the former case, it is not possible to reach the limit $S_{IJ} = 0$ without also reaching the limit $K_{IJ} = 0$. In the latter case, this is possible. This difference is reflected in state ordering.

4.2.1 Sigma Interactions

In the minimum basis set description, the H_2 molecule has three singlet states, which we shall label G, S, and D in the order of increasing energy, and three components of a triplet state T. All three singlets are bound, and the triplet is purely dissociative (Figure 4.3). It is interesting to ask why this is so. A qualitative answer can be found in an analysis of the nature of the state wave functions.

We shall start with the familiar simple MO picture in which each state is represented by a single configuration. We shall find it inadequate and shall then turn to configuration interaction and to the very pictorial and intuitive physical description provided by the VB method.

The MO view of the states of a sigma bond. In the simple MO picture (Figures 4.4 and 4.5), the ground state G is approximated by the ground configuration $^1|1^2\rangle$ with the bonding orbital $|1\rangle$ (HOMO) doubly occupied. The singly excited state S is represented by the singly excited configuration $^1|1\ -1\rangle$: One electron is in the bonding orbital HOMO $|1\rangle$ and one is in the antibonding orbital $|-1\rangle$ (LUMO), and they are singlet-coupled. The doubly excited state D is approximated by the doubly excited configuration $^1|-1^2\rangle$, with both electrons in LUMO. The triplet state T is represented by the triplet singly excited configuration $^3|1\ -1\rangle$, with the electrons triplet-coupled—one in HOMO, the other in LUMO. The MO energies shown in Figure 4.4 were obtained for the triplet configuration, since it alone shows proper dissociative behavior. The configurations are listed symbolically in Figure

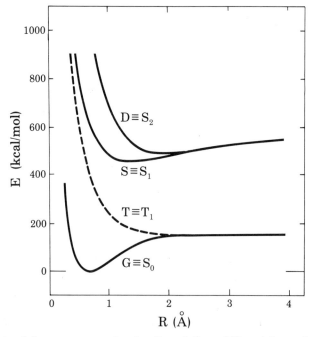

Figure 4.3 Potential energy curves for the dissociation of H_2: minimum basis set, full CI (3 × 3 CI model). AO exponent, 1.4. Singlet (solid lines) and triplet (dashed line) states.

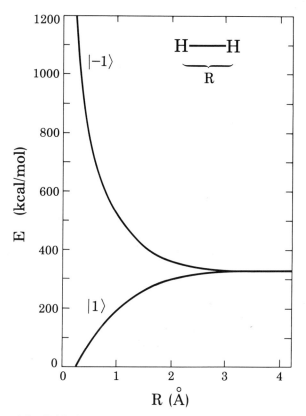

Figure 4.4 Orbital energies for H_2: minimum basis set, open shell.

4.1, and their energies are given in Figure 4.5. Using the label of the predominant configuration in each of the resulting states, we shall refer to them as the ground G state, the singly excited S state, the doubly excited D singlet state, and the triplet T state of the bond.

The mathematical representation of the state wave functions in the MO language, worked out in full detail in Appendix I, is

$$|G\rangle = (1/\sqrt{1 + \kappa^2})\,({}^1|1^2\rangle - \kappa\,{}^1|-1^2\rangle) \tag{4.34}$$

$$|S\rangle = {}^1|1\ -1\rangle \tag{4.35}$$

$$|D\rangle = (1/\sqrt{1 + \kappa^2})\,({}^1|-1^2\rangle + \kappa\,{}^1|1^2\rangle) \tag{4.36}$$

$$^{(1)}|T\rangle = {}^{3(1)}|1\ -1\rangle \tag{4.37}$$

$$^{(0)}|T\rangle = {}^{3(0)}|1\ -1\rangle \tag{4.38}$$

$$^{(-1)}|T\rangle = {}^{3(-1)}|1\ -1\rangle \tag{4.39}$$

where the configurations are defined by equations 4.1–4.6, with the MOs defined by $|b\rangle = |1\rangle$, $|a\rangle = |-1\rangle$.

The MOs are related to the AOs $|A\rangle$ and $|B\rangle$, located on hydrogen atoms A

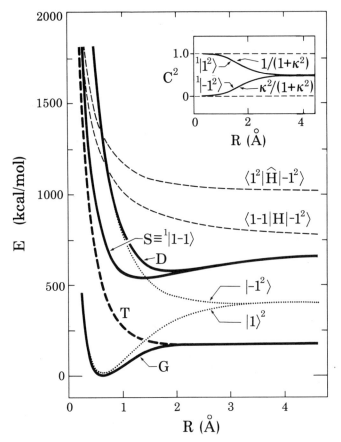

Figure 4.5 Potential energy curves for the dissociation of H_2: minimum basis set, full CI (3 × 3 CI model). AO exponent, 1.4. Singlet (solid lines) and triplet (dashed lines) states. The dotted lines show configuration energies, and the thin dashed lines show the magnitude of the off-diagonal matrix elements of the Hamiltonian that are responsible for configuration mixing. The insert indicates the weight of the $^1|1^2\rangle$ and $^1|-1^2\rangle$ configurations in the S_0 state.

and B, respectively, by

$$|1\rangle = \frac{1}{\sqrt{2 + 2S_{AB}}} [|A\rangle + |B\rangle] \quad (4.40)$$

$$|-1\rangle = \frac{1}{\sqrt{2 - 2S_{AB}}} [|A\rangle - |B\rangle] \quad (4.41)$$

where S_{AB} is the overlap of AOs $|A\rangle$ and $|B\rangle$. The mixing parameter κ is small relative to unity as long as the H—H bond has its normal ground-state length. This justifies the approximate identification of states with configurations used in the simple MO picture.

Difficulties with the simple MO view. In photochemistry, we are interested in identifying the geometries at which the lowest excited singlet S_1 and triplet T_1 states have minima, and Figures 4.4 and 4.5 suggest immediately that the simple MO picture will not be adequate. Two main problems are apparent.

First, the S_1 state is identical to the S state of our sigma bond, and the T_1 state is identical to its T state. The simple MO picture allows us to understand easily why the triplet has a minimum at infinite internuclear separation, that is, when the bond is completely broken. The electron in $|1\rangle$ provides a bonding interaction, but the electron in $|-1\rangle$ provides an even stronger antibonding interaction. The inequality of the magnitudes of the two interactions is a result of the inclusion of overlap in the calculation. In general, qualitative feeling for the dependence of MO energies on nuclear geometry can be obtained from considerations of the phase and magnitude of the overlap of the participating AOs. The origin of the bonding nature of $|1\rangle$ and of the antibonding nature of $|-1\rangle$ is described in Appendix I.

By the same token, one would expect the $|S\rangle$ state of the sigma bond to be dissociative as well. Figure 4.3 shows, however, that it is quite strongly bound. True, its minimum occurs at a much larger internuclear separation than that in the G state, roughly by a factor of two, but nowhere near the infinite separation where the T state has a minimum.

The simple MO picture does not provide any clue as to why this should be so. Even the D state of the bond, which only has antibonding electrons in the simple MO picture, is bound and not dissociative. Clearly, the simple-minded expectation that an understanding of MO energies as a function of internuclear separation (Figure 4.4) will provide an understanding of the location of minima in the excited states is not fulfilled.

The second serious problem occurs with the ground state. At large internuclear separations, the energy of the singlet ground configuration $^1|1^2\rangle$ actually is much higher than that of the triplet state (Figure 4.5). Thus, it certainly is a poor approximation for the true ground-state wave function. Consideration of the behavior of the mixing parameter κ as a function of internuclear separation provides a clue. As soon as one leaves the region of "normal" geometries of the ground-state molecule, i.e., as soon as one stretches the bond substantially, κ increases; and at infinite internuclear separation, the ground configurations $^1|1^2\rangle$ and the doubly excited configuration $^1|-1^2\rangle$ actually contribute equally to G (Figure 4.5). The energetic effect of their mixing (the correlation energy) is huge and certainly cannot be neglected. This is reasonable, given that the two configurations are nearly or exactly degenerate and can be expected whenever the σ and σ^* orbitals have comparable energies (biradicaloid geometries). While the need for configuration mixing can be accepted as an unfortunate necessity, it does not in itself provide much physical insight.

The VB view of the states of a sigma bond. In order to understand (i) the location of the minima in the singlet states and (ii) the large effect of configuration interaction in the ground state at large internuclear separations, it is very useful to adopt the alternative VB viewpoint. A transformation of the state wave functions into the VB language shows the following:

$$|G\rangle = (1 + 2\tau\kappa + \tau^2)^{-1/2}(^1|AB\rangle + \tau^1|Z_2\rangle)$$
$$\kappa = \langle ^1Z_2|^1AB\rangle = 2S_{AB}/(1 + S_{AB}^2) \tag{4.42}$$

$$|S\rangle = \frac{1}{\sqrt{2(1 - S_{AB}^2)}} [^1|A^2\rangle - {}^1|B^2\rangle] = {}^1|Z_1\rangle \qquad (4.43)$$

$$|D\rangle = (1 - 2\tau'\kappa + \tau'^2)^{-1/2}(\tau'^1|AB\rangle - {}^1|Z_2\rangle)$$

$$\tau' = (\tau + \kappa)/(1 + \tau\kappa) \qquad (4.44)$$

$$^{(1)}|T\rangle = {}^{3(1)}|AB\rangle$$

$$^{(0)}|T\rangle = {}^{3(0)}|AB\rangle \qquad (4.45)$$

$$^{(-1)}|T\rangle = {}^{3(-1)}|AB\rangle$$

where $^1|Z_1\rangle$ and $^1|Z_2\rangle$ stand for the "zwitterionic" combinations of hole–pair wave functions:

$$^1|Z_2\rangle = [2(1 + S_{AB}^2)]^{-1/2}[^1|A^2\rangle + {}^1|B^2\rangle] \qquad (4.46)$$

$$^1|Z_1\rangle = [2(1 - S_{AB}^2)]^{-1/2}[^1|A^2\rangle - {}^1|B^2\rangle] \qquad (4.47)$$

and where the VB structures are those defined in equations 4.10–4.15. They are represented symbolically in Figure 4.2. The mixing parameter τ or τ' is small relative to unity at all geometries of interest. The "zwitterionic" structures $^1|Z_1\rangle$ and $^1|Z_2\rangle$ are defined for convenience as symmetry-adapted linear combinations of the hole–pair VB structures $^1|A^2\rangle$ and $^1|B^2\rangle$ with separated charges and can also be referred to as hole–pair structures for brevity.

The energies of the VB structures are shown in Figure 4.6. As described in more detail in Appendix I, their behavior as a function of distance is intuitively clear.

First, we obtain information on the physical origin of the minima in S_1 and S_2. The high energy of the hole–pair structures $^1|A^2\rangle$ and $^1|B^2\rangle$ and its increase with increasing internuclear distance are readily understood as a result of charge separation, and the rise of all the curves at short internuclear distances is clearly due to poorly screened internuclear repulsion. Thus, the presence of minima in the excited singlet states is perfectly sensible. The presence of a minimum in the ground singlet state and the repulsive nature of the triplet state are also easily understood in terms of the addition or removal of electron density from the internuclear region resulting from the interference of the two AO wave functions (Appendix I).

The only concern might be whether the mixing of the VB structures, whose behavior we now understand individually, does not somehow change the picture. The energetic equivalence of $^1|A^2\rangle$ and $^1|B^2\rangle$ favors their interaction, which causes their combined occurrence in $^1|Z_1\rangle$ and $^1|Z_2\rangle$. However, the interaction element is proportional to S_{AB}^2 and is only large at short internuclear distances. This explains why the energies of the structures $^1|Z_1\rangle$ and $^1|Z_2\rangle$ do not differ much from that of $^1|A^2\rangle$ or $^1|B^2\rangle$ at extended internuclear distances. The interaction element between the low-energy dot–dot VB structure $^1|AB\rangle$ and the high-energy zwitterionic combination of hole–pair structures $^1|Z_2\rangle$ is roughly proportional to S_{AB}. However, $^1|AB\rangle$ and $^1|Z_2\rangle$ are far in energy, and their mixing does not lead to qualitative changes in the shapes of the potential energy curves. In the region of equilibrium ground-state geometry, this mixing has a significant effect on the energy of the higher of the two resulting states, D. Only a shallow minimum at a relatively large internuclear distance remains in D, compared with a deep minimum at a short

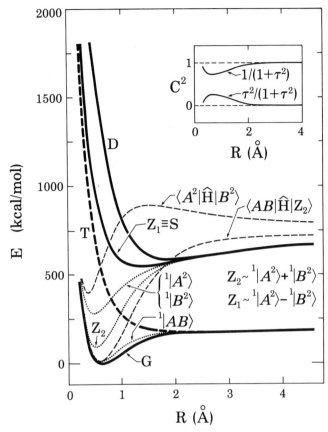

Figure 4.6 Potential energy curves for the dissociation of H_2: minimum basis set, full CI (3×3 CI model). AO exponent, 1.4. Singlet (solid lines) and triplet (dashed lines) states. The dotted lines show the energies of the dot–dot $^1|AB\rangle$ and hole-pair $^1|A^2\rangle$, $^1|B^2\rangle$ VB structures and of their linear combination $^1|Z_2\rangle$ ($^1|Z_1\rangle$ is identical with $|S\rangle$). The thin dashed lines show the magnitude of the off-diagonal matrix elements of the Hamiltonian that are responsible for VB structure mixing (resonance). The insert indicates a rough measure of the importance of the VB structures $^1|AB\rangle$ and $^1|Z_2\rangle$ in the S_0 state (see text).

internuclear separation in $^1|Z_1\rangle$. All in all, as Figure 4.6 shows, at medium and large internuclear distances, the energy curves for the VB structures are quite similar to those of the final states. In particular, those curves of the individual VB structures that have bound minima yield states that also have bound minima.

VB versus MO–CI. The second above-mentioned problem—namely, the origin of the improper dissociation behavior of the $^1|1^2\rangle$ configuration and the need for CI—can now also be qualitatively understood. When expanded in terms of VB structures, $^1|1^2\rangle$ is seen to be an in-phase combination of equal parts of $^1|AB\rangle$ and $^1|Z_2\rangle$:

$$^1|1^2\rangle = \frac{\sqrt{1 + S_{AB}^2}}{\sqrt{2(1 + S_{AB})}} [^1|Z_2\rangle + {}^1|AB\rangle] \quad (4.48)$$

If $^1|1^2\rangle$ is to represent the ground state, as simple MO theory insists, this is bad enough at normal geometries, since the hole–pair structures are so unfavorable. At intermediate and long bond lengths, it becomes totally untenable because the ground state of H_2 would then dissociate with equal likelihood into an ion pair and an atom pair.

In introducing the CI procedure, this situation is remedied by subtracting the undesirable charge-separated component from the $^1|1^2\rangle$ configuration. This is accomplished by subtracting the doubly excited $^1|-1^2\rangle$ configuration, which represents an out-of-phase combination of $^1|Z_2\rangle$ and $^1|AB\rangle$. In this way, the VB formulation helps us to understand how configuration interaction works.

An MO–VB comparison. In summary, the simple MO description is useful at short internuclear distances, where individual configurations represent reasonable approximations to electronic states. For example, it is useful to think of the pair of configurations $^1|1-1\rangle$ and $^3|1-1\rangle$ as producing a pair of related states where the triplet T lies below the excited singlet $|S\rangle$ by an amount approximately equal to $2K_{1,-1}$, twice the exchange integral between the two MOs $|1\rangle$ and $|-1\rangle$ (Figures 4.3 and 4.6).

However, at intermediate and large internuclear separations (biradicaloid geometries), the introduction of configuration interaction into the MO picture is absolutely necessary. In this region, the VB description provides a clearer and more intuitive interpretation of the nature of electron motion and an easier rationalization of the shape of the energy curves. Since excited states are of interest, inclusion of hole–pair structures in the VB picture is essential. At these geometries, it is useful to think of the pair of VB structures, $^1|AB\rangle$ and $^3|AB\rangle$, as producing a pair of related states (Figures 4.3 and 4.6). The triplet T lies above the ground-state singlet G by an amount approximately equal to minus twice the VB "exchange integral" between the two AOs $|A\rangle$ and $|B\rangle$, $2K_{AB}^{vb}$. The "exchange integral" of the VB method, K_{AB}^{vb}, should not be confused with the already familiar exchange integral $K_{1,-1}$ of the MO method. A detailed comparison is given in Appendix I, in which the MO and VB solutions for the H_2 molecule in the minimum basis set are described in their full glory.

Here, we only note that unlike $K_{1,-1}$, the VB "exchange integral" K_{AB}^{vb} contains contributions from the one-electron part of the Hamiltonian as well as an electron-repulsion term. The former contain overlap and vanish if $|A\rangle$ and $|B\rangle$ are orthogonal. If that happens, we obtain $K_{AB}^{vb} = K_{AB} \rangle 0$, and then the triplet T will lie below the singlet G by $2K_{AB}$. We shall encounter this situation in twisted ethylene described by a two-electron two-orbital model. In the present case of H_2, S_{AB} vanishes only at infinite internuclear separation and K_{AB} then vanishes as well, so that G and T are degenerate. At finite distances, the one-electron part in K_{AB}^{vb} dominates and gives this integral a negative value. Therefore, in H_2, T always lies above G. At a general biradicaloid geometry of many other molecules, either sign can occur, so that either state can be lower.

The fact that the triplet state T is simply related to the ground-state singlet G at some geometries and to the excited state singlet S at others, in the sense of having a similar energy and a roughly similar electron density distribution, is due to its dual nature. Like G, it is a "dot–dot" (covalent) state in which the electrons avoid each other (VB language). Like S, it is an excited state in which an anti-

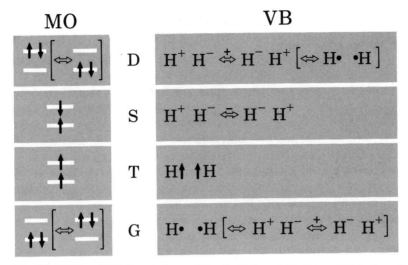

Figure 4.7 The MO and VB descriptions of the G, T, S, and D states of H_2 in the minimum basis set model.

bonding MO is occupied (MO language). At biradicaloid geometries, where both MOs are essentially nonbonding, only the former property matters. At ordinary geometries, where the antibonding orbital is high in energy, mostly the latter property matters.

A pictorial representation of the correspondence between the MO and VB descriptions of the states of the H_2 molecule in the minimum basis set is given in Figure 4.7. Table 4.1 summarizes the MO and VB expressions for their energies. The former are cast in terms of the one-electron energies $h_{\pm 1}$, MO energies $\varepsilon_{\pm 1}$, nuclear repulsion energy e^2/R, and electron-repulsion integrals between MOs, both of the Coulomb (J) and of the exchange (K) types. The latter are cast in terms of

TABLE 4.1 VB and MO Expressions for the State Energies of the H_2 Molecule

Energies: VB

$E(^1|AB\rangle) = E_A + E_B + (J^{vb}_{AB} + K^{vb}_{AB})/(1 + S^2_{AB})$

$E(^3|AB\rangle) = E_A + E_B + (J^{vb}_{AB} - K^{vb}_{AB})/(1 - S^2_{AB})$

$E(^1|A^2\rangle) = E(^1|B^2\rangle) = E_A + E_B + E^{CT}_{AB}$

$\langle A^2|\hat{H}|B^2\rangle = (E_A + E_B) S^2_{AB} + K^{vb}_{AB}$

Energies: MO

$E(^1|1^2\rangle) = 2h_1 + J_{11} + e^2/R = 2\varepsilon_1 - J_{1,1} + e^2/R$

$E(^3|1\,-1\rangle) = h_1 + h_{-1} + J_{1,-1} - K_{1,-1} + e^2/R = \varepsilon_1 + \varepsilon_{-1} - J_{1,1} - J_{1,-1} + e^2/R$

$E(^1|1\,-1\rangle) = h_1 + h_{-1} + J_{1,-1} + K_{1,-1} + e^2/R = \varepsilon_1 + \varepsilon_{-1} - J_{1,1} - J_{1,-1} + 2K_{1,-1} + e^2/R$

$E(^1|-1^2\rangle) = 2h_{-1} + J_{-1,-1} + e^2/R = 2\varepsilon_{-1} + J_{-1,-1} - 4J_{1,-1} + 2K_{1,-1} + e^2/R$

the energies of the constituent atoms (E_H), of the VB Coulomb (J^{vb}) and exchange (K^{vb}) integrals, and of the charge-transfer energy (E_{AB}^{CT}). All of these quantities are defined in detail in Appendix I.

The H_2 molecule: Accurate calculations and Rydberg states. The H_2 molecule represents a simple enough problem that an essentially exact solution is possible at the Born–Oppenheimer level. The results of such a calculation are in excellent agreement with all experimental observations. The computed energy curves are shown in Figure 4.8 along with their spectroscopic notation.

In spite of the simple nature of the minimum basis set description that we have given above, one has no trouble identifying our model states (G, T, and S) of Figure 4.3 with the real Born–Oppenheimer states of Figure 4.8. There are two main differences. The first one is trivial from our point of view: The absolute energies and energy curve shapes computed by the simple method are only qualitatively, but not quantitatively, correct. The second one is of more significance for photochemistry: Additional excited states are present at energies comparable with those of the S state. These are the so-called Rydberg states, with electron density in a very diffuse orbital, best described in terms of AOs of higher quantum numbers: 2s, 2p, and so on. The lowest excited singlet, $B^1\Sigma_u^+$, corresponds to our S state and is bound, with a minimum at 1.29 Å. This can be compared with the ground-state equilibrium bond length of 0.74 Å. However, the state does not dissociate to an ion pair as our result suggested. Although it initially behaves as if it were going to, it then undergoes an avoided crossing with one of the Rydberg states and proceeds to dissociate to an H atom in its ground 1s state and to another one in its 2s state. Such "Rydbergization" of a valence state along a bond dissociation coordinate is a fairly common occurrence. Along with the converse processes, "de-Rydbergization," it plays an important role in their photochemistry.

The state that corresponds to our D state is even more complicated. It is known

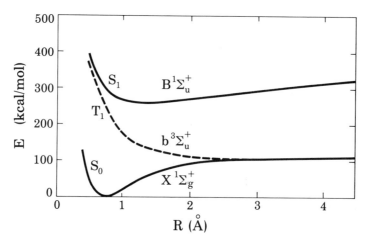

Figure 4.8 Accurate potential energy curves of H_2. Based on W. Kolos and L. Wolniewicz, *J. Chem. Phys.* **45**, 509 (1966); **49**, 404 (1968).

as the F state, with a minimum at 2.32 Å. Because of an avoided crossing with another Rydberg state, it has a second minimum at 1.01 Å, at which point it is known as the E state. Yet other Rydberg states, such as the C state, also occur in this general energy region.

The lesson that the comparison with exact calculations and with experiment teaches is that the simple concepts concerning the nature of the four electronic states, G, T, S, and D, are correct in principle but have to be applied with caution when it comes to higher energy excitations. In larger molecules, the lowest Rydberg state is generally found 2–3 eV below the ionization potential, and its possible involvement in singlet photochemical processes should always be kept in mind. This is particularly true in smaller molecules with relatively high lowest excitation energy. Fortunately, it appears that the S_1 wave functions in the regions of minima from which the return to S_0 occurs are usually of valence rather than of Rydberg nature, and this permits many qualitative conclusions even without an explicit consideration of Rydberg states.

4.2.2 Pi Interactions

Consider now a simple model of the π system in the ethylene molecule as a prototype of a π bond (Figure 4.9). The σ electrons are considered only at the SCF level, and their excitation is not included in the CI treatment. Thus, they are viewed only as a nonpolarizable time-averaged charge distribution (core). In many ways, this model is entirely analogous to the just-discussed model of a σ bond, since we only use a minimum basis set for the description of the π system; that is, we use only the (nonorthogonal) atomic orbitals $|A\rangle$ and $|B\rangle$. In our discussion, we shall concentrate on the differences.

The MO view of the states of a pi bond. The already familiar six states G, S, D, T_{+1}, and T_0, and T_{-1} again appear, with the triplet triply degenerate in the present approximation. A plot of their energies against the twist angle is shown in Figure 4.10. The representation of the states in terms of MO configurations, whose energies are shown in Figure 4.10 as well, is the same as in equations 4.34–4.39. The energies of the MOs and the weights of the configurations in the individual states are shown later, in Figure 4.16 (case q = 0). Once again, in the neighborhood of the ordinary planar geometry, the G state is well represented by the ground configuration $^1|1^2\rangle$, the S state by the singly excited configuration $^1|1\ -1\rangle$, the D state by the doubly excited configuration $^1|-1^2\rangle$, and the three components of the T state by $^{3(1)}|1\ -1\rangle$, $^{3(0)}|1\ -1\rangle$, and $^{3(-1)}|1\ -1\rangle$.

The energies of the MOs behave as expected from the correlation diagram in

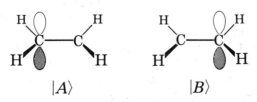

Figure 4.9 The minimum AO basis for the π system of ethylene.

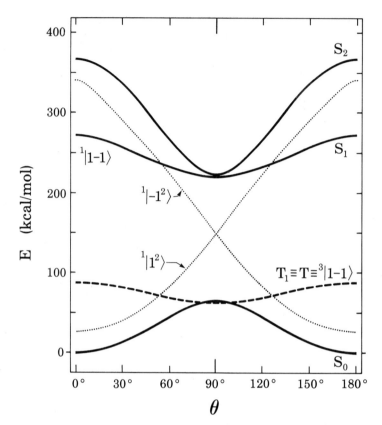

Figure 4.10 State diagram for ethylene as a function of the twist angle θ. Minimum π basis set, full CI (3 × 3 CI model). Singlet (solid lines) and triplet (dashed lines) states. The dotted lines show the energies of the $^1|1^2\rangle$ and $^1|-1^2\rangle$ configurations.

Figure 3.6. As ethylene is twisted towards 90°, the HOMO $|1\rangle$ rises in energy, while the LUMO $|-1\rangle$ drops. The energies of the two MOs become equal at 90°, where $|A\rangle$ and $|B\rangle$ are orthogonal, and the lines in the plot continue their respective climb or drop. Because of this orbital crossing, geometries near 90° twist are biradicaloid. However, we have kept the label $|1\rangle$ for the bonding HOMO and $|-1\rangle$ for the antibonding LUMO at all geometries so that we should properly refer to orbital touching rather than crossing.

From the consideration of the MO energies alone, it is possible to understand the course of the state energy curves, and in this sense the MO picture does better here than for the dissociation of H_2. The loss of the bonding nature of $|1\rangle$ and the even stronger loss of the antibonding nature of $|-1\rangle$ upon twisting lead one to expect the ground configuration $^1|1^2\rangle$ to rise and the doubly excited configuration $^1|-1^2\rangle$ to drop in energy, as observed.

Where the MO picture without configuration interaction fails is in the detailed description in the region of biradicaloid geometries, near 90°. In a better description, the G and D states do not become degenerate, but instead remain separated (in minimum basis set approximation followed by 3 × 3 CI, by ~80 kcal/mol), so that the crossing is strongly avoided. It is avoided over a fairly large range of twist angles, so that the change in the wave function is rather gradual (cf. Figure 2.11, p. 69), and at low vibrational energies, the probability of jumps from one surface to the other is much less than unity. As in the H_2 case, the S and T states remain well represented by single configurations.

The VB view of the states of a pi bond. The VB description again comes to the rescue: It does not predict a degeneracy at 90° twist, since the lower two states are represented by dot–dot structures without charge separation and the higher two states are represented by high-energy hole–pair separated-charge structures. The representation of states by VB structures is exactly the same as in equations 4.42–4.47. At all geometries, the dot–dot structure $^1|AB\rangle$ is a good approximation for the ground state G. As in the case of H_2, at the ordinary geometries an in-phase combination of the hole–pair structures $^1|A^2\rangle$ and $^1|B^2\rangle$ also contributes weakly to G. As the molecule deviates from planarity, their contribution decreases and at 90° twist the G state is purely of dot–dot nature, that is, $G = {}^1|AB\rangle$.

Again at all geometries, the triplet state T is represented by the VB structure $^3|AB\rangle$ and the singly excited state S is represented by the out-of-phase combination $^1|Z_1\rangle$ of the hole–pair VB structures. The doubly excited state D is predominantly zwitterionic (hole–pair $^1|Z_2\rangle$). Its small content of the dot–dot structure $^1|AB\rangle$ diminishes with an increasing angle of twist and vanishes at the perpendicular geometry.

At the 90° twisted geometry, ethylene is a perfect biradical (see Section 4.4.2). At this geometry the two AOs, $|A\rangle$ and $|B\rangle$, are orthogonal, so that $|A\rangle = |A\rangle$ and $|B\rangle = |B\rangle$, and $S_{AB} = S_{AB} = 0$, using our standard notation of roman capital characters A and B for localized orthogonal orbitals and italic capital letters A and B for atomic orbitals. The minimum basis set four-state description of the wave functions for the two nonbonding electrons, then, is

$$|G\rangle = (1/\sqrt{2})\, (^1|1^2\rangle - {}^1|-1^2\rangle) = {}^1|AB\rangle \tag{4.49}$$

$$|S\rangle = {}^1|1\ -1\rangle = (1/\sqrt{2})\, (^1|A^2\rangle - {}^1|B^2\rangle) \tag{4.50}$$

$$|D\rangle = (1/\sqrt{2})\, (^1|1^2\rangle + {}^1|-1^2\rangle) = (1/\sqrt{2})\, (^1|A^2\rangle + {}^1|B^2\rangle) \tag{4.51}$$

$$|T\rangle = {}^3|1\ -1\rangle = {}^3|AB\rangle \tag{4.52}$$

where the MO configurations are defined as in equations 4.1–4.6 and the MOs $|1\rangle$ and $|-1\rangle$ are the in-phase and out-of-phase combinations of the two p-type AOs $|A\rangle$ and $|B\rangle$ as defined in equations 4.40 and 4.41. If we take the energy of an electron in the p orbital as zero and ignore the constant nuclear repulsion term, the energies of both MOs, $|1\rangle$ and $|-1\rangle$, are also zero. The following results are obtained for the energies of the six states at the orthogonally twisted geometry (see Sections 4.4 and 4.5 and Appendix II for further detail):

$$E(G) = J_{1,1} - K_{1,-1} = J_{AB} + K_{AB} \tag{4.53}$$

$$E(S) = J_{1,-1} + K_{1,-1} = J_{AA} - K_{AB} \tag{4.54}$$

$$E(D) = J_{1,1} + K_{1,-1} = J_{AA} + K_{AB} \tag{4.55}$$

$$E(T) = J_{1,-1} - K_{1,-1} = J_{AB} - K_{AB} \tag{4.56}$$

We see that the changes in the nature of the ethylene wave functions upon going from an ordinary to a biradicaloid geometry are identical to those already analyzed in the case of H_2. In the present case, this transformation is accomplished by twisting the π bond; in the previous case, it is accomplished by stretching the σ bond. An understanding of the ordering of states at the 90° twisted geometry of ethylene can now be developed in terms analogous to those used for H_2.

The two covalent states are of low energy, and the two ionic states are of high energy. As before, the need for subtracting $^1|-1^2\rangle$ from $^1|1^2\rangle$ in the description of the ground state arises from the necessity to remove the zwitterionic (hole–pair) contribution $^1|Z_2\rangle$ from $^1|1^2\rangle$, which is given by equation 4.48.

The tendency of the G and T states of a σ or π bond to acquire purely dot–dot covalent nature once the bond is completely broken is perfectly general in the minimum basis set model, in which only two AOs are used for its description. Nothing is to be gained by mixing the high-energy charge-separated hole–pair structures into the ground state. When, then, do hole–pair VB structures mix into the ground-state wave functions of bonds, and why? The answer is not easy to give in the VB language but is clear in MO terms: At geometries where $|1\rangle$ lies substantially below $|-1\rangle$, it is energetically favorable to occupy the former and not the latter. This provides a tendency to give $^1|1^2\rangle$ a large weight and $^1|-1^2\rangle$ a small weight. Once the two weights are unequal, however, the hole–pair VB structures are inevitably not subtracted exactly from the total wave function.

Sigma–pi comparison. There are two interesting differences between the behavior of state energy curves for the σ bond and for the π bond. Both are related to the fact that the former is broken by separating the two centers to infinite distance, while in the latter their distances change only insignificantly. As a σ bond is stretched, the charge separation energy, and therefore also the energy of the hole–pair structures $^1|A^2\rangle$ and $^1|B^2\rangle$, increases monotonically. As the π bond is twisted, the charge separation energy hardly changes at all.

The first difference is now understandable: The hole–pair (zwitterionic) S and D states of H_2 have a minimum halfway between a normal and a biradicaloid geometry, in spite of simple MO arguments, while in ethylene the minimum occurs at the 90° twist biradicaloid geometry, in agreement with the simple MO arguments.

The second difference is subtle: In completely dissociated nonpolar σ bonds (both participating atoms of equal electronegativity), the G and T states are degenerate and the S and D states are degenerate. This is a consequence of the infinite separation of the two participating AOs $|A\rangle$ and $|B\rangle$, which leads to a zero overlap density everywhere and thus a zero exchange integral as well as a zero overlap S_{AB}. At finite separations, G is always below T (the VB "exchange integral" K_{AB}^{vb} is negative) and D is always above S. However, for the completely twisted nonpolar π bond, the G and T and the S and D state pairs are only approximately degenerate at the present level of approximation. The reason for this is the finite separation of $|A\rangle$ and $|B\rangle$: The overlap density between the orthogonally twisted orbitals $|A\rangle$ and $|B\rangle$ does not vanish, although their overlap S_{AB} does, as do all terms proportional to it. Thus, the VB "exchange integral" K_{AB}^{vb} does not vanish and instead becomes equal to the electron repulsion integral K_{AB}.

Since even at 90° twist the VB "exchange integral" $K_{AB}^{vb} = K_{AB} = K_{AB}$ is positive, albeit very small, the degeneracies noted above are removed. This places T below G and D above S by an amount equal to $2K_{AB}$. We shall see later that this result changes when the ethylene model is improved to allow for the presence of other electrons in the molecule. At the two-electron two-orbital four-state level for the two-center problem, however, the result represents a general difference between biradicaloid geometries at which $S_{AB} = K_{AB} = 0$ and those at which $S_{AB} = 0$, and $K_{AB} > 0$.

The notion that in a simple model for a molecule some states are purely of hole–pair nature (zwitterionic) while others have mixed dot–dot and hole–pair (covalent–zwitterionic) nature has far more generality than might appear at first sight. It should be noted that in some model cases, states of the mixed category can actually be purely dot–dot (covalent), for instance, the T state of H_2.

The concept is encountered in many simple models of all molecules containing only AOs of equal electronegativity. The condition for such separation of states into two categories is that the AOs of the basis set must be separable into two groups, say starred and unstarred, in such a way that the one-electron part of the Hamiltonian has zero matrix elements between two AOs if they belong to the same group. Such AO systems are known as alternant systems. For instance, simple models for the π systems of polyenes and aromatics are of this kind.

When simple models are refined, the strict distinction between the two kinds of states generally disappears. Often, however, enough of a difference remains that the "hole–pair" versus "mixed" classification retains its usefulness.

As in the discussion of H_2, we could now describe the results of more accurate calculations. These reflect the presence of other bonds in the molecule and lead to the reversal of the energetic order of the $|G\rangle$ and $|T\rangle$ states, and of that of the two next higher singlet states, at orthogonal geometries. The consideration of the existence of Rydberg orbitals introduces additional complications, particularly at planar geometry. These matters will be discussed in Sections 4.8.1 and 5.1.2.

Finally, we note that also the label C ("covalent") is used frequently in the literature for the G state, and the labels Z_1 and Z_2 ("zwitterionic") for the S and D states, respectively. We shall use these conventions interchangeably.

4.3 POLAR BONDS

We are now ready to compare the results of the previous section with those of models for two-electron two-center σ and π interactions of the polar kind in which the electronegativities of the two participating orbitals $|A\rangle$ and $|B\rangle$ are different.

4.3.1 Sigma Interactions

In order to model electronegativity differences between the two participating atoms in a bond, we have repeated calculations on the H_2 molecule using a minimum basis set, giving one of the hydrogen atoms a nuclear charge of $Z_A = 0.8$ or 0.6 and the other a nuclear charge of $Z_B = 1.2$ or 1.4 elementary units. This "polarizes" the H—H bond. Subsequently, we shall consider results for an ordinary H_2 molecule in the field of a nearby positive charge and results for a real molecule, LiH. In all of these instances, the already familiar three singlet states and a triplet state result. However, the singlets are no longer described by the simple wave functions $|G\rangle$, $|S\rangle$, and $|D\rangle$ but, instead, are described by their mixtures. The triplet wave function remains equal to $|T\rangle$.

A "polarized" H_2 molecule

The MO picture. In the polar case, the MOs are no longer determined by symmetry:

$$|1\rangle = \frac{1}{\sqrt{C_1^2 + 2C_1C_2S_{AB} + C_2^2}} [C_1|A\rangle + C_2|B\rangle] \quad (4.57)$$

$$|-1\rangle = \frac{1}{\sqrt{C_3^2 - 2C_3C_4S_{AB} + C_4^2}} [C_3|A\rangle - C_4|B\rangle] \quad (4.58)$$

The energies of these orbitals, calculated for the triplet configuration, are shown in Figure 4.11. The more electronegative orbital $|B\rangle$ predominates in the bonding MO $|1\rangle$, whereas the less electronegative orbital $|A\rangle$ predominates in the antibonding orbital $|-1\rangle$. This polarization of the MOs is particularly strong at large internuclear distances. The dependence of the mixing coefficients on internuclear distance shows that the in-phase character of the bonding orbital and the out-of-phase character of the antibonding MO remain what they were in the nonpolar case of H_2.

The state wave functions have the form

$$|S_0\rangle = C'_{1^2,S_0}|1^2\rangle + C'_{1-1,S_0}|1-1\rangle + C'_{-1^2,S_0}|-1^2\rangle \quad (4.59)$$

$$|S_1\rangle = C'_{1^2,S_1}|1^2\rangle + C'_{1-1,S_1}|1-1\rangle + C'_{-1^2,S_1}|-1^2\rangle \quad (4.60)$$

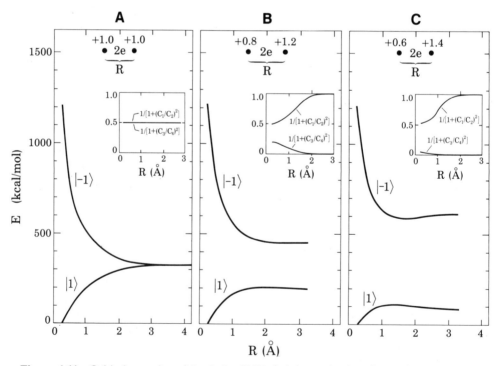

Figure 4.11 Orbital energies of "polarized" H_2 (minimum basis set). Nuclear charges as shown. The inserts show the polarization of the MOs $|1\rangle$ and $|-1\rangle$ (see text).

$$|S_2\rangle = C'_{1^2,S_2}{}^1|1^2\rangle + C'_{1-1,S_2}{}^1|1-1\rangle + C'_{-1^2,S_2}{}^1|-1^2\rangle \tag{4.61}$$

$$^{(1)}|T\rangle = {}^{3(1)}|1-1\rangle \tag{4.62}$$

$$^{(0)}|T\rangle = {}^{3(0)}|1-1\rangle \tag{4.63}$$

$$^{(-1)}|T\rangle = {}^{3(-1)}|1-1\rangle \tag{4.64}$$

The MO configurations are those of equations 4.1–4.6 and 4.16–4.21; the MOs are now denoted $|b\rangle = |1\rangle$ and $|a\rangle = |-1\rangle$, as defined in equations 4.57 and 4.58. The energies of the three states and of the three configurations, as well as their weights in each state, are plotted as a function of internuclear distance in Figure 4.12. The S_0 and T_1 states are affected only slightly by the polarization. The S_1 and S_2 states split widely apart, and the former becomes purely dissociative. The T_1 state dissociates to the same limit as the S_0 state in nonpolar and weakly polarized H_2, but it dissociates to the same limit as the S_1 state in the more strongly polarized H_2 model. This behavior may appear peculiar at first but is easily rationalized when the nature of the electronic wave functions is considered. This is generally difficult in the MO model, since both the shape of the MOs, as defined by their coefficients (Figure 4.11), and the weight of each configuration in the final state (Figure 4.12)

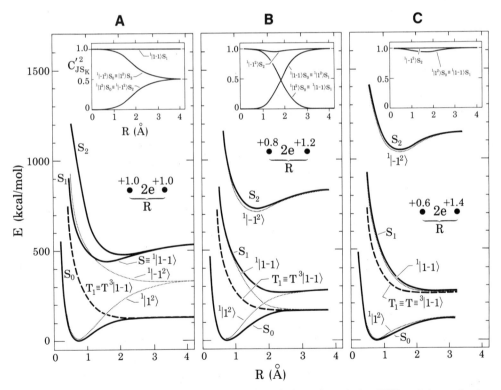

Figure 4.12 Potential energy curves for the dissociation of "polarized" H_2: minimum basis set, full CI (3 × 3 CI model). Singlet (solid lines) and triplet (dashed lines) states. Nuclear charges as shown. The insert shows the weights of the important configurations in the three singlet states. Configuration energies are shown by dotted lines.

change as the geometry changes. Fortunately, we only need to consider the behavior at large internuclear separations to understand the difference between the state energy curves in parts A, B, and C of Figure 4.12. At the dissociation limit, the MOs of "polar" H_2 are completely polarized: $|1\rangle$ is identical to the AO $|B\rangle$, and $|-1\rangle$ is equal to the AO $|A\rangle$.

The wide separation of the S_1 and S_2 curves in part B of Figure 4.12 is now seen to be due to the widely different energies of their dissociation limits, represented by the hole–pair structures $^1|B^2\rangle = {}^1|1^2\rangle$ for S_1 and $^1|A^2\rangle = {}^1|-1^2\rangle$ for S_2 (recall that the nuclear charges are $Z_B = 1.2$ and $Z_A = 0.8$). The common energy at the dissociation limit for S and T_1 is that of the singlet and triplet dot–dot structures $^{1,3}|AB\rangle$.

As Z_B increases further and Z_A is reduced accordingly, it eventually becomes energetically preferable to place both electrons on one of the fully separated atoms, that is, $^1|B^2\rangle \equiv {}^1|1^2\rangle$ is now not only lower in energy than $^1|A^2\rangle \equiv {}^1|-1^2\rangle$ but also lower than $^{1,3}|AB\rangle$. This situation has been reached for $Z_B = 1.4$, $Z_A = 0.6$ (Figure 4.12C). Now, the dissociation limit for the S_0 state is the hole–pair structure $^1|B^2\rangle$, the degenerate limit for S_1 and T_1 is $^{1,3}|AB\rangle$. Not surprisingly, S_2 is now described by $^1|A^2\rangle$ and is very high in energy.

This dramatic dependence of the qualitative shape of the energy curves for the excited states of a single bond has considerable implications for photochemistry, and we shall analyze the situation in far more detail in the remainder of this chapter. Also in ground-state chemistry, the information in Figure 4.12 is significant: While part A models a nonpolar covalent bond, part B models a polar covalent bond and part C models a dative bond.

The VB picture. The alternative construction of the four states from the VB starting point is shown in Figure 4.13. The wave functions now have the form

$$|S_0\rangle = C'_{A^2,S_0}{}^1|A^2\rangle + C'_{AB,S_0}{}^1|AB\rangle + C'_{B^2,S_0}{}^1|B^2\rangle \quad (4.65)$$

$$|S_1\rangle = C'_{A^2,S_1}{}^1|A^2\rangle + C'_{AB,S_1}{}^1|AB\rangle + C'_{B^2,S_1}{}^1|B^2\rangle \quad (4.66)$$

$$|S_2\rangle = C'_{A^2,S_2}{}^1|A^2\rangle + C'_{AB,S_2}{}^1|AB\rangle + C'_{B^2,S_2}{}^1|B^2\rangle \quad (4.67)$$

$$^{(1)}|T\rangle = {}^{3(1)}|AB\rangle \quad (4.68)$$

$$^{(0)}|T\rangle = {}^{3(0)}|AB\rangle \quad (4.69)$$

$$^{(-1)}|T\rangle = {}^{3(-1)}|AB\rangle \quad (4.70)$$

where the VB structures are those defined in equations 4.10–4.15. The energies of the VB structures and the energies of the final states are shown in Figure 4.13. The VB analysis is simpler than the MO analysis because we no longer face the problem of a geometry-dependent MO basis set, but the results are the same.

A comparison of parts A and B in Figure 4.13 again reflects the fact that the energies of the two hole–pair structures, $^1|A^2\rangle$ and $^1|B^2\rangle$, are no longer equal, as they were in the nonpolar case.

An H_2 molecule in the field of a charge. Another artificial way of making the two hydrogen atoms in H_2 differ in electronegativity is to locate a positive charge on the internuclear axis just outside of the molecule. Minimum basis set calculations

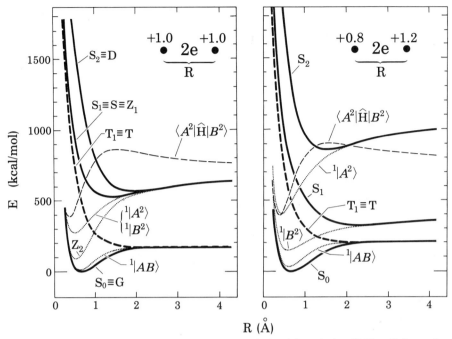

Figure 4.13 Potential energy curves for the dissociation of "polarized" H_2: minimum basis set, full CI (3 × 3 CI model). Singlet (solid lines) and triplet (dashed lines) states. Nuclear charges as shown. The dotted lines show the energies of the dot–dot $^1|AB\rangle$ and hole–pair $^1|A^2\rangle$, $^1|B^2\rangle$ VB structures. The thin dashed lines show the magnitude of the off-diagonal elements of the Hamiltonian.

for the S_0, S_1, S_2, and T states then produce the state energies displayed in Figure 4.14 along with those of an unperturbed H_2 molecule. The differences are clearly the same as those obtained with the artifice of changing the nuclear charges, and both sets of results provide an indication of the behavior of electrons in the various electronic states of a polar bond.

The LiH molecule: MO calculations and Rydberg states. Although the LiH molecule cannot be calculated with anywhere near the accuracy that the H_2 molecule can, it is still sufficiently small that high-quality results are obtainable. The results of such calculations are in good agreement with all experimental observations. The computed energy curves are shown in Figure 4.15A. In spite of the simple nature of the above model calculations for H_2, one has no trouble identifying the four model states (S_0, T_1, S_1, and S_2) of Figures 4.12–4.14 with the more realistic calculations of Figure 4.15A. If the calculation is repeated with a minimum basis set (Figure 4.15B), the excitation energies are too high, but qualitatively the results look exactly like those for the "polarized" H_2 molecule.

4.3.2 Pi Interactions

We shall now briefly consider a model of the polar π bond by inspecting the results for an ethylene molecule with a positive charge located on the C—C bond axis next to it.

166 TWO-CENTER REACTIONS: ONE ACTIVE ORBITAL PER ATOM

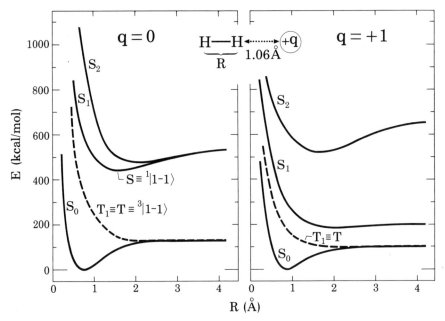

Figure 4.14 Potential energy curves for the dissociation of H_2 in the field of an outside charge $q|e|$ located on the molecular axis at a distance of 1.06 Å from one of the protons. Minimum basis set, full CI (3 × 3 CI model). Singlet (solid lines) and triplet (dashed lines) states.

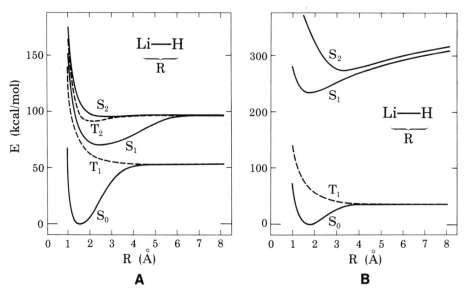

Figure 4.15 Potential energy curves for the dissociation of LiH: (A) *Ab initio* large-scale CI calculation, with Rydberg orbitals; (B) minimum basis set, full valence shell CI (3 × 3 CI model; note the twofold compression of the vertical scale). Singlet (solid lines) and triplet (dashed lines) states.

The MOs have the form given in equations 4.57 and 4.58. Their energies are plotted in Figure 4.16A as a function of the twist angle. The degree of localization of the HOMO on the more electronegative orbital $|B\rangle$ and of the LUMO on the less electronegative $|A\rangle$ is a strong function of the twist angle. In general, what matters is the ratio of the electronegativity difference to the bond strength. When the interaction between the orbitals $|A\rangle$ and $|B\rangle$ vanishes (90° twist), the localization is complete as long as there is any electronegativity difference between $|A\rangle$ and $|B\rangle$.

There is an interesting difference between the polar and nonpolar case. Because of the increased symmetry of the latter, the MO energies are degenerate at 90° twist, while this is not so in the former; that is, the perturbation of the symmetry causes the crossing of the MOs to be avoided.

The energies of the MO configurations, defined once again by equations 4.1–4.6, along with the energies of the final states, are given in Figure 4.17. The energies in parts B and C of Figure 4.17 should be compared with the state energies for the unperturbed ethylene without any adjacent charge in Figure 4.17A, calculated at

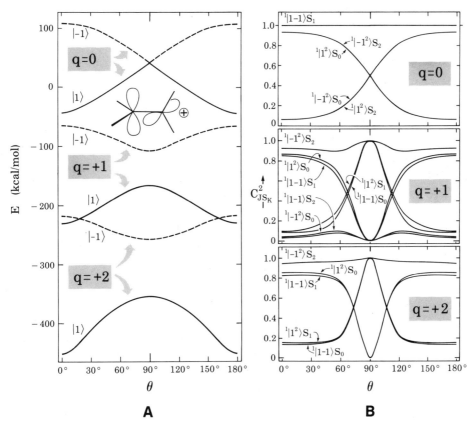

Figure 4.16 Ethylene in the field of an outside charge $q|e|$ located on the C—C axis 1.85 Å from one of the carbon atoms. (A) MO energies (open shell) and (B) weights of configurations in the state wave functions as a function of the twist angle θ. For $q = 2$, the squared coefficients $C^2_{S_2}|1-1\rangle$, $C^2_{S_1}|-1^2\rangle$, $C^2_{S_2}|1^2\rangle$, and $C^2_{S_0}|-1^2\rangle$ are less than 0.04 everywhere and are not shown. See also Figure 4.17.

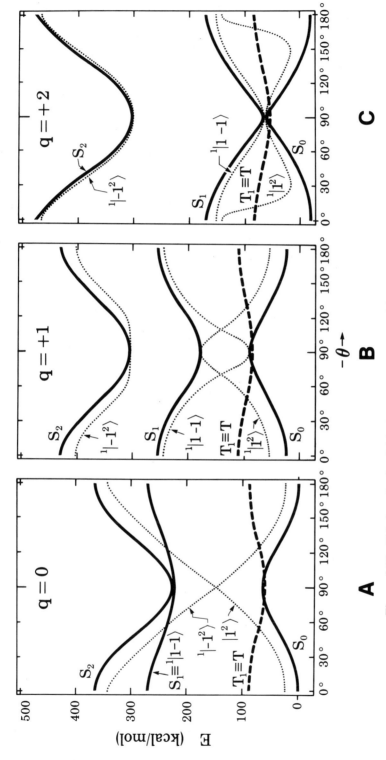

Figure 4.17 Ethylene in the field of an outside charge $q|e|$ located on the C—C axis 1.85 Å from one of the carbon atoms: *ab initio* π CI (3 × 3 CI model). Potential energy curves for the singlet (solid lines) and triplet (dashed lines) states, and configuration energies (dotted lines) as a function of the twist angle θ.

the same level of approximation. The weights with which the MO configurations enter into the final states S_0, S_1, S_2, and T (equations 4.59–4.64 are shown in Figure 4.16B.

The most striking difference between the two sets of curves is the move of the S state from near degeneracy with the D state at orthogonal geometries in the nonpolar case to near degeneracy with the ground state in the polar case. Indeed, for a suitable choice of charge, one obtains an exact degeneracy between the S_0 and S_1 surfaces (Section 4.5.3).

In the nonpolar case, the large gap between the ground state and the excited singlet state at orthogonal geometry was due to relatively strongly avoided crossing between the G and D states, so that an MO crossing did not lead to a crossing at the state level. Naively, one might have expected that an avoidance of a crossing that is already present at the MO level can only increase the separation between the ground state and the excited singlet state, but this is clearly not so. The reason can be seen in the presence of three, rather than just two, close-lying singlet states. The symmetry-breaking perturbation introduced in the polar case causes strong interaction between the S and D states of the nonpolar molecule at orthogonal geometries. This matter is discussed in greater detail in Section 4.5.3.

4.4 BIRADICALS

4.4.1 Fundamentals

A common elementary step in a chemical reaction is the breaking of a bond (e.g., a σ bond by stretching or a π bond by twisting). Since in the ground state the bonds generally correspond to electron pairs, it is therefore useful to investigate the electronic states of a system containing two electrons. In doing so, we shall consider the electrons present in the other bonds as inactive bystanders. This is clearly only an approximation ("frozen core approximation") and will have to be scrutinized subsequently, but it turns out to be quite useful in introducing all of the necessary basic concepts.

The 3 × 3 CI model. A system containing two electrons will, in general, have an infinite number of stationary electronic states available to it. However, only the lowest few will be of interest presently, and it is a reasonable approximation to consider only those states in which the two electrons are only allowed to move in two orbitals. This would correspond to the minimum basis set description of bonds in organic molecules. The resulting model gives rise to three singlet states and one triplet state and is also widely known as the 3 × 3 CI model. Since it really deals with four electronic states, it would perhaps be more correct but also much more cumbersome to refer to it as the 3 × 3 + 1 × 1 CI model in recognition of the block-diagonal nature of the CI matrix.

This model description is actually not only sufficient for the case of two electrons forming a normal bond at its equilibrium length, where two of the three singlet states would typically suffice, but is also sufficient for the description of the electronic states encountered along the dissociation path, and in the limit of complete bond dissociation. In the limit of a broken bond, we can think of the orbitals of a system as two localized and noninteracting atomic or hybrid orbitals. If these two orbitals are of equal energy (degenerate), the broken-bond system will be referred to as a perfect biradical. In such a case, the configuration $^1|A^2\rangle$ in which both electrons are assigned into one of the localized orbitals, $|A\rangle$, has the same energy as the configuration $^1|B^2\rangle$, in which they are both assigned to the other, $|B\rangle$. If the noninteracting localized orbitals differ in energy, we shall refer to the system as a biradicaloid. In such a case, the above two configurations differ in energy, and we shall adopt the convention that the configuration $^1|B^2\rangle$ is lower in energy than the configuration $^1|A^2\rangle$.

We have already indicated in Chapter 2 that biradicaloid geometries are likely points of return from an excited electronic surface to the ground-state surface, and it is therefore only fitting that we should consider electronic states of molecules at such geometries in some detail. We shall see that even the simple model provides quite useful information about the order of energies of the three singlet and one triplet states and, in particular, about the likely instances in which a touching of the S_1 and S_0 surfaces is to be expected.

The one-electron (orbital) basis set. In the following, we shall be dealing with the exact solution of the simple two-electron two-orbital model, in particular with its exact two-electron wave functions. For such wave functions it is immaterial whether they are built from one or another choice of the two orbitals used to define the model. So far, we have considered the localized form of the two orbitals, $|A\rangle$ and $|B\rangle$, but often it is more convenient to work with the delocalized form, $|a\rangle$, $|b\rangle$, and possibly even with some intermediate form. We shall label the general form of the two orbitals $|\mathscr{A}\rangle$, $|\mathscr{B}\rangle$ and develop the model in its general form. Subsequently, $|A\rangle$, $|B\rangle$ or $|a\rangle$, $|b\rangle$ can be substituted for $|\mathscr{A}\rangle$, $|\mathscr{B}\rangle$. Initially, we assume that the orbitals $|\mathscr{A}\rangle$, $|\mathscr{B}\rangle$ are orthogonal; that is,

$$S_{\mathscr{AB}} = \langle \mathscr{A} | \mathscr{B} \rangle = 0 \qquad (4.71)$$

This does not necessarily mean that these orbitals are noninteracting. For instance, an s orbital and a p orbital on the same atom are mutually orthogonal and do not interact through the one-electron part of the Hamiltonian, but two sp^3 orbitals on the same atom are also mutually orthogonal yet do interact.

The two-electron (configuration or structure) basis set without spin adaptation. The Hamiltonian matrix. If we ignore spin for the moment, there are four ways in which two electrons can be accommodated in the two orbitals $|\mathscr{A}\rangle$, $|\mathscr{B}\rangle$. These correspond to the configurations $|\mathscr{A}^2\rangle = |\mathscr{A}(1)\rangle|\mathscr{A}(2)\rangle$, $|\mathscr{B}^2\rangle = |\mathscr{B}(1)\rangle|\mathscr{B}(2)\rangle$, $|\mathscr{AB}\rangle = |\mathscr{A}(1)\rangle|\mathscr{B}(2)\rangle$, and $|\mathscr{BA}\rangle = |\mathscr{B}(1)\rangle|\mathscr{A}(2)\rangle$. The energy of the system can be thought of as being composed of the energies of each of the two electrons, both kinetic and potential. The potential energy is due to the interactions with the nuclei and with all other electrons in the molecule, and to the mutual electrostatic repulsion of the two electrons. The Hamiltonian operator for the two electrons can then be written in the form

$$\hat{H} = \hat{h}(1) + \hat{h}(2) + e^2/r_{12} \qquad (4.72)$$

In order to find the eigenstates of the system, we need to construct and diagonalize the Hamiltonian matrix. The diagonal elements are:

$$\langle \mathscr{A}^2|\hat{H}|\mathscr{A}^2\rangle = 2h_{\mathscr{AA}} + J_{\mathscr{AA}} \tag{4.73}$$

$$\langle \mathscr{B}^2|\hat{H}|\mathscr{B}^2\rangle = 2h_{\mathscr{BB}} + J_{\mathscr{BB}} \tag{4.74}$$

$$\langle \mathscr{AB}|\hat{H}|\mathscr{AB}\rangle = \langle \mathscr{BA}|\hat{H}|\mathscr{BA}\rangle = h_{\mathscr{AA}} + h_{\mathscr{BB}} + J_{\mathscr{AB}} \tag{4.75}$$

In this expression, the diagonal elements of the one-electron Hamiltonian are defined by

$$h_{\mathscr{AA}} = \langle \mathscr{A}|\hat{h}|\mathscr{A}\rangle \tag{4.76}$$

$$h_{\mathscr{BB}} = \langle \mathscr{B}|\hat{h}|\mathscr{B}\rangle \tag{4.77}$$

The Coulomb repulsion integrals $J_{\mathscr{AA}}$, $J_{\mathscr{AB}}$, $J_{\mathscr{BB}}$, as well as the exchange integral $K_{\mathscr{AB}}$, which will appear later, have already been defined (equations 1.4, and 1.5).

Equations 4.73–4.75 have a simple physical interpretation. For instance, the energy of the configuration $|\mathscr{A}^2\rangle$ is equal to twice the one-electron energy plus the time-averaged repulsion of the two electrons in orbital $|\mathscr{A}\rangle$. In a similar fashion, the energy of a configuration in which each orbital is occupied once is given by the sum of the one-electron energies for the two orbitals plus the electron-repulsion integral for the electrostatic interaction of the two time-averaged charge distributions corresponding to them.

Off-diagonal elements can be evaluated similarly, and the whole matrix is found to have the form

$$\begin{array}{l}|\mathscr{A}^2\rangle: \\ |\mathscr{B}^2\rangle: \\ |\mathscr{AB}\rangle: \\ |\mathscr{BA}\rangle: \end{array} \begin{pmatrix} 2h_{\mathscr{AA}} + J_{\mathscr{AA}} & K_{\mathscr{AB}} & h_{\mathscr{AB}} + (\mathscr{AA}|\mathscr{AB}) & h_{\mathscr{AB}} + (\mathscr{AA}|\mathscr{AB}) \\ K_{\mathscr{AB}} & 2h_{\mathscr{BB}} + J_{\mathscr{BB}} & h_{\mathscr{AB}} + (\mathscr{BB}|\mathscr{BA}) & h_{\mathscr{AB}} + (\mathscr{BB}|\mathscr{BA}) \\ h_{\mathscr{AB}} + (\mathscr{AA}|\mathscr{AB}) & h_{\mathscr{AB}} + (\mathscr{BB}|\mathscr{BA}) & h_{\mathscr{AA}} + h_{\mathscr{BB}} + J_{\mathscr{AB}} & K_{\mathscr{AB}} \\ h_{\mathscr{AB}} + (\mathscr{AA}|\mathscr{AB}) & h_{\mathscr{AB}} + (\mathscr{BB}|\mathscr{BA}) & K_{\mathscr{AB}} & h_{\mathscr{AA}} + h_{\mathscr{BB}} + J_{\mathscr{AB}} \end{pmatrix}$$

$$\tag{4.78}$$

Here, $h_{\mathscr{AB}}$ is given by

$$h_{\mathscr{AB}} = \langle \mathscr{A}|\hat{h}|\mathscr{B}\rangle \tag{4.79}$$

and the hybrid integral $(\mathscr{AA}|\mathscr{AB})$ is defined by

$$(\mathscr{AA}|\mathscr{AB}) = \langle \mathscr{A}(1)\mathscr{A}(2)|e^2/r_{12}|\mathscr{A}(1)\mathscr{B}(2)\rangle \tag{4.80}$$

Spin adaptation and the final form of the Hamiltonian matrix. While the wave functions $|\mathscr{A}^2\rangle$ and $|\mathscr{B}^2\rangle$ satisfy Pauli's principle and clearly represent singlet wave functions, the wave functions $|\mathscr{AB}\rangle$ and $|\mathscr{BA}\rangle$ are not antisymmetric with respect to the exchange of the two electrons. The properly adapted combinations are

$(|\mathscr{AB}\rangle \pm |\mathscr{BA}\rangle)/\sqrt{2}$, where the plus sign corresponds to a singlet and the minus sign to a triplet wave function. Since we also wish to treat the configurations $|\mathscr{A}^2\rangle$ and $|\mathscr{B}^2\rangle$ on the same footing, it is useful to introduce the linear combinations $(|\mathscr{A}^2\rangle \pm |\mathscr{B}^2\rangle)/\sqrt{2}$ as well, and we shall write these as $^1|\mathscr{A}^2 \pm \mathscr{B}^2\rangle$. In terms of this new basis set, the Hamiltonian matrix will be block-diagonal because the singlet part will not interact with the triplet part. It has the form

$$\begin{array}{l} ^1|\mathscr{A}^2 - \mathscr{B}^2\rangle: \\ ^1|\mathscr{A}^2 + \mathscr{B}^2\rangle: \\ ^1|\mathscr{AB}\rangle: \\ ^3|\mathscr{AB}\rangle: \end{array} \begin{pmatrix} E(T) + 2K'_{\mathscr{AB}} & \delta_{\mathscr{AB}} & \gamma^-_{\mathscr{AB}} & 0 \\ \delta_{\mathscr{AB}} & E(T) + 2(K'_{\mathscr{AB}} + K_{\mathscr{AB}}) & \gamma_{\mathscr{AB}} & 0 \\ \gamma^-_{\mathscr{AB}} & \gamma_{\mathscr{AB}} & E(T) + 2K_{\mathscr{AB}} & 0 \\ 0 & 0 & 0 & E(T) \end{pmatrix}$$

(4.81)

In writing this Hamiltonian matrix, we have introduced the abbreviations

$$\gamma_{\mathscr{AB}} = 2h_{\mathscr{AB}} + (\mathscr{AA}|\mathscr{AB})^* + (\mathscr{BB}|\mathscr{BA}) \qquad (4.82)$$

$$\gamma^-_{\mathscr{AB}} = (\mathscr{AA}|\mathscr{AB})^* - (\mathscr{BB}|\mathscr{BA}) \qquad (4.83)$$

$$\delta_{\mathscr{AB}} = h_{\mathscr{AA}} - h_{\mathscr{BB}} + (J_{\mathscr{AA}} - J_{\mathscr{BB}})/2 \qquad (4.84)$$

$$K'_{\mathscr{AB}} = [(J_{\mathscr{AA}} + J_{\mathscr{BB}})/2 - J_{\mathscr{AB}}]/2 \qquad (4.85)$$

$$E(T) = h_{\mathscr{AA}} + h_{\mathscr{BB}} + J_{\mathscr{AB}} - K_{\mathscr{AB}} \qquad (4.86)$$

The diagonal energies of the four wave functions of the new basis set, $^1|\mathscr{A}^2 - \mathscr{B}^2\rangle$, $^1|\mathscr{A}^2 + \mathscr{B}^2\rangle$, $^1|\mathscr{AB}\rangle$, and $^3|\mathscr{AB}\rangle$, form a regular pattern around the average energy E_0:

$$E_0 = E(T) + K'_{\mathscr{AB}} + K_{\mathscr{AB}} \qquad (4.87)$$

Since $K_{\mathscr{AB}}$ and $K'_{\mathscr{AB}}$ are always positive, the energy of the triplet state, $E(T)$, is the lowest of the four.

The significance of the off-diagonal mixing elements $\delta_{\mathscr{AB}}$ and $\gamma_{\mathscr{AB}}$ is relatively simple. The quantity $\delta_{\mathscr{AB}}$ is equal to half the energy difference of the configurations $|\mathscr{A}^2\rangle$ and $|\mathscr{B}^2\rangle$, so that it can be viewed as the electronegativity difference between the orbitals $|\mathscr{A}\rangle$ and $|\mathscr{B}\rangle$. We shall choose the orbital label so that $|\mathscr{A}\rangle$ is always the less electronegative orbital, so that $\delta_{\mathscr{AB}}$ is always a positive quantity or zero.

The actual physical meaning of $\delta_{\mathscr{AB}}$ cannot be specified until the choice of the orbitals $\mathscr{A}$ and $\mathscr{B}$ themselves is specified (Section 4.5.1).

The quantity $\gamma_{\mathscr{AB}}$ is a measure of the interaction between orbitals $|\mathscr{A}\rangle$ and $|\mathscr{B}\rangle$. Once again, its physical meaning depends on the choice of the orbitals $|\mathscr{A}\rangle$ and $|\mathscr{B}\rangle$ (Section 4.5.1). If these orbitals are localized, that is, $|\mathscr{A}\rangle = |A\rangle$ and $|\mathscr{B}\rangle = |B\rangle$, it corresponds quite closely to twice the resonance integral between the atomic orbitals $|A\rangle$ and $|B\rangle$ of semiempirical theories.

The physical significance of the quantity $\gamma'_{\mathscr{AB}}$ is less clear-cut. We shall see in the following that it represents a measure of the deviation of our choice of orbitals $|\mathcal{A}\rangle$ and $|\mathcal{B}\rangle$ from both the most and the least localized possible choice. In either of these limiting choices, $\gamma'_{\mathcal{AB}}$ vanishes.

Effect of orbital transformations. As noted above, the final state wave functions will be independent of the choice of the orbitals $|\mathcal{A}\rangle$ and $|\mathcal{B}\rangle$, since the exact solution of the model will be found. Since for practical reasons it is useful to adopt one or another choice of the orbitals at various times, say localized or delocalized, we shall consider this now briefly. A more complete description is given in Appendix II.

It may perhaps be confusing to see the basis set orbitals mixed in a completely arbitrary manner, as we are about to do now, given that the beginning students are taught rules such as "orbitals of different symmetries do not mix." Surely, in a species such as orthogonal ethylene the localized orbitals $|A\rangle$ and $|B\rangle$ of Section 4.2.2 are of different symmetry with respect to the two HCH planes, and yet we propose to work with their linear combinations $|b\rangle = |1\rangle$ and $|a\rangle = |-1\rangle$ of Section 4.2.2. This might be considered excusable because $|A\rangle$ and $|B\rangle$ have identical energies. However, we shall consider the mixtures $|b\rangle$ and $|a\rangle$ even in the presence of a positive charge next to the twisted ethylene (Figure 4.16), where the energies of $|A\rangle$ and $|B\rangle$ are different.

The origin of the apparent discrepancy with what one is taught about MOs is, as usual, semantic. Basis set orbitals of different symmetries indeed do not simultaneously enter into an MO, provided that it is required that the latter be an eigenfunction of a one-electron Hamiltonian operator, such as a Hückel or a Hartree–Fock operator (that is, a canonical MO). However, since we shall be working with the full CI solution of our model problem anyway, there is no particular virtue in using MOs that are such eigenfunctions; for biradicals, it can actually be positively misleading. Once the requirement is dropped, there is no reason why any orbital could not be mixed with any other on the way to the construction of the full CI solution.

The transformations among the various possible choices of orbital pairs can be characterized by a parameter ω which ranges from 0 to $\pi/2$ and can be written as

$$|\mathcal{A}_\omega\rangle = |\mathcal{A}\rangle \cos \omega + |\mathcal{B}\rangle \sin \omega \qquad (4.88)$$

$$|\mathcal{B}_\omega\rangle = -|\mathcal{A}\rangle \sin \omega + |\mathcal{B}\rangle \cos \omega \qquad (4.89)$$

and this can be abbreviated in matrix notation as

$$\begin{pmatrix} |\mathcal{A}_\omega\rangle \\ |\mathcal{B}_\omega\rangle \end{pmatrix} = \begin{pmatrix} \cos \omega & \sin \omega \\ -\sin \omega & \cos \omega \end{pmatrix} \begin{pmatrix} |\mathcal{A}\rangle \\ |\mathcal{B}\rangle \end{pmatrix}, \qquad \omega = [0, \pi/2] \qquad (4.90)$$

Starting with an arbitrary pair of orbitals $|\mathcal{A}\rangle$, $|\mathcal{B}\rangle$, we have an infinite number of choices of the transformation parameter ω, each of them defining a new pair of orbitals $|\mathcal{A}_\omega\rangle$, $|\mathcal{B}_\omega\rangle$. Such a transformation has the following effect on the wave functions of the two-electron basis set:

$$^3|\mathcal{A}_\omega, \mathcal{B}_\omega\rangle = {}^3|\mathcal{A}\mathcal{B}\rangle \qquad (4.91)$$

$$^1|\mathcal{A}_\omega^2 + \mathcal{B}_\omega^2\rangle = {}^1|\mathcal{A}^2 + \mathcal{B}^2\rangle \tag{4.92}$$

$$\begin{pmatrix} ^1|\mathcal{A}_\omega, \mathcal{B}_\omega\rangle \\ ^1|\mathcal{A}_\omega^2 - \mathcal{B}_\omega^2\rangle \end{pmatrix} = \begin{pmatrix} \cos 2\omega & -\sin 2\omega \\ \sin 2\omega & \cos 2\omega \end{pmatrix} \begin{pmatrix} ^1|\mathcal{AB}\rangle \\ ^1|\mathcal{A}^2 - \mathcal{B}^2\rangle \end{pmatrix} \tag{4.93}$$

It is seen that the form of the wave function of the triplet state and the form of the function $^1|\mathcal{A}^2 + \mathcal{B}^2\rangle$ are not changed by the orbital transformation. On the other hand, a wave function that has the form $^1|\mathcal{AB}\rangle$ for one orbital choice will have the form $^1|\mathcal{A}^2 - \mathcal{B}^2\rangle$ for an orbital choice that has been transformed by $\omega = \pi/4$.

The lack of symmetry between the three singlet wave functions is only apparent and is caused by restriction to real orbital transformations. If complex transformations are admitted as well, all three singlet wave function forms can mix upon orbital transformation.

The degree of freedom represented by the choice of the transformation parameter ω can be used to cast the Hamiltonian matrix in a form that contains at least one vanishing off-diagonal matrix element. In the following, we shall choose this to be the quantity $\gamma_{\mathcal{AB}}$. This can be accomplished in two ways. First, $|\mathcal{A}\rangle$ and $|\mathcal{B}\rangle$ can be the most localized possible pair of orbitals, $|A\rangle$, $|B\rangle$. In this case, K_{AB} will be as small as possible and K'_{AB} will be as large as possible. Second, $|\mathcal{A}\rangle$ and $|\mathcal{B}\rangle$ can be the most delocalized possible set of orbitals, $|a\rangle$, $|b\rangle$. In this case, K'_{ab} will be as small as possible and K_{ab} will be as large as possible (the sum of $K_{\mathcal{AB}}$ and $K'_{\mathcal{AB}}$ is constant).

The transformation between $|A\rangle$, $|B\rangle$ and $|a\rangle$, $|b\rangle$ is associated with the choice $\omega = \pi/2$:

$$|a\rangle = \frac{1}{\sqrt{2}}(|A\rangle - |B\rangle) \tag{4.94}$$

$$|b\rangle = \frac{1}{\sqrt{2}}(|A\rangle + |B\rangle) \tag{4.95}$$

The relation between the exchange integrals $K_{\mathcal{AB}}$ defined for the localized orbital set and those defined for the delocalized orbital set is given by

$$K'_{AB} = K_{ab} \tag{4.96}$$

$$K'_{ab} = K_{AB} \tag{4.97}$$

Since $K_{\mathcal{AB}} - J_{\mathcal{AB}}$ is invariant with respect to the transformation of orbitals $\mathcal{A}$ and $\mathcal{B}$, the definitions of the most localized and most delocalized orbitals correspond to the usual convention, according to which the former should minimize and the latter should maximize the interorbital Coulomb repulsion integral $J_{\mathcal{AB}}$. The fact that the exchange and Coulomb repulsion integrals between the two orbitals acquire their minimum values K_{AB} and J_{AB} for the most localized choice and the maximum values K_{ab} and J_{ab} for the most delocalized choice thus makes physical sense, since the localized orbitals $|A\rangle$, $|B\rangle$ avoid each other in space to the maximum degree possible, whereas the opposite is true for $|a\rangle$ and $|b\rangle$. An algorithm that permits the construction of the most localized ($|A\rangle$, $|B\rangle$) and most delocalized ($|a\rangle$, $|b\rangle$) pair of orbitals from an arbitrary starting pair of orbitals is given in Appendix II.

In order to simplify further discussion, we shall assume from now on that the orbitals chosen are the most localized choice $|A\rangle$, $|B\rangle$, although these are often not the ones that come out from computations using standard programs. For this choice, γ_{AB}^- vanishes and the Hamiltonian matrix is characterized by the two off-diagonal elements δ_{AB} and γ_{AB} and by the quantities K_{AB} and K'_{AB} which enter the diagonal elements ($K_{AB} \leq K'_{AB}$).

Special cases. We shall now consider the various possible cases individually. The simplest is that of a perfect biradical in which δ_{AB} and γ_{AB} both vanish. In such a case, the two localized orbitals do not interact and they have equal energies, so that this two-electron two-orbital system clearly is of biradical character. Such systems are discussed in Section 4.4.2.

Systems in which δ_{AB} or γ_{AB} (or both) are nonvanishing also may have some biradical character or may be quite ordinary molecules in which all electrons either participate in bonding or form lone pairs. We shall refer to systems in which δ_{AB} or γ_{AB} or both are different from zero as biradicaloids and shall cover them in Sections 4.5 and 4.6. There are two special kinds of biradicaloids. In homosymmetric biradicaloids, δ_{AB} vanishes but γ_{AB} does not (Section 4.5.2). In heterosymmetric biradicaloids, γ_{AB} vanishes and δ_{AB} does not (Section 4.5.3). In the general case, nonsymmetric biradicaloids, both of these quantities are nonzero (Section 4.5.3).

Natural orbital occupancies. Perhaps the least ambiguous measure of the degree of biradical character of a system are the natural orbital occupancies. Natural orbitals of an electronic system are defined as those orbitals that diagonalize its exact first-order density matrix (the bond order and charge density matrix). These orbitals generally have fractional occupancies n (orbital "charge densities") and vanishing interactions ("bond orders") between orbitals. The natural orbitals of a state can be readily calculated exactly once its exact wave function is known. This will be so in the simple 3×3 model. In general, only an approximate wave function is known, permitting an only approximate calculation of natural orbitals and their occupancies.

In the ground states of ordinary "closed-shell" molecules, all natural orbital occupancies are either close to two, describing a nearly perfectly coupled electron pair ("occupied" natural orbitals), or close to zero ("vacant" natural orbitals). In biradicals and biradicaloids, this is true for all but two of the natural orbitals, whose occupancies are close to one, describing a nearly perfectly uncoupled electron pair ("open-shell"). For additional detail, see Appendixes II and III.

4.4.2 Perfect Biradicals

In a perfect biradical (Appendix II), the localized orbitals $|A\rangle$ and $|B\rangle$ are degenerate ($\delta_{AB} = 0$) and do not interact ($\gamma_{AB} = 0$). Then, the Hamiltonian matrix (4.81) is already diagonal in the representation of the states $^3|AB\rangle$, $^1|AB\rangle$, $^1|A^2 - B^2\rangle$, and $^1|A^2 + B^2\rangle$, and the state energies are given by the diagonal elements of the matrix (4.98). Figure 4.18 shows the resulting energy pattern of the four states and the associated wave functions expressed in both sets of starting orbitals. A few examples of perfect biradicals are given in Figure 4.19A.

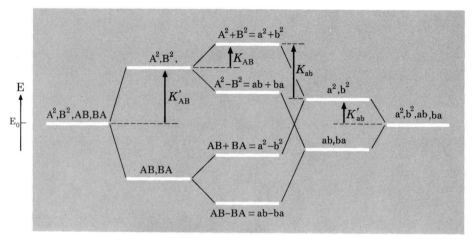

Figure 4.18 Wave functions and energies of a perfect biradical constructed from the most localized (A,B) and most delocalized (a,b) set of degenerate orbitals.

$$
\begin{array}{l}
^1|A^2 - B^2\rangle: \\
^1|A^2 + B^2\rangle: \\
^1|AB\rangle: \\
^3|AB\rangle:
\end{array}
\begin{pmatrix}
E(T) + 2K'_{AB} & 0 & 0 & 0 \\
0 & E(T) + 2(K'_{AB} + K_{AB}) & 0 & 0 \\
0 & 0 & E(T) + 2K_{AB} & 0 \\
0 & 0 & 0 & E(T)
\end{pmatrix}
$$

(4.98)

The triplet $^3|AB\rangle$ is the lowest of the four states. The lowest of the three singlets is the $^1|AB\rangle$ state, located $2K_{AB}$ above the triplet. If one wished to use the delocalized orbitals, $|a\rangle$, $|b\rangle$, an equivalent way of writing this singlet wave function would be $^1|a^2 - b^2\rangle$. The average energy of these lower two states is $E_0 - K'_{AB}$. The remaining two singlets are located at higher energies. Their average energy is $E_0 + K'_{AB}$, they are again separated by $2K_{AB}$, and the lower of them has the wave function $^1|A^2 - B^2\rangle$. In terms of delocalized orbitals, this can be written equally well as $^1|ab\rangle$. The highest energy singlet has the wave function $^1|A^2 + B^2\rangle$, which can also be written as $^1|a^2 + b^2\rangle$.

As noted above, $K'_{AB} \geq K_{AB} \geq 0$, since we have chosen to work in the localized orbital basis, $|A\rangle$, $|B\rangle$. In the general case of a perfect biradical, $K_{AB} \neq 0$ and $K'_{AB} \neq K_{AB}$. There are, however, two important limiting categories of perfect biradicals:

(i) In *pair biradicals*, $K_{AB} = 0$. This condition can be strictly satisfied only if the separation between the localized orbitals A, B is infinite so that the biradical consists of a pair of distant radicals (e.g., a completely dissociated H_2 molecule). In practice, the condition can be nearly satisfied when the distance between the centers of the localized orbitals $|A\rangle$, $|B\rangle$ is only a little more than 1 Å [e.g., in Figure 4.19A, orthogonally twisted ethylene (**1**), trimethylene (**2**), tetramethylene (**3**), tetramethyleneethane (**4**), but also square cyclobutadiene (**5**) and the isoelec-

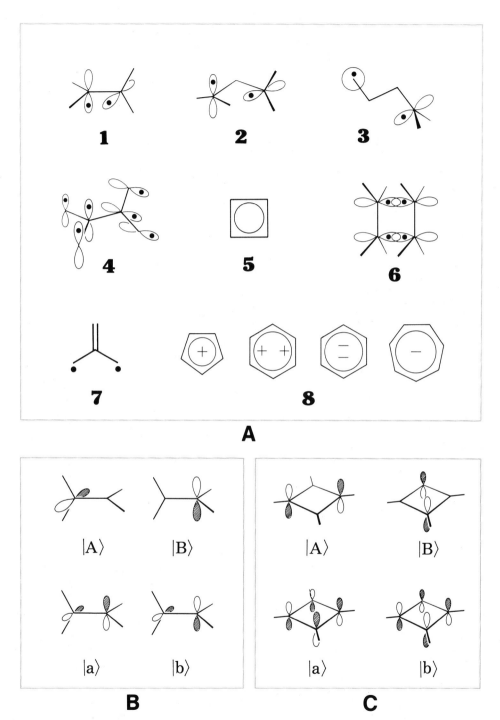

Figure 4.19 (**A**) Examples of perfect biradicals. (**B**) Localized (A,B) and delocalized (a,b) nonbonding orbital set for twisted ethylene. (**C**) Localized (A,B) and delocalized (a,b) nonbonding orbital set for cyclobutadiene.

tronic $2_s + 2_s$ cycloaddition transition state (**6**)]. In general, the pericyclic transition states of ground-state "forbidden" processes are pair biradicals if the cyclic orbital array contains as many electrons as it contains interacting orbitals (for olefin $2_s + 2_s$ cycloaddition, both numbers equal 4). This is of considerable importance for photochemical pericyclic processes. According to the simple model we are considering presently, in pair biradicals the lowest singlet S_0 is degenerate with the triplet state T, and the two upper singlets S_1 and S_2 are also degenerate. We shall see in Chapter 5 that in pericyclic orbital arrays this is only approximately so, but the trend still imposes important consequences.

(ii) In *axial biradicals*, $K'_{AB} = K_{AB}$. This condition can be imposed by symmetry in the presence of a threefold or higher order rotational symmetry axis, e.g., in O, O_2, linear CH_2, trimethylenemethane (**7** in Figure 4.19A), or in cyclic molecules with $4N$ π electrons in a charged perimeter which has the shape of a regular polygon (**8** in Figure 4.19A). In general, the pericyclic transition states of ground-state "forbidden" processes are axial biradicals if the number of orbitals in the cyclic array does not equal the number of electrons they hold ("charged perimeter").

Within the present model, in axial biradicals the singlet–triplet splitting $2K_{AB}$ is large so that T_1 lies far below S_0, and S_0 and S_1 are degenerate. This degeneracy is model-independent if imposed by symmetry. The large T–S_0 and S_1–S_2 separations have important consequences for photochemical pericyclic processes in ions and the isoconjugate heterocycles.

In a perfect biradical, any pair of orthogonal real orbitals $|\mathscr{A}\rangle$, $|\mathscr{B}\rangle$ represents the natural orbitals for the four states of the two-orbital two-electron model, S_0, S_1, S_2, and T. The density and the occupation numbers for both orbitals are equal to unity in each of the states. This displays quite clearly the "open-shell" or "perfect-biradical" nature of the four states, which is totally independent of the choice of the orbital basis set, $|\mathscr{A}\rangle$, $|\mathscr{B}\rangle$. The open-shell character of all four wave functions, obvious from this consideration, is less clear when one merely inspects the form of the wave functions as given in Figure 4.18 and on the left of the matrix 4.98.

It is interesting for a photochemist to consider the conditions under which the S_0 and S_1 surfaces touch in a perfect biradical. We have seen from the above that within the two-electron two-orbital model the condition for this degeneracy in a perfect biradical is $K_{AB} = K'_{AB}$.

In nonlinear molecules, Jahn–Teller distortion will prevent the equilibrium geometry in the lowest singlet state S_0 from coinciding exactly with the high-symmetry geometry needed for the perfect biradical to be axial, but we are primarily interested in the S_1 surface, for which the high-symmetry geometry will just be very favorable.

In most perfect biradicals the equality of K_{AB} and K'_{AB} is not forced by symmetry, so that S_0 and S_1 are split by $2(K'_{AB} - K_{AB})$, which typically amounts to several dozen kilocalories per mole (e.g., in orthogonally twisted ethylene and the trimethylene and tetramethylene biradicals).

Although the S_1 surface still lies relatively low in these perfect biradicals, it does not even come close to touching S_0. With regard to the S_1–S_0 gap, the worst cases are pair biradicals formed by stretching a single bond to infinite length. In these the gap is equal to the repulsion integral J_{AB}, typically hundreds of kilocalories per mole.

As we shall see below (see also Appendix II), the degeneracy of the lowest two singlet states is generally far more readily reached or at least approached in biradicaloids than in perfect biradicals. On the other hand, in perfect biradicals, a triplet has the best chance to follow Hand's rule and to represent the ground state of the system.

4.5 BIRADICALOIDS

4.5.1 Fundamentals

In a biradicaloid two-electron two-orbital system the two localized nonbonding orbitals $|A\rangle$, $|B\rangle$ either interact ($\gamma_{AB} \neq 0$) or differ in energy ($\delta_{AB} \neq 0$), or both. Thus, at least one of the off-diagonal matrix elements, γ_{AB} and δ_{AB}, is different from zero. Examples of biradicaloids related to ethylene are shown in Figure 4.20.

The wave functions for the states of the system and their energies are obtained by the diagonalization of the Hamiltonian matrix. The wave functions will have a

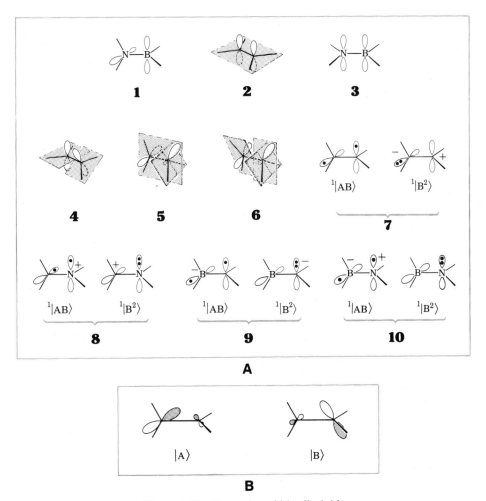

Figure 4.20 Examples of biradicaloids.

form in which all three basis functions, $^1|\mathscr{AB}\rangle$, $^1|\mathscr{A}^2 - \mathscr{B}^2\rangle$, and $^1|\mathscr{A}^2 + \mathscr{B}^2\rangle$ are mixed:

$$|S_i\rangle = C_{i,-}\,^1|\mathscr{A}^2 - \mathscr{B}^2\rangle + C_{i,+}\,^1|\mathscr{A}^2 + \mathscr{B}^2\rangle + C_{i,0}\,^1|\mathscr{AB}\rangle, \quad i = 0, 1, 2$$

(4.99)

The coefficients $C_{i,+}$ are invariant with respect to real orbital basis set transformations, since the function $^1|\mathscr{A}^2 + \mathscr{B}^2\rangle$ itself is, but the coefficients $C_{i,0}$ and $C_{i,-}$ are not. The natural orbitals $|\mathscr{A}_\omega\rangle$, $|\mathscr{B}_\omega\rangle$ are obtained from the localized orbitals $|A\rangle$, $|B\rangle$ by a transformation characterized by the parameter ω (cf. equations 4.88–4.90), whose actual value can be computed by diagonalizing the Hamiltonian matrix 4.81. Their occupation numbers $n_{\mathscr{A}_\omega}$ and $n_{\mathscr{B}_\omega}$ are (see Appendix II)

$$n_{\mathscr{A}_\omega} = 1 - 2C_{i,+}\sqrt{1 - C_{i,+}^2}$$
$$n_{\mathscr{B}_\omega} = 1 + 2C_{i,+}\sqrt{1 - C_{i,+}^2}$$

(4.100)

An infinitely strongly perturbed biradicaloid. In the limiting case of an infinitely strong perturbation $\delta_{AB}^2 + \gamma_{AB}^2$, the value of the parameter ω can be obtained simply. In this limit,

$$C_{0,+} = 1/\sqrt{2}, \ C_{0,-} = -1\sqrt{2}, \ C_{0,0} = 0, \ n_{\mathscr{A}_\omega} = 0, \ n_{\mathscr{B}_\omega} = 2, \ |S_0\rangle = {}^1|\mathscr{B}_\omega^2\rangle$$

$$\omega = (1/2)\tan^{-1}(\gamma_{AB}/\delta_{AB})$$

(4.101)

While the total strength of the perturbation, $\delta_{AB}^2 + \gamma_{AB}^2$, is invariant to an orbital transformation, we see that the form of the doubly occupied orbital $|\mathscr{B}_\omega\rangle$ and of the vacant orbital $|\mathscr{A}_\omega\rangle$ is determined by the ratio of γ_{AB} to δ_{AB} in the localized basis. If $\gamma_{AB} = 0$ (no bonding interaction between localized orbitals $|A\rangle$ and $|B\rangle$), we have $\omega = 0$ and $|\mathscr{B}_\omega\rangle = |B\rangle$ and $|\mathscr{A}_\omega\rangle = |A\rangle$, so that in S_0 both electrons are in the more stable localized orbital $|B\rangle$ and form a more or less nonbonding electron pair, normally referred to as a "lone pair" if orbital $|B\rangle$ is mostly localized on a single atom. This situation is approached in orthogonally twisted aminoborane (**1** in Figure 4.20A). If $\delta_{AB} = 0$ (no electronegativity difference between orbitals $|A\rangle$ and $|B\rangle$) and $\gamma_{AB} < 0$ as usual, we have $\omega = -\pi/4$ and $|\mathscr{B}_\omega\rangle = |b\rangle$, $|\mathscr{A}_\omega\rangle = |a\rangle$, so that in S_0 both electrons are in the bonding perfectly delocalized orbital $|b\rangle$ and form a nonpolar bond. This situation is approached in planar ethylene (**2** in Figure 4.20A). In a general case, $|\mathscr{A}_\omega\rangle$ and $|\mathscr{B}_\omega\rangle$ are partially delocalized on both centers and the two electrons in $|\mathscr{B}_\omega\rangle$ form a polar bond. This situation is approached in planar aminoborane (**3** in Figure 4.20A).

Real biradicaloids. In reality, neither the bonding perturbation γ_{AB} nor the electronegativity difference δ_{AB} can be infinitely large. The occupation number of the "occupied" orbital $|\mathscr{B}_\omega\rangle$ is then somewhat less than two ($n_{\mathscr{B}_\omega} < 2$), and that of the "vacant" orbital $|\mathscr{A}_\omega\rangle$ is somewhat larger than zero ($n_{\mathscr{A}_\omega} > 0$). This is the actual situation in cases **1–3** in Figure 4.20A.

The gradual change from the perfect-biradical situation ($n_{\mathscr{A}} = n_{\mathscr{B}} = 1$) to the ordinary closed-shell bond or lone-pair situation ($n_{\mathscr{B}_\omega} \simeq 2$, $n_{\mathscr{A}_\omega} \simeq 0$) illustrates very nicely the continuous nature of the conversion of the biradical into the biradicaloid, and eventually into an ordinary molecule, by the introduction of a suitable perturbation.

Orbital transformations and the perturbation parameters γ and δ. We have already noted that $\delta_{\mathscr{AB}}$ is a measure of the energy difference of orbitals $|\mathscr{A}\rangle$ and $|\mathscr{B}\rangle$, while $\gamma_{\mathscr{AB}}$ is a measure of the degree to which they interact. The detailed physical significance of these quantities depends upon the choice of the orbitals $|\mathscr{A}\rangle$ and $|\mathscr{B}\rangle$. While $\delta^2_{\mathscr{AB}} + \gamma^2_{\mathscr{AB}}$ is invariant to this choice, for any perturbation there is a choice that makes $\delta_{\mathscr{AB}}$ vanish and another choice that makes $\gamma_{\mathscr{AB}}$ vanish (Appendix II). In general, these choices correspond neither to the localized orbitals $|A\rangle$, $|B\rangle$ nor to the delocalized orbitals $|a\rangle$, $|b\rangle$.

Since we wish to work with the localized orbital set $|A\rangle$, $|B\rangle$, or sometimes with the delocalized set $|a\rangle$, $|b\rangle$, we have to accept the existence of two independent perturbation prameters, δ_{AB} and γ_{AB}, but matters simplify in that for these two orbital set choices, $\gamma^-_{AB} = \gamma^-_{ab} = 0$ in the Hamiltonian matrix. The descriptions provided by the two orbital choices are related by

$$\delta_{AB} = \gamma_{ab}$$
$$\gamma_{AB} = \delta_{ab}$$
(4.102)

An illustrative example. A simple illustration of the orbitals of a perfect biradical is provided by those of orthogonally twisted ethylene (Figure 4.19B), and another by those of square cyclobutadiene (Figure 4.19C). Both twisted ethylene and cyclobutadiene are perfect biradicals, but a suitable perturbation converts them into biradicaloids.

Two types of perturbation of the perfect biradical, twisted ethylene, will be considered: return to planarity and pyramidalization on one of the carbon atoms. The localized orbital set, $|A\rangle$, $|B\rangle$, is shown on the top and the delocalized one, $|a\rangle$, $|b\rangle$, on the bottom in Figure 4.19B at the perfect biradical geometry. Symmetry dictates that the energies of the two localized orbitals be equal, $\delta_{AB} = 0$, and that the orbitals cannot interact, $\gamma_{AB} = 0$. The same is true of the delocalized orbitals, $\delta_{ab} = \gamma_{ab} = 0$. When the orthogonal ethylene is distorted towards planarity (formula **4** in Figure 4.20A), the energies of the localized orbitals remain equal by symmetry, $\delta_{AB} = 0$, but their interaction no longer vanishes, $\gamma_{AB} \neq 0$. The energies of the two delocalized orbitals are no longer equal, since one is now bonding and one antibonding, $\delta_{ab} \neq 0$, but symmetry demands that they cannot interact, $\gamma_{ab} = 0$.

In contrast, when one of the carbon atoms in orthogonal ethylene is pyramidalized and the molecule is allowed to remain orthogonally twisted (formula **5** in Figure 4.20A), the energies of the two localized orbitals begin to differ, $\delta_{AB} \neq 0$, because one of them is no longer a pure p orbital but is, instead, a hybrid with some s character. However, symmetry still prevents them from interacting, $\gamma_{AB} = 0$. The energies of the delocalized orbitals remain the same because both are equally distributed over both carbon atoms, $\delta_{ab} = 0$. However, $\gamma_{ab} \neq 0$ because $|a\rangle$ and $|b\rangle$ are clearly distinct from the usual canonical MOs, which have unequal coefficients on the two carbon atoms.

Upon simultaneous pyramidalization and deviation from orthogonality (formula **6** in Figure 4.20A), δ_{AB}, γ_{AB}, δ_{ab}, and γ_{ab} all become nonzero.

Along similar lines, a consideration of the behavior of the orbitals $|A\rangle$, $|B\rangle$ of cyclobutadiene (Figure 4.19C) shows that the introduction of a heteroatom such as an aza nitrogen into a single position or into positions 1 and 3 in the square geometry ($\delta_{AB} \neq 0$) produces a heterosymmetric biradicaloid.

In the perfect biradical, orthogonally twisted unpyramidalized ethylene and even in the nonorthogonal but unpyramidalized form, the delocalized orbitals $|a\rangle$, $|b\rangle$ are identical to the usual canonical MOs. However, at singly pyramidalized geometries, the canonical MOs correspond to the localized orbitals, $|A\rangle$, $|B\rangle$, if the geometry is still orthogonal and correspond to neither $|a\rangle$, $|b\rangle$ nor $|A\rangle$, $|B\rangle$ if it is not. The same is true of the canonical orbitals of an azacyclobutadiene with alternating bond lengths. Since standard computer programs normally produce canonical MOs, in nonsymmetric biradicaloids the algorithm described in the Appendix II is needed to convert them to the set $|A\rangle$, $|B\rangle$ or $|a\rangle$, $|b\rangle$.

4.5.2 Homosymmetric Biradicaloids

When two orthogonal orbitals on two different atoms interact through the one-electron part of the Hamiltonian ($\gamma_{AB} \neq 0$), they generally cannot be perfectly localized. When they reside on two neighboring atoms, each of them is localized mostly on one and partly on the other partner atom. An example is the $|A\rangle$, $|B\rangle$ orbital set on partially twisted ethylene (Figure 4.20B).

The wave functions and their energies. The singlet part of the Hamiltonian matrix of the homosymmetric biradicaloid has the block-diagonal form

$$\begin{array}{c} {}^1|A^2 - B^2\rangle: \\ {}^1|A^2 + B^2\rangle: \\ {}^1|AB\rangle \quad : \end{array} \begin{pmatrix} E(T) + 2K'_{AB} & 0 & 0 \\ 0 & E(T) + 2(K'_{AB} + K_{AB}) & \gamma_{AB} \\ 0 & \gamma_{AB} & E(T) + 2K_{AB} \end{pmatrix} \quad (4.103)$$

The antisymmetric wave function ${}^1|A^2 - B^2\rangle$ remains an eigenstate as it was in the perfect biradical. Since it corresponds to ${}^1|ab\rangle$, it can be referred to as the "singly excited" S_1 state.

The symmetric functions ${}^1|A^2 + B^2\rangle$ and ${}^1|AB\rangle$ are mixed by the off-diagonal element γ_{AB} and produce the ground state S_0 and the "doubly excited" state S_2. If the phases of the orbitals $|A\rangle$ and $|B\rangle$ have been chosen so as to make the overlap S_{AB} positive, the "resonance integral" γ_{AB} will normally be negative if the interaction is primarily directly through space. The sign of γ_{AB} can be the same as the sign of S_{AB} when the interaction between $|A\rangle$ and $|B\rangle$ is primarily mediated through bonds. We shall see an example of such a situation below (trimethylene, Figure 4.21). For γ_{AB} negative, the functions ${}^1|AB\rangle$ and ${}^1|A^2 + B^2\rangle$ are mixed in-phase in S_0 and out-of-phase in S_2; for positive γ_{AB} the opposite holds.

The expressions for the energies and wave functions of the singlet states are

	Wave function	Energy
S_2:	$\cos \alpha \, {}^1|A^2 + B^2\rangle + \sin \alpha \, {}^1|AB\rangle$	$K'_{AB} + 2K_{AB} + \sqrt{K'^2_{AB} + \gamma^2_{AB}}$
S_1:	${}^1|A^2 - B^2\rangle$	$2K'_{AB}$
S_0:	$-\sin \alpha \, {}^1|A^2 + B^2\rangle + \cos \alpha \, {}^1|AB\rangle$	$K'_{AB} + 2K_{AB} - \sqrt{K'^2_{AB} + \gamma^2_{AB}}$

(4.104)

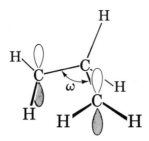

Figure 4.21 The trimethylene biradical.

$$\alpha = (1/2)\tan^{-1}(\gamma_{AB}/K'_{AB}) \tag{4.105}$$

Figure 4.22A demonstrates the dependence of the state energies on the strength of the perturbation γ_{AB} for a particular choice of K_{AB} and K'_{AB}.

The strength of the perturbation is given by the ratio γ_{AB}/K'_{AB}. For $\gamma_{AB} = 0$, the states S_0, S_1, and S_2 obviously are those of the perfect biradical. In the limit of very large negative γ_{AB}, the state S_0 goes to $(^1|AB\rangle + {}^1|A^2 + B^2\rangle)/\sqrt{2}$, and this can be written in terms of delocalized orbitals as $^1|b^2\rangle$. The simplest description of the ground state of an ordinary molecule, with both electrons occupying the bonding orbital, is then correct.

The S_0–S_1 gap can only be increased with respect to a perfect biradical when γ_{AB} is introduced as a perturbation. For instance, it increases rapidly when orthogonally twisted ethylene is brought back towards planarity or when two radicals are brought into interaction with sigma-type overlap of their orbitals.

As the absolute value $|\gamma_{AB}|$ increases, S_0 is stabilized relative to T. Their energies are equal when $|\gamma_{AB}| = 2\sqrt{K_{AB}(K_{AB} + K'_{AB})}$. For even larger absolute values of $|\gamma_{AB}|$, S_0 lies below T.

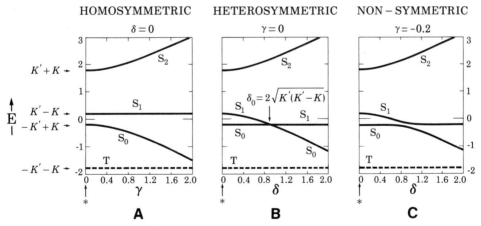

Figure 4.22 3 × 3 CI model: Generation of a biradicaloid by perturbation of a perfect biradical (indicated by asterisks). Singlet (solid lines) and triplet (dashed lines) state energies for examples of homosymmetric (A), heterosymmetric (B), and nonsymmetric (C) biradicaloids. The subscripts A, B on K_{AB}, K'_{AB}, δ_{AB} and γ_{AB} have been dropped to improve legibility. Reproduced by permission from V. Bonačić-Koutecký, J. Koutecký, and J. Michl, *Angew. Chem. Int. Ed. Engl.* **26**, 170 (1987).

Natural orbital occupancies. In all four states of the homosymmetric biradicaloid the natural orbitals coincide with the delocalized orbitals $|a\rangle$ and $|b\rangle$. Expressions for their occupation numbers are given in Appendix II. In the S_1 and T_1 states, they are equal to unity in both orbitals at all values of γ_{AB}, $n_a = n_b = 1$, demonstrating clearly their open-shell nature. In the S_0 and S_2 states, they are equal to unity in the limit of perfect biradical, $\gamma_{AB} = 0$, and then gradually reach the values $n_a \simeq 0$, $n_b \simeq 2$ as γ_{AB}/K'_{AB} increases.

An illustrative example. An instructive example of the effect of the variation of γ_{AB} on the S_0–T gap are π,π-1,3-biradicals. In these, γ_{AB} contains two opposed contributions that nearly cancel and whose relative weight depends critically on the CCC valence angle. With the orbital phase choice indicated in Figure 4.21, the negative contribution to γ_{AB} is due to direct through-space interaction, and the positive one is due to interaction through the hyperconjugating central CH$_2$ group. The former is fairly sensitive to the CCC valence angle, becoming more negative as the angle decreases. The latter does not depend much on this angle. As a results, γ_{AB} is positive for large CCC angles and negative for small ones, and there is a small range of CCC angles where γ_{AB} is so close to zero that the system is a nearly perfect biradical and possesses a triplet ground state.

Once S_0 drops significantly below T_1, the two-electron two-orbital system would normally no longer be called a biradicaloid, at least not in its ground state.

4.5.3 Heterosymmetric Biradicaloids

The wave functions and their energies. In this case, the singlet part of the Hamiltonian matrix has the block-diagonal form

$$\begin{array}{l} {}^1|A^2 - B^2\rangle: \\ {}^1|A^2 + B^2\rangle: \\ {}^1|AB\rangle \quad : \end{array} \begin{pmatrix} E(T) + 2K'_{AB} & \delta_{AB} & 0 \\ \delta_{AB} & E(T) + 2(K'_{AB} + K_{AB}) & 0 \\ 0 & 0 & E(T) + 2K_{AB} \end{pmatrix} \quad (4.106)$$

In this instance, $^1|AB\rangle$ remains as an eigenstate and corresponds to the dot–dot, or "covalent", structure with one electron in each of the two localized orbitals. The hole–pair unpolarized functions $^1|A^2 - B^2\rangle$ and $^1|A^2 + B^2\rangle$, with equal electron densities in orbitals $|A\rangle$ and $|B\rangle$, mix out-of-phase to produce a lower energy polarized state with excess electron density in orbital $|B\rangle$ and in-phase to produce an oppositely polarized higher energy state. As shown in formulas **7–10** in Figure 4.20A, the dot–dot $^1|AB\rangle$ structures and the hole–pair $^1|B^2\rangle$ structures differ in the attribution of formal charges. Depending on the nature of the atoms in the π bond (e.g., B, C, or N), either one can be "zwitterionic." For this reason, we prefer the unambiguous notation "dot–dot" and "hole–pair."

The expressions for the energies and wave functions of the three states are

	Wave function	Energy
S_2 :	$|2\rangle = \cos\beta \, {}^1|A^2 + B^2\rangle + \sin\beta \, {}^1|A^2 - B^2\rangle$	$2K'_{AB} + K_{AB} + \sqrt{K'^2_{AB} + \delta^2_{AB}}$
S_1 or S_0:	$|1\rangle = -\sin\beta \, {}^1|A^2 + B^2\rangle + \cos\beta \, {}^1|A^2 - B^2\rangle$	$2K'_{AB} + K_{AB} - \sqrt{K'^2_{AB} + \delta^2_{AB}}$
S_0 or S_1:	$^1|AB\rangle$	$2K_{AB}$

(4.107)

$$\beta = (1/2)\tan^{-1}(\delta_{AB}/K_{AB}) \qquad (4.108)$$

Figure 4.22B displays the state energies as a function of the perturbation δ_{AB} for a particular choice of K_{AB} and K'_{AB}. In addition to the wave function $^1|AB\rangle$, which represents either S_0 or S_1, we have the two functions $|2\rangle$ and $|1\rangle$, where the latter represents either S_1 or S_0. These two functions originate from the mixing of $^1|A^2 - B^2\rangle$ and $^1|A^2 + B^2\rangle$ as a result of the perturbation by δ_{AB}. For $\delta_{AB} = 0$ the state wave functions are those of a perfect biradical, and the configurations $^1|A^2 - B^2\rangle$ and $^1|A^2 + B^2\rangle$ are not mixed. This situation obtains with orthogonal ethylene (**7** in Figure 4.20A). For a very large δ_{AB}, the state $|1\rangle$ goes from $^1|A^2 - B^2\rangle$ to $(^1|A^2 - B^2\rangle - {}^1|A^2 + B^2\rangle)/\sqrt{2} = {}^1|B^2\rangle$, while the state $|2\rangle$ goes from $^1|A^2 + B^2\rangle$ to $^1|A^2\rangle$. This process can be described as a polarization of the unpolarized states $^1|A^2 + B^2\rangle$ and $^1|A^2 - B^2\rangle$ into the polar configurations containing both electrons in one or the other localized orbital. In this limit, the simple MO description of the ground state of an ordinary molecule is correct, with both electrons occupying the lower energy orbital $|B\rangle$. The simple MO and VB descriptions are then identical. This situation is approached in orthogonal aminoborane (**10** in Figure 4.20A). Note that only the ratio δ_{AB}/K_{AB} counts; that is, the magnitude of δ_{AB} required for a given degree of polarization increases as K_{AB} increases.

Natural orbital occupancies. In all four electronic states of the heterosymmetric biradicaloid, the natural orbitals coincide with the localized orbitals $|A\rangle$, $|B\rangle$. The expressions for the occupation numbers are given in Appendix II. For the $^1|AB\rangle$ and $^3|AB\rangle$ states, they are $n_A = n_B = 1$ for any δ_{AB}, and these states are of "open-shell" nature. For the states $|1\rangle$ and $|2\rangle$, they go from $n_A = n_B = 1$ in the perfect biradical to $n_A \simeq 0$, $n_B \simeq 2$ as the perturbation δ_{AB}/K_{AB} increases (see the discussion of "charge-transfer" biradicaloids and "TICT" states in Section 6.4.2).

Singlet state: Degeneracy and polarizability. In biradicals with a small K_{AB} and therefore with nearly degenerate S_1 and S_2 states (approximately "pair" biradicals), such as a 90° twisted ethylene, even a weak polarizing perturbation δ_{AB} such as pyramidalization on one center or the presence of a nearby charge causes an essentially complete polarization of the functions $^1|A^2 + B^2\rangle$ and $^1|A^2 - B^2\rangle$ into the hole–pair structures $^1|A^2\rangle$ and $^1|B^2\rangle$. A very high polarizability of the excited state results; this has been long recognized, for example, for twisted ethylene in its S_1 state ("sudden polarization").

In biradicals with large values of K_{AB}, a much larger orbital energy difference δ_{AB} between orbitals $|A\rangle$ and $|B\rangle$ will have to be introduced in order to produce a similar degree of polarization in the excited state. In an axial biradical, $K_{AB} = K'_{AB}$, and for the degenerate S_0 and S_1 states, the polarization of $^1|A^2 + B^2\rangle$ and $^1|A^2 - B^2\rangle$ into $^1|A^2\rangle$ and $^1|B^2\rangle$ by the action of a perturbation δ_{AB} tends to be incomplete even if the energies of the orbitals $|A\rangle$ and $|B\rangle$ are quite different. Thus the bending of linear carbene leaves the 2p orbital $|A\rangle$ intact but stabilizes $|B\rangle$ very much by giving it some s character, and yet numerical calculations show that $|A^2\rangle$ and $|B^2\rangle$ are still extensively mixed in the description of the S_0 state of even strongly bent carbene.

S_0–S_1 degeneracy. Since a suitable choice of δ_{AB} will cause S_0 and S_1 to be degenerate starting with any general perfect biradical whose K'_{AB} is larger than K_{AB}, it is possible to design a large number of systems in which S_0 and S_1 surfaces touch or nearly touch. The twisted structures **8** and **9** in Figure 4.20A represent simple examples.

The point at which the S_0–S_1 degeneracy occurs is given by

$$\delta_0 = 2\sqrt{K'_{AB}(K'_{AB} - K_{AB})} \qquad (4.109)$$

and this expression clearly displays the limiting cases of a perturbed axial biradical in which $K_{AB} = K'_{AB}$ and $\delta_0 = 0$ and of a perturbed pair biradical in which $K_{AB} = 0$ and $\delta_0 = 2K'_{AB} = J_{AA}$.

We shall distinguish three classes of heterosymmetric biradicaloids: (i) weakly heterosymmetric ones in which δ_{AB} is smaller than δ_0; (ii) strongly heterosymmetric ones in which δ_{AB} is larger than δ_0; and (iii) critically heterosymmetric ones in which $\delta_{AB} = \delta_0$ (Figure 4.22B).

Weak heterosymmetry. If $\delta_{AB} < \delta_0$, the lowest singlet S_0 is represented by $^1|AB\rangle$ similarly as in a perfect biradical, and S_1 is represented by a mixture of the polarized hole–pair structures $^1|A^2\rangle$ and $^1|B^2\rangle$, with the latter dominating. This situation is usually encountered in uncharged biradicaloids, where the covalent structure $^1|AB\rangle$ involves no formal separation of charges, while the hole–pair structures $^1|A^2\rangle$ and $^2|B^2\rangle$ do. Examples are 90° twisted unsymmetrical double bonds such as orthogonally twisted propene or twisted ethylene pyramidalized at one carbon atom (**5** in Figure 4.20A).

Strong heterosymmetry. If $\delta_{AB} > \delta_0$, $^1|AB\rangle$ describes S_1, and the lower singlet S_0 is represented by a mixture of the hole–pair structures $^1|A^2\rangle$ and $^1|B^2\rangle$, with the latter much more stable than the former and usually strongly dominating. If the heterosymmetry is very pronounced, that is, if δ_{AB} is very large, both electrons are thus kept virtually exclusively in orbital $|B\rangle$ in the ground state, often as a lone pair. Once δ_{AB} has exceeded the value $2\sqrt{K'_{AB}(K'_{AB} + K_{AB})}$, S_0 lies below T—often far below it. For both reasons, strongly heterosymmetric biradicaloids are normally not considered to be biradicaloids at all, at least not in the ground state. If the hole–pair structure $^1|B^2\rangle$ involves formal separation of charge, such species are usually referred to as zwitterions or ion pairs, with a positive charge on center A and a negative charge on center B.

However, really large δ_{AB} values are normally reached in systems in which it is the covalent $^1|AB\rangle$ structure that carries separated formal charges, negative on A and positive on B, while the polarized structure $^1|B^2\rangle$ does not. It is then the excited state S_1 that has both the charge separation and the spatially separated odd electrons and is sometimes referred to as a charge-transfer biradicaloid. Examples of strongly heterosymmetric biradicaloids are molecules containing a noninteracting donor–acceptor pair, such as the 90° twisted aminoborane H_2NBH_2 (**10** in Figure 4.20A), and the TICT states of compounds such as *p-N,N*-dimethylaminobenzonitrile, discussed in Section 6.4.2.

Critical heterosymmetry. If $\delta_{AB} = \delta_0$, we expect a degeneracy of S_0 and S_1 from the simple model, and this is by far the most interesting case for a photochemist. Now, the covalent configuration $^1|AB\rangle$ has the same energy as the out-of-phase combination of $^1|A^2 + B^2\rangle$ and $^1|A^2 - B^2\rangle$, which can, at this value of δ_{AB}, usually be approximated by $^1|B^2\rangle$ alone. This situation is most readily obtained if neither $^1|AB\rangle$ nor $^1|B^2\rangle$ involves a formal charge separation. Then the two configurations differ by a shift of formal charge—either a positive charge as in **8** or a negative charge as in **9** (Figure 4.20A). Charged biradicaloids, therefore, have a particularly good chance of exhibiting S_0-S_1 degeneracies.

In summary, the way in which the typically large S_0-S_1 gap in a perfect biradical can be brought to zero exactly or approximately by a perturbation that turns it into a biradicaloid can be summarized as follows: Orbital $|A\rangle$ needs to be destabilized and/or orbital $|B\rangle$ needs to be stabilized to such a degree as to make the energies of the configurations $^1|AB\rangle$ and $^1|B^2\rangle$ approximately equal.

4.5.4 Nonsymmetric Biradicaloids

In this general case, the singlet part of the Hamiltonian matrix has the form

$$\begin{array}{l} ^1|A^2 - B^2\rangle: \\ ^1|A^2 + B^2\rangle: \\ ^1|AB\rangle \quad : \end{array} \begin{pmatrix} E(T) + 2K'_{AB} & \delta_{AB} & 0 \\ \delta_{AB} & E(T) + 2(K'_{AB} + K_{AB}) & \gamma_{AB} \\ 0 & \gamma_{AB} & E(T) + 2K_{AB} \end{pmatrix} \quad (4.110)$$

and although the expressions for state energies can be written in an explicit manner, they are complicated and not instructive. An example is given in Figure 4.22C. A closer analysis shows that S_0 and S_1 can only be degenerate when γ_{AB} vanishes and that the general shape of the function $E(S_1) - E(T)$ is as shown in Figure 4.23 for one particular choice of constant values for K' and K.

The conclusion to be drawn from Figure 4.23 is that the choice $\delta_{AB} \rightarrow \delta_0$ and $\gamma_{AB} \rightarrow 0$ is optimal when biradicaloids with a small S_0-S_1 energy gap are sought.

The one case of nonsymmetric biradicaloid in which a solution can be written simply is a perturbed axial biradical for which $K_{AB} = K'_{AB}$ (Appendix II).

Sudden polarization. As a general nonsymmetric biradicaloid approaches the limit $\delta_{AB} = 0$, that is, becomes a homosymmetric biradicaloid, the S_1 and S_2 states remain perfectly balanced with respect to charge as γ_{AB} is changed from large negative to large positive values, e.g., by twisting ethylene from the cis to the trans configuration. As soon as $\delta_{AB} \neq 0$, however, this balance is lost because $^1|A^2 + B^2\rangle$ and $^1|A^2 - B^2\rangle$ mix. If δ_{AB} is comparable to or smaller than the S_1-S_2 energy gap, the imbalance appears only for very small values of γ_{AB}. In ethylene, where the S_1-S_2 energy gap is very small as long as δ_{AB} is small, since it is approximately a pair biradical ($K_{AB} \simeq 0$), this polarization of the S_1 and S_2 states appears and disappears quite suddenly as γ_{AB} sweeps through zero. This effect has been referred to as sudden polarization. For larger values of δ_{AB} in pair biradicals and for all values of δ_{AB} in axial biradicals and for other biradicals with large K_{AB} values, this polarization is calculated to develop and disappear much more gradually.

4.6 FROM BIRADICALS TO BONDS AND LONE PAIRS: SURFACE SHAPES

Figure 4.23 is useful for a discussion of the classification of the effects of perturbations on a perfect biradical for any one given particular choice of K_{AB} and K'_{AB},

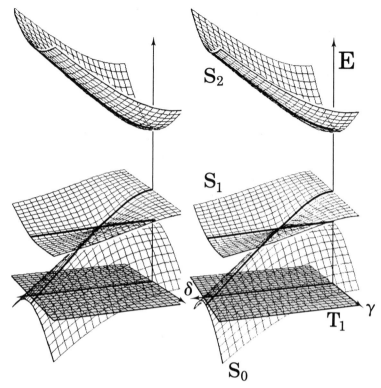

Figure 4.23 3 × 3 CI model: A stereo view of the energies of the singlet states of a biradicaloid relative to that of the triplet, plotted as a function of the perturbation parameters γ_{AB} and δ_{AB}. Reproduced by permission from V. Bonačić-Koutecký, J. Koutecký, and J. Michl, *Angew. Chem., Int. Ed. Engl.* **26**, 170 (1987).

but it does not in itself provide a useful guide for the shape of the potential energy surfaces for a two-electron two-orbital system as a function of its geometry and of the nature and of the organization of its environment, which would be of prime interest in a photochemical process. In order to obtain such guidance, it is necessary to express our reference energy, E(T), as a function of molecular geometry and environmental effects.

We shall again consider separately the case of sigma interaction, in which the dissociated fragments are infinitely far apart so that not only the resonance integral but also electron repulsion integrals vanish, and the case of pi interaction, in which the bond is broken while keeping substantial values for the electron repulsion integrals.

In Section 4.6.1 we shall consider the sigma interaction in a qualitative way. In Section 4.6.2, we shall provide a similarly qualitative picture for a pi interaction, and follow it up with a more quantitative treatment using a simple elaboration of the 3 × 3 CI model to obtain the potential energy surfaces. This is easier to do for a pi interaction because the magnitude of the electron repulsion terms K_{AB} and K'_{AB} can be considered approximately independent of the twist angle. In the case of sigma interaction, these terms clearly cannot be considered independent of the A to B separation.

In Section 4.6.3, we shall discuss the limitations of the simple 3 × 3 CI model, and in Section 4.6.4, we shall treat spin-orbit coupling in biradicals and biradicaloids.

4.6.1 Sigma Interactions

We shall now use Figure 4.23 for a qualitative analysis of the dissociation of a sigma bond between orbitals $|A\rangle$ and $|B\rangle$ located on species $\mathfrak{A}$ and $\mathfrak{B}$, respectively.

The electronegativity difference between the two orbitals, or, more accurately, half the energy difference between the structures $^1|A^2\rangle$ and $^1|B^2\rangle$, will be an important variable characterizing the bond and is given by δ_{AB}. It is determined primarily by the nature of the atoms that carry orbitals $|A\rangle$ and $|B\rangle$, by their bonding situation, which determines formal charges, and by the charge-stabilizing ability and organization of the surrounding solvent. We shall consider δ_{AB} as a continuous variable, although, in practice, only relatively minor variations in its value can be accomplished by continuous changes in hybridization or in the orientation of solvent molecules. In practice, then, a substantial change in δ_{AB} would most readily be accomplished by a change in the solvent. We shall use three prototype sigma bonds as concrete realizations of the so far fairly abstract concept: A covalent sigma bond between like atoms (e.g., Si—Si in Si_2H_6) is a typical example of the case of small δ_{AB} ($\delta_{AB} = 0$); a charged sigma bond such as C—N in CH_3—NH_3^+ is a typical example of the case of $\delta_{AB} = \delta_0$; and a dative sigma bond such as B—N in BH_3NH_3 is a typical example for a very large δ_{AB} value.

The degree of interaction between the orbitals $|A\rangle$ and $|B\rangle$, γ_{AB}, as well as the exchange integral K_{AB}, will decrease approximately exponentially in absolute value as the separation R of the orbitals $|A\rangle$ and $|B\rangle$ increases. At the same time, the electron repulsion integrals J_{AB} will decrease more slowly, approximately as R^{-1}.

If we wish to consider the potential energy surfaces S_0, S_1, and T as a function of δ_{AB} and R on the basis of the 3 × 3 CI model considered so far, we need to consider the variation in γ_{AB}, K_{AB}, and J_{AB} (and therefore K'_{AB}) with R, so that Figure 4.23, in which K_{AB} and K'_{AB} are assumed constant, is not directly applicable. Such analysis leads to the conclusion embodied in Figures 4.24 and 4.25.

Figure 4.24 shows schematically the energies of the dot–dot and hole–pair structures as a function of δ_{AB} in the dissociated limit $R = \infty$ (pair biradical, $K_{AB} = 0$). Only the lower-energy hole–pair structure is shown (at $\delta_{AB} = 0$ the two are degenerate). The energy of the dot–dot structure does not depend on δ_{AB}, since the average electronegativity is assumed constant. On the left, where $\delta_{AB} = 0$, both hole–pair structures are much higher in energy because they involve charge separation. As δ_{AB} increases, the more stable hole–pair structure decreases in energy until at $\delta_{AB} = \delta_0$ it becomes degenerate with the dot–dot structure. For large δ_{AB}, the hole–pair structure lies below the dot–dot structure. This result is immediately obvious from Figure 4.23 when one sets γ_{AB} and K_{AB} equal to 0 and K'_{AB} relatively large.

In the dissociated limit, the VB structures of Figure 4.24 will represent states except in the limit $\delta_{AB} = 0$, where $\mathfrak{A}^+:\mathfrak{B}^-$ and $\mathfrak{A}:^-\mathfrak{B}^+$ need to be mixed equally. Since at $R = \infty$ the singlet and the triplet dot–dot structures are exactly degenerate, we obtain for $\delta_{AB} < \delta_0$ a covalent S_0 degenerate with T_1 (singlet or triplet radical pair) and an ionic S_1 (ion pair), and we obtain for $\delta_{AB} > \delta_0$ a covalent S_1 degenerate with T_1 (singlet or triplet radical pairs) and an ionic S_0 (ion pair).

The dissociation of the NaCl molecule in the gas phase is an example of the

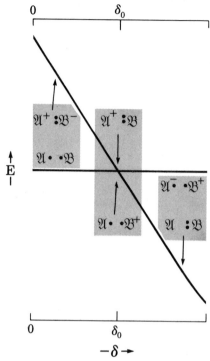

Figure 4.24 Energies of the dot–dot and hole–pair configurations of heterosymmetric biradicaloids as a function of the perturbation parameter δ_{AB}. Schematic.

former situation (in the ground state, Na and Cl atoms are produced), and its dissociation in aqueous solution is an example of the latter situation (in the ground state, hydrated Na^+ and Cl^- ions are produced).

The whole range of possibilities for δ_{AB} and R is displayed in Figure 4.25, in which the S_0, S_1 and T_1 energies are plotted as a function of R and δ_{AB} in a schematic fashion. The lines shown in Figure 4.24 for the limit $R = \infty$ appear in the rear right face of Figure 4.25. Cuts through the surfaces along lines of constant δ_{AB} correspond to ordinary bond dissociation curves, with those for the limit $\delta_{AB} = 0$ appearing in the rear left face of Figure 4.25. These are qualitatively the same as the dissociation curves of H_2 (Figure 4.3). As δ_{AB} increases, the shape of the S_0 and T_1 curves does not change much, but the S_1 energy rapidly approaches that of T_1 until at $\delta_{AB} = \delta_0$ the two differ very little and merge in the fully dissociated limit. Up to this point, the bonding in the ground state can be properly described as covalent.

For still larger values of δ_{AB}, the nature of bonding in the S_0 state is better described as dative, and the minimum in S_0 gradually becomes shallower (an extreme case of such a situation is encountered in HeH^+). The S_1 and T_1 surfaces remain close together and gradually become flatter until they show almost no dependence on R. This can be understood using NH_3BH_3 as an example. At large R values in both S_1 and T_1, an electron is transferred from the HOMO, located predominantly on the N atom, to the LUMO, located predominantly on the B atom, so that a "charge-transfer biradicaloid" is produced. An increase in R is

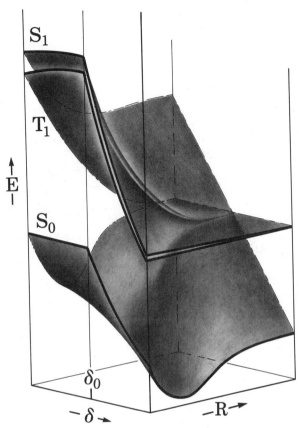

Figure 4.25 Energies of the S_0, S_1, and T_1 states of a polar sigma bond as a function of bond length R and of the polarity parameter δ in the 3 × 3 CI model. Schematic.

unfavorable since it increases charge separation. However, a decrease in R is also unfavorable because it increases the antibonding nature of the singly occupied LUMO orbital.

The information contained in Figure 4.25 can now be summarized on the three examples chosen, although the exact value of δ_{AB} for BH_3NH_3 is not obvious without further calculations. In the case of ordinary covalent bonds (small δ_{AB}) we expect excitation into T_1 to produce dissociation to the loose biradicaloid geometry (R = ∞) and excitation into S_1 to produce bond weakening and stretching to a tight biradicaloid geometry (R ≈ $2R_e$). In the case of a charged covalent bond, we expect both excitation into T_1 and excitation into S_1 to produce dissociation to the loose biradicaloid geometry. In the case of a dative bond, we expect both excitation into T_1 and into S_1 to break the dative bond present in S_0 but to provide little, if any, driving force towards actual dissociation, so that radiationless return to S_0 may well compete with the drifting of the $\mathfrak{A}^+$ and $\mathfrak{B}^-$ ions apart. Since in this case the S_1 and T_1 curves are expected to be so flat, even a minor structural variation may cause a shallow minimum to occur either at small or large R values, with possibly important photophysical or photochemical consequences.

4.6.2 Pi Interactions

Next we return to Figure 4.23 in order to provide a qualitative analysis of the breaking of a pi bond between orbitals $|A\rangle$ and $|B\rangle$ by twisting.

General analysis. The electronegativity difference δ_{AB} remains as one of our variables. Since the $|A\rangle$ to $|B\rangle$ separation is not affected much by the twisting, we shall assume that K_{AB} and K'_{AB} remain constant, so that we can use Figure 4.23 directly as soon as a proper functional form is obtained for the T_1 energy. This will be done below. In the way of introduction, however, we can first consider Figure 4.26, in which the twist angle dependence of the energies was drawn in a purely qualitative fashion. In the limit $\delta_{AB} = 0$ (front right face) we see the familiar S_0, S_1 and T_1 state energy curves for the twisting of ethylene. As δ_{AB} increases, the S_0–S_1 gap at 90° twist decreases until at $\delta_{AB} = \delta_0$ it vanishes. For even larger δ_{AB} values, it increases again and the barrier to twist in the S_0 surface decreases. The S_1–T_1 separation is large for $\delta_{AB} < \delta_0$, but the two surfaces are nearly parallel for $\delta_{AB} > \delta_0$. The S_0–T_1 separation is large everywhere except for $\delta_{AB} < \delta_0$ at twist angles near 90°.

For $\delta_{AB} < \delta_0$, the S_0 state is best described as containing a covalent pi bond and for $\delta_{AB} > \delta_0$, a dative pi bond, except at 90°, where the bond is fully broken. The T_1 state is well described by the dot–dot structure at all geometries. The S_1 state is described by a hole–pair structure (at $\delta_{AB} = 0$, a combination of two such structures) as long as $\delta_{AB} < \delta_0$. For $\delta_{AB} > \delta_0$, it is described by the singlet "dot–dot" structure ("charge-transfer biradicaloid").

The expected consequence for photochemical behavior would be a twisting from

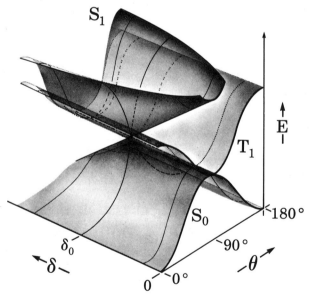

Figure 4.26 Energies of the S_0, S_1, and T_1 states of a polar pi bond as a function of the twist angle θ and of the polarity parameter δ_{AB}. 3 × 3 CI model. Schematic. Adapted by permission from V. Bonačić-Koutecký, J. Köhler, and J. Michl, *Chem. Phys. Lett.* **104**, 440 (1984).

the planar to the orthogonal geometry in both the S_1 and T_1 states, regardless of the value of δ_{AB}.

A crude model. A somewhat more detailed analysis of the model results will be presented next, based on a specific assumption for the functional form of E(T) in the 3×3 CI model. This will permit us to convert Figure 4.23 into a series of increasingly more realistic drawings of potential energy surfaces.

We start from nonorthogonal atomic orbitals $|A\rangle$ and $|B\rangle$. Within the two-electron two-orbital model, one obtains

$$E(T) = \frac{1}{1 - S_{AB}^2} \left(h_{AA} + h_{BB} + \frac{J_{AA} + J_{BB}}{2} - 2K'_{AB} - K_{AB} - 2h_{AB}S_{AB} \right)$$

$$= E_T^0 + \frac{1}{1 - S_{AB}^2} [(h_{AA} - h_{AA}^0) + (h_{BB} - h_{BB}^0) + (J_{AB} - J_{AB}^0) \quad (4.111)$$

$$- (K_{AB} - K_{AB}^0) + S_{AB}^2 E_T^0 - 2h_{AB}S_{AB}]$$

where the superscript ^{0} refers to the orthogonal reference geometry, for which $\gamma_{AB} = 0$:

$$E_T^0 = h_{AA}^0 + h_{BB}^0 + J_{AB}^0 - K_{AB}^0 \quad (4.112)$$

For the purposes of a qualitative discussion, the difference terms in parentheses can be neglected. Since h_{AB} is approximately proportional to S_{AB}, it is reasonable to approximate E_T by

$$E(T) = E_T^0 + cS_{AB}^2/(1 - S_{AB}^2) \quad (4.113)$$

where c is a constant that reflects the average electronegativity of orbitals $|A\rangle$ and $|B\rangle$. Its value must be positive if a deviation from the perfect-biradical geometry produces bonding between orbitals $|A\rangle$ and $|B\rangle$ in the ground state, as would ordinarily be the case.

Since the resonance integral between orthogonalized localized orbitals $|A\rangle$ and $|B\rangle$ is approximately proportional to overlap between the atomic orbitals $|A\rangle$ and $|B\rangle$, we can finally write

$$E(T) = E_T^0 + f[\gamma_{AB}^2/(1 - g\gamma_{AB}^2)] \quad (4.114)$$

where the constants f and g are positive. The shape of this function defines the shape of the T_1 surface in Figure 4.27. Thus, E_T equals E_T^0 if γ_{AB} vanishes, and it increases gradually as γ_{AB} increases. This result reproduces the well-known general fact that triplet energies are particularly low at geometries at which the two singly occupied orbitals do not interact. After all, upon such interaction the antibonding combination is destabilized more than the bonding one is stabilized.

A more realistic model. The last item to consider is the effect of changes in geometry and solvent environment on δ_{AB} and on E_T^0. To a first approximation, the value of δ_{AB} in the orthogonalized basis is determined by the value of δ_{AB} in

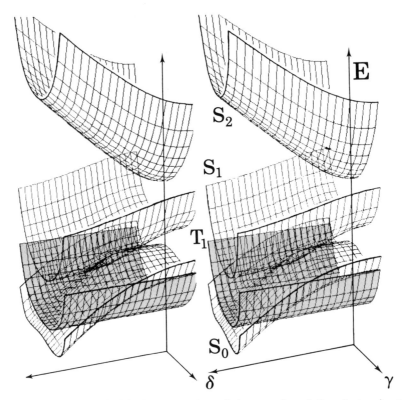

Figure 4.27 3 × 3 CI Model: A stereo view of the energies of the electronic states of biradicaloids, plotted as a function of the perturbation parameters γ and δ. See text. Reproduced by permission from V. Bonačić-Koutecký, J. Koutecký, and J. Michl, *Angew. Chem., Int. Ed. Engl.* **26**, 170 (1987).

the nonorthogonal basis, which, in turn, depends primarily on the nature of the atoms on which $|A\rangle$ and $|B\rangle$ are located and on the substituents they carry. The shapes of singlet potential energy surfaces for a series of such choices of δ_{AB} are shown in Figure 4.28. To obtain a final qualitatitive view of the shapes of the singlet surfaces starting from Figure 4.27, one needs to choose an appropriate value of δ_{AB}, say δ'_{AB}, as dictated by the chemical nature of the localized orbitals $|A\rangle$ and $|B\rangle$ and by the charge-stabilizing or charge-destabilizing effects of the solvent environment. Geometry changes are generally of three kinds: (1) those that cause the orbitals $|A\rangle$ and $|B\rangle$ to interact (twisting in orthogonal ethylene), (2) those that cause changes in the relative electronegativities of orbitals $|A\rangle$ and $|B\rangle$, and (3) those that change the average electronegativity of the orbitals $|A\rangle$ and $|B\rangle$ and, thus, change E_T^0. The latter two are normally associated with rehybridization at the centers that carry orbitals $|A\rangle$ and $|B\rangle$. However, rearrangement of solvent molecules may have similar and even larger effect, particularly if the solvent is polar. Any particular geometry change may, of course, contribute in more than one category.

The effect of the changes of type (1) is already incorporated in Figures 4.23 and 4.27 in the form of the dependence on γ_{AB}. The effects of the changes of types (2)

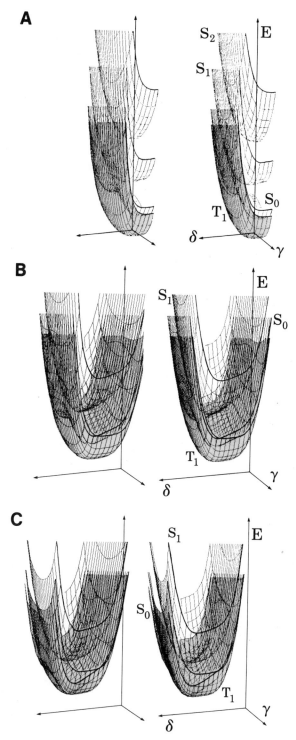

Figure 4.28 3 × 3 CI model: A stereo view of the energies of the electronic states of a biradicaloid, plotted as a function of the perturbation parameters γ_{AB} and δ_{AB}. (A) Weakly heterosymmetric. (B) Critically heterosymmetric. (C) Strongly heterosymmetric. See text. Reproduced by permission from V. Bonačić-Koutecký, J. Koutecký, and J. Michl, *Angew. Chem. Int. Ed. Engl.* **26**, 170 (1987).

and (3) are limited by the steric and other problems which other bonds in the molecule will experience upon excessive rehybridization and which solvent molecules will feel when their charges are brought too close to the substrate molecule. Their effects on the potential energy surfaces of Figure 4.23 can then be simulated by adding suitable empirical potentials. This was done in Figure 4.28, which displays the energies as a function of γ_{AB} and δ_{AB} for propene ($\delta'_{AB} \simeq 0$), protonated formaldimine ($\delta'_{AB} \leq \delta_0$), and aminoborane ($\delta'_{AB} > \delta_0$). In all three examples, minima in S_1 and T_1 are located at orthogonal geometries, but the S_1 surface of protonated formaldimine clearly is most interesting, due to the accessibility of the geometry at which S_1 and S_0 are degenerate. The effects of the variations in the average electronegativity could be represented similarly, but they call for the use of yet another dimension in the graph.

The transformation of the general shape of the singlet and triplet surfaces of Figure 4.27 into model surfaces of particular molecules in which the range of δ_{AB} is limited, exemplified in Figure 4.28, can be generalized for other cases. For instance, for molecules with double bonds constrained to partially twisted geometries, such as bridgehead (anti-Bredt) olefins, an empirical potential restricting the accessible range of values of γ_{AB} could be added.

The generalized singlet and triplet surfaces of Figure 4.27 permit a visualization of the particular surfaces for all kinds of π-interacting two-electron systems. In the center ($\delta_{AB} = \gamma_{AB} = 0$), one finds the orthogonal geometry and the perfect biradical. Motion along the γ_{AB} coordinate introduces a π bond between $|A\rangle$ and $|B\rangle$. This is nonpolar if $\delta_{AB} = 0$ and increasingly polar as δ_{AB} is allowed to grow. Finally, motion along the δ_{AB} coordinate converts the system into a donor–acceptor pair, with a lone pair as the donor and a vacant orbital as the acceptor. The degree of interaction between the donor and acceptor orbitals increases as γ_{AB} is allowed to grow. A similar analysis can be performed for σ-interacting two-electron systems. The main difference is due to the fact that now the bond-breaking process is associated with a very large variation of K_{AB} as well, while in Figure 4.27 we assumed that K_{AB} is not significantly affected by bond twisting.

In a sense then, Figure 4.27 provides a qualitative view of the main features of the triplet and singlet surfaces of two-electron systems ranging from perfect biradicals to nonpolar bonds, polar bonds, and lone-pair–acceptor combinations.

4.6.3 Limitations of the Simple Model

Up to this point, the states of the two-electron system were treated by means of a very simplified model. The electrons were permitted to move only in a very limited space, dictated by the two orbitals $|\mathscr{A}\rangle$, $|\mathscr{B}\rangle$, and the presence of all the other electrons of the molecule was felt only through their time-averaged fields, which means that all mutual correlation of the motion of the selected two electrons with that of the remaining ones was neglected. It is important to inquire what effect the removal of the approximations will have on the results.

Basis set limitation. The limitation to only two orbitals $|\mathscr{A}\rangle$, $|\mathscr{B}\rangle$ turns out not to be very serious for low-energy states. Sure enough, the introduction of large orbital basis sets for the description of the states of the two electrons improves the description of the low-lying states and adds a large number of highly excited states to the four states present in the simple model. However, in photochemical processes we shall typically only be concerned with the lowest triplet and the lowest two singlets, and there will rarely be a qualitatively significant difference when the larger basis set is introduced. Perhaps the largest effects are observed in the smallest molecules. Thus, in our example of H_2 given in Section 4.2.1, the S_0 and T_1 states of the H_2 molecule were described reasonably well, but the S_1 and S_2 states had energies that were unreasonably high. With the use of a large basis set, these two latter states are seen to have considerable Rydberg character. As the H_2 molecule dissociates, the combination of a Rydberg state of one hydrogen atom with the ground state of the other H atom in reality becomes the lowest excited singlet state of the system, so that the dissociation in the S_1 surface does not actually produce a H^+H^- ion pair but, instead, produces a ground-state hydrogen atom plus a Rydberg excited hydrogen atom.

As noted in Section 1.3.3, in larger molecules Rydberg states still tend to mix with singlet valence states at ordinary geometries, but this mixing tends to be less important at biradicaloid geometries where the valence states are particularly low in energy. Since these are of particular interest to photochemists as the points of return from the excited to the ground states, the qualitative picture obtained without the use of Rydberg orbitals tends to be qualitatively useful.

The fixed core approximation. The other serious shortcoming of the simple model, the use of a fixed core to represent all the other electrons and nuclei of the molecule, is more serious. When the motion of the two electrons of the subsystem of interest is allowed to correlate with the motion of the other electrons in the molecule, two important changes are brought about in the description provided by the simple model. Both of them have to do with states that are predicted to be split by $2K_{AB}$ in the simple model, and they are therefore particularly important when K_{AB} is a small number.

The S_0–T ordering. In perfect biradicals that approach the pair biradical limit of $K_{AB} = 0$, T is expected to lie only a little below S_0 according to the simple model, since $2K_{AB}$ is small. A correlation with the motion of additional electrons in the molecule provides a mechanism that leads to a preferential stabilization of S_0 relative to T, so that the order is frequently reversed. This happens, for instance, in orthogonally twisted ethylene and in square cyclobutadiene. The prediction of a T–S_0 crossing, which, according to the simple model, should occur in homosymmetric biradicaloids when $|\gamma_{AB}|$ reaches the value of $2\sqrt{K_{AB}(K'_{AB} + K_{AB})}$ and in heterosymmetric biradicaloids when δ_{AB} reaches the value of $2\sqrt{K'_{AB}(K'_{AB} + K_{AB})}$, is therefore worthless in these cases, although it may still be of some use in those biradicals in which K_{AB} is large. Although the consideration of the correlation with the motions of other electrons in the molecule thus can be critical for the determination of the nature of the ground state of the molecule, singlet or triplet, it actually has relatively little effect on the considerations relevant to photochemistry because in these cases, S_0 and T still tend to be very close in energy and both are thermally accessible.

The S_1–S_2 ordering. The second consequence of the correlation with the motion of other electrons in the molecule is potentially more serious for photochemical considerations. When K_{AB} is small, the simple model predicts the S_1–S_2 gap in a perfect biradical to be small, with the $^1|A^2 - B^2\rangle$ state below $^1|A^2 + B^2\rangle$. In the presence of correlation effects from the other electrons present, the $^1|A^2 + B^2\rangle$ state is differentially stabilized relative to $^1|A^2 - B^2\rangle$, so that the order may be reversed (see Section 4.8). This happens both in twisted double bonds and in square cyclobutadiene. Since we shall see later that the location of minima in the states represented by $^1|A^2 + B^2\rangle$ and by $^1|A^2 - B^2\rangle$ in the simple model may be different, the order of the states may be critical for determining the geometry at which a minimum in S_1 is located.

Fortunately, the mechanism provided by the simple model for finding S_0–S_1 degeneracy in heterosymmetric biradicaloids by polarization of the $^1|A^2 + B^2\rangle$ and $^1|A^2 - B^2\rangle$ states into states described approximately by the $|A^2\rangle$ and $|B^2\rangle$ configurations, and the preferential stabilization of the latter, is quite independent of the initial order of the $^1|A^2 + B^2\rangle$ and $^1|A^2 - B^2\rangle$ states in the parent perfect biradical. The results of the simple model have been confirmed by good-quality calculations. Of course, the exact values of δ_{AB} at which S_0 and S_1 touch are then not given accurately by the simple expressions of the model, but the qualitative conclusions remain unchanged.

4.6.4 Spin–Orbit Coupling in Biradicals and Biradicaloids

Spin–orbit coupling is the main mechanism for intersystem crossing in biradicals in which the localized orbitals $|A\rangle$ and $|B\rangle$ are in reasonable proximity, such as twisted alkenes and 1,3- and 1,4-biradicals. Only in biradicals with quite distant radical centers and in radical pairs that are not in direct contact is the S_0–T_1 splitting sufficiently small for the hyperfine mechanism to play a generally dominant role.

In the present section, we consider briefly S_0–T_1 and S_1–T_1 coupling in biradicals and biradicaloids by the spin–orbit mechanism, relying on the general discussion given in Section 1.3.2. References listed in Section 4.9 should be consulted for additional detail.

Within the framework of the two-electron two-orbital model, the singlet state wave functions $|S_i\rangle$ are written in the form (4.99). The triplet functions are written as $|^uT\rangle$, u = x, y, z, using the triplet spin eigenfunctions Θ_x, Θ_y, and Θ_z defined in Appendix II (equation II.13). Using the definitions introduced in Section 1.3.2 and dropping the electron index j on $\mathbf{r}_{\alpha j}$ and $\hat{l}_{uj}$, we obtain

$$\langle ^uT|\hat{H}_{SO}|S_i\rangle = C_{i,+} \frac{e^2\hbar}{2m^2c^2} \langle \mathscr{A} | \sum_\alpha \frac{Z_\alpha}{|\mathbf{r}_\alpha|^3} \hat{l}_u|\mathscr{B}\rangle \qquad (4.115)$$

This result shows that in the 3 × 3 CI model, the spin–orbit coupling matrix element depends on three factors: the $^1|A^2 + B^2\rangle$ character of the singlet state, the spatial disposition of the orbitals $|\mathscr{A}\rangle$ and $|\mathscr{B}\rangle$, and nuclear charges (heavy atom effect). The actual intersystem crossing rate will also depend on the energy gap.

According to equation 4.115, the triplet spin–orbit couples with singlet states only to the extent that the latter contain a contribution from the function $^1|\mathscr{A}^2 + \mathscr{B}^2\rangle$. In a perfect biradical, such as the species listed in Figure 4.19, this wave function represents the S_2 state, and spin–orbit coupling with both the S_0 and the S_1 states then vanishes in this approximation.

4.6 FROM BIRADICALS TO BONDS AND LONE PAIRS: SURFACE SHAPES

Recall that the coefficient $C_{i,+}$ is invariant with respect to orthogonal orbital transformations; equation 4.115 holds for any choice of $\mathscr{A}$ and $\mathscr{B}$. It is convenient to continue further discussion in terms of the most localized orbitals $|A\rangle$ and $|B\rangle$, since the action of the angular momentum operator $\hat{l}$ on these is most readily visualized.

In a homosymmetric biradicaloid, $|S_0\rangle$ acquires some $^1|A^2 + B^2\rangle$ character ($C_{0,+} \neq 0$) and therefore becomes capable of mixing with $|T\rangle$ in the present approximation. This happens, for instance, in partially twisted ethylene and in rectangular cyclobutadiene. While an increase in γ_{AB} and the concomitant increase in $C_{0,+}$ upon distortion from a perfect biradical enhance the spin–orbit coupling matrix element, this also increases the S_0–T energy gap. These two factors will work against each other, with the former enhancing and the latter reducing the rate of intersystem crossing. Qualitatively, one can conclude that intersystem crossing, say, in ethylene, will be the fastest at some large but not orthogonal degree of twisting.

In a weakly heterosymmetric biradicaloid, $|S_1\rangle$ acquires some $^1|A^2 + B^2\rangle$ character and therefore will mix with $|T\rangle$. Now, an increase of δ_{AB} and the concomitant increase in $C_{0,+}$ upon distortion from a perfect biradical not only enhances the spin–orbit coupling matrix element but also decreases the S_1–T energy gap. At $\delta_{AB} = \delta_0$, the S_0 and S_1 surfaces touch; beyond that, it is S_0–T coupling that is favorable. One can expect intersystem crossing to S_0 to be particularly fast when $\delta_{AB} \simeq 2\sqrt{K'_{AB}(K'_{AB} + K_{AB})}$, that is, in heterosymmetrical biradicaloids perturbed somewhat beyond the critical point δ_0. In even more strongly heterosymmetric biradicaloids, $|S_0\rangle$ has nearly exactly 50% $^1|A^2 + B^2\rangle$ character, and spin–orbit coupling to $|T\rangle$ is still very favorable, but of course, as δ_{AB} increases further, the S_0–T gap increases and the intersystem crossing rate drops.

Next to the content of the $^1|A^2 + B^2\rangle$ character in the singlet wave function, the magnitude of the matrix element $\langle A|\Sigma_\alpha (Z_\alpha/|\mathbf{r}_\alpha|^3)\hat{l}_u|B\rangle$ is of critical importance for the magnitude of the spin–orbit coupling matrix element (4.115). As discussed in Section 1.3.2, the three components of the angular momentum operator $\hat{l}_u$ (u = x,y,z) annihilate s and p_u orbitals and rotate the remaining two p orbitals around the x, y, or z axis, respectively. The weighting by $|\mathbf{r}_\alpha|^{-3}$ means that in the sum over nuclei, the nucleus on which the p orbital is located contributes much more than the others, which can be neglected. It then becomes obvious that a requirement for the matrix element to be sizable is that the p components of the orbital $|B\rangle$, after rotation by 90° about the axis u, should overlap well with orbital $|A\rangle$. Thus, a parallel arrangement of the axes of the p components of the two orbitals on atoms contained in a plane is worst of all, since they will be orthogonal after one of them is rotated by 90° around either of the directions perpendicular to the orbital axes. An initial inclination of 90° is fine, provided that the orbital centers are not far apart. The best arrangement is for $|A\rangle$ and $|B\rangle$ to be two different p orbitals on the same center, e.g., on a nitrogen atom in a nitrene. If they are located at different centers, the overlap of A with rotated B falls off rapidly as one goes beyond 1,3-biradicals.

However, even in the longer chains there will be special geometries at which the overlap of orbital $|A\rangle$ with a rotated orbital $|B\rangle$ will be large, namely, those in which the two radical centers are in close proximity, nearly closing a cycle. These also are the geometries at which $|A\rangle$ and $|B\rangle$ can have a substantial resonance integral and thus a substantial γ_{AB}, so that $C_{i,+}$ does not vanish. A geometry reminiscent of the arrangement normally drawn for a disrotatory or conrotatory ring closure would appear ideal. Of course, as γ_{AB} increases, the energy of the $|T\rangle$ state increases as well, and the geometrical requirement described here may well cause the presence of an activation barrier in the intersystem crossing process (the geometries of the fastest crossing are energetically somewhat above the triplet minimum).

Once the crossing to the $|S_0\rangle$ state occurs, the biradical will find itself at a geometry ideally suited for immediate bond closure, looking down into an ~80 kcal/mol abyss and having essentially no choice other than closing a bond. This would then, in effect, make the process of intersystem crossing and bond closure in such biradicals kinetically and stereochemically concerted.

One additional factor affecting the size of the matrix element in equation 4.115 and affecting the rate of intersystem crossing in biradicals and biradicaloids remains to be mentioned (it was already noted in Chapter 1): The presence of the factor $Z_k/|\mathbf{r}_k|^3$ is responsible for the heavy-atom effect.

Finally, it should be mentioned that in many biradicals it may not be desirable to approximate the most localized orbitals $|A\rangle$ and $|B\rangle$ by a single AO each. When the radical centers are not close together, hyperconjugative delocalization of $|A\rangle$ and $|B\rangle$ into a structural region of common overlap may make an important contribution to the matrix element in equation 4.115.

4.7 BEYOND THE SIMPLE MODEL: PHOTODISSOCIATION OF THE SINGLE BOND

In previous sections of Chapter 4, we have considered the relation between a biradical or a noninteracting donor–acceptor combination on the one hand and a nonpolar bond, polar bond, or a dative bond on the other hand, using a simple two-electron two-orbital model (3 × 3 CI for singlet states). Although we also pointed out briefly the modifications introduced by more sophisticated theoretical descriptions, we did not describe the result in any degree of detail. The results presented in Sections 4.7 and 4.8 are based on adequate quality *ab initio* calculations, mostly large-scale CI with a double-zeta-quality basis set.

4.7.1 Covalent Sigma Bond

In this section the simple concepts derived within the two-electron two-orbital model for the dissociation of a prototype covalent bond with $\delta_{AB} = 0$ in its various electronic states will be tested on more realistic model cases. We have chosen as an illustration the dissociation of the Ge—Ge bond in digermane, Ge_2H_6, and of the C—C bond in cyclopropane.

Digermane. Figure 4.29 shows the lowest few potential energy curves obtained from an *ab initio* large-scale CI calculation using the pseudopotential method, without inclusion of Rydberg configurations. The result is qualitatively indistinguishable from those represented in Figures 4.3 and 4.8, as is the nature of the resulting wave functions, except, of course, for quite small values of R. In the dissociated limit, the S_0 and T_1 states are covalent and are represented by the G

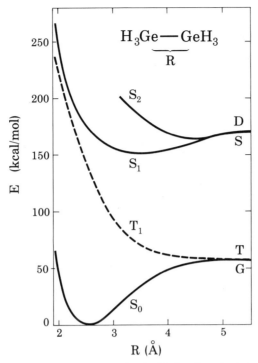

Figure 4.29 Potential energy curves for singlet (solid lines) and triplet (dashed line) states for the dissociation of the Ge—Ge bond in digermane. The geometry is that of digermane, and the Ge—Ge distance is the only geometrical parameter varied. *Ab initio* large-scale CI calculation with pseudopotentials for Ge.

and T wave functions of a biradical, respectively. The much higher energy S_1 and S_2 states are "zwitterionic," and their wave functions can be described approximately as $^1|Z_1\rangle$ and $^1|Z_2\rangle$, as defined in equations 4.46 and 4.47. These states have shallow minima at relatively large bond lengths, and they rise in energy upon further increase in the Ge—Ge separation. Locally excited states of the type GeH_3 + GeH_3^* are calculated at somewhat higher energies. It is possible that a calculation with a larger basis set would place them below the Z_1 and Z_2 states, in which case the situation would be just what it is in H_2, where the true correlation of the analogous excited states upon dissociation is not with an ion pair but rather with H(1s) and H(2s).

Either way, the conclusions we arrived at from the consideration of the simple model remain intact. Photochemical dissociation of the Ge—Ge bond should proceed without a significant barrier in the triplet state T_1 and should yield a radical pair (the product has a "loose" biradicaloid geometry). In the singlet state S_1, the bond should not dissociate but should only stretch by nearly a factor of two and become much weaker (a "tight" biradicaloid geometry). There is a fair chance that some fluorescence might be observed (fluorescence from singlet excited alkanes is indeed known), and there is an excellent chance that other unimolecular or bimolecular reactions might take place by motion towards other minima along paths other than Ge—Ge bond dissociation (say, H_2 elimination with the formation of a germylene). Of course, since Ge has a relatively high atomic number, intersystem crossing followed by triplet photochemistry may be a serious competitor. More definitive conclusions could only be reached after inclusion of Rydberg configurations. A general conclusion, then, is that saturated molecules can be expected to undergo simple covalent bond dissociation reactions in their T_1 state and to either be stable (fluoresce) or undergo more complex reactions in their S_1 state.

Cyclopropane. Figure 4.30 shows the energies of the few lowest states of cyclopropane along a ring-opening path which leads to the "90°, 90°" geometry of the trimethylene biradical. Only the S_0 ground state of cyclopropane is bound; in the S_1 and T_1 states the ring should open spontaneously without having to overcome a barrier. Once the trimethylene biradical geometry is reached, the surfaces become quite flat and the S_0 and T_1 states become nearly degenerate. The S_1 state lies much higher. There is no local minimum in the S_0 state that would correspond to a trimethylene biradical geometry, and the closure to cyclopropane will proceed without a barrier once the S_0 state is reached.

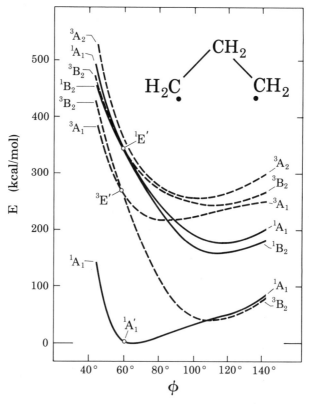

Figure 4.30 State energy diagram for the cyclopropane-to-trimethylene conversion. All bond lengths and angles are held at their cyclopropane values except that the CCC angle ϕ is varied. *Ab initio* calculation. Reproduced by permission from R. J. Buenker and S. D. Peyerimhoff, *J. Phys. Chem.* **73**, 1299 (1969).

4.7 BEYOND THE SIMPLE MODEL: PHOTODISSOCIATION OF THE SINGLE BOND 203

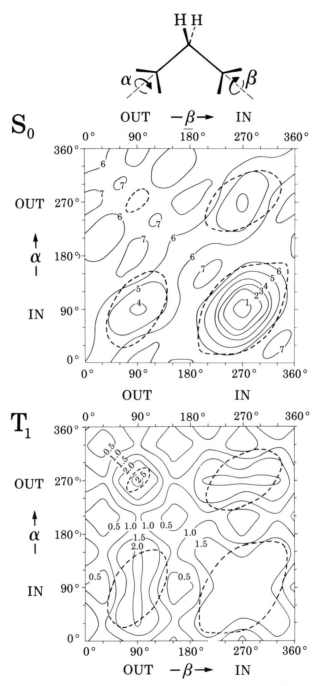

Figure 4.31 Contour diagram for the energies (kcal/mol) of the S_0 (top) and T_1 (bottom) states of a trimethylene biradical pyramidalized by 20° at both ends, as a function of rotation of the methylene groups (angles α, β). CCC angle: 114°. IN: Inward pyramidalization, OUT: Outward pyramidalization. The dashed lines indicate S_0–T_1 intersections. *Ab initio* two-configuration MCSCF. Reproduced by permission from C. D. Doubleday, Jr., J. W. McIver, Jr., and M. Page, *J. Am. Chem. Soc.* **104**, 6533 (1982).

The minimum in the T_1 surface is hard to locate exactly, because the surface is so flat. It occurs at a CCC valence angle of about 114°; moderate pyramidalization of the two CH_2 groups and their rotation about the C—C bonds have very little effect on the energy. A section of the S_0 and T_1 surfaces, with several S_0—T_1 intersections, is shown in Figure 4.31. It is hardly surprising that extensive stereorandomization of the substituents occurs in triplet reactions that proceed through this biradical.

It would then appear that the two radical centers in trimethylene hardly know about each other's presence. However, there is strong experimental evidence from ESR and IR spectra of substituted triplet trimethylenes that the central CH_2 group hyperconjugates strongly with the radical centers. Indeed, the through-bond coupling of the two radical centers via the CH_2 group forms the basis of an ingenious explanation proposed for the dependence of the relative energies of the S_0 and T_1 states on the CCC valence angle (Figure 4.32).

At the planar (0°,0°) geometry of trimethylene, the orbitals of the two radical centers combine in a symmetric (S) and an antisymmetric (A) fashion. Interaction with the doubly occupied bonding orbital of the CH_2 group component of the out-of-phase combination of the two C—H bonding orbitals destabilizes the S combination relative to the A combination. As the CCC valence angle decreases, the two atomic orbitals overlap more strongly and the S combination is stabilized. In a relatively narrow range of CCC angles, the two effects just cancel and the S and

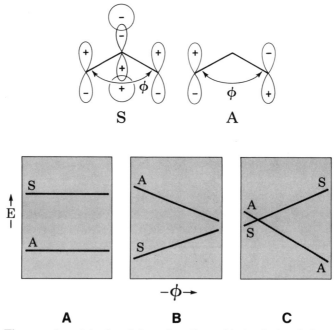

Figure 4.32 The energies of the S and A nonbonding orbitals of trimethylene as a function of the CCC valence angle ϕ, illustrating the opposed effects of through-bond (A) and through-space (B) interaction as well as their sum (C). Adapted by permission from C. D. Doubleday, Jr., J. W. McIver, Jr., and M. Page, *J. Am. Chem. Soc.* **104**, 6533 (1982); see also A. H. Goldberg and D. A. Dougherty, *J. Am. Chem. Soc.* **105**, 284 (1983).

A orbitals are essentially degenerate. For these values of the CCC angle, T_1 lies below S_0. For smaller or larger values, T_1 lies above S_0.

4.7.2 Charged Sigma Bond

Next we compare the results obtained in the simple two-electron two-orbital model for the case $\delta_{AB} = \delta_0$ with those obtained from an *ab initio* large-scale CI calculation on protonated methylamine, $CH_3NH_3^+$. Rydberg configurations are again ignored. The hole–pair (CH_3^+ :NH_3, $^1|B^2\rangle$) and dot–dot ($CH_3\cdot$ ·NH_3^+, $^{1,3}|AB\rangle$) structures are not quite isoenergetic, and their exact order depends on the geometries assumed. In the calculation, the geometries have been assumed tetrahedral throughout, but at the equilibrium ground-state geometries of the products, $^1|B^2\rangle$ would lie below

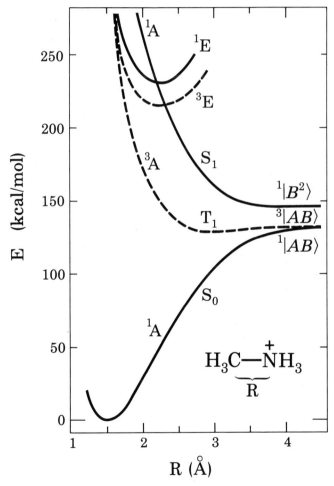

Figure 4.33 Potential energy curves for the dissociation of the C—N bond in protonated methylamine. All geometric parameters are held at standard $CH_3NH_3^+$ values except that the C—N distance R is varied. *Ab initio* large-scale CI calculation.

$^{1,3}|AB\rangle$ because the ionization potential of CH$_3\cdot$ is lower than that of NH$_3$. Regardless of these details, the S$_0$, S$_1$, and T$_1$ curves clearly come close enough together at the dissociation limit to provide us with some feeling for the limiting case of degeneracy of S$_0$ and S$_1$ at infinite internuclear separation.

Once again, the result (Figure 4.33) is qualitatively the same as that obtained for the simple model. As expected from Figures 4.12–4.14, both S$_1$ and T$_1$ are strongly dissociative, and the nature of the wave functions also is as expected. The calculated energy at the dissociated limit is too high because its geometry has not been optimized, but this is immaterial for the present argument. The *ab initio* results reinforce our previous conclusion that charged sigma bonds should be prone to photodissociation in both the T$_1$ and S$_1$ states, finding little, if any, chance for fluorescence.

4.7.3 Dative Sigma Bond

Our final comparison of a good-quality *ab initio* calculation with the simple two-electron two-orbital model deals with the case of a large electronegativity difference, $\delta_{AB} > \delta_0$.

The large-scale CI results for the dissociation of the ammonia–borane adduct, BH$_3$NH$_3$, in its lowest electronic states are shown in Figure 4.34.

The results are only schematic, since the geometry at both B and N has been kept tetrahedral along the whole path; this artificially increases the energy of the ground ($^1|B^2\rangle$) and charge-transfer excited ($^{1,3}|AB\rangle$) states at the dissociation limit. Not surprisingly, the bond in the S$_0$ state is weaker than a charged or an uncharged covalent bond. The picture of the excited states is complicated by the presence of a low-energy locally excited state of the BH$_3$ fragment ($^{1,3}E$). The ^1E state curve shows clear evidence of an avoided crossing with a higher state that is not shown. The states of interest are the dot–dot excited states $^{1,3}A'$, well described by the charge-transfer biradicaloid configurations $^{1,3}|AB\rangle$. The increase of their energies with increasing separation is readily understood in terms of classical electrostatics of the BH$_3^-$ NH$_3^+$ ion pair. With full geometry optimization, these states might actually become the S$_1$ and T$_1$ states of the molecule. Even at the present level of calculation, however, they provide a clear illustration of the behavior qualitatively expected from the simple model for a δ_{AB} value well in excess of δ_0.

4.8 BEYOND THE SIMPLE MODEL: PHOTOISOMERIZATION OF THE DOUBLE BOND

4.8.1 Covalent Pi Bond

In this section we discuss photochemical reaction paths that proceed through a twisted π-bond geometry. The geometrical isomerization processes in ethylene, protonated formaldimine, and aminoborane can be considered as classical prototypes for cases involving no electronegativity difference between the two termini of the double bond, a significant difference, and an overwhelming difference, respectively. Thus, they are prototypes of covalent, charged, and dative π bonds. The qualitative aspects of the fundamental unity of these three apparently quite dissimilar cases have been discussed on simple models in Part A of this book. Many

4.8 BEYOND THE SIMPLE MODEL: PHOTOISOMERIZATION OF THE DOUBLE BOND

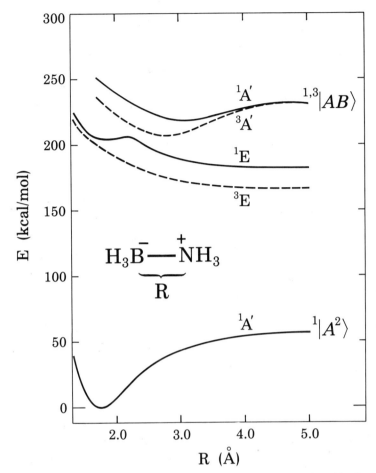

Figure 4.34 Potential energy curves for the dissociation of the B—N bond in the ammonia-borane adduct. All geometric parameters are held fixed; only the B—N distance R is varied; all angles are tetrahedral). *Ab initio* large-scale CI calculation.

related systems in which $\pi\pi^*$ excited states play a dominant role can be derived conceptually by manipulation of the nature of the double-bonded atoms (as in Si=Si) as well as by attachment of substituents at termini of the double bond (as in stilbene).

Unless specified otherwise, the results shown presently were obtained from calculations in which Rydberg states were not considered. This simplification is justified by the fortunate circumstance that at the photochemically important twisted geometries the lowest excited states are all of purely valence character while the Rydberg states are high in energy. At planar geometries this is not so, as mentioned already.

If heteroatoms carrying lone pairs are introduced, $n\pi^*$ excited states may be of low energy and may complicate the situation considerably (Chapter 7).

Ethylene. Figure 4.35 shows a one-dimensional cut through the potential energy hypersurfaces of ethylene along a twisting path of D_2 symmetry. The distance

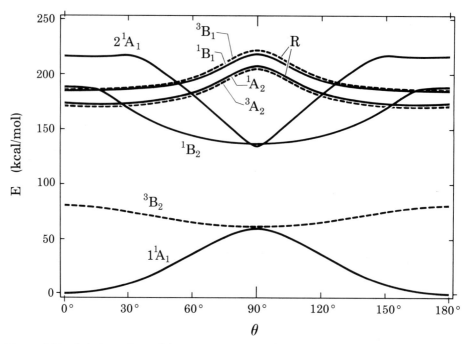

Figure 4.35 Ethylene: Potential energy curves as a function of the twist angle θ. *Ab initio* large-scale CI calculation including Rydberg configurations. R: Rydberg states.

between the two carbon atoms was kept constant at the value calculated to be optimal in the lowest singlet excited state at orthogonal geometry. The HCH angles and the CH bond lengths were those of planar ground-state ethylene at equilibrium geometry. Small variations in these geometrical parameters have little effect. For instance, a calculation in which the separation of the carbon atoms is chosen to vary simultaneously with the twist, starting with the ground-state equilibrium value, yields comparable shapes for the potential energy curves and modifies primarily the absolute values of the excitation energies at nearly planar geometries.

Valence states. The results for the ground state, the lowest two singlet excited states, and the lowest triplet states are similar to those obtained from the correlation diagrams of Chapter 3 and the simpler model calculations of Sections 4.2.2 (Figure 4.10) and 4.6.2. At 90° twist, the computed energy gap between the ground and lowest excited singlet state is 85 kcal/mol. The main differences are that the present calculation shows (i) a crossing between the states S and D (i.e., a touching between the lowest two excited singlet states S_1 and S_2) at a twist angle of 82° and (ii) no crossing between the states $C = S_0$ and $T = T_1$. This reversal of the order of states within the S, D and C, T pairs relative to the simple 3×3 CI picture can be easily understood qualitatively when it is recognized that a double bond contains four electrons and when the correlation of all four is considered. The simplest way to proceed is to construct the double bond from two carbenes.

As will be shown in more detail in Section 5.1.1 (Figure 5.8), two covalent (dot–dot) singlet structures $^1|\overparen{AB}\overparen{CD}\rangle$ and $^1|\overparen{A\underline{BC}D}\rangle$ can be written for four electrons in four orbitals. The former contains singlet-coupled electron pairs ("bonds") in orbital pairs $|A\rangle$, $|B\rangle$ and $|C\rangle$, $|D\rangle$; the latter, in orbital pairs $|A\rangle$, $|C\rangle$ and $|B\rangle$, $|D\rangle$. An example is the two VB structures of cyclobutadiene. Consider the limit of ethylene molecule dissociated into two carbene molecules. The four orbitals in question are the two nonbonding orbitals of each carbene, one of p and one of sp^2 character. Those on one carbene shall be labeled $|A\rangle$ (p) and $|B\rangle$ (sp^2); those on the other, $|C\rangle$ (p) and $|D\rangle$ (sp^2).

The $^1|\overline{AB}\overline{CD}\rangle$ singlet structure now corresponds to the combination of two $^1B_1$ CH_2 molecules, and the $^1|\overline{ABCD}\rangle$ structure corresponds to the combination of two $^3B_1$ CH_2 molecules coupled into an overall singlet. As the carbenes are brought together to form a 90° twisted ethylene, their sp^2 orbitals $|B\rangle$ and $|D\rangle$ interact end-on, and the two spin-coupling schemes behave quite differently. In $^1|\overline{AB}\overline{CD}\rangle$, the electrons engaged in the new sigma interaction of $|B\rangle$ with $|D\rangle$ are singlet-paired and represent a single C—C bond. Each of the other two electrons occupied one of the mutually orthogonal p orbitals $|A\rangle$ and $|C\rangle$ on the twisted double bond. They are mutually coupled into a singlet as well, but this makes little difference energetically because $|A\rangle$ and $|C\rangle$ do not interact. The VB structure $^1|\overline{AB}\overline{CD}\rangle$ dominates in the S_0 state of twisted ethylene.

In the $^1|\overline{ABCD}\rangle$ structure, the electrons engaged in the sigma interaction of $|B\rangle$ with $|D\rangle$ are not singlet-coupled but represent a mixture of a sigma bond (singlet coupling) a sigma antibond (triplet coupling). Since $^3\sigma \rightarrow \sigma^*$ excitation of a C—C bond requires energy, this structure is much less favorable. The other two electrons are again located in each of the two mutually orthogonal nonbonding p orbitals $|A\rangle$ and $|C\rangle$. Now they are coupled into a triplet, but once again this is without much consequence for the energy because the orbitals $|A\rangle$ and $|C\rangle$ do not interact.

At the geometry of twisted ethylene, the low-energy covalent structure $^1|\overline{AB}\overline{CD}\rangle$ keeps a vestige of the triplet spin coupling between electrons in the orbitals $|A\rangle$ and $|B\rangle$ and the orbitals $|C\rangle$ and $|D\rangle$ that made each $^3B_1$ carbene much more stable than $^1B_1$ carbene. The resulting saving in electron repulsion energy thus occurs on both carbon atoms of the twisted ethylene molecule in its S_0 state. This can be contrasted with the situation in the T_1 state, which correlates with a $^1B_1 + {}^3B_1$ carbene combination, in which only one of the carbenes minimizes the repulsion between electrons in its singly occupied orbitals. The resulting lowering of S_0 below T_1 is often referred to as due to "dynamic spin polarization."

In addition, even a weak mixing between the two now energetically quite separated VB structures described by the two spin-coupling schemes will help stabilize the lower-energy one and thus contribute to establishing the S_0, T_1 state order.

The high-energy covalent structure of 90° twisted ethylene $^1|\overline{ABCD}\rangle$ is totally symmetric, like the $^1|Z_2\rangle$ combination of the hole–pair structures, and probably lies only a little above it in energy. The mixing of the two lowers the $^1|Z_2\rangle$ energy below $^1|Z_1\rangle$, which is of the wrong symmetry and cannot benefit from a similar interaction.

Rydberg states. Figure 4.35 also shows the lowest two singlet and the lowest two triplet Rydberg states of ethylene. Although at planar and nearly planar geometries the singlet Rydberg state is the lowest excited singlet state, at highly twisted geometries its energy is well above those of the valence S_1 and S_2 states shown, since it runs closely parallel to the ground-state surface. This behavior is fairly typical of Rydberg states and provides a justification for the use of simpler computational models in which Rydberg states are ignored.

Although the presence of low-lying Rydberg states is not essential for the cis–trans isomerization under discussion presently, it may cause a barrier in the S_1 surface on the way to the orthogonal geometry. These states may also offer access to a variety of other photochemical processes. The relative energies of the lowest Rydberg and the lowest valence state will be affected by the presence of a solvent as well as of substituents that are found in the more complicated olefins.

In the present section, we concentrate on cis–trans isomerization, and for this purpose the photochemical path can be viewed as an initial excitation into a vertical S_1 or T_1 state followed by twisting until the minimum in S_1 or T_1 at orthogonal geometry is reached. Return to the S_0 state and final relaxation to either the cis or the trans planar isomer at equilibrium in the ground state follow.

The nature of the valence-state wave functions. The nature of the four valence states critical for the cis–trans isomerization mechanism (S_0, S_1, S_2, and T_1) is worthy of a more detailed examination. In this respect, the large-scale CI calculations

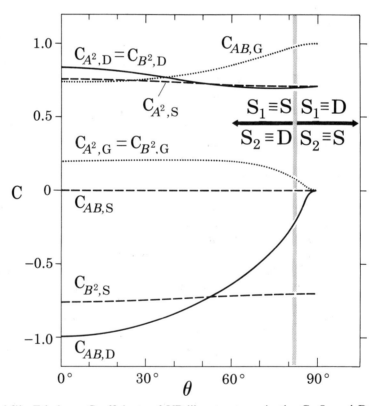

Figure 4.36 Ethylene: Coefficients of VB-like structures in the G, S, and D states as a function of the twist angle θ. *Ab initio* large-scale CI calculation. Near 90° twist, D represents S_1 and S represents S_2. Elsewhere, S represents S_1 and D represents S_2. Reproduced by permission from V. Bonačić-Koutecký, L. Pogliani, M. Persico, and J. Koutecký, *Tetrahedron* **38**, 741 (1982).

4.8 BEYOND THE SIMPLE MODEL: PHOTOISOMERIZATION OF THE DOUBLE BOND

under consideration presently yield a picture very similar to that already developed in the simple model calculations of Section 4.6.2. The primary importance of the detailed consideration of the nature of the wave functions lies in the insight it provides into the course of the potential energy curves.

The nature of the valence-state wave functions will be discussed in VB-like terms that more readily provide a simple physical insight. This will be done in a way similar to that introduced in Section 4.2.2, but it is now much more difficult to transform the *ab initio* CI wave function based on canonical MOs into a VB-like form because it contains contributions from thousands of configurations. The task is made possible by the circumstance that three singlet and one triplet configuration together provide at least 80% of the weight in the total correlated wave function. These are just the configurations that were considered in the simple two-electron two-orbital model of Sections 4.2.2 and 4.6.2. Therefore, we shall assume in the following that an understanding of these dominant contributions represents an approximate understanding of the overall wave function. In this set of calculations, Rydberg states were not considered.

Using procedures outlined in Section 4.1, we can use the knowledge of the coefficients of the configurations $|a^2\rangle$, $|b^2\rangle$, and $|ab\rangle$ in the CI wave function of each state to find the coefficient of the configurations $^1|A^2\rangle$, $^1|B^2\rangle$ and $^1|AB\rangle$ based on the orthogonalized localized orbitals A and B (equations 4.31–4.33). The results are shown in Figures 4.36 and 4.37. Figure 4.36 shows the coefficients of the three VB-like configurations in the ground state, singly excited state, and doubly excited

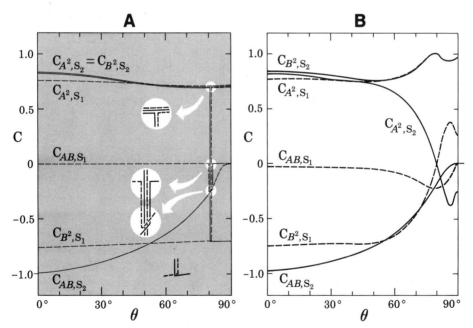

Figure 4.37 Ethylene: Coefficients of VB-like structures in the S_0, S_1, and S_2 states as a function of the twist angle θ. (A) Both CH$_2$ groups planar (the discontinuities in the values of the coefficients are caused by the change in the state order as θ approaches 90°). (B) One CH$_2$ group pyramidalized by 20°. *Ab initio* large-scale CI calculation. Reproduced by permission from V. Bonačić-Koutecký, L. Pogliani, M. Persico, and J. Koutecký, *Tetrahedron* **38**, 741 (1982).

state as a function of the twist angle (these are not true VB configurations, since they are built from orthogonal orbitals $|A\rangle$, $|B\rangle$ rather than the atomic orbitals $|A\rangle$, $|B\rangle$). The ground state, G, is of A_1 symmetry and is represented primarily by the covalent structure $^1|AB\rangle$. The coefficients of the two ionic configurations, $^1|A^2\rangle$ and $^1|B^2\rangle$, are equal ($C_{A^2,G} = C_{B^2,G}$), and their contribution decreases with the twist angle and vanishes at the orthogonal geometry ($C_{AB,G}$). The singly excited state (S) is of B_2 symmetry, and this precludes any contributions of covalent character. It is represented by an out-of-phase combination of the $^1|A^2\rangle$ and $^1|B^2\rangle$ configurations, $C_{A^2,S} = C_{B^2,S}$. This state is highly zwitterionic but not as purely so as suggested by the simplest description of Sections 4.2.2 and 4.6.2, since we are now analyzing only about 80% of the total wave function.

The doubly excited state (D) is of A_1 symmetry and is represented by the in-phase combination of the localized configurations $^1|A^2\rangle$ and $^1|B^2\rangle$, combined with the covalent configuration $^1|AB\rangle$. The weight of the latter diminishes with increasing degree of twist and vanishes at orthogonality. The factor that makes the doubly excited state lower in energy than the singly excited state at angles very close to 90° is increased mixing with configurations other than the three considered here.

In the case of the triplet state the symmetry is B_1. This state is represented essentially purely by the covalent configuration $^3|AB\rangle$ at all twist angles.

It should be noted that the localized orbitals $|A\rangle$ and $|B\rangle$ used in the VB-like description are not mutually orthogonal for twist angles other than 90°. This makes the direct comparison of the numerical magnitudes of the coefficients shown in Figure 4.36 between different states difficult because the squares of the coefficients add up to numbers that are different for every state. This is of particular concern at twist angles close to zero.

Figure 4.37A shows the expansion coefficients once again, this time for the states S_1 and S_2. Because of the surface touching at the twist angle of 82°, the plots for these two states are discontinuous. It may appear arbitrary and unnecessary to plot the coefficients for the S_1 and S_2 states in view of these unpleasant discontinuities, but as soon as symmetry is lowered and the crossing of the singly and doubly excited states at 82° is avoided, the singly excited state S and the doubly excited state D are mixed in the final wave functions of the S_1 and the S_2 states, and plots such as that of Figure 4.37B can only be meaningfully constructed for the S_1 and S_2 surfaces but not for the S and D states.

The effect of pyramidalization. An example of a one-dimensional cut through the potential energy hypersurfaces for the twisting of ethylene which is of lower symmetry than the D_2 case considered above is shown in Figure 4.38 (Rydberg states have been omitted here). The path chosen shows results for a geometry in which one of the termini of the double bonds has been pyramidalized to the extent of 20°. The pyramidalization induces a difference in the electronegativity of the two centers A and B, with the pyramidalized center B now being more electronegative. As expected, in Figure 4.38 the crossing of the singly and doubly excited singlet state S and D is avoided. This can also be traced in the plot of the coefficients of the three leading configurations shown in Figure 4.37B. The coefficients of the three configurations in the ground state S_0 are very similar to what they were in unpyramidalized ethylene (Figure 4.36) and are not shown. The coefficients in the two excited states S_1 and S_2 differ in the case of the unpyramidalized (Figure 4.37A)

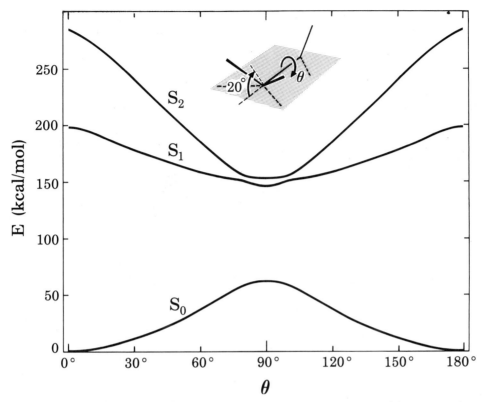

Figure 4.38 Ethylene with one CH$_2$ group pyramidalized by 20°: Potential energy curves as a function of the twist angle θ. *Ab initio* large-scale CI calculation without Rydberg configurations. Adapted by permission from R. J. Buenker, V. Bonačić-Koutecký, and L. Pogliani, *J. Chem. Phys.* **73**, 1836 (1980).

and pyramidalized (Figure 4.37B) ethylene in a way that is easily understood. The sharp discontinuities observed for the case of an unavoided crossing (e.g., for the coefficient C_{AB,S_1}) become rounded when the crossing is avoided.

The most interesting aspect of Figure 4.37 is the behavior of the polar hole-pair configurations $^1|A^2\rangle$ and $^1|B^2\rangle$, which dominate the excited states S_1 and S_2. Because of the difference in the electronegativity of the two centers, the configuration $^1|B^2\rangle$ dominates the lower S_1 state while $^1|A^2\rangle$ dominates the upper state S_2. In each state, the minor polar component is admixed to the major one in a way that changes in the vicinity of the avoided crossing. For S_1 the mixing is out-of-phase for smaller angles and in-phase for angles close to 90°, while for S_2 the opposite holds. This clearly reflects the memory of the S and D states of the planar ethylene.

Because of the imbalance in the contributions from the two hole-pair configurations $|A^2\rangle$ and $|B^2\rangle$, both the S_1 and S_2 states are highly polar and the polarity increases substantially with an increasing twist angle. Since $^1|B^2\rangle$ dominates in S_1 while $^1|A^2\rangle$ dominates in S_2, the directions of the dipoles are opposite in the two states and remain constant for all angles of twist.

The photochemical cis–trans isomerization path. In summary, the photochemical cis–trans isomerization path in the singlet state involves downhill travel from the planar geometry, possibly after overcoming a barrier imposed by the presence of the Rydberg state. This is followed by a brief sojourn at orthogonal geometry, at which S_1 and S_2 are strongly diabatically coupled (Section 1.2.2, Figures 2.10 and 2.11). Coupling to S_0 is much weaker, but after a fair number of vibrations, radiationless decay to S_0 will follow. The exact nature of the geometrical distortions that facilitate this return is still under discussion, and it has been proposed recently that this may be a CCH valence angle distortion. Return to a planar geometry, either cis or trans, on the S_0 surface completes the reaction.

The isomerization in the triplet state is similar except that T_1 is sufficiently lower in energy that Rydberg states are unlikely to impose barriers. The $T_1 \rightarrow S_0$ return at orthogonal geometry is spin-forbidden, but it is facilitated by the very small energy separation and by partial untwisting (Section 4.6.4).

Hydrogen motion. CCH valence angle distortion in singlet orthogonally twisted ethylene represents a beginning of a migratory path of a hydrogen atom from one to the other CH_2 group. A completion of the path would produce methylcarbene, but only the beginning is of immediate interest. The plane defined by the originating carbon, the terminating carbon, and the migrating hydrogen in twisted ethylene (**1**) can be used to distinguish σ and π symmetry. The nonbonding orbital $|B\rangle$ on the originating carbon (on the left in **1–3**) is of π symmetry, and the nonbonding orbital $|A\rangle$ on the terminating carbon (on the right in **1–3**) is of σ symmetry (**1A–3A**). Initially, their energies are equal. As the hydrogen migrates, $|B\rangle$ continues to be well approximated as a $2p_z$ π orbital on the originating carbon. It cannot interact with $|A\rangle$ as long as the symmetry plane is kept and the system is a heterosymmetric biradicaloid in the sense of Section 4.5 ($\gamma_{AB} = 0$), and its energy remains nearly constant. However, $|A\rangle$, which starts as an approximately $2p_z$ σ orbital on the terminating carbon in **1A**, is first destabilized and becomes the C–C antibonding orbital of the triangular CCH array in **2A** (i.e., an orbital analogous to the one whose double, single, or zero occupancy dictates the familiar forbiddenness of anionic and radical 1,2-sigmatropic hydrogen shifts and the contrasting allowedness of cationic ones). Subsequently, however, $|A\rangle$ is stabilized again and finally ends up considerably more stable than $|B\rangle$, since it becomes a nearly pure 2s σ orbital on the originating carbon atom in **3A**.

According to numerical calculations, translated into the simple 3×3 CI model of Section 4.5, the electronegativity difference parameter δ_{AB}, which vanishes in-

4.8 BEYOND THE SIMPLE MODEL: PHOTOISOMERIZATION OF THE DOUBLE BOND

itially (**1A**), becomes sufficiently positive to exceed the critical value δ_0 on the way to **2A**, then recrosses this value again, becomes zero, and finally its absolute value increases again above δ_0 as the geometry of **3A** is reached. If this is correct, an S_0–S_1 touching (conical intersection) is expected each of the three times that δ_{AB} reaches the value δ_0, and the energies of the dot–dot and one of the hole–pair structures become equal. The dot–dot structures are shown in **1–3**. The hole–pair structures are generated by replacing either the dot in the π orbital or the dot in the σ orbital by a plus sign and the other by a minus sign.

The presence of at least one S_0–S_1 touching along the symmetric migration path is guaranteed by the order of the states at the initial (dot–dot singlet lowest) and final (hole–pair singlet lowest) geometries.

Two funnels have been located more recently; the expected region of the third was not searched. It was proposed that the first funnel, located between the structures **1** and **2**, is responsible for the very fast $S_1 \to S_0$ return of simple twisted olefins. It is apparent that at least some paths from excited twisted ethylene to methylcarbene may well be strewn with funnels and that it need not be simple for the molecule to find its way all the way to the carbene product. However, there is experimental evidence that in some cases it actually does.

Propene. A second example of alkene twisting is provided by propene, in which the two termini of the double bond differ in electronegativity. Figures 4.39 and 4.40 represent a cut through the potential energy hypersurfaces taken at a constant

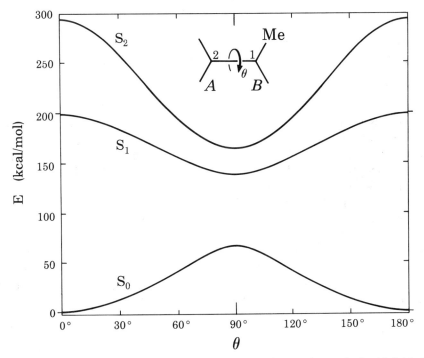

Figure 4.39 Singlet states of propene as a function of the twist angle θ. *Ab initio* large-scale CI calculation. Reproduced by permission from V. Bonačić-Koutecký, L. Pogliani, M. Persico, and J. Koutecký, *Tetrahedron* **38**, 741 (1982).

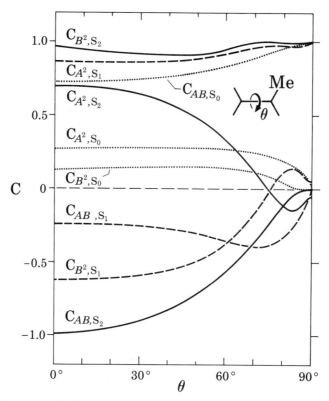

Figure 4.40 Propene: Coefficients of VB-like structures in the S_0, S_1, and S_2 states as a function of the twist angle θ. Ab initio large-scale CI calculation. Reproduced by permission from V. Bonačić-Koutecký, L. Pogliani, M. Persico, and J. Koutecký, *Tetrahedron* **38**, 741 (1982).

CC bond length and constant CH_2 angles. The resulting energy curves for the lower three singlet states, shown in Figure 4.39, are very similar to those shown in Figure 4.38 for ethylene pyramidalized at one end. The behavior of the coefficients in the VB-like description of the large-scale CI wave function is also similar in the two cases, as shown by comparison of Figures 4.37B and 4.40.

The dipole moments of the S_1 and S_2 states of propene are shown as a function of the twist angle in Figure 4.41. The results for singly pyramidalized ethylene are very similar and are not shown. The large increase in the dipoles as the twist angle approaches 90° has been referred to as "sudden polarization" of the S_1 and S_2 states. The suddenness of the onset of the polarization is a function of the degree of the electronegativity difference between the two centers, which dictates the extent to which the crossing of the S and D states is avoided. Although the phenomenon of sudden polarization does not seem to have any consequence for the cis–trans isomerization mechanism as such, it may make other paths available for reactions such as an attack by a polar solvent on the highly polarized alkene. The possibility that the positive end of the polarized twisted olefin is attacked by a nucleophilic solvent will be difficult to distinguish from an attack of the same solvent on a planar Rydberg state of the alkene.

4.8 BEYOND THE SIMPLE MODEL: PHOTOISOMERIZATION OF THE DOUBLE BOND

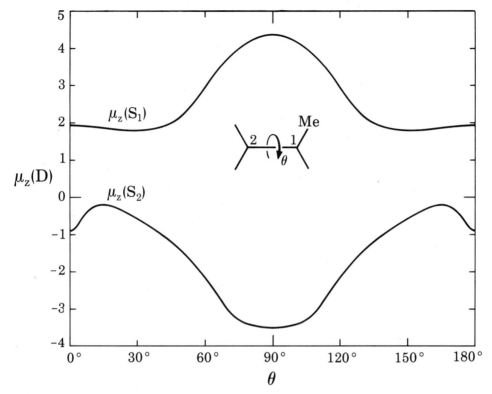

Figure 4.41 Propene: The component of the dipole moment in the S_1 and S_2 states along the C—C axis as a function of the twist angle θ. *Ab initio* large-scale CI calculation. Reproduced by permission from V. Bonačić-Koutecký, L. Pogliani, M. Persico, and J. Koutecký, *Tetrahedron* **38**, 741 (1982).

4.8.2 Charged Pi Bond ($R_2C=N^+R_2$)

Continuing in the direction of an increasing electronegativity difference between the two termini of a double bond, we arrive at charged π bonds. Perhaps the best-known representatives are the protonated Schiff bases, of which formaldiminium is the simplest example. A cut through the hypersurfaces calculated for the twisting of this ion around the CN bond by the same *ab initio* large-scale CI method that was used to obtain the results for ethylene and perturbed ethylenes is shown in Figure 4.42. The particular one-dimensional path displayed was generated by keeping the bond lengths and bond angles constant. For a general twist angle, the molecule is of C_2 symmetry; at planarity and at 90° twist, it is of C_{2v} symmetry.

The trends that were noted upon perturbing the ethylene double bond slightly are strongly accentuated presently. The crossing between the S and D states, which is not avoided in D_2 symmetry ethylene and was only weakly avoided in singly pyramidalized ethylene and in propene, is now avoided very strongly, so that the $^1|B^2\rangle$ configuration dominates the S_1 states at all torsion angles, except that at planar and nearly planar geometries a $\sigma\pi^*$ state is calculated to be the lowest singlet (a Rydberg state may lie below the $^1|B^2\rangle$ state as well). Note that the $^1|B^2\rangle$ structure, which can be represented by $\rangle C^+ — \ddot N\langle$, involves charge relocation relative to the

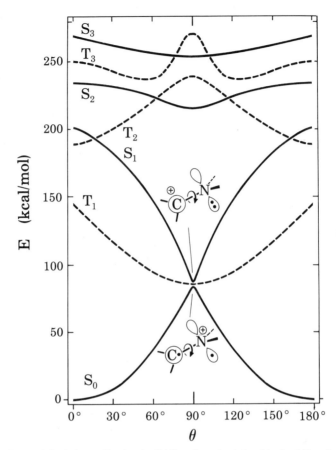

Figure 4.42 Formaldiminium: Singlet (solid lines) and triplet (dashed lines) state energies as a function of the twist angle θ. Ab initio large-scale CI calculation. Reproduced by permission from V. Bonačić-Koutecký, K. Schöffel, and J. Michl, *Theor. Chem. Acta* **72**, 459 (1987).

$^1|AB\rangle$ structure, which can be represented by $\rangle C=N^+\langle$, but not charge separation, as it did in ethylene. This is a consequence of the overall positive charge on the ion due to the presence of an additional proton in nitrogen compared to carbon. On the other hand, the ionic structure $^1|A^2\rangle$ involves an accumulation of two charges on the nitrogen atom and a negative charge on carbon and is so high in energy that it plays no role of interest in photochemistry.

An S_0–S_1 touching. The state characterized by the hole–pair structure $^1|B^2\rangle$ is of comparable stability to the state characterized by the dot–dot structure $^1|AB\rangle$ at the 90° twist angle where their mixing is prevented by symmetry. As the twist angle deviates towards planarity, the two structures can mix and the resulting states separate rapidly. As a consequence, the avoided touching between the S_0 and S_1 surfaces is still present as it was in ethylene but is only very weakly avoided. The degree to which the crossing is avoided and the actual order of the energies of the $^1|B^2\rangle$ and $^1|AB\rangle$ structures at 90° twist are a quite sensitive function of the quality of calculation and of the molecular geometry and will undoubtedly also be quite sensitive to substitution and solvent effects. Clearly, in the methyleneimonium ion we have arrived at just the right electronegativity difference $\delta_{AB} = \delta_0$ of the two termini of the double bond that corresponds to the crossing point in the diagram displaying the energies of the $^1|B^2\rangle$ and $^1|AB\rangle$ configurations as a function of δ (Figure 4.24, cf. Figures 4.22 and 4.26).

Two possible consequences of an S_0–S_1 touching. The touching or near touching of the S_1 and S_0 surfaces may have some interesting consequences (cf. Section 2.2.2). First, it may have observable effects on the quantum yields of the photochemical syn–anti or cis–trans isomerization. Excitation of a planar molecule at 0° twist to the S_1 state will be followed by a twisting motion that will bring it to the region of the crossing, where the nature of the S_1 wave function changes very rapidly. A memory effect may result if the crossing is avoided sufficiently weakly, such that the motion will continue on the ground-state surface towards 180° twist with a high probability. The quantum yield of the cis–trans isomerization may thus reach the limiting value of 1. At the same time, excitation of the planar molecule at 180° twist to S_1 will be followed by motion directed in a mirror fashion, bringing possibly all of the molecules to 0° twist in the ground state, so that this quantum yield might have a value of unity as well.

When the multidimensional nature of the potential energy surface is recognized, it becomes apparent that the memory effect may decrease the quantum yields just as easily as it could increase them, depending on the trajectories taken. In this other limit, both yields could be as low as zero. Either way, interesting deviations from simple expectations could result. On the other hand, the photochemical isomerization in the triplet state should be quite normal and analogous to that in simple alkenes, with the two quantum yields adding to unity if no other processes intervene. It is interesting to note that the $^3|AB\rangle$ configuration is degenerate or nearly degenerate with the $^1|B^2\rangle$ and $^1|AB\rangle$ singlet configurations.

Second, the energetic near-equivalence of the two lowest singlet states at orthogonal geometries and their very different charge distribution, once $\rangle C^+\!\!-\!\!\ddot{N}\langle$ and once $\rangle C\!\!=\!\!N^+\langle$, suggest that it might be possible to stabilize either state preferentially by suitable solvation. If planarization could be prevented, this would lead to two ground-state "isomers" differing in the position of charge and in the organization of the solvent, and these might possibly be interconvertible photochemically. In such a system, light absorption would lead to long-lived ground-state charge translocation, and we shall return to this topic in Sections 6.3.3 and 6.3.4.

4.8.3 Dative Pi Bond ($R_2B^-\!\!=\!\!N^+R_2$)

It is possible to increase the electronegativity difference δ_{AB} between the two termini of a double bond further by removing a proton from the nucleus B and thus arriving at aminoborane (Figure 4.43). Like ethylene, this molecule is again electroneutral, but now the dot–dot structure $^1|AB\rangle$ involves charge separation and the hole–pair structure $^1|B^2\rangle$ does not. Thus the former dominates the excited state S_1, and the latter dominates the ground state S_0. At 90° twist, where the two structures cannot mix, this description is the most accurate. Because of the large difference in energy between the $^1|B^2\rangle$ and $^1|AB\rangle$ structures at 90°, the crossing of the S_0 and S_1 surfaces is again avoided very strongly as was the case in ethylene. The lowest triplet state

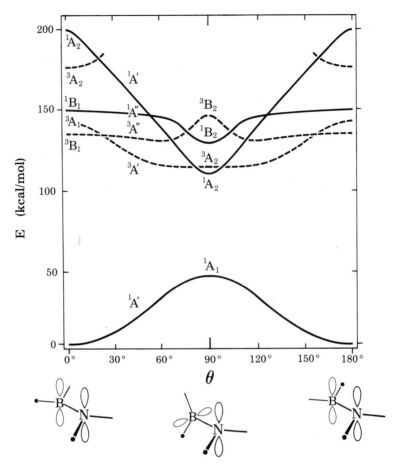

Figure 4.43 Aminoborane: Singlet (solid lines) and triplet (dashed lines) state energies as a function of the twist angle θ. Ab initio large-scale CI calculation including Rydberg configurations. Reproduced by permission from V. Bonačić-Koutecký and J. Michl, *J. Am. Chem. Soc.* **107**, 1765 (1985).

is dominated by the $^3|AB\rangle$ configuration, and at 90° twist this again is close in energy to the corresponding singlet $^1|AB\rangle$ configuration. The latter charge-transfer biradicaloid structure characterizes the lowest excited state S_1. Thus the 90° twisted aminoborane does not behave like a biradical in the sense that its lowest triplet state is not nearly degenerate with the ground state as was the case in twisted molecules whose electronegativity difference δ_{AB} is smaller, but, instead, it is nearly degenerate with its much higher energy excited singlet state.

The reversal of the energies of the dot–dot $^1|AB\rangle$ and hole–pair $^1|B^2\rangle$ structures also has important consequences for the properties of the ground state, which now has only a weak double-bond character with a low bond order even at planar geometry. Indeed, aminoborane is not normally considered a doubly bonded molecule.

Twisted internal charge-transfer (TICT) states. The 90° twisted aminoborane in its dipolar S_1 state is a simple prototype of the so-called twisted internal charge transfer (TICT) state discussed in Section 6.4.2. So far, such states have been primarily of spectroscopic interest, but they can also be viewed as intermediates in cis–trans isomerization of molecules containing a bond with a partial double-bond character.

Photochemical cis–trans isomerization of aminoborane can be expected to proceed by absorption to the S_1 state, followed by twisting to the orthogonal geometry. The charge-transfer biradical twisted S_1 state may proceed to the S_0 state by radiationless transition as is common in all of the twisted bonds discussed so far. However, in view of the large S_1–S_0 separation, it is also quite possible that detectable radiative transition will occur. Either way, vibrational equilibration to either the 0° or 180° geometry should follow. This photochemical isomerization has not received much attention experimentally because compounds with partial double bonds of this kind usually cis–trans isomerize fairly rapidly even in the ground state, since the double-bond character is only weakly pronounced in S_0. The ground-state barrier shown in Figure 4.43 is relatively high, but it is likely to be quite exaggerated because the geometry has not been optimized. If it can be observed, the photochemical reaction may be of an interesting kind where the sum of the quantum yields of fluorescence and of product formation exceeds unity.

The reason why emission from TICT states is not observed more frequently is that often the TICT state is not the lowest excited state of the twisted molecule. Rather, some other excited state, usually of locally excited character, is the lowest excited singlet and is responsible for the observed fluorescence. Frequently, the use of a highly polar solvent can preferentially lower the energy of the dipolar state

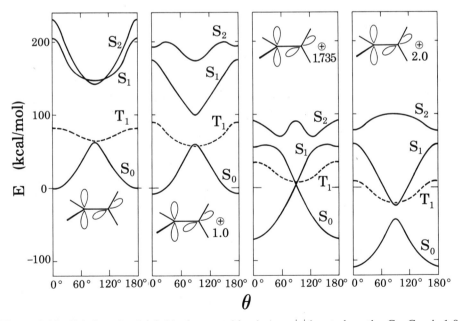

Figure 4.44 Ethylene in the field of an outside charge q|e| located on the C—C axis 1.85 Å from one of the carbon atoms. From left to right, q = 0, +1, +1.735, +2. Energies of singlet (solid lines) and triplet (dashed lines) states are shown as a function of the twist angle θ. *Ab initio* large-scale CI calculation. Reproduced by permission from E. Lippert, W. Rettig, V. Bonačić-Koutecký, F. Heisel, and A. Miehé, *Adv. Chem. Phys.* **68**, 1 (1987).

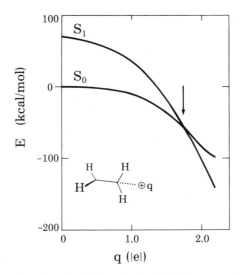

Figure 4.45 Orthogonally twisted ethylene in the field of an outside charge q located on the C—C axis 1.85 Å from one of the carbon atoms. Energies of the S_0 and S_1 states as a function of q. *Ab initio* large-scale CI calculation. Reproduced by permission from V. Bonačić-Koutecký, J. Koutecký, and J. Michl, *Angew. Chem. Int. Ed. Engl.*, **26**, 170 (1987).

and make it the lowest excited singlet. In this case, the TICT fluorescence may be observable even if it is not seen in a nonpolar solvent.

4.8.4 Pi Bond in the Field of a Charge

In Sections 4.8.1–4.8.3 we have compared the *ab initio* computed potential energy surfaces of ethylene, formaldiminium, and aminoborane, and noted that their dependence on the twist angle qualitatively corresponds to expectations based on the simple 3 × 3 CI model (Figure 4.26).

Unlike the simple model calculation, this type of calculation does not permit a continuous variation of the electronegativity difference between the two bond termini. However, this can be achieved even in a numerical computation for a particular molecule, using the artifice of placing a charge next to the molecule. A positive charge placed on the C—C axis will, in effect, make the closer carbon atom more electronegative than the more distant one, and the electronegativity difference is continuously adjustable either by a variation of the magnitude of the charge or by a variation of its distance from the molecule.

The system consisting of ethylene plus point charge can be viewed as modeling either a solvated or an adsorbed ethylene molecule, but in a more general sense it can be viewed as modeling any perturbed double bond with two termini of different electronegativity.

Figures 4.44 and 4.45 show several one-dimensional cuts through the potential energy surfaces. Figure 4.44 shows the state energies as a function of the twist angle for four choices of charge size, while Figure 4.45 shows them at orthogonally twisted geometries as a continuous function of charge size.

The behavior of the energies and of the associated wave functions is that expected from the qualitative arguments. In particular, the S_0–S_1 touching in a critically heterosymmetric biradicaloid is exhibited clearly.

4.9 COMMENTS AND REFERENCES

The first discussions of the theory of **electronic states of biradicals and biradicaloids** with application to photochemistry appeared in 1972: L. Salem and C. Rowland, *Angew. Chem. Int. Ed. Eng.* **11**, 92 (1972); J. Michl, *Mol. Photochem.* **4**, 257 (1972). The subject has been treated in books: W. T. Borden, *Diradicals*, Wiley, New York, 1982; L. Salem, *Electrons in Chemical Reactions: First Principles*, Wiley, New York, 1982. The subject has been updated and summarized recently, with particular emphasis on the concept of critically heterosymmetric biradicaloids and S_0-S_1 degeneracies: V. Bonačić-Koutecký, J. Koutecký, and J. Michl, *Angew. Chem. Int. Ed. Engl.* **26**, 170 (1987). The role of biradicaloid geometries in organic photochemistry has been surveyed repeatedly: J. Michl, *Topics Curr. Chem.* **46**, 1 (1974); J. Michl, *Photochem. Photobiol.* **25**, 141 (1977). Spectroscopy of conjugated triplet biradicals is reviewed in J. Wirz, *Pure Appl. Chem.* **56**, 1289 (1984), nonconjugated flexible chain biradicals are discussed in C. Doubleday, Jr., N. J. Turro, and J.-F. Wang, *Accounts Chem. Res.* **22**, 199 (1989) and G. L. Closs, R. J. Miller, and O. D. Redwine, *Accounts Chem. Res.* **18**, 196 (1985). A recent survey of time-resolved studies of biradicals is L. J. Johnston and J. C. Scaiano, *Chem. Rev.* **89**, 521 (1989). Spin–orbit coupling in biradicals is covered in both of the above references to Salem and in: M. Kasha and H. R. Rawls, *Photochem. Photobiol.* **7**, 561 (1968); T. Lee, *J. Am. Chem. Soc.* **99**, 3909 (1977); A Halevi and C. Trindle, *Israel J. Chem.* **16**, 283 (1977); S. S. Shaik, *J. Am. Chem. Soc.* **101**, 3184 (1979).

The change in the **order of nearly degenerate states** that occurs frequently when the simple 3 × 3 CI model is elaborated requires explicit recognition of the role of other electrons present. Their role is similar in pericyclic systems such as square H_4 or square cyclobutadiene and in twisted ethylene, and this has been the subject of much discussion. The case of a cyclic array was analyzed by: W. T. Borden, *J. Chem. Soc. Chem. Commun.*, 881 (1969); *J. Am. Chem. Soc.* **97**, 5968 (1975); W. Gerhartz, R. D. Poshusta, and J. Michl, *J. Am. Chem. Soc.* **98**, 6427 (1976); H. Kollmar and V. Staemmler, *Theor. Chim. Acta.* **48**, 223 (1978). The case of a twisted double bond was analyzed by J. C. Mulder, *Nouv. J. Chim.* **4**, 283 (1980), who emphasized the formal similarity of the two cases.

In addition to calculations on the **twisting of ethylene** [R. J. Buenker, V. Bonačić-Koutecký, and L. Pogliani, *J. Chem. Phys.* **73**, 1836 (1980)] and propene [V. Bonačić-Koutecký, L. Pogliani, M. Persico, and J. Koutecký, *Tetrahedron* **38**, 741 (1982)], a calculation of the role of distortions of twisted ethylene towards methylcarbene geometry has appeared: I. Ohmine, *J. Chem. Phys.* **83**, 2348 (1985), who revised the results of an earlier calculation with a small basis set [E. M. Evleth and A. Sevin, *J. Am. Chem. Soc.* **103**, 7414 (1981)]. For a recent survey of the photochemistry of simple alkenes, see M. G. Steinmetz, in *Organic Photochemistry*, Vol. 8 (A. Padwa, Ed.), Marcel Dekker, New York, 1987, p. 67.

The concept of **"sudden polarization"** was introduced by: V. Bonačić-Koutecký, P. Bruckmann, P. Hiberty, J. Koutecký, C. Leforestier, and L. Salem, *Angew. Chem. Int. Ed. Engl.* **14**, 575 (1975); V. Bonačić-Koutecký, *J. Am. Chem. Soc.* **100**, 396 (1978) and has been analyzed repeatedly since: for example, V. Bonačić-Koutecký, R. J. Buenker, and S. D. Peyerimhoff, *J. Am. Chem. Soc.* **101**, 5927 (1979); V. Bonačić-Koutecký, J. Čížek, D. Döhnert, and J. Koutecký, *J. Chem. Phys.* **69**, 1168 (1978); J. Koutecký, V. Bonačić-Koutecký, J. Čížek, and D. Döhnert, *Int. J. Quant. Chem.* **12**, 357 (1978). For reviews, see: L. Salem, *Acc. Chem. Res.* **12**, 87 (1979); J.-P. Malrieu, *Theor. Chim. Acta* **59**, 251 (1981). Sudden polarization in planarized allene was investigated by B. Lam and R. D.

Johnson, *J. Am. Chem. Soc.* **105**, 7479 (1983), and, in twisted methylenecyclopropene, by R. D. Johnson and M. W. Schmidt, *J. Am. Chem. Soc.* **103**, 3244 (1981).

The effects of the **introduction of an electronegativity difference** on the ground-state energy and wave function of a biradical were noted by E. M. Evleth, *Chem. Phys. Lett.* **3**, 122 (1969). The general treatment was worked out in: V. Bonačić-Koutecký, J. Köhler, and J. Michl, *Chem. Phys. Lett.* **104**, 440 (1984); V. Bonačić-Koutecký and J. Michl, *J. Am. Chem. Soc.* **107**, 1765 (1985). References to twisted internal charge transfer states are given in Section 6.5.

Spin-orbit coupling in biradicals and biradicaloids was qualitatively analyzed in: L. Salem and C. Rowland, *Angew. Chem. Int. Ed. Engl.* **11**, 92 (1972). Some of the conclusions have been confirmed recently by *ab initio* calculations: R. A. Caldwell, L. Carlacci, C. E. Doubleday, Jr., T. R. Furlani, H. F. King, and J. W. McIver, Jr., *J. Am. Chem. Soc.*, **110**, 6901 (1988).

CHAPTER 5

Cyclic Multicenter Reactions: One Active Orbital Per Atom

Many organic photochemical reactions are believed to follow geometric paths along which new bonding interactions are generated and old ones destroyed in a cyclic array, more or less in concert. Although the formation or breaking of some of the bonds may lag behind others, as long as the cyclic interactions are important, they tend to keep any initially present stereochemical information intact, and this has practical consequences in that those reactions whose whole course is "concerted" yield products in a stereospecific manner.

A commonly used designation for a pericyclic process states the number of participating orbitals in each component: 2 + 2, 2 + 4, 2 + 2 + 2, and so on. Subscripts s and a are used to differentiate between suprafacial and antarafacial participation of a component: $2_s + 2_s, 2_s + 2_a$, and so on. In a suprafacially reacting partner, the lobes of the orbitals involved in overlap with the orbitals of the neighboring partner or partners are separated by an even number of intra-orbital nodal surfaces (usually zero). An example is the two lobes on the AOs in the butadiene positions 1 and 4 that are located on the same side of the molecular plane. In a Diels–Alder 2 + 4 cycloaddition, these are the lobes used to make the new σ bonds. In an antarafacially reacting partner, the lobes of the orbitals involved in overlap with the neighboring partner or partners are separated by an odd number of intraorbital nodal surfaces (usually one). An example is the two lobes on the AOs in ethylene positions 1 and 2 that are located on opposite sides of the molecular plane, which coincides with the intra-orbital node of all of the carbon 2p orbitals of π symmetry. If the new bonds are formed on opposite sides of the plane, the ethylene is said to participate in 2_a fashion.

Sometimes it is desirable to indicate whether the participating orbitals of a component interact with each other in a sigma or a pi manner. This can be shown by adding the subscript σ or π in front: $_\pi 2_s, _\sigma 2_a$, and so on.

In the present chapter, we concentrate on processes of the $2_s + 2_s$ kind, which do not suffer from steric problems and in which the number of electrons a calculation has to contend with is still relatively small. However, the concepts have a more general validity. As usual, we shall use the qualitative tool of correlation diagrams to bring out the generality of the results and to indicate their possible limitations.

The chapter is limited to processes involving one active orbital per participating atom, analogous to the bitopic two-center processes of Chapter 4. Reactions involving atoms that participate through two or more orbitals are analogous to polytopic two-center processes and will be handled jointly with the latter in Chapter 7.

Finally, we do not consider cycloadditions that only appear to be photochemical but actually occur on a ground-state surface. In these, the photochemical step merely generates the requisite reactive intermediates, such as an ion pair or a small-ring *trans*-cycloalkene.

5.1 ELECTRONS IN A CYCLIC ARRAY OF ORBITALS

Reaction paths proceeding through geometries in which the reacting species contain electrons in an array of n orbitals with cyclic overlap are common in organic photochemistry. Antiaromatic arrays are of particular interest because of their biradicaloid nature. These are 4N-electron arrays of the Hückel type, where all overlaps between neighboring orbitals can be chosen positive, and (4N + 2)-electron arrays of the Möbius type, where an odd number of overlap integrals is negative for any choice of basis orbital phases. These are the situations commonly labeled "excited-state-allowed" and "ground-state-forbidden" in what we refer to as the usual orbital symmetry rules (Woodward–Hoffmann rules, formulated in this particular fashion by Dewar and by Zimmerman). The rules also state that aromatic arrays, (4N + 2)-electron Hückel and 4N-electron Möbius, correspond to "excited-state-forbidden" and "ground-state-allowed" paths, but these are of less interest in photochemistry. At any rate, in order to understand photochemical pericyclic processes, we need to go well beyond this elementary formulation of the rules.

The electronic states of antiaromatic systems have been quite well understood from π-electron theory for a long time. Two distinct cases are important: (i) "uncharged perimeters," Hückel 4N = n or Möbius 4N + 2 = n, and (ii) "charged perimeters," Hückel 4N $\neq$ n or Möbius 4N + 2 $\neq$ n. Of these, the uncharged perimeter case is the more important in practice, in particular the Hückel N = 1 and N = 2 cases, which correspond to suprafacial $2_s + 2_s$ and $4_s + 4_s$ cycloadditions, respectively. In Section 5.1.1, we shall consider in detail the nonpolar Hückel N = 1 case, which can be viewed as a cyclic interaction of two nonpolar bonds. In Section 5.1.2 we shall examine in a less detailed manner the polar Hückel N = 1 case, that is, a cyclic interaction of two polar bonds. An extension to other uncharged perimeters and a brief discussion of the new aspects presented by the charged perimeters will be found in Section 5.1.3.

5.1.1 Two Nonpolar Bonds

We shall consider first the simplest model case as a prototype: a collection of four protons and four electrons, in the minimum basis set approximation. This is a rather poor model for bimolecular photochemistry of molecular hydrogen, in which Rydberg orbitals must be important. However, it will permit us to illustrate many concepts important for the cyclic interaction of two nonpolar chemical bonds in various states of excitation and thus for organic photocycloadditions. Along the "concerted" reaction path of approach of two H_2 molecules, the four H atoms lie at the vertices of a rectangle or a trapezoid, and this will be investigated presently. The "nonconcerted" approach will be addressed in Section 6.1. A few other approaches, less interesting in the present context, have also been investigated in the literature. A full exploration of the six-dimensional nuclear configuration space of H_4 has never been undertaken, although it might bring additional interesting insights.

We shall analyze the results of *ab initio* calculations using a minimum basis set. The exact results for state energies obtained within this simple model will be compared with three approximate descriptions: (i) a simple MO picture based on MOs of the total H_4 system followed by 3×3 configuration interaction, (ii) a simple VB picture without dot–dot ↔ hole–pair (covalent ↔ ionic) resonance, and (iii) a subsystem picture based on the states of two H_2 molecules. This comparison provides a clear picture of the nature of electronic wave functions.

We have initially chosen more or less arbitrarily a rectangular reaction path for the transformation H_A—H_C + H_B—H_D ⇆ H_A—H_B + H_C—H_D and shall consider the more general trapezoidal path later (Figure 5.1). Figure 5.2 presents the state energies obtained in the four-orbital (minimum basis set) approximation. The choice of the path was guided by more complete calculations of the energies of the three lowest-energy singlet states for all trapezoidal geometries. Although the detailed shape of the state energy curves changes as the path is adjusted, the general features and the nature of the electronic states have been found to remain unaffected, and the curves shown in Figure 5.2 are schematic but representative.

The exact solution (four orbitals, four electrons). In the lowest singlet state G (C), the reaction is seen to have a large activation barrier, apparently resulting from an avoided touching with the state labeled D. This is not a surprise, and the reader has undoubtedly already identified the ground-state $2_s + 2_s$ reaction as a Woodward–Hoffmann forbidden process. Several of the low-energy excited states have a minimum in the region of the ground-state activation barrier. For two of them, labeled S (Z_1) and D (Z_2), the minima are particularly deep.

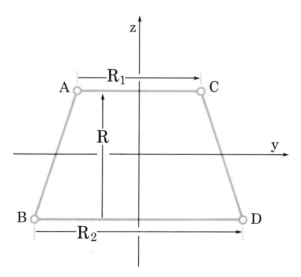

Figure 5.1 Coordinate system and geometric parameters of trapezoidal H_4.

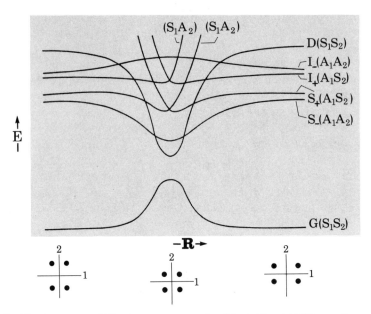

Figure 5.2 Singlet states of H_4 along a rectangular $H_2 + D_2 \rightleftarrows 2HD$ reaction path. Schematic, adapted from an *ab initio* full CI minimum basis set calculation with optimized orbital exponents. Symmetry labels shown refer to the two symmetry planes indicated at the bottom.

Our next task is to rationalize the shapes of the energy curves for the low-lying excited states in simple terms and to understand the nature of their electronic wave functions.

The simple four-state MO picture

The molecular orbitals. In the minimum basis set, the four canonical MOs of a rectangular H_4 molecule are determined by symmetry (Figure 5.3). In the D_{2h} group, they belong to the a_g and b_{1u} (bonding) and b_{2u} and b_{3g} (antibonding) representations. At infinite separation they are pairwise degenerate, and their energies are identical to those in an isolated H_2 molecule. As the H_A—H_C and H_B—H_D molecules approach, the mutual interaction of their MOs grows and the degenerate pairs split. The energy of the b_{1u} MO increases, and that of the b_{2u} decreases. As a square geometry is reached, the delocalized MOs become those familiar from cyclobutadiene (cf. Figure 4.19C). In the D_{4h} symmetry group of the square, we have a bonding MO of a_g symmetry, the b_{1u} and b_{2u} orbitals of the rectangle now form a degenerate pair of nonbonding MOs of e_u symmetry, and there is an antibonding MO of b_g symmetry. If the square is now deformed into a rectangle with the H_A—H_B and H_C—H_D pairs on the short sides, the degeneracy of the nonbonding MO pair is lifted again. The b_{1u} MO continues to be destabilized, and the b_{2u} MO can be stabilized. At infinite separation of the H_A—H_B and H_C—H_D molecules, the orbitals a_g and b_{2u} are degenerate (bonding MO of H_2) and b_{3g} and b_{1u} are

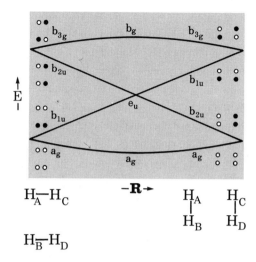

Figure 5.3 Rectangular reaction path for $H_2 + D_2 \rightleftarrows 2HD$. MO correlation diagram.

degenerate (antibonding MO of H_2). The crossing of the b_{1u} and b_{2u} MOs along the rectangular reaction path is thus demanded by symmetry. In the vicinity of the square geometry, the geometry is biradicaloid, since two approximately or exactly nonbonding MOs are available for two electrons in the ground state.

The states. From what we have already learned on the case of H_2, it is clear that it would be hopeless to try to describe the singlet electronic states of H_4 at the biradicaloid geometries by individual configurations without configuration interaction. The minimum extent of configuration interaction which has any chance of success for singlet states is 3×3 CI; this has been used repeatedly in the literature as a basis for numerical computations.

The 3×3 CI approximation is strictly analogous to the full treatment of two electrons in two orbitals, which we have already discussed in Chapter 4 in considerable detail. These are the nearly degenerate b_{1u} and b_{2u} MOs. Excitations from and into other MOs (a_g, b_{3g}) are not considered. The three singlet configurations then have MO occupancies $a_g^2 b_{1u}^2$, $a_g^2 b_{1u}^1 b_{2u}^1$ and $a_g^2 b_{2u}^2$. An analogous treatment of the triplets takes into account a single set of three triplet configurations with MO occupancy $a_g^2 b_{1u}^1 b_{2u}^1$ and three different spin functions.

Although the 3×3 CI description of the singlet states is fairly commonly used, we have already noted that it has important shortcomings. These can be overcome by inclusion of additional configurations. A total of 20 singlet configurations, 15 triply degenerate triplet configurations, and one quintuply degenerate quintet configuration are possible. It is usually not necessary to perform full CI, i.e., to consider all of the configurations of a given multiplicity, in order to obtain good results, but we have nevertheless listed all 20 singlet configurations in Figure 5.4. Note that two of the configurations, $^1|\overline{2\ 1}-\overline{1\ -2}\rangle$ and $^1|\overline{2\ 1}-1\ -2\rangle$, do not differ in orbital occupancy but only in the spin assignment, i.e., in spinorbital occupancy. They are linearly independent, since there are two linearly independent singlet spin functions for four single electrons in four orbitals. In $^1|\overline{2\ 1}-\underline{1\ -2}\rangle$ the electrons in orbital

230 CYCLIC MULTICENTER REACTIONS: ONE ACTIVE ORBITAL PER ATOM

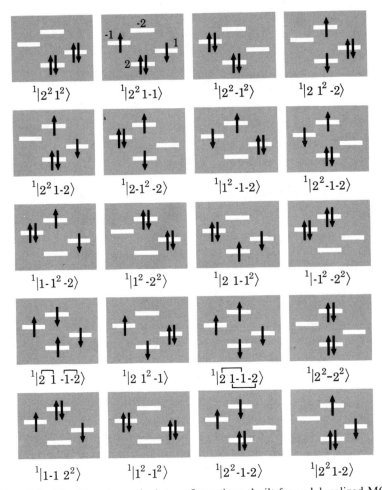

Figure 5.4 The H_4 system: singlet configurations, built from delocalized MOs.

pairs 2, 1 and -1, -2 have singlet-coupled opposite spins and the spin function is $(1/2)[(2\ \bar{1}) - (\bar{2}\ 1)][(-1\ -\bar{2}) - (-\bar{1}\ -2)]$, where an occupancy of an orbital with an electron of spin β is indicated by a bar, and occupancy with an electron of spin α is denoted by the absence of a bar (cf. Section 4.1). In $^1|2\ 1\ -1\ -2\rangle$, the electrons in orbital pairs 2, -1 and 1, -2 are singlet-coupled and the spin function is $(1/2)[(2\ -\bar{1}) - (\bar{2}\ -1)][(1\ -\bar{2}) - (\bar{1}\ -2)]$. In an actual calculation one would need to recognize that these spin functions are not mutually orthogonal, but for our present purposes this is immaterial.

The three times 15 triplet configurations are obtained from the 20 singlet configurations by (i) deleting the six with closed-shell orbital occupancies, (ii) combining each of the remaining ones with three triplet spin functions leading to the $+\hbar$, 0, and $-\hbar$ components, and (iii) recognizing that there are three linearly independent triplet spin functions for four single electrons in four orbitals for each of the three values of spin projection into the z axis ($+\hbar$, 0, and $-\hbar$), so that the 2, 1, -1, -2 occupancy now leads not to two but to nine independent configurations.

The quintet configurations are obtained by combining the orbital occupancy 2, 1, −1, −2 with the five possible quintet spin functions that correspond to the five values of spin projection into the z axis, $+2\hbar$, $+\hbar$, 0, $-\hbar$, $-2\hbar$.

The simple four-state MO picture, i.e., the approximation of the states of H_4 based on 3 × 3 CI for singlets and including only the lowest triplet, yields the energies shown in Figure 5.5B. Comparison with the full CI curves along the same reaction path, shown in Figure 5.5A, is instructive. Aside from a general decrease of the energy of all states in the better description, which is of little consequence for the understanding of the nature of the electronic wave functions and of photochemical processes, we note that the four states of interest (G or C, S or Z_1, D or Z_2, and T) are actually described quite well in the simple four-state MO picture. Except in the vicinity of the square geometries, the ground state G is primarily represented by the ground configuration $^1|2^21^2\rangle$, the S state by the singly excited configuration $^1|2^21-1\rangle$, and the D state by the doubly excited configuration $^1|2^2-1^2\rangle$. As a square geometry is approached, the simple description of G as $^1|2^21^2\rangle$ and D as $^1|2^2-1^2\rangle$ becomes unacceptable because the two configurations mix strongly. At square geometries, the simple MO description depicts the G state as $|G (3 \times 3, sq.)\rangle = (1/\sqrt{2})(^1|2^21^2\rangle - {}^1|2^2-1^2\rangle)$ and the D state as $|D (3 \times 3, sq.)\rangle = (1/\sqrt{2})(^1|2^21^2\rangle + {}^1|2^2-1^2\rangle)$, in complete analogy to the already familiar results for a fully dissociated H_2 molecule. For the S state, one has $|S (3 \times 3, sq.)\rangle = {}^1|2^21-1\rangle$, the same as at rectangular geometries. Obviously, the configurations $^1|a_g^2 b_{1u}^2\rangle$ and $^1|a_g^2 b_{2u}^2\rangle$ undergo an avoided crossing along the reaction path since the orbitals b_{1u} and b_{2u} cross. As a consequence, the configurations $^1|2^21^2\rangle$ and $^1|2^2-1^2\rangle$ undergo an avoided touching (note that b_{1u} is 1 and b_{2u} is −1 on one side of the square geometries, while b_{1u} is −1 and b_{2u} is 1 on the other side).

The resulting avoided touching of the G and D states satisfactorily mimics the behavior of the full CI solutions and accounts for the barrier in the G state and a minimum in the D state. The central minimum in the S state also is reproduced qualitatively correctly. However, the comparison of the simple four-state MO picture with the full CI solution reveals two serious difficulties. One of these has to do with the state ordering at square geometries, and the other with the limiting behavior upon dissociation to two H_2 molecules. We shall now consider the nature and origin of the two difficulties.

Square geometries and the MO–VB dichotomy.

The MO picture. In the simple four-state MO picture, i.e., 3 × 3 CI for singlets and a single triplet, the D state lies above the S state at all geometries, as it did in the H_2 molecule. Also, the G state lies above the T state at all square geometries (and, indeed, at any other geometries at which the two nonbonding orbitals of a biradical are exactly degenerate). At square geometries, the two energy differences, $E(D) - E(S)$ and $E(G) - E(T)$, are equal and are given by twice the value of the exchange integral between the localized MOs $|B'\rangle = (1/\sqrt{2})(|1\rangle - |-1\rangle)$ and $|A'\rangle = (1/\sqrt{2})(|-1\rangle + |1\rangle)$. Making the identification $|1\rangle = |b\rangle$, $|-1\rangle = |a\rangle$, this agrees with the results given in Chapter 4 (equations 4.94 and 4.95). Here, we use primes on the labels of the localized orbitals $|A'\rangle$ and $|B'\rangle$ in order to avoid confusion with the atomic orbitals $|A\rangle$ and $|B\rangle$ located on H_A and H_B.

The localized MOs are shown in Figure 5.6 (cf. Figure 4.19C) along with their overlap density. The self-repulsion of this density, which defines the exchange

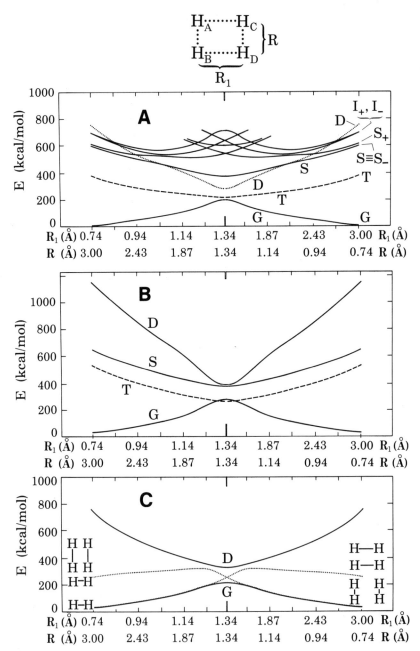

Figure 5.5 Rectangular H_4 in minimum basis set ("$2_s + 2_s$ cycloaddition paths"). (A) Full CI; (B) 3 × 3 CI; (C) energies of the dot–dot VB structures (dotted lines) and of the two covalent states resulting from their mixing (solid lines).

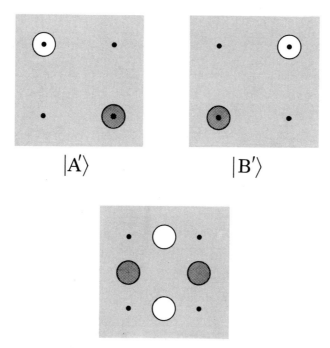

Figure 5.6 The localized MOs $|A'\rangle$ and $|B'\rangle$ and their overlap density (schematic).

integral $K_{A'B'}$, will clearly be small if the square is large. In the limit of an infinitely large square, it is zero and the G and T states are degenerate, as are the S and D states. As the square becomes smaller, the exchange integral increases and the degeneracy disappears. For the square size selected in Figure 5.5, the degeneracy is already quite clearly removed.

Compared with the full CI solution given in Figure 5.5A, the prime difficulty with the simple four-state MO picture lies in the description of the D state. A minor difficulty is encountered with the G state.

Instead of remaining above the S state, in the full CI solution the D state quite clearly dips below it at square geometries, so that it now represents the lowest excited singlet state, S_1. The difficulty is particularly clearly revealed in a plot of state energies for square geometries as a function of the square size (Figure 5.7A). The state ordering anticipated from the simple MO picture (Figure 5.7B) is correct only in the limit of very small squares. For larger squares, the behavior of the D state obtained from the simple MO description is completely wrong, since it keeps its "zwitterionic" hole–pair nature and does not approach the proper limit at infinite square size. This situation is quite analogous to that encountered with the simplest MO description of the dissociation of the H_2 molecule. It is obvious that it can be remedied by an increase in the number of configurations used in the CI. The configurations that need to be added are $^1|1^2 - \bar{1}^2\rangle$, $^1|2^2 - \bar{2}^2\rangle$, $^1|2\,\bar{1} - \bar{1}\,\bar{2}\rangle$ and $^1|\bar{2}\,1 - 1\,\bar{-2}\rangle$. It is not immediately obvious why these particular configurations are required and why they are needed so badly for the D state and yet not for the S state. Similarly as in the case of H_2, we shall find answers to these questions in the alternative view of the electronic wave function provided by the VB method.

The difficulty with the G state also occurs at square geometries. In the simple

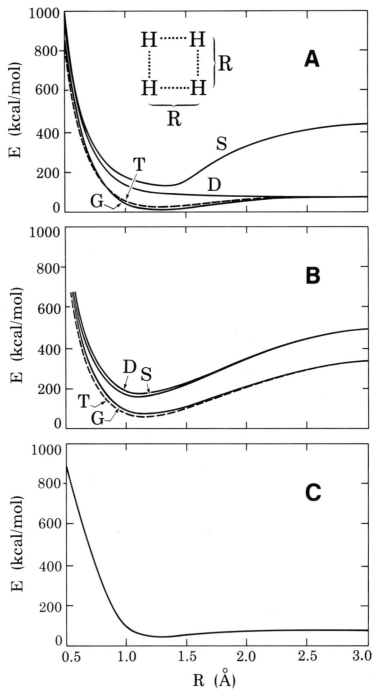

Figure 5.7 Square H_4 in minimum basis set. Singlet-state energies. (A) Full CI; (B) 3 × 3 CI; (C) energies of the two degenerate dot–dot VB structures, whose mixing yields the G and D states in the simple VB picture. In a larger basis set calculation, the ionic S state will be differentially stabilized.

four-state MO picture, G is predicted to lie above T, while in the full CI solution the opposite order is found; however, the energy differences involved are quite small.

The VB picture: The dot–dot (covalent) states. The 20 singlet VB structures of H_4 are shown in Figure 5.8. The two dot–dot (covalent) structures, $^1|\overgroup{AB\overgroup{CD}}\rangle$ and $^1|\overgroup{A\overgroup{BC}D}\rangle$, which differ only in spin assignment, involve no charge separation. The former, $^1|\overgroup{AB\overgroup{CD}}\rangle$, involves a singlet coupling between H_A and H_B and between H_C and H_D. Its energy as a function of rectangular geometry is shown as a dotted line in Figure 5.5C. According to our prior discussion of the states of H_2, it should represent an approximation to the ground state of two noninteracting diatomics, H_AH_B and H_CH_D, at intermediate and long internuclear separations A—B and C—D. It represents exactly such a ground state in the limit of fully dissociated H_AH_B and H_CH_D molecules. The energy of this VB structure will be low if H_AH_B and H_CH_D have ordinary bond lengths and do not interact much, that is, on the right-hand side of Figure 5.5C.

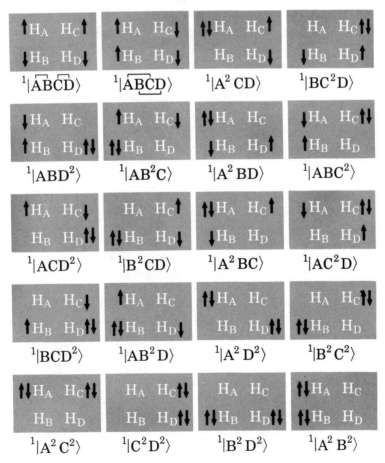

Figure 5.8 The H_4 system: singlet VB structures, built from AOs.

In this structure, $^1|\overline{AB}\overline{CD}\rangle$, there is mixed singlet-like and triplet-like interaction between H_A and H_C and between H_B and H_D. Its energy will therefore be high if the distances H_AH_C and H_BH_D are short and the distances H_AH_B and H_CH_D long, i.e., on the left-hand side of Figure 5.5C. The second covalent structure, $^1|\overline{AB\overline{CD}}\rangle$, involves singlet coupling between H_A and H_C and between H_B and H_D, and mixed singlet-like and triplet-like coupling between H_A and H_B and between H_C and H_D. Its energy is represented in Figure 5.5C by a dotted curve which mirrors that of the other covalent VB structure, and it is high on the right-hand side. Mixing of the two structures yields a higher-energy singlet state that represents a combination of the triplet states of two noninteracting diatomics H_AH_C and H_BH_D, that is, a state of double triplet excitation, coupled into an overall singlet, and a lower-energy singlet state that represents an approximation to the ground state of a pair of noninteracting molecules H_AH_C and H_BH_D.

At square geometries, the two covalent VB structures are degenerate. Their crossing in Figure 5.5C will be avoided when they are allowed to interact, since they are both totally symmetric. This interaction will convert the crossing of $^1|\overline{AB}\overline{CD}\rangle$ with $^1|\overline{AB\overline{CD}}\rangle$, shown in Figure 5.5C, into an avoided crossing that will look very much like the avoided touching of the G and D states obtained from the full CI calculation. Thus, like the simple four-state MO description, the simple VB description accounts correctly for the presence of a reaction barrier in the ground-state surface and for the deep central minimum in the D state. Moreover, it provides a pictorial physical significance for the D state. From the simple four-state MO description, we already know that it is a doubly excited singlet state which accounts for our choice of label. The VB analysis tells us that it is that doubly excited singlet state in which each of the two H_2 molecules is in an excited triplet state.

The VB picture: The hole–pair (zwitterionic) states. It is much harder to derive the course of the energy curves for the other states by inspection of the VB structures of Figure 5.8. Since there were only two dot–dot (covalent) VB structures, and since these were needed to approximate the G and D states, all other states must be at least partly hole–pair (zwitterionic) in character. One can generally expect the VB structures with a single charge separation, such as $^1|A^2CD\rangle$, to have energies below those with a double charge separation, such as $^1|A^2C^2\rangle$. All of the ionic structures can be expected to lie high in energy at large internuclear separations between the charged centers and to be stabilized as the separation decreases. Thus, it is hardly a surprise to see minima for at least some of the energy curves at square geometries. Experience from H_2 suggests, however, that in the region of compact geometries the interaction between the individual ionic VB structures may be strong and may have a large effect on the final shapes of the state energy curves. This suspicion is corroborated by inspection of Figure 5.5A.

It is obvious that symmetry-adapted combinations of the individual hole–pair VB structures have to be used to approximate the states of the H_4 system, and the knowledge of the VB picture of the states of an H_2 molecule then permits an assignment of the physical nature of the states. For instance, on the right-hand side of Figure 5.5, where the distances A—B and C—D are short and A—C and B—D long, the combination $^1|A^2CD\rangle - ^1|B^2CD\rangle$ represents approximately a singlet excitation in the H_AH_B molecule, with the H_CH_D molecule in the ground state. The combination $^1|ABC^2\rangle - ^1|ABD^2\rangle$ represents approximately a singlet excitation in the H_CH_D molecule, with the H_AH_B molecule in the ground state. The in-phase and out-of-phase mixtures of these two combinations are then symmetry-adapted and correspond to the two exciton states. When we need to refer to them individually, we shall use the labels S_+ and S_-, respectively. On the left-hand side of Figure 5.5, the same configurations have a different physical significance. Here, the A—C and B—D distances are short, so that it is sensible to speak in terms of molecules $H_A H_C$ and $H_B H_D$. Then, $^1|A^2CD\rangle$, $^1|ABC^2\rangle$, $^1|AC^2D\rangle$, and $^1|A^2BC\rangle$ are charge-transfer structures in which an electron has been transferred from $H_B H_D$ to $H_A H_C$, and these need to be combined with analogous configurations describing electron transfer in the opposite sense. The energy of the charge-transfer structures is particularly sensitive to the distance of the two H_2 molecules. These structures contribute significantly to low-lying excited singlets only at quite compact geometries. It is qualitatively resonable that the lowest among the zwitterionic excited states should have a minimum at square geometries: One expects it to contain a contribution from VB structures describing the stabilized exciton state as well as contributions from various charge-transfer structures. At large intermolecular distances, the charge-transfer structures combine to produce charge-transfer (ion-pair) states. The lowest two of these, which are then nearly degenerate, correspond to the in-phase and out-of-phase combinations of the structures $H_2^+ H_2^-$ and $H_2^- H_2^+$. When needed, we shall refer to them as the I_+ and I_- states, respectively.

The large and small square limits. Clearly, the description by VB structures provides far more physical insight than the previous description in terms of configurations built on MOs that are delocalized over the whole system. Still, the existence of a large number of ionic VB structures confuses the issue. We shall see below that a happy compromise can be reached by describing the states of H_4 in terms of the states of the individual H_2 molecules.

Before proceeding to that, however, we need to return to the case of square geometries. There, the simple four-state MO description provided a singlet state order G, S, D, with S and D becoming degenerate in the limit of an infinitely large square (Figure 5.7B). The simple VB description yields two covalent states, G and D, which are degenerate in the limit of an infinite square, and higher lying ionic configurations, a suitable mixture of which must stand for the S state (Figure 5.7C). While neither simple picture agrees exactly with the full CI description (Figure 5.7A), the simple VB picture comes much closer if the square is large, whereas the simple four-state MO picture does not do too badly if the square is small. Thus, the situation is reminiscent of that in the H_2 molecule again. As before, the intermolecular distances of interest in real molecules are such that neither simple picture applies.

In order to understand what is wrong with the way in which the simple four-state MO approach describes the D state, we need to express the three singlet configurations $^1|2^2 1^2\rangle$, $^1|2^2 1\ -1\rangle$, and $^1|2^2 - 1^2\rangle$ in terms of VB structures. It is then found that G is a largely dot–dot (covalent) state but S and D are purely and largely hole–pair (ionic) states, respectively, for a square of any size. Yet we know from the analysis of bonding in VB terms, as well as from the full CI calculation, that for an infinitely large square the D state is purely dot–dot (covalent) and degenerate with the ground state. After all, the two states then differ only by the orientation of electron spins on infinitely separated hydrogen atoms, and this clearly cannot affect the energy. It is then obvious why much additional configuration interaction will be needed beyond the 3 × 3 level to subtract the inappropriate hole–pair (ionic) character from the wave function of the D state. This additional CI effort will be the largest at the infinite square geometry, where the resulting wave function must be purely covalent, but it is still substantial at geometries of actual interest. This "memory" of the covalent nature which the D state has in the limit of large squares causes it to be of mixed nature, dot–dot and hole–pair (i.e., covalent ↔ zwitterionic), at intermediate square geometries. The addition of covalent character lowers the energy of the D state, and we can now understand that at intermediate and large square geometries, it will lie below the S state.

In summary, then, both methods agree that the G state is predominantly of dot–dot (covalent) nature and the S state of hole–pair (zwitterionic) nature at the biradicaloid square geometries. The simple VB method without covalent-ionic mixing is right in stating that the D state is covalent and lies below the S state for very large squares, and the simple four-state MO method is right in stating that the D state is zwitterionic and lies above the S state for very small squares. For most square sizes of interest, such as those near the S_1 minimum, neither simple answer is right. For these square geometries, the D state is of mixed hole–pair and dot–dot (i.e., ionic–covalent) nature, and it lies below the S state (Figure 5.7A–C).

Somewhat similar qualitative arguments can be used to understand the reversal of the G–T state ordering between the four-state MO picture and the full CI solution ("dynamic spin polarization").

Rectangular geometries and the subsystem configuration picture. A second difficulty with the simple four-state MO picture is that it does not yield the proper dissociation behavior for the S and D states in the limit of separation into two H_2 molecules, as is seen in the extreme left and extreme right of Figure 5.5. Here, once again, the flexibility of the 3×3 CI procedure is insufficient for proper rendition of the nature of these electronic states. From the VB description as well as from the full CI results, we know that the G state correlates with two H_2 molecules in their ground states, that the S state correlates with a combination of one ground state and one singly excited H_2 molecule, and that the D state correlates with two H_2 molecules, each in its triplet state, coupled into an overall singlet.

While it is possible to understand the behavior of the H_4 states in terms of the VB structures, a useful alternative view is provided by an analysis based on the electronic states of the two H_2 subsystems. Then, the zero-order states for a discussion of H_4 states at any rectangular (or square) geometry are the various possible combinations of the G, T, S, and D states of the individual H_2 molecules and of charge-transfer states in which electrons have been transferred from one H_2 system to the other. This description is guaranteed to yield the proper limiting behavior for the dissociation into two H_2 molecules, but it is more cumbersome in the central region of biradicaloid geometries.

Since the MO and VB expressions for the G and D states contain mixing coefficients that depend on the bond length in the H_2 molecule, the translation of the "subsystem state" description into the already familiar MO or VB description is a function of the A—B and C—D separations. This difficulty can be avoided if configurations based on the MOs of the H_2 subsystems rather than on their states are used as the basis of the description. We shall use a and a' for bonding and antibonding MO for the $H_A H_B$ molecule and b and b' for those of the $H_C H_D$

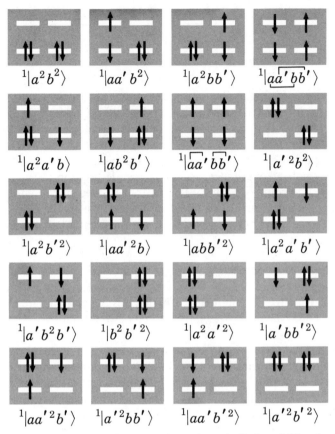

Figure 5.9 The H_4 system: singlet configurations, built from MOs of subunits.

molecule. The 20 singlet configurations built on these MOs are collected in Figure 5.9. One of them corresponds to a subsystem state that represents the "double triplet" state, namely, $^1|aa'bb'\rangle$. Most need to be combined in a simple fashion, using geometry-dependent coefficients, in order to produce pure states of the subsystems. This is easily done, since the composition of subsystem states in terms of their configurations is already known. Thus, the representation of the states of H_4 in the limit of dissociation to two H_2 molecules remains simple. If the A—B and C—D distances are short, so that the ground state of H_2 can be reasonably approximated by its ground configuration, and the A—C and B—D distances long, we have the following approximate relations for the six lowest singlet states:

$$|G\rangle \cong {}^1|a^2b^2\rangle \tag{5.1}$$

$$|S_+\rangle \cong \frac{1}{\sqrt{2}} ({}^1|aa'b^2\rangle + {}^1|a^2bb'\rangle) \tag{5.2}$$

$$|S_-\rangle \cong \frac{1}{\sqrt{2}} ({}^1|aa'b^2\rangle - {}^1|a^2bb'\rangle) \tag{5.3}$$

$$|D\rangle = {}^1|aa'bb'\rangle \tag{5.4}$$

$$|I_+\rangle = (1/\sqrt{2})({}^1|a^2a'b\rangle + {}^1|ab^2b'\rangle) \tag{5.5}$$

$$|I_-\rangle = (1/\sqrt{2})({}^1|a^2a'b\rangle - {}^1|ab^2b'\rangle) \tag{5.6}$$

Here, it is recognized that in the limit of large internuclear separations the "singly excited" state S is nearly degenerate, as is the "charge-transfer" state I. The two components of each of these states are split by amounts related to the rate of exchange of excitation (S_+, S_-) or of charge switching (I_+, I_-) between the two molecules, respectively. The former is known as exciton splitting (see Section 5.4.2).

As the intermolecular distance decreases, the description of the states of H_4 in terms of subsystem configurations of Figure 5.9 becomes more complex. At square geometries, the wave function of the S state clearly shows that the minimum in the S state is due to a combination of stabilization by exciton interaction and charge-transfer interaction.

General trapezoidal geometries. Both a rectangular and a symmetric collinear (Section 6.1) arrangement of the four H atoms can be viewed as special cases of a trapezoidal geometry. In the former, the lengths of the two parallel sides of the trapezoid are equal; in the latter, the distance between the two sides is zero.

A visual representation. A representation of the energies of the G, D, and S states at all trapezoidal geometries in a very simple *ab initio* approximation is shown in Figures 5.10–5.14. The coordinate axes are the lengths of the two parallel sides, R_1 and R_2, and their separation is R (Figure 5.1). The scale for the vertical axis R is twice as large as the others in order to improve visualization. The figures show perspective views of nested equipotential surfaces. These contain the points representing geometries for which a state has a constant value of energy. The values between neighboring surfaces differ by 0.05 atomic units (~30 kcal/mol) and are listed in Figures 5.10–5.14. Results for the G state are displayed in Figures 5.10 and 5.11, which show two different views of the same set of equipotential surfaces. In the C_{2v} group of a trapezoid, the G state is of symmetry A_1. Figure 5.12 shows the results for the D state, which also is of symmetry A_1. Figures 5.13 and 5.14 shows two views of the nested surfaces for the S state, which is of symmetry B_2. Since the D and S states differ in symmetry, their surfaces cross freely. The points at which they have the same energy constitute a surface that is shown in Figures 5.13 and 5.14. Its cross-section with the planes that limit the half-octant displayed is shown with a thick line. In the region between this cross-section surface and the R_1R_2R axes, the S state is below D in energy. In the rest of the space, D is lower.

In principle, all three coordinates range from $-\infty$ to $+\infty$. Since the energy is symmetric across all coordinate axes, only the first octant needs to be shown. Moreover, the energy is unchanged when R_1 and R_2 are interchanged, so that the half-octant shown contains all the relevant information. The plane $R_1 = R_2$ corresponds to all rectangular geometries, the line $R_1 = R_2 = R$ corresponds to all squares, and the plane $R = 0$ corresponds to all symmetric linear geometries.

Figures 5.10–5.14 contain a tremendous amount of information about the three lowest states of H_4 in the minimum basis set approximation (in this case, AO basis exponents were optimized separately for each state and each geometry). This is

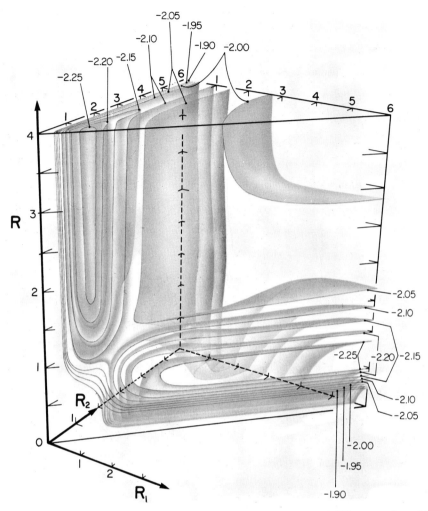

Figure 5.10 Equipotential surfaces of trapezoidal H_4 in the ground (G) state, viewed from the "front." *Ab initio* minimum basis set full CI calculation. Orbital exponents were optimized for each state and each geometry. Reproduced by permission from W. Gerhartz, R. D. Poshusta, and J. Michl, *J. Am. Chem. Soc.* **98**, 6427 (1976).

best appreciated when it is realized that the three reaction paths considered so far, namely, a selected set of rectangular geometries, square geometries, and a selected set of symmetrical linear geometries, only contain information along three lines through the half-octant. We shall not discuss all the available information here in detail, and we refer the interested reader to the original article (Section 5.5). Rather, we shall only indicate the qualitative significance of the displayed shapes briefly.

The G, S, and D states. The surfaces for the G state represent a purely repulsive situation for the approach of two H_2 molecules. The paths of easiest approach lead through the two tubes, a vertical one corresponding to rectangles formed by two

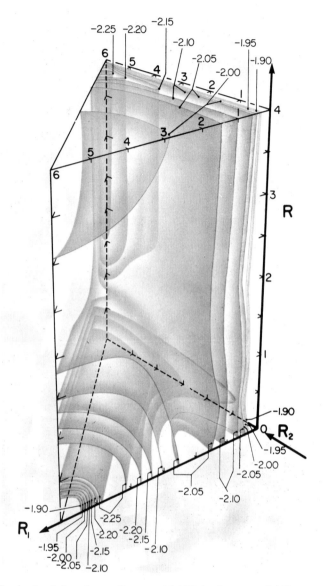

Figure 5.11 Equipotential surfaces of trapezoidal H_4 in the ground (G) state, viewed from the "rear." *Ab initio* minimum basis set full CI calculation. Orbital exponents were optimized for each state and each geometry. Reproduced by permission from W. Gerhartz, R. D. Poshusta, and J. Michl, *J. Am. Chem. Soc.* **98**, 6427 (1976).

parallel molecules $H_A H_C$ and $H_B H_D$, and a horizontal one corresponding to trapezoids formed by two molecules of equal length, $H_A H_B$ and $H_C H_D$. The two tubes are separated by a region of high energy which occurs along a surface of points (geometries) at which the two VB structures, $^1|\overline{ABCD}\rangle$ and $^1|A\overline{BCD}\rangle$, undergo an avoided crossing. The surface contains the line of all squares ($R_1 = R_2 = R$) and cuts the bottom plane ($R = 0$) in a curved line that runs from near the origin

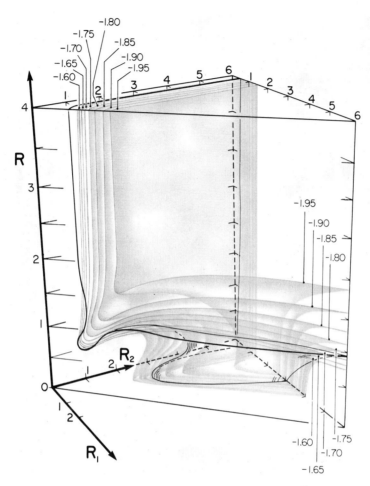

Figure 5.12 Equipotential surfaces of trapezoidal H_4 in the "doubly excited" (D) state. *Ab initio* minimum basis set full CI calculation. Orbital exponents were optimized for each state and each geometry. Reproduced by permission from W. Gerhartz, R. D. Poshusta, and J. Michl, *J. Am. Chem. Soc.* **98,** 6427 (1976).

toward the point $R_1 = 2$ Å, $R_2 = 6$ Å. In between the two tubes, the surface of avoided crossing follows a series of deep furrows that represent relatively high energies in the G state (Figure 5.11) and relatively low energies in the D state (Figure 5.12). The D state is purely repulsive in all respects. The S state contains a minimum that lies inside the egg-shaped region shown in Figures 5.13 and 5.14 ("excimer minimum").

It is a depressing thought to realize that all of this complexity reflects only the behavior of the three lowest singlet states in a single three-dimensional subspace of the total six-dimensional space of H_4. One other three-dimensional subspace and a few other geometries have been investigated. Among the more interesting findings is the behavior of square H_4 upon distortion into a diamond. In the G state, the energy increases because the H atoms on a diagonal are triplet-coupled. In the D state, the energy decreases because now they are singlet-coupled; as the

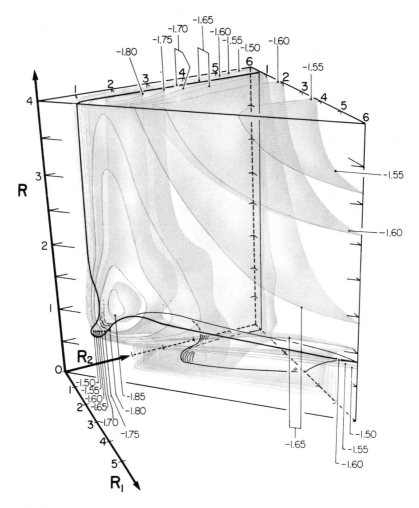

Figure 5.13 Equipotential surfaces of trapezoidal H_4 in the "singly excited" (S) state, viewed from the "front." *Ab initio* minimum basis set full CI calculation; Orbital exponents were optimized for each state and each geometry. Reproduced by permission from W. Gerhartz, R. D. Poshusta, and J. Michl, *J. Am. Chem. Soc.* **98**, 6427 (1976).

motion proceeds, a ground-state H_2 molecule results from the shorter diagonal, and the two remaining H atoms are singlet-coupled. When they, too, are brought together, a second ground-state H_2 molecule results. Thus, the D state of a square H_4 correlates with the G state of two H_2 molecules placed across each other, via a G-D touching attributable to critical heterosymmetry (Section 4.6). If the distortion of the square is performed by an out-of-plane deformation that shortens both diagonals equally, it proceeds through a regular tetrahedron, at which the D and G states are degenerate by symmetry.

5.1.2 Two Polar Bonds

In preparation for discussion of substituent effects on pericyclic processes we shall now consider briefly the changes that occur when two interacting bonds are po-

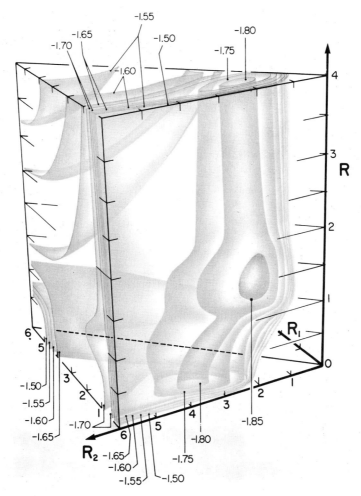

Figure 5.14 Equipotential surfaces of trapezoidal H_4 in the "singly excited" (S) state, viewed from the "rear." *Ab initio* minimum basis set full CI calculation. Orbital exponents were optimized for each state and each geometry. Reproduced by permission from W. Gerhartz, R. D. Poshusta, and J. Michl, *J. Am. Chem. Soc.* **98**, 6427 (1976).

larized. This could again be done by using the artifice of nonintegral nuclear charges or by considering constructs such as Li_2H_2, but we shall satisfy ourselves with the concepts introduced in Chapter 4 and shall test their qualitative validity by comparison with *ab initio* calculations on isoelectronic π systems, namely, perturbed square cyclobutadienes.

As usual in photochemistry, we search for minima in the excited state surfaces—in this case, specifically for regions of S_0–S_1 touching. We have just seen in the nonpolar cases of square H_4 and face-to-face ethylene cycloaddition that the separation between the S_0 and S_1 states remains considerable at the geometry of the square pericyclic minimum. We have also seen it is possible to reduce this S_0–S_1 gap to zero by introducing a difference in the energies of the localized nonbonding orbitals $|A\rangle$ and $|B\rangle$ by making the diagonals unequal.

In the unperturbed square biradical the S_0 state was well described by the dot–dot structure $^1|AB\rangle$, and the S_1 and S_2 states were well described by an equal mixture of the hole–pair structures $^1|A^2\rangle$ and $^1|B^2\rangle$ ($^1|Z_1\rangle$ and $^1|Z_2\rangle$, defined in equations 4.46 and 4.47); however, in the resulting heterosymmetrically perturbed biradical the exact balance will be lost, the S_1 state will be dominated by $^1|B^2\rangle$, and its energy will drop rapidly as $|B\rangle$ becomes more electronegative.

In the four-electron four-orbital case the nature of the excited singlets is not quite that described by the 3×3 model, as we have just discussed in Section 5.1.1 on the example of H_4, and the D state contains a significant admixture of structures other than $^1|B^2\rangle$. Still, under conditions when the 3×3 model is still applicable, e.g., for moderately sized or small H_4 squares, the S_1 and S_2 (D and S) states are still fairly close in energy and an increase in the electronegativity of the orbital $|B\rangle$ would still be expected to convert the S_1 wave function into an essentially pure $|B^2\rangle$ structure and to stabilize it significantly. Then, we would expect a qualitatively identical behavior to that in the two-electron two-orbital case.

In order to stabilize $|B\rangle$ relative to $|A\rangle$, we also can make the two diagonally opposed atomic orbitals of the square array on which $|B\rangle$ resides more electronegative than the other two, on which $|A\rangle$ is located (Figure 5.15A). If we were to increase the electronegativity of two adjacent atomic orbitals instead, we would still be lifting orbital degeneracy, but now by introducing $\gamma_{AB} \neq 0$ instead of $\delta_{AB} \neq 0$ in the sense of Section 4.5, and we would be increasing the S_0–S_1 gap instead of decreasing it.

These concepts have been verified by an *ab initio* large-scale CI calculation on perturbed square cyclobutadienes whose formulas are listed in Figure 5.15B. The observed behavior of the state energies is exactly that predicted by the simple 3×3 CI model. The S_0–S_1 splitting is large in the perfect biradical, cyclobutadiene (**3**, ~45 kcal/mol), vanishes almost completely in the critically heterosymmetric biradicaloids **4** and **5**, and is again large in the strongly heterosymmetric biradicaloids **6** (~50 kcal/mol) and **7** (~90 kcal/mol). The triplet lies close to the $^1|AB\rangle$ singlet, i.e., close to S_0 in **3–5** and close to S_1 in **4–7**. These results ($\delta_{AB} \neq 0$) can be contrasted with those computed for the opposite substitution pattern ($\gamma_{AB} \neq 0$, **1** and **2**). Now, there is no S_0–S_1 crossing but, instead, a strong stabilization of S_0 and destabilization of S_2.

5.1.3 Other Pericyclic Ring Sizes

Ground-state "forbidden" pericyclic reactions involve an "antiaromatic" biradicaloid geometry at the reaction midpoint, where the G and D configurations undergo an avoided crossing. This provides an opportunity for the existence of a pericyclic minimum in the S_1 state, as just discussed in detail for the four-electron four-orbital case of H_4. The presence of this minimum, in turn, provides a driving force for pericyclic photochemical reaction, and these reactions are said to be excited-state "allowed." Whether barriers are present on the way to such a minimum cannot be determined from the mere knowledge that the pericyclic array is aromatic or antiaromatic, and actual construction of correlation diagrams is required. Information on their presence or absence is essential before a prediction can be made as to whether an excited-state "allowed" process is truly allowed; even then, it need not be competitive.

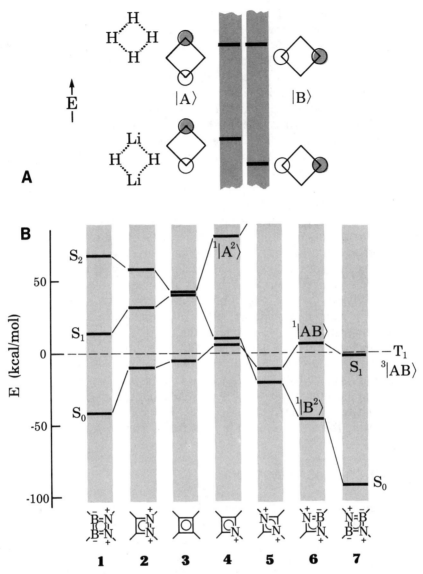

Figure 5.15 (A) The localized form of nonbonding orbitals of square H_4 and Li_2H_2. Schematic. (B) The singlet and triplet energies of a series of push–pull perturbed cyclobutadienes at square geometries. *Ab initio* large-scale CI calculations. Reproduced by permission from V. Bonačić-Koutecký, J. Koutecký, and J. Michl, *Angew. Chem. Int. Ed. Engl.* **26**, 170 (1987).

It is important to know whether the D state dips below the S state in the region of the avoided crossing, since only then will it produce a minimum in the S_1 surface. We have just seen that this is so for realistic geometries of H_4 and shall see below that it also holds for other four-electron four-orbital systems. This is understandable considering that for 4N-electron 4N-orbital Hückel perimeters the exchange integral K_{AB} between the localized nonbonding MOs is minuscule (zero in the differential overlap approximation, "pair" biradical). To the first approximation, D lies above

S by $2K_{AB}$, and the mixing of covalent character into the D wave function which takes place in a better approximation easily lowers the energy of the D state below that of S. The same situation is expected for all uncharged perimeters [4N-electron 4N-orbital Hückel or $(4N + 2)$-electron $(4N + 2)$-orbital Möbius systems].

However, as we have seen in Section 4.4, when the number of electrons and the number of participating orbitals in the antiaromatic pericyclic array are different (charged perimeters such as $C_3H_3^-$, $C_5H_5^+$, $C_6H_6^{2+}$), the exchange integral K_{AB} is large ("axial" biradical). Then S lies far below D in the simple 3×3 CI model, and T far below G, and even a large degree of admixture of covalent character into D as the model is improved does not suffice to lower D below S, so that no pericyclic minimum in the S_1 surface is present. The large S–D separation should also make it relatively harder to polarize the S wave function by introducing $\delta_{AB} \neq 0$ and to decrease the S_0–S_1 gap in the process.

At the same time, the relatively favorable energy of the triplet state at the pericyclic geometry optimizes the chances that a triplet-state reaction will proceed along a concerted pathway and that the minimum in T_1 will be at such a "tight" cyclic geometry and not the "loose" open-chain geometry chracteristic of the T_1 minima of species derived from an uncharged perimeter.

Only relatively few pericyclic reactions of ions have been of experimental interest. However, many important pericyclic reactions involve the heterocyclic analogs of such charged species. For example, the known disrotatory interconversion of aziridine and its zwitterion, $CH_2=\overset{+}{NH}-CH_2^-$, is isoelectronic with the interconversion of the cyclopropyl anion and the allyl anion, and the known conrotatory interconversion of phenylvinylamines and zwitterions derived from indoles is isoelectronic with the interconversion of the pentadienide anion and the cyclopenten-3-ide anion. These and related processes occur in the triplet state in a concerted fashion, following the orbital symmetry rules (see Section 5.2.3).

5.2 ELECTROCYCLIC REACTIONS

Two fundamentally different stereochemical courses, disrotatory and conrotatory, are possible for electrocyclic processes. In the former, the sense of rotation of the two terminal groups of a conjugated π system that eventually yield the new σ bond is opposite. In the latter, the sense is the same for both groups. The former motion preserves a (possibly only approximate) plane of symmetry and the latter a (possibly only approximate) twofold axis of rotation. A more subtle distinction, which will not be addressed here, is that between the two possible senses of rotation available for each of these modes.

These two pericyclic processes obey the usual orbital symmetry rules stated simply in the introduction to Section 5.1. A nonconcerted ring-opening reaction, in which only one terminal group rotates, is also possible and will be discussed in Section 6.2.5.

5.2.1 Interconversion of Butadiene and Cyclobutene

The concerted disrotatory interconversion of butadiene and cyclobutene is a classical prototype of a ground-state "forbidden" pericyclic process involving an "uncharged" perimeter in the cyclic orbital array. In this reaction, we shall see how both the VB and the MO approaches to correlation diagrams can be used to arrive at this conclusion; we subsequently shall compare the state correlation diagram with the results of numerical calculations.

VB structure and state correlation. This example (Figure 5.16) lies on the borderline of what is still qualitatively sensible with the VB approach and could hardly have been worked out originally without help from actual numerical calculations. The active part of the molecule contains four atomic orbitals, but they are all simultaneously in interaction. Two covalent singlet structures can be formed, depending on how the individual electron spins are coupled. In Figure 5.16 we have indicated singlet and (approximate) triplet local coupling as usual by the letters s and t. In the lower of the two covalent structures on the left-hand side we have singlet coupling between two pairs of strongly overlapping orbitals and only one unfavorable triplet coupling within such a pair, since the terminal two atomic orbitals do not overlap significantly. The other possibility corresponds to the symmetric excited state of butadiene in which there are two unfavorable triplet couplings and only one favorable singlet coupling. The singlet coupling of the paired electrons on atoms 1 and 4 does not lower the energy, since these two orbitals do not overlap significantly. The nature of this higher-energy structure corresponds

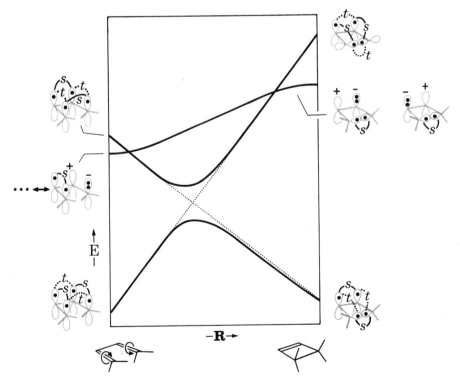

Figure 5.16 Disrotatory cyclobutene–butadiene conversion. A VB structure and state correlation diagram.

to a combination of two triplet ethylenes into an overall singlet. The corresponding singlet state is occasionally referred to as the doubly excited state or the "bitriplet" state of butadiene. Its exact location is still under dispute because it carries no intensity in optical absorption from the ground state. In our diagram we have placed it slightly above the lowest optically allowed state of butadiene, which is represented by VB structures of zwitterionic nature. A representative structure of this kind is shown in the diagram, but many others contribute as well.

On the right-hand side the spin coupling scheme in the lower of the two covalent configurations is exactly the opposite of the one which lies lower on the left. This is because very strong bonding is possible between the orbitals located on atoms 1 and 4 which now overlap, and it is thus favorable to place singlet-coupled local pairs between atoms 1 and 4 and atoms 2 and 3. The lowest excited VB structures now correspond to those of ethylene, whereas the bitriplet type of VB structure is very high in energy because it involves triplet excitation in both the double bond and the newly formed σ bond; the singlet coupling between the orbitals on atoms 1 and 2 and atoms 3 and 4 does not provide any stabilization because one of the orbitals involved is of π and one of σ symmetry. Representing the symmetric states by their dominant configurations and drawing the correlation lines as indicated produces a crossing that will be avoided when interaction between structures is considered, since their symmetries are the same. Because of the antisymmetric nature of the optically allowed excited state, its crossing with the others is not avoided as long as the plane of symmetry that relates atoms 1 with 4 and 2 with 3 is preserved. Overall, these final interactions will convert the zero-order approximations to molecular states into state correlation lines in the diagram.

The type of argument just presented for butadiene is difficult to make in the absence of calculations. For instance, it is not obvious why the analogous conrotatory path that begins with a similar crossing of configurations in a simple-minded diagram leads to a significantly different result, since this crossing is very strongly avoided so that only a small barrier remains in the ground-state surface after interactions of VB structures are considered. In cases like this, and in those even more complicated, it is preferable to use the MO approach.

MO correlation. In order to construct the MO correlation diagram, we need to consider the four π orbitals in butadiene that correlate with the π and π^* orbitals of the double bond and with the σ and σ^* orbitals of the newly formed σ bond of cyclobutene.

Along the disrotatory pathway (Figure 5.17), the symmetry element preserved throughout is the symmetry plane perpendicular to the plane of the molecule. The classification of orbitals according to their symmetry is shown on both sides of Figure 5.17. When levels of like symmetry are connected, we obtain the correlation diagram shown in the figure. The form of the four orbitals part-way through the reaction path is also shown schematically in the diagram. It indicates how the pure π orbitals of the left-hand side develop. Two remain π in character, and the other two gradually lose the contributions of π symmetry and become pure σ orbitals.

This can be contrasted with the conrotatory path (Figure 5.18), where the symmetry element preserved throughout is a twofold axis of rotation passing through the midpoint of the C_2—C_3 bond. The symmetry of each MO relative to this operation is again shown on each side of the diagram. When lines for connecting levels of like symmetry are drawn, the MO correlation diagram results as is shown

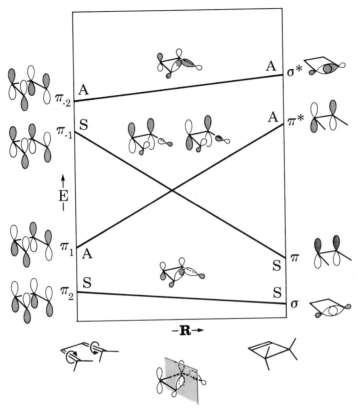

Figure 5.17 Disrotatory cyclobutene–butadiene interconversion. An MO correlation diagram.

in Figure 5.18. Once again, the course of the orbital shape changes along the reaction path is indicated schematically.

In both cases, first-order perturbation theory can be used readily to provide an understanding of the slopes of the correlation lines. The difference between the disrotatory and conrotatory case is then observed to originate in the sign of the original overlap between the two atomic orbitals which are being twisted in order to eventually produce the σ and σ^* orbitals of the new single bond.

The use of noninteracting subunits. Interesting insight into the origin of the difference in the MO correlation diagrams for the disrotatory and the conrotatory path is obtained when the total four-electron four-orbital system is decomposed into two subunits. This approach makes it particularly easy to see why perfect symmetry is not required for conclusions to be drawn, since the orbital nodal properties rather than orbital symmetries are what actually matters. The two atomic orbitals located at the terminal atoms 1 and 4 of butadiene interact with their neighbors on atoms 2 and 3 in the π system of butadiene. These interactions are lost gradually as the terminal methylene groups rotate. The rotation permits the two atomic orbitals to establish a new interaction of the σ type.

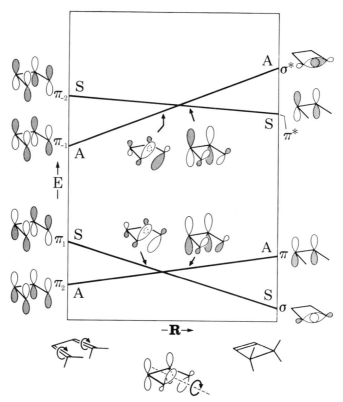

Figure 5.18 Conrotatory cyclobutene–butadiene interconversion. An MO correlation diagram.

Let us first consider these two orbitals as one subunit (I) and consider the 2p orbitals on atoms 2 and 3 of butadiene as another subunit (II). Let us assume for the moment that the interactions between the two subunits have been turned off. The correlation diagram would then have the appearance shown on the left in Figure 5.19: At the geometry of cyclobutene, the two AOs of subunit I combine into a σ, σ^* orbital pair, whereas at the geometry of butadiene, they hardly interact at all and represent a pair of nonbonding orbitals. The two AOs of subunit II are combined into a π, π^* orbital pair and remain essentially constant along the whole reaction path because the two subunits are assumed not to interact. There is no difference between the disrotatory and the conrotatory path.

When the interaction of the subunits is introduced, this will no longer be true. At any general rotation angle, each of the AOs on the methylene groups of subunit I contains partly π and partly σ character. Their relative amounts can be obtained by projecting the direction of the orbital axis into the plane of the carbon atoms and into a direction perpendicular to this plane. The π component will interact with the π and π^* orbitals of subunit II, and the effect of this interaction on orbital energies is indicated by arrows on the left-hand side of Figure 5.19.

Along the disrotatory path, the π-symmetry part of the original σ orbital of the single bond that is being destroyed by the rotation has the same sign on both

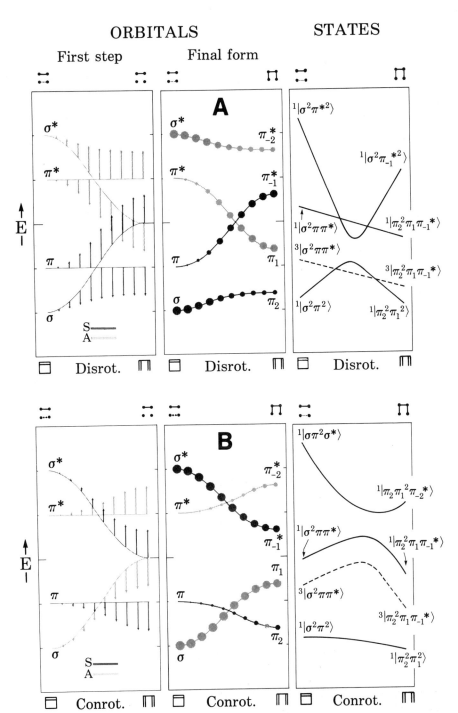

Figure 5.19 Disrotatory (A) and conrotatory (B) cyclobutene–butadiene interconversion. A stepwise derivation of the MO correlation diagram: (Left) Orbital correlation in the absence of interaction between orbitals of the double bond and those of the original single bond; the effect of such interaction is shown by arrows. (Center) Interaction is allowed for, and the MO correlation diagram results (the size of the circles gives the coefficients of the terminal carbon AOs in each MO, and their shading indicates their phase relationship). (Right) The resulting state correlation diagram. Symbols on top: The array of interacting orbitals. Sold lines denote π interaction; dashed lines denote σ interaction. Reproduced by permission from J. Michl, *Mol. Photochem.* **4**, 287 (1972).

termini, so that it will interact with the π orbital of the cyclobutene double bond. In a similar fashion, the π-symmetry part of the original σ^* orbital will interact with the π^* orbital because they have like nodal properties. Along the conrotatory path, however, the rotation of the AOs on the terminal atoms is such that the nodal properties of the π-symmetry parts of the original σ and σ^* orbitals of the single bond of cyclobutene are interchanged. Then, the original σ orbital of subunit I will interact with the π^* orbital of subunit II, and the original σ^* orbital of subunit I will interact with the π orbital of subunit II. As a result, the resulting two correlation diagrams, shown in the center in Figure 5.19, acquire a completely different character. In these, the weights of the contributions of the AOs that originally formed the σ bond to the MOs are indicated by shaded circles.

It is seen that in the absence of assistance from interaction with the cyclobutene π bond, the bond-breaking rotation of the two methylene groups would produce a biradical. Because of the assistance by the π system, the product will be a much more stable molecule, namely, butadiene. Along the disrotatory path, the assistance is inefficient so that a biradicaloid geometry still has to be passed. Along the conrotatory path, the assistance is efficient and the appearance of nonbonding orbitals is avoided altogether. The former is favorable for a photochemical, and the latter is favorable for a thermal transformation.

The degree of assistance that the more complicated π systems provide towards the breaking of a single bond by rotation of the termini into conjugation with π orbitals is not always immediately apparent but can be estimated easily using the "subunits" approach. We shall provide an illustration of such a more complex case below, but we need to discuss the derivation of configuration and state correlation diagrams first.

Configuration and state correlation. The conversion of the MO correlation diagrams of Figure 5.19 into configuration correlation diagrams is straightforward and follows the guidelines given in Chapter 3. Along the disrotatory path, the ground-state "forbidden" MO correlation with a normal orbital crossing results in a configuration correlation with the HOMO → LUMO characteristic configuration going at an approximately constant energy across the diagram. Because the HOMO → LUMO excitation energy is substantially lower for the more extended π system of butadiene than it is for the isolated double bond of cyclobutene, the correlation for the characteristic configuration runs somewhat uphill from butadiene to cyclobutene. This tends to impose a barrier when starting with the former but none when starting with the latter (Figure 5.19).

The correlation of the ground configuration of either starting material with a doubly excited configuration of its isomer produces the expected configuration crossing at the biradicaloid geometry halfway along the reaction path. After the introduction of configuration interaction, the configuration crossing is avoided and the anticipated barrier in S_0 and minimum in S_1 state appear. The singlet electrocyclic phototransformation of both butadiene and cyclobutene can thus be expected to proceed via excitation into a spectroscopic allowed excited state followed by travel to the pericyclic minimum in S_1 (in the case of butadiene, over a barrier; in the case of cyclobutene, over none), followed by return to the S_0 surface and formation of butadiene or cyclobutene.

The triplet configuration correlation diagram and the essentially identical state correlation diagram do not suggest the presence of a minimum anywhere along the disrotatory path, other than at the geometry of the more conjugated starting material, butadiene.

This is somewhat misleading, for two reasons. First, drawing the correlation lines straight is only a rough approximation and we have seen in the case of twisted ethylene (Chapter 4) how a minimum tends to be produced in the T surface at a biradicaloid geometry. Second, along the concerted pericyclic path the two roughly nonbonding orbitals characterizing the biradicaloid geometries are kept unnecessarily close together, and this is undesirable for the dot–dot structures characterizing the S_0 and T_1 states of an uncharged perimeter at biradicaloid geometries. Both the "forbidden" thermal process and the triplet reaction will therefore tend to follow the nonconcerted "linear" reaction paths, which we shall discuss in Chapter 6.

Only when steric constraints make it impossible for the molecule to escape from the pericyclic "concerted" path for an uncharged perimeter is the triplet reaction likely to follow it. Then, it would be expected to proceed from the isomer with the less extensive π system (cyclobutene) to the one with the larger π system (butadiene) in an adiabatic process. An example of such a situation is the conversion of Dewar aromatics to aromatics, as demonstrated in Section 2.7 on the photochemical reactions of 1,4-dewarnaphthalene. The case of charged perimeters is discussed in Section 5.2.3.

Along the conrotatory path, correlation-imposed barriers in S_1 and T_1 appear in Figure 5.19B, as expected from the general treatment in Chapter 3. A singlet or triplet photochemical reaction would generally not be expected to take such a path unless the overall exothermicity of the transformation lowers the barrier significantly or removes it altogether; however, then it would be expected to be adiabatic and to yield an excited product, since no correlation-imposed minimum or funnel is expected along the reaction path.

Numerical calculations for the disrotatory path. Figure 5.20 shows an example of a one-dimensional cut through the multidimensional potential energy surfaces for this type of reaction path. From left to right, the C=C—C angles were varied from their values in the ground state of cyclobutene (95°) to the value appropriate for butadiene (120°). Simultaneously, other geometrical parameters were adjusted in a linear fashion. The calculations utilized a minimum basis set of valence orbitals to which diffuse atomic orbitals were added, and a moderate amount of configuration interaction.

The potential energy diagram shown in Figure 5.20 looks very much like the simple correlation diagram shown in Figure 5.19. In particular, the ground state and the doubly excited state undergo an avoided crossing that results in a barrier in S_0 and a minimum in S_1, whereas the singlet and triplet singly excited states correlate more or less straight across. They lie at higher energy for cyclobutene than for butadiene because $\pi\pi^*$ excitation energies generally decrease with the extended length of the conjugated system. The T_1 state is well represented by the HOMO → LUMO excitation. At the present level of calculation, the singly excited state S represents the lowest excited singlet state S_1, whereas the doubly excited state D represents the second excited state S_2. Since the D state descends rapidly in energy towards the avoided crossing, a small barrier in the S_1 state results.

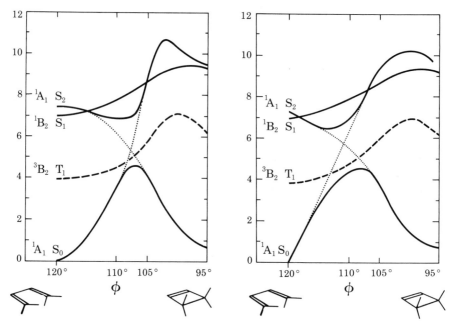

Figure 5.20 Disrotatory cyclobutene–butadiene interconversion for two reasonable but otherwise arbitrary reaction paths (left and right). Singlet (solid lines) and triplet (dashed lines) state energies calculated by an *ab initio* CI procedure as a function of the rotation angle ϕ of the CH$_2$ groups. Dotted lines indicate the crossing of configurations. Reproduced by permission from D. Grimbert, G. Segal, and A. Devaquet, *J. Am. Chem. Soc.* **97,** 6629 (1975).

Whether this would remain upon thorough geometry optimization or in a better-quality calculation is uncertain. At any rate, it appears that the vertically excited singlet butadiene molecule should have little trouble in finding its way into the pericyclic minimum that results from the avoided crossing. Approach from cyclobutene side should also be easy. A search of lower-symmetry geometries is needed to locate the true minimum (cf. critical heterosymmetry, Section 4.6). The triplet curve shown is probably irrelevant to triplet photochemistry, which is likely to follow a nonconcerted path (Chapter 6).

5.2.2 Electrocyclic Processes in Polycyclic Systems

The extension of the concepts that govern the electrocyclic interconversion of butadiene and cyclobutene to larger ring systems derived from an uncharged perimeter is straightforward. Section 5.1.3 and correlation diagrams make it clear that the stereochemical nature of the ground-state "forbidden" path which proceeds through a biradicaloid geometry, and thus will have a pericyclic minimum in the S_1 surface, changes as one proceeds along the series cyclobutene, cyclohexadiene, cyclooctatriene, etc. The expected photochemical path is disrotatory in cyclobutene, cyclooctatriene, etc., and is conrotatory in cyclohexadiene, cyclodecatetraene, etc.

A more challenging situation is encountered in polycyclic systems. Here, the construction of the appropriate correlation diagrams is frequently facilitated by the use of the "interacting subunits" method outlined in Section 3.3. We shall illustrate

5.2 ELECTROCYCLIC REACTIONS 257

this on a simple example, the electrocyclic ring opening of 1,2-dihydrocyclobut[a]acenaphthylene; similar principles apply in the case of 1,4-dewarnaphthalene discussed in Section 2.7.

Taking the π system of naphthalene, the π system of the isolated double bond, and the σ bond to be broken as three noninteracting subunits, we construct first the MO correlation diagram in Figure 5.21A. When the interaction between the subunits is introduced, orbitals of like nodal properties will interact as indicated by arrows, producing the diagram shown in Figure 5.21B. This MO correlation diagram clearly is of the "abnormal" MO crossing kind, in which the characteristic configuration is not obtained by the HOMO → LUMO excitation but by another one. As a result, a barrier along the reaction path both in the S_1 and T_1 states is immediately anticipated. The configuration correlation diagram shown in Figure 5.21C indeed exhibits such a barrier, in addition to the quite clearly present pericyclic minimum. The introduction of configuration interaction does not change the picture for the T_1 state but complicates it further for the singlet states, since the L_b state of naphthalene rather than its L_a (HOMO → LUMO) state represents S_1. Clearly, at least two of the lowest excited states of the starting material correlate with product states that are quite high in energy, and barriers are expected both in S_1 and in T_1. This result accounts for the experimental observations according to which the photochemical process cannot be triggered by excitation into the low-lying vibrational levels of the S_1 state or into the T_1 state, and both of these states

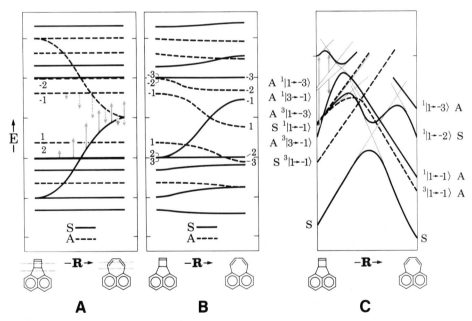

Figure 5.21 Disrotatory electrocyclic ring closure in pleiadiene. Stepwise derivation of the correlation diagram. (A) In the absence of interaction between the AOs of the original σ bond and those of the π systems. The effect of such interaction is indicated by arrows (1, HOMO of naphthalene; 2, HOMO of ethylene; -1, LUMO of naphthalene; -2, LUMO of ethylene). (B) MO correlation diagram in the presence of the interaction. (C) Configuration and state correlation diagram.

258 CYCLIC MULTICENTER REACTIONS: ONE ACTIVE ORBITAL PER ATOM

are deactivated primarily by luminescence and vertical intersystem crossing and, thus, are not photochemically reactive. However, upon the use of high-energy photons or upon the successive absorption of a second photon by the T_1 state, the photochemical reaction can be effected and pleiadiene formation is observed.

5.2.3 Electrocyclic Ring Opening in Oxirane

The concerted disrotatory interconversion of oxirane and the oxonium ylide zwitterion, $CH_2\!=\!\overset{+}{O}\!-\!CH_2^-$, will serve as a prototype of a classical ground-state "forbidden" pericyclic process involving a "charged" perimeter in the cyclic orbital array. This process, as well as the interconversion of aziridine and its zwitterion, $CH_2\!=\!\overset{+}{N}H\!-\!CH_2^-$, is formally derived from a similar interconversion of the cyclopropyl anion and the allyl anion, which is less convenient to investigate in practice.

The derivation of the appropriate correlation diagrams is quite analogous to what we have already seen in Section 5.2.1 and, thus, will not be presented. The orbital symmetry rules are followed in their usual form for a four-electron system, the conrotatory process is ground-state-allowed, and the disrotatory process is ground-state-forbidden. An aspect of these ring-opening reactions that is somewhat out of the ordinary is the fact that the product is very much higher in energy than the starting material and can be described as a biradicaloid.

An example of a one-dimensional cut through the potential energy surface is shown in Figure 5.22 for oxirane. The reaction path is defined by a stretching of the C—C bond with simultaneous rotation of the CH_2 groups into the plane of the two carbon atoms and the heteroatom. The calculations used a simple *ab initio* model (limited CI, minimum basis set). They start from the three-membered ring geometry ($\phi = 62°$) and proceed to the ring-opened form ($\phi = 109°$) of oxirane. The biradicaloid nature of the open form is reflected both in its S_0 energy, which is considerably above that of the closed form, and in the closeness of its S_0 and T_1 energies.

Both the barrier in the S_1 surface for the ground-state-allowed conrotatory mode, and a more or less straight correlation of the S_1 surface across the diagram for the ground-state-forbidden disrotatory mode are standard hallmarks of Woodward–Hoffmann correlation diagrams. There are two significant differences between the present results for the photochemically allowed disrotatory mode and the results shown earlier for cyclobutene going to butadiene (Section 5.2.1). Both are characteristic of the charged–uncharged perimeter dichotomy. First, the calculated shape of the avoided crossing between the S_0 ground (G ≡ C) and S_2 doubly excited (D ≡ Z_2) states is such that Z_2 does not dip below the singly excited state (S ≡ Z_1); and second, in the region of the avoided crossing, T_1 dips below S_0. Both of these features are expected of 4N-electron n-center perimeters when $4N \neq n$ (Section 5.1.3).

The curves in Figure 5.22 suggest that the singlet excited-state reaction will not proceed along the conrotatory path but will follow the disrotatory path. Moreover, if the doubly excited surface Z_2 really does not dip below the singly excited state Z_1, it is fairly probable that along the concerted path the product will be formed in the excited singlet state. However, such a conclusion cannot be drawn with certainty at the present level of calculation.

A similar result is obtained for the triplet state (Figure 5.22). The T_1 state of oxirane correlates with T_1 of the open form along the disrotatory path but not along the conrotatory path. Along the latter, the T_1 of the product originates in a

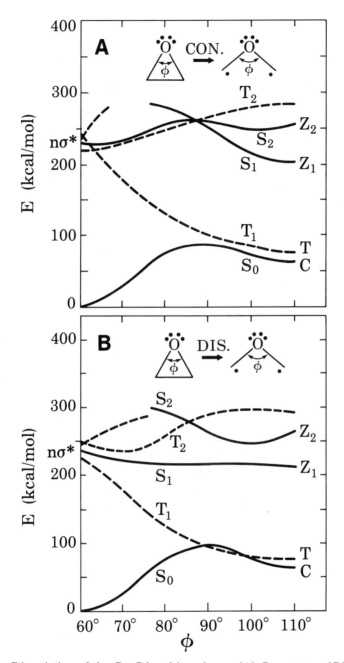

Figure 5.22 Dissociation of the C—C bond in oxirane. (A) Conrotatory (C_2); (B) disrotatory (C_s). *Ab initio* limited CI state energies (minimum basis set) for singlets (solid lines) and triplets (dashed lines) as a function of the COC valence angle ϕ. Reproduced by permission from B. Bigot, A. Sevin, and A. Devaquet, *J. Am. Chem. Soc.* **101,** 1095 (1979).

higher triplet state of oxirane. However, the resulting barrier is calculated to be very small, and a much better level of approximation, as well as proper geometry optimization, is needed to determine whether one may indeed conclude that disrotation will be the preferred process in the triplet reaction. In addition, an examination of the nonconcerted ring-opening path, in which only one CH_2 group rotates (Section 7.5.1), is needed as well before it can be stated reliably that theory favors the concerted process.

It appears, however, that even the present primitive level of theory is correct, in that, experimentally, the reaction appears to proceed in disrotatory fashion both in the singlet and the triplet excited states. This has been particularly well documented in the case of the next larger pericyclic ring size with a charged perimeter, a class of photochemical reactions that has acquired some synthetic importance.

The hydrocarbon prototype is the pentadienide–cyclopentenide interconversion:

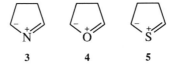

Numerous heterocyclic analogs of this process have been investigated, the zwitterionic products containing rings such as

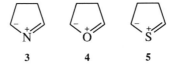

They are readily oxidized to pyrroles, furans, and thiophenes. As expected from the orbital symmetry rules, the photochemical conversion follows the conrotatory path. Once again, this is true starting both in the singlet and in the triplet excited states; in the latter case, the products are formed initially in their triplet states.

A tentative conclusion, then, is that unlike triplet reactions of formally uncharged perimeters, which generally prefer to proceed in a nonconcerted manner in the absence of specific constraints, those of formally charged perimeters proceed in a true pericyclic concerted fashion and give stereospecific products in accordance with orbital symmetry rules.

5.3 BOND-SHIFT REACTIONS

Bond-shift reactions are another important family of photochemical processes. In these, two stereochemical considerations are important.

One is the point of attachment of the migrating single bond in the starting material and in the product. If the attachment in the two species is on the same side of the surface defined by the π-electron system, the migration is termed suprafacial. If it is on the opposite sides, the migration is termed antarafacial. Antarafacial migrations are hardly possible sterically in the small cyclic array required by a 1,3 shift, but they can occur in shifts involving larger cycles.

The second stereochemical point deals with the presence or absence of an umbrella-type inversion at the migrating center. If the same lobe of the migrating orbital that was used to make the migrating allylic sigma bond in the starting material is used to make the bond in the product, the reaction proceeds with retention of configuration. If hydrogen is the migrating group, it uses its nodeless 1s orbital for bonding, and only migration without inversion is possible. With migrating groups such as methyl or trimethylsilyl, an inversion can occur in which it is the opposite lobe of the orbital which is used to form the allylic migrating bond that is used for bonding in the starting material and in the product.

There are two important types of photochemical bond-shift reactions. The usual kind, familiar from ground-state chemistry, is characterized by a pericyclic array of overlapping orbitals that encompasses both termini of a conjugated π system and the two orbitals of the migrating bond in a single ring (Figure 3.8). This array is isoelectronic with an annulene, and the usual rules for "allowed" and "forbidden" processes apply. These shifts are called sigmatropic (e.g., **6** → **7**) and are covered in Section 5.3.1. Their reaction path is clearly of the concerted type, with all new bonds being formed and all old bonds being broken approximately concurrently. A special type of such a photochemical shift is presumed to occur in orthogonally twisted olefins and leads to carbenes. It is discussed briefly in Section 4.8.1.

Sigmatropic reaction

Pseudosigmatropic reaction

The other kind of shift, which we shall call pseudosigmatropic (e.g., **6** → **8**), involves a cyclic array of interacting orbitals containing an odd number of orbitals. The missing orbitals are the one or more orbitals at the end of the conjugated chain that is opposite to the migrating allylic bond. This appendage to the ring then interacts through one of its ends with only one member of the ring (Figure 5.23, structures at the bottom). The orbital array is then isoelectronic with a generalized fulvene, and the reaction is always ground-state "forbidden" because the product is a biradical. It is therefore of interest as a possible photochemical process, but of course the final isolable material would normally be the thermal transformation products from the biradical, such as a cyclopropane. Pseudosigmatropic shifts are covered in Section 5.3.2. Their reaction path is concerted when the biradical is viewed as the product, since the old bond to the migrating group is broken and the new bond to this group is formed concurrently. However, when the cyclopropane is viewed as the product, the reaction path may be termed nonconcerted, since the migration step and the ring-closure step are separated in time. Either way, the path involves a cyclic array of interacting orbitals and therefore is discussed in this chapter.

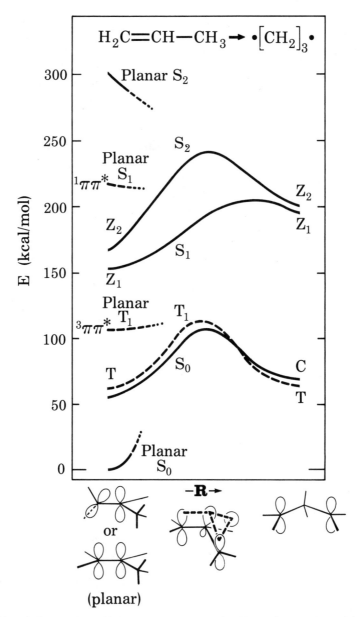

Figure 5.23 Hydrogen 3,2 shift in propene. Singlet (solid lines) and triplet (dashed lines) state energies from an *ab initio* large-scale CI calculation along a path that converts orthogonally twisted propene to the trimethylene biradical by concurrently adjusting all geometric parameters from adduct to product by proportional amounts.

5.3.1 Sigmatropic Reactions

The usual orbital correlation rules determine which among these pericyclic processes are ground-state "forbidden", that is, which ones run through a biradicaloid geometry and therefore are likely to have a pericyclic minimum in their S_1 surface, providing a driving force for a photochemical reaction. The rules are the opposite for migrations with and without inversion, and they also are the opposite for suprafacial and antarafacial migrations. A process that is relatively common in the singlet excited state is a 1,3-suprafacial hydrogen shift in an alkene.

The MO correlation diagram for this reaction is shown in Figure 3.8. The corresponding configuration and state correlation diagrams are given in Figure 3.11 and show that a pericyclic minimum in S_1 is indeed likely to be present, making the excited singlet reaction quite analogous to the disrotatory interconversion of butadiene and cyclobutene discussed in Section 5.2.1.

The pericyclic minimum is more likely to be accessible in alkenes in which the double bond is prevented from twisting. Once the bond has twisted, the travel to the pericyclic minimum may be too endothermic to compete efficiently with alternative processes such as vertical return to S_0.

5.3.2 Pseudosigmatropic Reactions

The simplest example of this type of a process is the 1,2 hydrogen shift in propene to yield the trimethylene biradical. The biradical can close to cyclopropane or return back to propene in a thermal process.

Figure 5.23 shows the result of *ab initio* large-scale CI calculations for two one-dimensional cuts through the potential energy surfaces relevant to this process. In one of these, the reaction starts at the relaxed twisted geometry of propene; in the other, at its planar geometry. The states on the left of the diagram are the already familiar states of a double bond, with S_0 greatly destabilized and T_1, S_1, and S_2 greatly stabilized by twisting. The states on the right-hand side are those expected for a simple biradical. The dot–dot structures C and T dominate the S_0 and T_1 states, and the much higher energy hole–pair structures dominate the much less stable S_1 and S_2 states.

Starting at the twisted geometry, the reaction is quite strongly endothermic in all states except T_1, where it is roughly thermoneutral. However, even here, a relatively large barrier is calculated. Although it would undoubtedly be reduced upon optimization of the reaction path, possibly to a quite small value, there is little doubt that the reaction will have a far better chance to occur if the twisting can be prevented, say by a suitable steric constraint.

Starting at the planar geometry, the reaction becomes strongly exothermic in T_1, where it may well proceed without a significant barrier, and mildly exothermic even in S_1. Considering that the alternative ordinary 1,3 sigmatropic shift has a good driving force in S_1 but not particularly in T_1, one can understand the experimental result that 1,2 shifts tend to occur in the triplet state and that 1,3 shifts tend to occur in the excited singlet state.

5.4 CYCLIC ADDITION AND REVERSION REACTIONS

The third pericyclic process that we shall consider is the most important one from the synthetic chemist's viewpoint, namely, concerted cycloaddition. We shall simultaneously discuss its reverse, namely, concerted cycloreversion. The usual correlation rules (Section 5.1) are applicable.

There are two stereochemically distinct ways in which a component can participate. If the orbital lobes located on the same side of the surface defined by the carbon atoms of the conjugated system are used in the process of making the two new bonds to the cycloaddition partner, the component is said to participate in a suprafacial manner. If the two lobes are located on opposite sides of this surface, it is said to participate in an antarafacial manner.

It is common to refer to cycloaddition reactions by labels such as $4_s + 2_a$, where the numbers 4 and 2 indicate the number of participating π centers in each reaction partner, and the letters s and a indicate the mode of participation of each partner (suprafacial and antarafacial, respectively). The rules that dictate in which cases the MO correlation diagram is of the ground-state "forbidden" type reverse when the participation mode of a component is changed between supra and antara. In most photochemical cycloadditions that proceed in the singlet state and are pericyclic, both components participate in a suprafacial manner, and in the following we shall limit our attention to this case.

A nonconcerted alternative to the presently considered pericyclic path is also available. In this, the two new bonds are formed one at a time, and an intermediate biradicaloid geometry is passed regardless of the number of electrons and the stereochemical arrangement. Paths of this kind will be considered in Section 6.2.6.

5.4.1 Cycloaddition and Cycloreversion

Cycloadditions that are suprafacial in both components lead to an MO correlation diagram of the ground-state "forbidden" type, and thus they pass through a biradicaloid geometry and offer a potential for a pericyclic minimum in S_1 and photochemical reactivity, when the total count of participating electrons is a multiple of four. The most common cases are $2_s + 2_s$ and $4_s + 4_s$ cycloadditions.

$2_s + 2_s$ Cycloaddition: A correlation diagram. The MO correlation diagram for such a face-to-face cycloaddition of two ethylene molecules to produce cyclobutane is shown in Figure 5.24.

On the left-hand side we have shown the energies of the π and π^* MOs of ethylene. Both levels are degenerate, since the two ethylene molecules are assumed to be infinitely far apart. On the right-hand side the same four orbitals representing the active group are shown as they appear in the cyclobutane product, namely, as two σ and two σ^* orbitals of the two new single bonds. If cyclobutane were assumed perfectly symmetric, these orbitals would of course mix with those of the other two C—C bonds, but we can ignore this complication for the present purposes. Numerous symmetry elements are preserved along the reaction path. As usual, those symmetry planes cutting across the bonds that are either being formed or destroyed are the important ones to consider. They are shown in Figure 5.24 and labeled σ_1 and σ_2. The symmetry or antisymmetry of the MOs relative to these two planes is shown by the usual labels on the left-hand side and the right-hand side of Figure 5.24 in the order σ_1, σ_2. Levels representing orbitals of like symmetry are connected in the usual fashion, producing the correlation diagram shown. Once again the reasons for orbitals increasing or decreasing in energy are obvious from the inspection of the signs of overlaps between AOs which are being brought into interaction.

Figure 5.25 shows the configuration correlation diagram derived from the MO correlation diagram of Figure 5.24, and it also shows the state correlation diagram

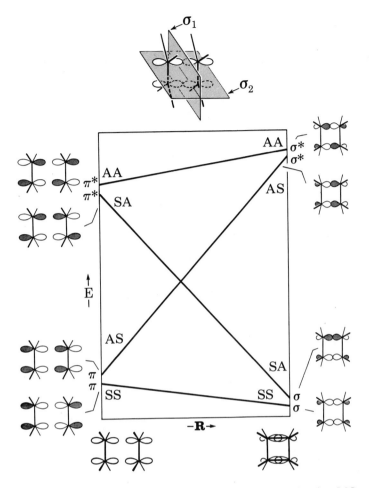

Figure 5.24 Face-to-face ($2_s + 2_s$) cycloaddition of two ethylene molecules. MO correlation diagram.

that results upon the subsequent introduction of configuration mixing. The latter displays clearly both the barrier in the S_0 surface and the pericyclic minimum in the S_1 surface.

The great similarity of the MO, configuration, and state correlation diagrams to the analogous diagrams obtained for the $H_AH_B + H_CH_D \rightleftarrows H_AH_C + H_BH_D$ reaction path in Section 5.1.1 is striking. The detailed analysis available for the H_4 case then leads to the more realistic state correlation diagram shown in Figure 5.26. At infinite separation of the two components, this analogy leads to the identification of the ground-state G of the whole system as corresponding to a combination of the ground states of the two components, of the singly excited singlet (S) state of the whole system as corresponding to a combination of one singlet excited component with one ground-state component (since either component can be excited, this state is doubly degenerate), and of the doubly excited (D) state of the whole system as corresponding to a singlet combination of two components in their respective triplet states. Moreover, the analogy leads us to expect an additional doubly degenerate singlet excited state (I), in which an electron has been transferred from

266 CYCLIC MULTICENTER REACTIONS: ONE ACTIVE ORBITAL PER ATOM

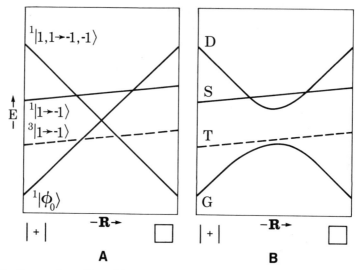

Figure 5.25 Face-to-face ($2_s + 2_s$) cycloaddition of two ethylene molecules. Configuration (A) and state (B) correlation diagram.

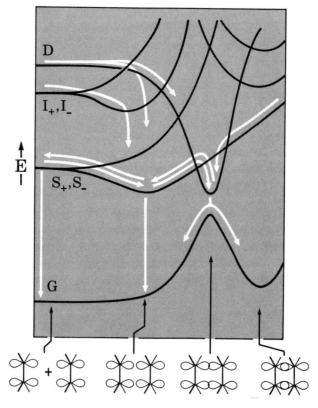

Figure 5.26 State correlation diagram for a ground-state forbidden concerted cycloaddition.

one component to the other. It would be possible to add lines for singly excited triplet (T) states of the whole system, which correspond to a combination of a triplet excited component with a ground-state component. Like the S state, the T state is doubly degenerate on the left-hand side of the diagram where the two components are infinitely far apart. An additional triplet state, as well as a quintet state, will originate in the combination of the two components in their respective triplet states, and both will be degenerate with the singlet D state.

In Figure 5.26, we have labeled the components of the degenerate states by using + and − to indicate the in-phase and the out-of-phase combination of the locally excited or charge-transfer states. At finite separations of the two components, the degeneracies are split.

The singlet pericyclic cycloaddition path is normally entered by the approach of one of the components in its singlet excited state to the other in its ground state, i.e., on the singly excited initially degenerate S_+, S_- surface. Three other possibilities exist, however: The components could approach each other as an ion pair, that is, on the initially degenerate charge-transfer excited I_+, I_- surface, or could both be excited into their respective triplet states, that is, on the doubly excited D surface, or, finally, the reaction could be a cycloreversion, that is, could start on the right-hand side in Figure 5.26.

In addition, the path could be entered on a triplet surface, either by bringing together a triplet excited component with a ground-state component or by triplet excitation of the adduct (triplet cycloreversion). Also, the triplet–triplet encounter statistically leads to an overall singlet coupling to the D state only in one out of nine times (in the absence of perturbations), whereas it yields an excited triplet in three out of nine times. The concerted path is generally not especially favorable for triplet cycloaddition, since it is endothermic. Approached in the cycloreversion direction, it does not suffer from this disadvantage; either way, there is no reason to expect a deep minimum in T_1 along this path, and triplet molecules are likely to proceed in a nonconcerted fashion to form an open-chain biradical (Section 6.2.6). There probably often is a very shallow minimum in the T surface along the concerted path, corresponding to a triplet excimer (Section 5.4.2). However, even after a sojourn in this minimum the T_1 molecules are still likely to stay off the concerted path and to yield an open-chain biradical.

Singlet cycloaddition

Approach on the S_+, S_- surface. In this most common case, reversible formation of an excimer is the first step. Energy transfer between the two components, as well as the nature of the weak bonding in the excimer complex, will be discussed in Section 5.4.2. Here we note only that the depth of the excimer well in the S state is a sensitive function of the nature of the two components and of the geometry of approach. In general, the head-to-head syn arrangement that maximizes the overlap of like orbitals of both partners should be the most favorable.

In the second step, the excimer rearranges to the geometry of the pericyclic minimum in the D surface. Since the S surface will generally slope uphill in this region in the direction from the two components to the cycloadduct, the correlation diagram suggests that a small barrier is to be expected for this motion. A lowering of symmetry from the maximum symmetry possible should make the S–D crossing

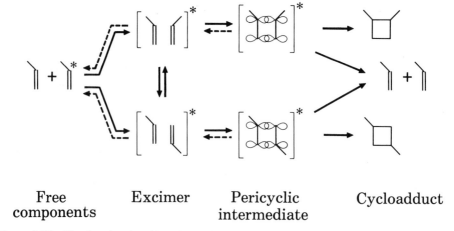

Figure 5.27 Head-to-head and head-to-tail regiochemistry in concerted photocycloaddition.

avoided and lower the barrier. Return from the pericyclic minimum in the D surface to the ground surface G is likely to be exceedingly fast, not giving much chance to the competing step of returning to the exciter.

The location of the pericyclic minimum in S_1 and its depth are of considerable interest. In general, more than one such minimum will be available due to the existence of various steric arrangements such as head-to-head and head-to-tail. Each will be most readily accessible from a different excimer precursor (Figure 5.27). The relative rates of the return to S_0 at the geometries of the various possible pericyclic minima will then be dictated by the outcome of a quite complicated competition affected primarily by the stabilities of the various excimers and by the heights of the barriers separating each of them from its respective associated pericyclic minimum.

If the return to the excimer geometry from a pericyclic intermediate is capable of competing with the return to S_0, also the depth of the pericyclic minimum will affect the relative yields of the various products.

Arguments of Chapter 4 and Section 5.1.2 suggest that the actual minimum of the pericyclic well in D will not be located at the most highly symmetric possible geometry but will, instead, be located at one whose symmetry is low enough to permit efficient mixing of the S and D states. If it is possible to arrange for the dot–dot and hole–pair configurations $^1|AB\rangle$ and $^1|B^2\rangle$ built from the localized orbitals $|A\rangle$ and $|B\rangle$ of the biradicaloid pericyclic intermediate to have the same energies, S_1 might actually touch S_0 or at least come close to it (a critically heterosymmetric biradicaloid geometry). These considerations are particularly relevant if substituents are present, and they will enter into the discussion of the regiochemistry and stereochemistry below.

Approach on the I_+, I_- surface. When a radical cation of one of the components in its ground state begins to approach a radical anion of the other component in its ground state, the total system of two components finds itself in the doubly degenerate excited state I_+, I_-. As the ions approach closer, the two states are weakly split; a discussion of their mutual interaction, as well as their interaction with locally excited states, is given in Section 5.4.2.

In solution, the location of the ion-pair states I_+ and I_- on the energy scale depends strongly on the solvating ability of the solvent. In a solvent that solvates both cations and anions well, it could possibly lie even below the excimer states S_+ and S_-, but most often it will lie above them. In such a case, rapid internal conversion to the lower excimer state will occur as soon as the two components begin to interact significantly. From then on, events that we have already discussed under "approach on the S_+, S_- surface" take over so that excimer emission, monomer emission from components that have diffused away from their partners, or product formation can all occur, as can intersystem crossing.

When the two oppositely charged doublet radical ions initially diffuse together, there is statistically only one chance out of four that the total system will be in a singlet state, and thus on the $I_\pm$ surface that we have discussed so far. In the absence of further perturbations, three-quarters of the time the "supermolecule" will be born in a triplet state, whose photochemistry is discussed in Chapter 6.

When the two components are still barely interacting, the singlet and triplet states into which the two doublets combine are very close in energy and hyperfine interactions have a significant effect on the rate of intersystem crossing, which then becomes dependent on the presence of an outside magnetic field. These effects will be further mentioned in Section 6.4.1, which deals with exciplexes, where they are more commonly observed.

The shape of the I_+ and I_- surfaces shown in Figure 5.26 is that expected in the gas phase. In solution, ions are solvated, and there will typically be barriers on the path of approach of the two components, caused by a decrease or total loss of solvation in the region between them (cf. loose versus tight ions pairs, etc.).

Approach on the D *surface.* When each of the two approaching components is in the triplet state, there will statistically be one chance in nine that the overall system will be in its singlet state and that the approach is on the D surface of Figure 5.26. There will also be one chance in three that the overall system is in its triplet state and that triplet photophysics or photochemistry will follow (Chapter 6). The remaining collisions, five out of nine, yield an overall quintet state and are usually purely repulsive.

Molecules that approach on the D surface find themselves on an increasingly steep slope towards the pericyclic minimum. Although some may reach the minimum directly and proceed to the S_0 state and thus to ground-state products or ground-state starting materials, it appears that many will undergo a rapid internal conversion to the lower excited singlet states of the supermolecule and eventually to the excimer surface S. From then on, events already discussed above will take over. Weak fluorescent emissions from upper excited singlet states, competing with internal conversion, have been observed in such triplet–triplet annihilation experiments. Since they were virtually unperturbed by the presence of the other component, it appears that the internal conversion from the D state to locally excited singlet states occurs at quite large distances of contact between the two triplet components.

Triplet–triplet annihilation can also be viewed as triplet energy transfer in which one of the components acts as a triplet energy donor and emerges in its ground singlet state, whereas the other acts as a triplet energy acceptor; a switch of its multiplicity produces a highly excited singlet state.

Cycloreversion. The cycloadduct has a smaller π system and therefore usually also a higher-lying S_1 (and T_1) state than the individual components. This tends to disfavor barriers along the reaction path on the S_1 surface entering from the right on Figure 5.26, provided that the S_1 state of the cycloadduct corresponds to the HOMO → LUMO excitation and that the orbital crossing diagram is of the normal type (Chapter 3). In aromatic molecules, this is often not fulfilled, and indeed the cycloreversion of the dimer of acenaphthylene seems to have been the first case in which an abnormal MO crossing was recognized.

If entry into the pericyclic minimum both from the right (cycloreversion) and from the left (cycloaddition) is possible without competing processes, and if the molecules stay in the pericyclic minimum long enough to lose memory of their origin, so that after return to S_0 they partition between cycloadduct and free components in a way which is independent of their original structure, the sum of the quantum yields of cycloaddition and cycloreversion will equal unity. This has indeed been verified on several cases in which the addition is intramolecular, so that there is no need for the relatively slow diffusion process that offers opportunities for other competing processes such as fluorescence and intersystem crossing. These observations provided good evidence in favor of the existence of a common intermediate in the cycloaddition and in the cycloreversion process, although in themselves they provided no evidence for the structure of this intermediate. Both its pericyclic nature and its electronically excited nature postulated in Figure 5.26 are based primarily on theoretical arguments, as well as on the stereospecific nature of singlet photocycladditions corresponding to "ground-state-forbidden" path. In some cases a triplet path cycloaddition competes, and this generally leads to the formation of some nonstereospecific products via open-chain biradical geometries (see Chapter 6). There is some direct evidence against the general involvement of such open-chain biradicals in the singlet process. However, in sterically constrained systems, the pericyclic minimum may well be forced to adopt a very nonsymmetric and possibly even an "open-chain" structure.

Molecules that enter the pericyclic minimum from along the cycloreversion path, from the right in Figure 5.26, may have some probability of escaping from it towards the excimer minimum and possibly even further along the S surface, to a free cycloreversion product in an electronically excited state. Small yields of fluorescent emission from free component products have been observed and represent examples of adiabatic organic photochemical processes.

$2_s + 2_s$ Cycloaddition: A computation. An *ab initio* large-scale CI calculation for a model path for the face-to-face cycloaddition of two ethylene molecules gave the results shown in Figure 5.28. They bear a striking resemblance to the correlation digram in Figure 5.25 and the other diagrams just discussed, and they support the interpretations given. For simple alkenes, this reaction path is particularly advantageous since the D state lies quite close to the S state in energy (the $^1\pi\pi^*$ excitation energy in planar ethylene is nearly exactly equal to twice is $^3\pi\pi^*$ excitation energy). While Figure 5.28 does show a well-developed excimer minimum in S located directly over the pericyclic minimum in D, so that only one minimum is present in the S_1 surface, it is possible that a second minimum in S_1 would be found at larger intermolecular separations upon geometry optimization. However, it is also possible that the quite high $^1\pi\pi^*$ excitation energy of simple alkenes really shifts

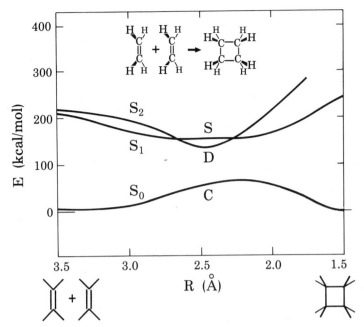

Figure 5.28 Face-to-face ($2_s + 2_s$) cycloaddition of two ethylene molecules. Singlet state energies from an *ab initio* large-scale CI calculation along a path in which standard cyclobutane geometry is approached with gradual pyramidalization of the CH_2 groups.

the excimer minimum towards the cycloadduct side so much that it lies above the pericyclic minimum in S_1. Then, the S_1 surface would contain no excimer minimum. Since the return from the pericyclic minimum to S_0 is undoubtedly very fast, this will provide essentially no opportunity for a reversal to the excited components on the S_1 surface, which normally occurs to some extent from excimers. In this event, the singlet 2 + 2 cycloaddition of alkenes may not be truly representative of other 2 + 2 or 4 + 4 cycloadditions of components with much lower energy S_1 states and/or higher T_1 states, for which excimers will have separate existence on the S_1 surface and a relatively long lifetime.

If only one stereochemical outcome is possible, we can then assume the following general kinetic scheme for singlet photocycloaddition initiated by an encounter of singlet excited alkene (or arene) with a ground-state alkene (or arene) (cf. Figure 5.26):

alkene + alkene* $\rightleftarrows$ excimer* $\rightleftarrows$ pericyclic intermediate* $\longrightarrow$

adduct or alkene + alkene

The Caldwell equation. Caldwell has proposed a simple equation that permits one to estimate whether a singlet photocycloaddition of an alkene or an arene $\mathfrak{A}$ with an alkene or arene $\mathfrak{B}$ has a chance to occur if other processes do not interfere:

$$\gamma = [E_T(\mathfrak{A}) + E_T(\mathfrak{B}) - E_S(\mathfrak{A})]/C^2 \qquad (5.7)$$

where

$$C^2 = 2(C_{1,\mathfrak{A}}^{HOMO}C_{1,\mathfrak{B}}^{HOMO} + C_{2,\mathfrak{A}}^{HOMO}C_{1,\mathfrak{B}}^{HOMO} + C_{1,\mathfrak{A}}^{LUMO}C_{1,\mathfrak{B}}^{LUMO} + C_{2,\mathfrak{A}}^{LUMO}C_{2,\mathfrak{B}}^{LUMO}) \quad (5.8)$$

is a sum of products of MO coefficients of the frontier orbitals at the positions of attack (1,2) in the two partners. Here, γ is minus the resonance integral for end-on interaction of carbon 2p orbitals at the distance corresponding to the transition state on the S_1 surface between the excimer minimum and the pericyclic minimum, defined by the S–D crossing in Figure 5.26. The larger the value of γ, the closer is the approach needed to reach the transition state, and the larger is the expected activation energy, since the S configuration is assumed to rise in energy towards the cycloadduct side of the reaction coordinate. The difference between $E_T(\mathfrak{A})$ + $E_T(\mathfrak{B})$, the sum of the triplet excitation energies of the partners $\mathfrak{A}$ and $\mathfrak{B}$, and $E_S(\mathfrak{A})$, the singlet excitation energy of the excited partner, represents the D–S energy difference at large separations, while C^2 represents a first-order perturbation theory estimate of the slope with which the D state descends from the separated partners towards the pericyclic minimum. Equation 5.7 is obtained by neglecting any changes in the energy of the S state as the two partners approach, approximating the stabilization of the D state by $C^2\gamma$, and asking for this stabilization to just compensate the initial D–S energy difference (Figure 5.26).

This simple equation has been remarkably successful. For $\gamma \leq 20$ kcal/mol, dimerizations are facile, for $20 \leq \gamma \leq 25$, reactivity is moderate, and arenes with $\gamma > 25$ kcal/mol have not been reported to dimerize.

The Caldwell equation is based strictly on reactant properties. This is both a strength in its applicability and a weakness in that the detailed aspects of the electronic structure in the intermediate reaction steps are not taken into account. Foremost among these are the depths of the two minima, which, we suspect, largely control the regiochemistry and stereochemistry of concerted singlet cycloadditions.

Regiochemistry and stereochemistry. In the presence of differentiating substituents, several cycloaddition modes are frequently possible, such as head-to-head versus head-to-tail, or exo versus endo. Each of these possibilities may be characterized by its own excimer minimum in addition to its pericyclic minimum. Even if only two isomers can be formed, the kinetic scheme is likely to be quite complicated (Figure 5.27). It is not easy to predict unequivocally how the change in a few rate constants will affect the yields of the products, but it seems clear that the relative stabilities of the two excimers and of the two pericyclic intermediates, as well as the size of the barriers that separate them, are likely to be important.

Simple theory (Section 5.4.2) suggests that the most highly symmetric excimer (head-to-head, endo) will be most stable as long as all other factors such as steric hindrance are assumed to be the same. This will also be the approach that will yield the largest value of C^2 in the Caldwell equation.

On the other hand, our consideration of the S_1 energies in biradicals perturbed in a polar fashion (Section 5.1.2) suggests that the head-to-tail arrangement leads to the deepest pericyclic minimum, with all other factors again being the same. This is due to the distribution of the localized nonbonding orbitals $|A\rangle$ and $|B\rangle$ on alternating atoms in the pericyclic orbital array, as shown in Figure 5.29A. In the head-to-tail arrangement, $\delta_{AB} \neq 0$, and there is an opportunity for the dot–dot configuration $^1|AB\rangle$ and the hole–pair configuration $^1|B^2\rangle$ built from the localized orbitals $|A\rangle$ and $|B\rangle$ to have similar energies. In the head-to-head arrangement, $\delta_{AB} = 0$, and the S_0–S_1 gap will be larger.

We are then faced with a situation in which the favored excimer geometry (head-to-head) is different from the favored pericyclic intermediate geometry (head-to-tail). The barrier separating the two will respond to a change in the energy of either one. When the head-to-tail geometry is compared to the head-to-head arrangement, the increase in the stability of the pericyclic minimum should lower the barrier in the direction from left to right (Figure 5.29), but the excimer minimum should be less deep and less effective as a reservoir feeding molecules to the pericyclic minimum. A definitive prediction of the relative quantum yields of the possible isomeric adducts is therefore hard to make. Indeed, examples of both types of regiochemistry

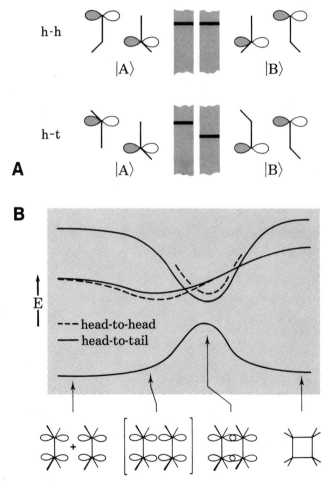

Figure 5.29 Concerted photodimerization. (A) The members of the nonbonding orbital pair in the pericyclic intermediate remain degenerate for head-to-head but not for head-to-tail regiochemistry. (B) The head-to-head and the head-to-tail arrangements have opposite effects on the stability of the excimer and of the pericyclic intermediate. Adapted from V. Bonačić-Koutecký, J. Koutecký, and J. Michl, *Angew. Chem. Int. Ed. Engl.* **26,** 170 (1987).

in singlet photocycloadditions are known. There is, however, no contest as to the expectation of a syn (endo) as opposed to anti (exo) geometry; in the former, there is a large area of overlap between the faces of the two components. In general, the energy differences between the different possible orientations are small, and the regiochemistry of photocycloaddition can be changed by relatively minor influences, such as performing a reaction in a micelle instead of an ordinary solution.

The actual quantum yields of the cycloadducts will also depend on the behavior of the molecules after return to S_0: A fraction of them will continue to the cycloadduct, the remainder will return to the separated components. Serious competition to cycloadduct formation may also come from unproductive decay of the excimer, particularly by emission or by intersystem crossing, and from intersystem crossing and fluorescence in the isolated components before diffusion brings them together. We did not show the intersystem crossing steps in Figures 5.26, 5.27, and 5.29 in order to keep them simple.

5.4.2 Excimers

Next, we shall address in somewhat more detail the electronic structure of excimers. In these, two molecules $\mathfrak{A}$ and $\mathfrak{B}$, usually but not necessarily both aromatic, are brought into fairly close contact (3.0–3.5 Å) so that they weakly overlap without losing their molecular identity. A species in such a minimum is referred to as an excimer if the two partners are identical, a heteroexcimer if they are similar but not identical (e.g., anthracene plus 9-methylanthracene), and an exciplex (Section 6.4.1) if they are different. In typical exciplexes, one partner is a good electron donor, the other a good electron acceptor (e.g., pyrene and N,N-dimethylaniline). In the ground state, the two components are normally not significantly bound in solution, so that $S_0 \rightarrow S_1$ emission or internal conversion is usually followed by dissociation.

The formation of these complexes is favored enthalpically but unfavorable entropically. In some cases, the complex can be intramolecular in that the two partners are linked by a chain or set of chains into a single molecule. Then, the unfavorable entropy factor is reduced. These excimer and exciplex minima in S_1 and T_1 are not always present for all pairs of partners. When they are, they are relatively shallow, with depths of up to about 10 kcal/mol, and relatively flat, with a variety of geometrical arrangements being usually thermally accessible at room temperature. Unless other factors intervene, for aromatics a parallel arrangement of the two planes is preferred in the excited singlet state, with the two molecules arranged symmetrically on both sides (head-to-head). This justifies the inclusion of singlet excimers and exciplexes, which lie on the least-motion path for face-to-face cycloaddition, in a chapter on cyclic multicenter reactions. In triplet excimers and perhaps also exciplexes, the two molecular planes are believed to form a large angle (110°). This places them on the least-motion path for linear multicenter reactions. Formally, then, these triplet species do not belong into this chapter. They have been included nevertheless, in order to keep the discussion of the wave functions of weak molecular complexes compact. The linear multicenter reactions in which triplet excimers appear as intermediates will be discussed in Chapter 6.

What is the reason for the presence of these minima in the S_1 and T_1 surfaces?

It is understandable that in the ground state, two closed-shell species normally do not bind together except for very weak van der Waals bonding. Why is this often different in the excited state?

A qualitative answer to this question at the one-electron level is provided by a consideration of MO energies. In the ground state, both partners have a doubly occupied HOMO and a vacant LUMO. In the degenerate S_1 state, one or the other has a singly occupied HOMO and LUMO. When the two partners are in close interaction, the HOMO and the LUMO are both split. In the ground state, both orbitals resulting from the HOMOs are doubly occupied, leading to a small net energy increase (closed-shell repulsion). In the excited state, the upper of these two orbitals contains only one electron, as does the lower of the two orbitals resulting from the interaction of the two LUMOs. Hence, a net energy lowering occurs.

The beginning of a more satisfactory answer was already provided in Section 5.1.1, dealing with the interaction of two diatomics in various states of excitation. There, we have seen first, that the approach of two closed-shell molecules in their ground state is purely repulsive, except for a shallow van der Waals minimum, and second, that there are three reasons why the approach is attractive if one of the molecules is excited. These are the interactions between the locally excited and charge-transfer excited configurations: the exciton interaction, the charge-transfer interaction, and the interaction of a locally excited configuration with a charge-transfer configuration. The same factors are responsible for the presence of excimer and exciplex minima in general. In H_4, all four electrons were involved in the game so that a square geometry was optimal. In complexes composed of two aromatics, only the few most loosely bound electrons are involved, whereas most electrons still form closed shells that resist close approach. It is then not surprising that the minima occur at relatively large distances of approach, where intermolecular overlap is very small. The five singlet and four triplet configurations involved are shown symbolically in Figure 5.30. In the following, we shall need to consider the matrix elements of the Hamiltonian between them.

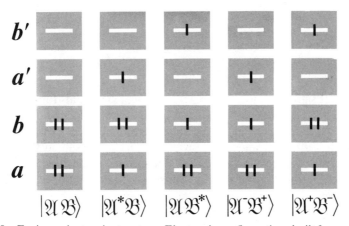

Figure 5.30 Excimer electronic structure: Electronic configurations built from subunit MOs.

Exciton interaction. The origin of the exciton stabilization can be understood as follows, assuming at first for simplicity that intermolecular overlap is negligible. Let us assume that the excited state of the partner $\mathfrak{A}$, which we shall denote by $^1\mathfrak{A}^*$ or $^3\mathfrak{A}^*$, is represented by a single configuration in which an electron has been promoted from the MO a to the MO a'. Similarly, let the excitation in the partner $\mathfrak{B}$, denoted by $\mathfrak{B}^*$, be represented by the promotion from b to b'. At infinite separation, the two partners do not interact, and a system composed of the two molecules has a ground state represented by $|\mathfrak{A}\mathfrak{B}\rangle = {}^1|\cdots a^2b^2\rangle$. It has two excited states of each multiplicity, represented by $^{1,3}|\cdots b^2aa'\rangle$ and $^{1,3}|\cdots a^2bb'\rangle$, where inactive closed shells are marked by dots. The energies are $E(\mathfrak{A}\mathfrak{B})$, $E(^{1,3}|\mathfrak{A}^*\mathfrak{B}\rangle)$, and $E(^{1,3}|\mathfrak{A}\mathfrak{B}^*\rangle)$, respectively:

$$E(^{1,3}|\mathfrak{A}^*\mathfrak{B}\rangle) - E(\mathfrak{A}\mathfrak{B}) = \varepsilon(a') - \varepsilon(a) - J_{aa'} + K_{aa'} \pm K_{aa'} \quad (5.9)$$

$$E(^{1,3}|\mathfrak{A}\mathfrak{B}^*\rangle) - E(\mathfrak{A}\mathfrak{B}) = \varepsilon(b') - \varepsilon(b) - J_{bb'} + K_{bb'} \pm K_{bb'} \quad (5.10)$$

where the upper signs hold for the singlet and the lower ones hold for the triplet states, and ε stands for SCF orbital energy.

At closer distances, the three zero-order states will interact. We shall now neglect the mixing of the zero-order ground state with the zero-order excited states and shall concentrate on the mutual mixing of the two excited singlets and of the two excited triplets. Its magnitude will be given by the size of the matrix element

$$H_1(S,T) = \langle \mathfrak{A}^*\mathfrak{B}|\hat{H}|\mathfrak{A}\mathfrak{B}^*\rangle = \langle ^{1,3}\cdots b^2aa'|\hat{H}|^{1,3}\cdots a^2bb'\rangle \quad (5.11)$$

Writing out the important part of the two wave functions explicitly for the four electrons outside of closed shells, we have

$$H_1(S,T) = \frac{1}{2}\langle\{|b\bar{b}a\bar{a}'| \pm |b\bar{b}a'\bar{a}|\}\hat{H}\{|a\bar{a}b\bar{b}'| \pm |a\bar{a}b'\bar{b}|\}\rangle$$

$$= -\langle a(1)a'(2)|\frac{e^2}{r_{12}}|b(1)b'(2)\rangle + \langle a(1)b(2)|\frac{e^2}{r_{12}}|a'(1)b'(2)\rangle$$

$$\pm \langle a(1)b(2)|\frac{e^2}{r_{12}}|a'(1)b'(2)\rangle$$

$$= -(ab|a'b') + (aa'|bb') \pm (aa'|bb') \quad (5.12)$$

where the upper sign holds for the singlet and lower sign holds for the triplet state. Here, we use both the standard Dirac notation for two-electron integrals and an abbreviated notation in which electron labels are not shown explicitly. The latter will be used in Tables 5.1 and 5.2. However, since we assumed that the overlap between orbitals of the partners is negligible, and since the partners are approaching in a face-to-face fashion, both overlap densities $a(1)b(1)$ and $a'(2)b'(2)$ must be negligible as well; for the interaction elements, we obtain

$$H_1(S) = 2\langle a(1)b(2)|\frac{e^2}{r_{12}}|a'(1)b'(2)\rangle \quad (5.13)$$

5.4 CYCLIC ADDITION AND REVERSION REACTIONS

TABLE 5.1 Excimer: The Hamiltonian Matrix Elements for Singlets

$\mathfrak{AB}$	$\mathfrak{A}^*\mathfrak{B}$	$\mathfrak{AB}^*$	$\mathfrak{A}^-\mathfrak{B}^+$	$\mathfrak{A}^+\mathfrak{B}^-$
0	$\sqrt{2}(aa'\|bb)$	$\sqrt{2}(bb'\|aa)$	$\sqrt{2}\varepsilon(b,a')$	$\sqrt{2}\varepsilon(a,b')$
	$\varepsilon(a') - \varepsilon(a)$ $-(aa\|a'a')$ $+2(aa'\|aa')$	$2(aa'\|bb')$	$-\varepsilon(a,b)$ $-(ab\|a'a')$ $+2(aa'\|ba')$	$\varepsilon(a',b')$ $-(a'b'\|aa)$ $+2(aa'\|ab')$
		$\varepsilon(b') - \varepsilon(b)$ $-(bb\|b'b')$ $+2(bb'\|bb')$	$\varepsilon(a',b')$ $-(a'b'\|bb)$ $+2(bb'\|ba')$	$-\varepsilon(a,b)$ $-(ab\|b'b')$ $+2(bb'\|ab')$
			$\varepsilon(a') - \varepsilon(b)$ $-(a'a'\|bb)$	0
				$\varepsilon(b') - \varepsilon(a)$ $-(b'b'\|aa)$

$$H_1(T) = 0 \tag{5.14}$$

In the zero overlap approximation, then, the interaction matrix element is given by twice the electrostatic interaction energy between the transition density of the $a \to a'$ transition on partner $\mathfrak{A}$ and the $b \to b'$ transition on partner $\mathfrak{B}$. The interaction vanishes for the triplet states because singlet–triplet transitions have zero transition densities.

If the interaction of the two transition densities is approximated by the interaction of their respective dipoles $\mathbf{M}_\mathfrak{A}$ and $\mathbf{M}_\mathfrak{B}$, that is, the transition moments, one obtains the usual dot-product expression:

$$H_1(S) = \mathbf{M}_\mathfrak{A} \cdot \mathbf{M}_\mathfrak{B} / R^3 \tag{5.15}$$

where R is the distance of the two aromatic planes. The splitting is large if the

TABLE 5.2 Excimer: The Hamiltonian Matrix Elements for Triplets

$\mathfrak{A}^*\mathfrak{B}$	$\mathfrak{AB}^*$	$\mathfrak{A}^-\mathfrak{B}^+$	$\mathfrak{A}^+\mathfrak{B}^-$
$\varepsilon(a') - \varepsilon(a)$ $-(aa\|a'a')$	0	$-\varepsilon(a,b)$ $-(ab\|a'a')$	$\varepsilon(a',b')$ $-(a'b'\|aa)$
	$\varepsilon(b') - \varepsilon(b)$ $-(bb\|b'b')$	$\varepsilon(a',b')$ $-(a'b'\|bb)$	$-\varepsilon(a,b)$ $-(ab\|b'b')$
		$\varepsilon(a') - \varepsilon(b)$ $-(a'a'\|bb)$	0
			$\varepsilon(b') - \varepsilon(a)$ $-(b'b'\|aa)$

transitions $a \rightarrow a'$ and $b \rightarrow b'$ have large dipole moments, that is, are strongly allowed. An example is provided by the ^1L_a states of aromatic molecules. To the contrary, ^1L_b states, allowed only very weakly, hardly provide any exciton stabilization. The value of $H_1(S)$ depends strongly on the orientation of the two partners (cf. equation 2.16): When the transition moments are parallel, their interaction is strong; when they are perpendicular, it vanishes. This factor plays an important role in determining the favored geometry of the excimers.

If the two partners are identical, the two excited states are degenerate at infinite separation and separated by twice $H_1(S)$ at finite separations. Using the notation S_+ for the in-phase and S_- for the out-of-phase combination, as before, we see that S_- lies below S_+ if $H_1(S)$ is positive. If $H_1(S)$ is negative, S_+ lies below S_-. The sign of $H_1(S)$ depends on the relative orientation of the two reaction partners, which determines the relative orientation of the interacting transition dipoles $\mathbf{M}_\mathfrak{A}$ and $\mathbf{M}_\mathfrak{B}$.

In both S_+ and S_-, the excitation is distributed equally between both identical partners. These are the stationary states. Under the usual experimental conditions, excimers are initially formed in a nonstationary state in which the excitation is fully localized on one or the other partner. Such a state corresponds to an in-phase or out-of-phase combination of S_+ and S_- and develops in time from one to the other combination, i.e., the excitation resonates between the two partners. The rate of this excitation exchange depends on $H_1(S)/\hbar$. This phenomenon forms the basis for the Förster mechanism of energy transfer.

In concluding, it needs to be emphasized once again that the simple results (5.13) and (5.14) hold only when the mutual overlap of the MOs of the two partners is negligible. At shorter separations, the triplet excited states will interact as well (and there will be energy transfer between them, too).

Charge resonance interaction. Consider now the states that result from the transfer of an electron from one partner to the other. Let us assume that these states are represented by the configurations $^{1,3}|\cdots a^2ba'\rangle$ which represents charge transfer to $\mathfrak{A}$, $\mathfrak{A}^-\mathfrak{B}^+$ and $^{1,3}|\cdots b^2ab'\rangle$, which represents charge transfer to $\mathfrak{B}$, $\mathfrak{A}^+\mathfrak{B}^-$. Their energies are:

$$E(^{1,3}|\mathfrak{A}^-\mathfrak{B}^+\rangle) - E(\mathfrak{A}\mathfrak{B}) = \varepsilon(a') - \varepsilon(b) - J_{a'b} \quad (5.16)$$

$$E(^{1,3}|\mathfrak{A}^+\mathfrak{B}^-\rangle) - E(\mathfrak{A}\mathfrak{B}) = \varepsilon(b') - \varepsilon(a) - J_{ab'} \quad (5.17)$$

where we have omitted integrals that contain the overlap density between MOs located on different partners. These excitation energies are equal to the ionization potential of one of the partners, minus the electron affinity of the other partner, reduced by the energy of the mutual electrostatic attraction of the resulting ion pair. At infinite separation, the two charge-transfer configurations do not interact. At finite separations, they will interact because of the existence of the matrix element

$$H_2(S,T) = \langle \mathfrak{A}^-\mathfrak{B}^+|\hat{H}|\mathfrak{A}^+\mathfrak{B}^-\rangle = \langle ^{1,3}\cdots a^2ba'|\hat{H}|^{1,3}\cdots b^2ab'\rangle \quad (5.18)$$

We shall assume again that the separation of the two partners is large enough for overlap to be negligible, and we shall use the shorthand notation introduced for the exciton states. Then,

$$H_2(S,T) = \frac{1}{2}\langle\{|a\bar{a}b\bar{a}'| \pm |a\bar{a}a'\bar{b}|\}\hat{H}\{|b\bar{b}ab'| \pm |b\bar{b}b'\bar{a}|\}\rangle \quad (5.19)$$

$$= \langle a(1)a'(2)|\frac{e^2}{r_{12}}|b(1)b'(2)\rangle$$

As long as the overlap between the orbitals of the two partners is negligible, the overlap densities $a(1)b(1)$ and $a'(2)b'(2)$ are both negligible and the interaction element vanishes for both multiplicities; that is, $H_2(S,T) = 0$.

Interaction between charge-transfer and locally excited configurations. In order to complete the picture of interactions that contribute to the existence of excimer minima, we also need to consider the interaction elements between a locally excited configuration, say $|\mathfrak{A}^*\mathfrak{B}\rangle$, and one or the other of the charge-transfer configurations, $|\mathfrak{A}^-\mathfrak{B}^+\rangle$ and $|\mathfrak{A}^+\mathfrak{B}^-\rangle$:

$$H_3(S,T) = \langle\mathfrak{A}^-\mathfrak{B}^+|\hat{H}|\mathfrak{A}^*\mathfrak{B}\rangle = \frac{1}{2}\langle\{|a\bar{a}b\bar{a}'| \pm |a\bar{a}a'\bar{b}|\}\hat{H}\{|b\bar{b}a\bar{a}'| \pm |b\bar{b}a'\bar{a}|\}\rangle$$

$$= -\varepsilon(a,b) + \langle a(1)b(2)|\frac{e^2}{r_{12}}|a'(1)a'(2)\rangle \pm \langle a(1)b(2)|\frac{e^2}{r_{12}}|a'(1)a'(2)\rangle$$

$$- \langle a(1)a'(2)|\frac{e^2}{r_{12}}|b(1)a'(2)\rangle \quad (5.20)$$

$$H_4(S,T) = \langle\mathfrak{A}^-\mathfrak{B}^+|\hat{H}|\mathfrak{A}\mathfrak{B}^*\rangle = \frac{1}{2}\langle\{|a\bar{a}b\bar{a}'| \pm |a\bar{a}a'\bar{b}|\}\hat{H}\{|a\bar{a}b\bar{b}'| \pm |a\bar{a}b'\bar{b}|\}\rangle$$

$$= \varepsilon(a',b') - \langle a'(1)b(2)|\frac{e^2}{r_{12}}|b'(1)b(2)\rangle + \langle b(1)b(2)|\frac{e^2}{r_{12}}|b'(1)a'(2)\rangle$$

$$\pm \langle b(1)b(2)|\frac{e^2}{r_{12}}|b'(1)a'(2)\rangle \quad (5.21)$$

Here, $\varepsilon(a,b)$ and $\varepsilon(a',b')$ stand for off-diagonal elements of the Fock operator. They would vanish if a and b were the canonical SCF MOs of the whole system. However, since a, a', b, and b' are MOs of the individual partners, these off-diagonal elements must be kept. Note also that we have kept the two-electron integrals in which only one of the two charge densities is due to overlap of MOs located on different partners. The significance and evaluation of the matrix elements $H_3(S,T)$ and $H_4(S,T)$ will be discussed below.

Interactions involving the ground state. Finally, the matrix elements of the Hamiltonian between the ground state on the one hand and a locally excited or a charge-transfer singlet state on the other hand are given by

$$H_5 = \langle \mathfrak{AB}|\hat{H}|\mathfrak{A}^*\mathfrak{B}\rangle = \frac{1}{\sqrt{2}}\langle |a\bar{a}b\bar{b}|\hat{H}\{b\bar{b}a\bar{a}'| + |b\bar{b}a'\bar{a}|\}\rangle \qquad (5.22)$$

$$= \sqrt{2}\,(\langle a(1)b(2)|\frac{e^2}{r_{12}}|a'(1)b(2)\rangle - \langle a(1)a'(2)|\frac{e^2}{r_{12}}|b(1)b(2)\rangle)$$

With the usual neglect of the integral containing two small overlap densities, this can be simplified to

$$H_5 = \sqrt{2}\,\langle a(1)b(2)|\frac{e^2}{r_{12}}|a'(1)b(2)\rangle \qquad (5.23)$$

Finally,

$$H_6 = \langle \mathfrak{AB}|\hat{H}|\mathfrak{A}^-\mathfrak{B}^+\rangle = \frac{1}{\sqrt{2}}\langle |a\bar{a}b\bar{b}|\hat{H}|\{|a\bar{a}b\bar{a}'| + |a\bar{a}a'\bar{b}|\}\rangle = \sqrt{2}\,\varepsilon(b,a') \qquad (5.24)$$

A summary of the interaction elements is given in Tables 5.1 (singlets) and 5.2 (triplets), using the abbreviated notation.

Interactions with other configurations. Frequently, other excited configurations lie low enough in energy and need to be included. One singlet and one triplet state result from double triplet excitations involving the four orbitals considered:

$$^1|^3\mathfrak{A}^{*3}\mathfrak{B}\rangle = {}^1|\cdots \overline{aa'bb'}\rangle = (1\sqrt{12})|\,2|a\bar{b}a'\bar{b}'| + 2|b\bar{a}b'\bar{a}'| \qquad (5.25)$$
$$+ |a\bar{a}'b\bar{b}'| + |a'\bar{a}b'\bar{b}| - |a\bar{a}'b'\bar{b}| - |a'\bar{a}b\bar{b}'|\rangle$$

$$^3|^3\mathfrak{A}^{*3}\mathfrak{B}^*\rangle = (1/\sqrt{2})|\,|a\bar{b}a'\bar{b}'| - |b\bar{a}b'\bar{a}'|\rangle \qquad (5.26)$$

At times, other MOs of one or both partners need to be used in order to include configurations describing simultaneous excitation and charge transfer such as $^{1,3}|\mathfrak{A}^{-*}\mathfrak{B}^+\rangle$ or $^{1,3}|\mathfrak{A}^-\mathfrak{B}^{+*}\rangle$ and other locally excited states of the partners. In aromatics, L_b and L_a states are often of comparable energy and both need to be considered. The description we have used here to represent $\mathfrak{A}^*$ and $\mathfrak{B}^*$ corresponds to the wave functions of the L_a states.

For illustrative purposes, the incorporation of all this complexity is not necessary, and we shall avoid it. However, it needs to be emphasized that this may no longer be admissible if quantitative results are desired.

Inclusion of intermolecular overlap. At typical intermolecular separations of 3.0–3.5 Å, the overlap between a π MO on one of the partners and a π MO on the other partner in an excimer or exciplex is much less than unity, of the order of 0.01–0.05. Therefore, it is possible to treat the effects of nonvanishing overlap by addition of correction terms to the framework already established. Since the overlap is so small, it is reasonable to neglect its second and higher powers. It then becomes attractive to express as many interactions as possible in terms of experimentally accessible quantities and the four overlap integrals $S_{a,b}$, $S_{a'b}$, $S_{ab'}$, and $S_{a'b'}$. This can be done by the use of suitable (although sometimes quite drastic) approximations. Only two interaction elements are not related to overlap. The first is due to the long-range exciton interaction, given for the singlet state by

$$H_1(S) = \langle {}^1\mathfrak{A}^*\mathfrak{B}|\hat{H}|{}^1\mathfrak{A}\mathfrak{B}^*\rangle = 2\langle a(1)b(2)|\frac{e^2}{r_{12}}|a'(1)b'(2)\rangle \quad (5.27)$$

and vanishing for the triplet,

$$H_1(T) = \langle {}^3\mathfrak{A}^*\mathfrak{B}|\hat{H}|{}^3\mathfrak{A}\mathfrak{B}^*\rangle = 0 \quad (5.28)$$

The magnitude of $H_1(S)$ can be estimated from known transition intensities using equation 5.15. The second interaction element that is not related to overlap is due to the charge resonance interaction and is negligible:

$$H_2(S,T) = \langle \mathfrak{A}^-\mathfrak{B}^+|\hat{H}|\mathfrak{A}^+\mathfrak{B}^-\rangle \cong 0 \quad (5.29)$$

Using the ground-state energy $E(\mathfrak{A}\mathfrak{B})$ as the energy zero, the most important interaction elements involving the charge-transfer configurations ${}^{1,3}|\mathfrak{A}^-\mathfrak{B}^+\rangle$ are

$$\langle {}^{1,3}\mathfrak{A}^-\mathfrak{B}^+|\hat{H}|{}^{1,3}\mathfrak{A}^*\mathfrak{B}\rangle \cong -S_{ab}[\Gamma_\mathfrak{B} + E({}^{1,3}|\mathfrak{A}^*\mathfrak{B}\rangle)] \quad (5.30)$$

$$\langle {}^{1,3}\mathfrak{A}^-\mathfrak{B}^+|\hat{H}|{}^{1,3}\mathfrak{A}\mathfrak{B}^*\rangle \cong S_{a'b'}\Gamma_\mathfrak{B} \quad (5.31)$$

$$\langle {}^1\mathfrak{A}^-\mathfrak{B}^+|\hat{H}|{}^1\mathfrak{A}\mathfrak{B}\rangle \cong \sqrt{2}\, S_{a'b'}\Gamma_\mathfrak{B} \quad (5.32)$$

$$\langle {}^1\mathfrak{A}^-\mathfrak{B}^+|\hat{H}|{}^1({}^3\mathfrak{A}^{*3}\mathfrak{B}^*)\rangle \cong -\frac{1}{2}\sqrt{6}\, S_{ab'}[\Gamma_\mathfrak{B} + E({}^1\mathfrak{A}^*\mathfrak{B})] \quad (5.33)$$

$$\langle {}^3\mathfrak{A}^-\mathfrak{B}^+|\hat{H}|{}^3({}^3\mathfrak{A}^{*3}\mathfrak{B}^*)\rangle \cong S_{ab'}[\Gamma_\mathfrak{B} + E({}^3|\mathfrak{A}^*\mathfrak{B}\rangle)] \quad (5.34)$$

and the matrix elements for ${}^{1,3}|\mathfrak{A}^+\mathfrak{B}^-\rangle$ are analogous, for example,

$$\langle {}^{1,3}\mathfrak{A}^+\mathfrak{B}^-|\hat{H}|{}^{1,3}\mathfrak{A}^*\mathfrak{B}\rangle \cong S_{a'b'}\Gamma_\mathfrak{A} \quad (5.35)$$

The symbols $\Gamma_\mathfrak{A}$ and $\Gamma_\mathfrak{B}$ stand for

$$\Gamma_\mathfrak{A} = -\gamma + IP_\mathfrak{A} - \tfrac{1}{2}J_\mathfrak{A} - \tfrac{1}{2}C \quad (5.36)$$

$$\Gamma_\mathfrak{B} = -\gamma + IP_\mathfrak{B} - \tfrac{1}{2}J_\mathfrak{B} - \tfrac{1}{2}C \quad (5.37)$$

where $IP_\mathfrak{A}$ and $IP_\mathfrak{B}$ are the ionization potentials of the partners $\mathfrak{A}$ and $\mathfrak{B}$ and where

$$J_\mathfrak{A} = (aa|aa) \simeq (aa|a'a') \quad (5.38)$$

$$J_\mathfrak{B} = (bb|bb) \simeq (bb|b'b') \quad (5.39)$$

$$C = (aa|bb) \simeq (aa|b'b') \simeq (a'a'|bb) \quad (5.40)$$

and

$$\gamma = A_C - \tfrac{1}{2}\langle I|U_I^*|I\rangle - 2\langle I+1|U_I^*|I+1\rangle \qquad (5.41)$$

where A_C is the electron affinity of the carbon atom in its valence state, $\langle I|U_I^*|I\rangle$ is the one-center penetration integral, and $\langle I+1|U_I^*|I+1\rangle$ is the nearest-neighbor penetration integral for an aromatic carbon atom. The value of γ is about 16 eV. Usually, $J_\mathfrak{A}$ and $J_\mathfrak{B}$ are of the order of 5 eV, and C is about 3 eV. Typical values of $\Gamma_\mathfrak{A}$ and $\Gamma_\mathfrak{B}$ then are about -12 eV.

The overlap integrals are a sensitive function of the distance of the partners and of the pairwise matching of the AO centers in the partners. Perfect matching of equivalent atoms can occur for an excimer. If the two partners differ greatly in size, the overlap integral will be small. The matching of nodal planes present in each of the MOs also is important.

Singlet excimers. Equality of the partners introduces an element of symmetry into many of the possible geometric arrangements of the complex. In particular, in the perfectly matched geometry favored by singlet excimers (Figure 5.31), the two locally excited configurations $^1|\mathfrak{A}^*\mathfrak{B}\rangle$ and $^1|\mathfrak{A}\mathfrak{B}^*\rangle$ produce an in-phase and an out-of-phase combination, $^1|\mathfrak{A}^*\mathfrak{B}\rangle \pm {}^1|\mathfrak{A}\mathfrak{B}^*\rangle$. Likewise, the charge-transfer configurations combine into $^1|\mathfrak{A}^-\mathfrak{B}^+\rangle \pm {}^1|\mathfrak{A}^+\mathfrak{B}^-\rangle$. Only configurations of equal symmetry can interact when final states are computed. The lowest of the resulting excited states is then the excimer state. Because of the stabilization by the interaction, its energy is generally lower than that of the separated pair of molecules $\mathfrak{A} + \mathfrak{B}^*$ or $\mathfrak{A}^* + \mathfrak{B}$, and an "excimer minimum" in the S_1 surface results. The symmetry of the electronic wave function in the excimer state tends to be such that the radiative transition to the ground state is forbidden, and excimers generally have long radiative lifetimes. This is important because it provides an opportunity for motion into other minima in the S_1 surface while the two partners are in close vicinity, as well as an opportunity for intersystem crossing. The fluorescent emission is facilitated by displacements of the exciter geometry to less symmetric forms.

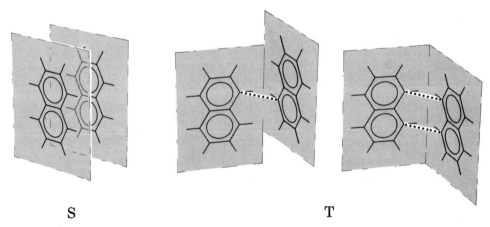

Figure 5.31 Preferred geometries of singlet (S) and triplet (T) excimers.

When the two aromatic hydrocarbons are linked by an alkane chain, the length of the chain is quite critical. Among short chains, only the trimethylene link permits the partners to attain a sandwich-like configuration and form a singlet excimer.

A classical illustration is provided by the excimer of pyrene. In this molecule, the lowest excited singlet state is L_b, with L_a only 0.42 eV higher in energy. Both states therefore need to be considered. The L_a state on each pyrene is approximated by the configuration $^1|\cdots 1\,-1\rangle$, and the L_b state is approximated by the configuration $^1|\cdots 2\,1^2\,-1\rangle - ^1|\cdots 2^21\,-2\rangle$. Thus, the orbitals a, a', b, and b' needed for the description of the excimer are the orbitals 2, 1, -1, and -2 on each pyrene. These can be labeled $2_\mathfrak{A}$, $1_\mathfrak{A}$, $-1_\mathfrak{A}$, $-2_\mathfrak{A}$, $2_\mathfrak{B}$, $1_\mathfrak{B}$, $-1_\mathfrak{B}$, and $-2_\mathfrak{B}$. However, a calculation of overlap integrals at the most symmetric geometry (D_{2h}) shows that only those between MOs $1_\mathfrak{A}$ and $1_\mathfrak{B}$ and MOs $-1_\mathfrak{A}$ and $-1_\mathfrak{B}$ do not vanish. Therefore, in the approximation adopted, interaction occurs only between the L_a states $^1|\mathfrak{A}^*\mathfrak{B}\rangle \pm ^1|\mathfrak{A}\mathfrak{B}^*\rangle$ and the charge-transfer states $^1|\mathfrak{A}^-\mathfrak{B}^+\rangle \pm ^1|\mathfrak{A}^+\mathfrak{B}^-\rangle$, and the only orbitals needed are $a \equiv 1_\mathfrak{A}$, $b \equiv 1_\mathfrak{B}$, $a' \equiv -1_\mathfrak{A}$, and $b' \equiv -1_\mathfrak{B}$. The nonvanishing overlap integrals are $S_{ab} = 0.059$ and $S_{a'b'} = 0.025$ when Slater AOs with exponent 1.625 are used as a first approximation.

Using IP = 7.53 eV, EA = 0.59 eV, $E(^1|\mathfrak{A}^*L_a\rangle) = 3.72$ eV, and $E(^3|\mathfrak{A}^*L_a\rangle) = 2.09$ eV and calculating $C = 3.29$ eV, one obtains $J = 4.85$ eV, $E(^1|\mathfrak{A}^-\mathfrak{B}^+\rangle) = E(^1|\mathfrak{A}^+\mathfrak{B}^-\rangle) = 3.65$ eV, and $\Gamma = -12.5$ eV, so that

$$H_3(S) = \langle ^1|\mathfrak{A}^-\mathfrak{B}^+|\hat{H}|^1\mathfrak{A}^*\mathfrak{B}\rangle = \langle ^1|\mathfrak{A}^+\mathfrak{B}^-|\hat{H}|^1\mathfrak{A}\mathfrak{B}^*\rangle \cong 0.53 \text{ eV} \quad (5.42)$$

$$H_4(S) = \langle ^1|\mathfrak{A}^-\mathfrak{B}^+|\hat{H}|^1\mathfrak{A}\mathfrak{B}^*\rangle = \langle ^1|\mathfrak{A}^+\mathfrak{B}^-|H|^1\mathfrak{A}^*\mathfrak{B}\rangle \cong -0.31 \text{ eV} \quad (5.43)$$

Estimating

$$H_1(S) = \langle ^1|\mathfrak{A}^*\mathfrak{B}|\hat{H}|^1\mathfrak{A}\mathfrak{B}^*\rangle \cong 0.1 \text{ eV} \quad (5.44)$$

for the exciton interaction element, the secular problem can be solved. The result for the singlet excimer emission energy is 2.58 eV and compares favorably with the experimental value of 2.60 eV. This perfect agreement was obtained by optimizing the value for the intermolecular distance within the generally accepted range of 3.0–3.5 Å. The resulting wave function is

$$0.50\{^1|\mathfrak{A}^*\mathfrak{B}\rangle - ^1|\mathfrak{A}\mathfrak{B}^*\rangle\} - 0.50\{^1|\mathfrak{A}^-\mathfrak{B}^+\rangle - ^1|\mathfrak{A}^+\mathfrak{B}^-\rangle\} \quad (5.45)$$

and demonstrates the simultaneous importance of the locally excited and charge-transfer excited states within the framework of the simple model.

Accepting this wave function as being qualitatively correct, it is possible to use experimental values of excitation energies and excimer enthalpy of formation to produce a rough picture of the relevant potential energy surfaces. The shift of the excimer emission energy (2.60 eV) relative to the excitation energy of the L_a transition of isolated pyrene (3.72 eV) is 1.12 eV, that is, 26.4 kcal/mol. This difference, of course, does not represent the depth of the excimer minimum in S_1 (Figure 5.32). First, the S_1 state of pyrene is ^1L_b, and this lies only 3.30 eV above the ground state in the isolated molecule. Second, the ground electronic state of the excimer lies above that of two separated molecules because the complex is not bound in the ground state. Experimentally, the depth of the pyrene singlet excimer

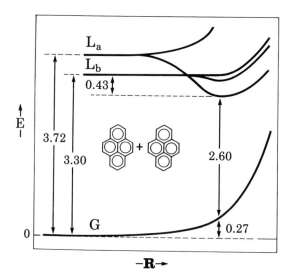

Figure 5.32 State energy diagram for pyrene excimer. Based on experimental data. Energies are expressed in electronvolts.

minimum is 9.8 kcal/mol (0.43 eV), so that the destabilization of the ground-state energy at this distance of approach is 6.4 kcal/mol (0.27 eV).

Triplet excimers. Triplet excimers can be generally expected to be less stable than singlet excimers: The exciton stabilization is missing because the transition density is almost exactly zero, and the locally excited states are now energetically farther removed from the charge-transfer states. Moreover, their observation has been more difficult. Long radiative lifetimes of phosphorescence make it very sensitive to quenching if diffusion can occur. If diffusion cannot occur, the excimer cannot be formed unless the two partners are associated to start with.

Thus, conclusive direct evidence for the existence of triplet excimers of aromatic hydrocarbons from emission and transient absorption spectroscopy is much more recent than that for their singlet counterparts. Studies containing both partners linked and thus predisposed for excimer formation have been particularly important. Unlike singlet excimers, which require a three-carbon link, triplet excimers are not sensitive to the link length and can be formed even from diarylmethanes. This alone shows that a sandwich-like arrangement of the two aromatic planes is not needed. This is understandable, given that the exciton interaction between the local transition moments is now negligible, so that it does not particularly matter if they are not parallel. The geometries believed to be optimal for triplet excimers are shown in Figure 5.31. The two molecular planes lie at about a 70° angle to each other, resembling the optimal arrangement in a gas-phase ground-state van der Waals dimer.

In a sense, the difference between the optimal singlet and triplet excimer geometries anticipates the energetically more strongly developed difference between the optimal excited singlet ("tight") and triplet ("loose") biradicaloid geometries that lie further along the concerted and nonconcerted reaction paths, respectively. In the excited singlet, the overlap of the interacting localized orbitals on the two touching molecules is maximized; in the triplet and in the ground state, it is minimized. The one close C—C distance in the optimal triplet excimer geometry shown in the center of Figure 5.31 suggests an incipient bonding interaction on the way to a triplet biradical in which each of the two unpaired electrons is localized in one of the aromatic rings, and there is a distinct similarity to a radical attack on a ground-state aromatic molecule.

It is likely that similar very weakly bound triplet excimers generally occur as intermediates in nonconcerted triplet dimerization reactions, discussed in Chapter 6. In those cases in which there is an excessive energy barrier separating the triplet excimer from the triplet biradical in which a new C—C bond between the aromatic rings has been formed, the cycloaddition will not take place, but the triplet excimer may still play an important role in quenching. This is due to its nonplanar geometry, which enhances spin–orbit coupling relative to that in an isolated planar aromatic molecule. The radiative lifetime of the phosphorescence of triplet excimers is indeed shortened, and there is evidence for a much enhanced rate of $T_1 \rightarrow S_0$ intersystem crossing.

5.5 COMMENTS AND REFERENCES

The states of a system consisting of **four electrons in a cyclic array of four orbitals** and their relation to organic photochemistry were analyzed in W. Th. A. M. van der Lugt and L. J. Oosterhoff, *J. Am. Chem. Soc.* **91**, 6042 (1969) and W. Gerhartz, R. D. Poshusta, and J. Michl, *J. Am. Chem. Soc.* **98**, 6427 (1976) and, most recently, in A. F. Voter and W. A. Goddard, III, *J. Am. Chem. Soc.* **108**, 2830 (1986) and V. Bonačić-Koutecký, J. Koutecký, and J. Michl, *Angew. Chem. Int. Ed. Engl.* **26**, 170 (1987). The electronic states of polar cyclobutadiene analogs (critical heterosymmetry) are also discussed in V. Bonačić-Koutecký, K. Schöffel and J. Michl, *J. Am. Chem. Soc.,* **111**, 6140 (1989). For a dissenting opinion on a four-orbital four-electron case, see K. Morihashi, O. Kikuchi, and K. Suzuki, *Chem. Phys. Lett.* **90**, 346 (1982), whose results on the photochemical electrocyclic ring closure of butadiene differ from those obtained earlier by van der Lugt and Oosterhoff quoted above and by D. Grimbert, G. A. Segal, and A. Devaquet, *J. Am. Chem. Soc.* **97**, 6629 (1975), with whose views we tend to agree. S_0-S_1 touchings at critically heterosymmetric biradicaloid geometries of the diamond-shaped cyclohetadiene type have recently been located numerically in calculations for 1,3-butadiene ring closure and 2+2 cycloaddition of two ethylenes: F. Bernardi, S. De, M. Olivucci, and M. A. Robb, *J. Am. Chem. Soc.* **112**, 1737 (1990). Recent experimental evidence suggests that the reverse reaction, vacuum UV photoreaction of simple cyclobutenes to produce butadienes, may proceed in part adiabatically: K. B. Clark and W. J. Leigh, *J. Am. Chem. Soc.* **109**, 6068 (1987). The adiabatic course of a similar reaction of a more complicated cyclobutene, 1,4-dewarnaphthalene, has been discussed in Section 2.7. Recent experimental evidence [M. O. Trulson, G. D. Dollinger, and R. A. Mathies, *J. Chem. Phys.* **90**, 4274 (1989)] and computational results [P. E. Share, K. L. Kompa, S. D. Peyerimhoff, and M. C. van Hemert, *Chem. Phys.* **120**, 411 (1988)] support the van der Lugt-Oosterhoff picture for the conrotatory opening of 1,3-cyclohexadiene.

A comparison with the cyclization of 1,4-diazabutadiene is available in Y. Jean and A. Devaquet, *J. Am. Chem. Soc.* **99**, 1949 (1977). Application to photochemical cycloadditions is also discussed in J. Michl, *Photochem. Photobiol.* **25**, 141 (1977) and E. M. Evleth and E. Kassab, *Can. J. Chem.* **61**, 306 (1983). The relation of these results to the cis–trans isomerization of double bonds has been emphasized by J. J. C. Mulder, *Nouv. J. Chim.* **4**, 283 (1980).

For **calculations on the electrocyclic ring opening of oxirane,** see B. Bigot, A. Sevin, and A. Devaquet, *J. Am. Chem. Soc.* **101,** 1095 (1979); for **aziridine,** see B. Bigot, A. Devaquet, and A. Sevin, *J. Org. Chem.* **45,** 97 (1980); for a summary of experimental results, see N. R. Bertoniere and G. W. Griffin, in *Organic Photochemistry,* (O. L. Chapman, Ed.), Marcel Dekker, New York, 1973, Vol. 3, p. 115, and A. Padwa and G. Grifin, in *Photochemistry of Heterocyclic Compounds* (O. Buchardt, Ed.), Wiley, New York, 1976, p. 41. Experimental evidence for stereospecific singlet and triplet cyclization of divinylamines, divinylethers, and divinylsulfides to **five-membered rings** is summarized in A. G. Schultz, *Acc. Chem. Res.* **16,** 210 (1983) and A. G. Schultz and L. Motyka, in *Organic Photochemistry* (A. Padwa, Ed.), Marcel Dekker, New York, 1983, Vol. 6, p. 1, see also K.-H. Grellmann, U. Schmitt, and H. Weller, *Chem. Phys. Lett.* **88,** 40 (1982) and references therein. An observation of adiabatic formation of triplet product was reported in E. W. Förster and K.-H. Grellmann, *Chem. Phys. Lett.* **14,** 536 (1972).

For a derivation of the **Caldwell equation** (equation 5.7) for **photocycloaddition,** see R. A. Caldwell, *J. Am. Chem. Soc.* **102,** 4004 (1980). For an earlier discussion in terms of PMO theory, see W. C. Herndon, *Topics Curr. Chem.* **46,** 141 (1974). The concerted cycloaddition mechanism proceeding via an excimer or exciplex and then a pericyclic minimum has gained general acceptance and agrees with the stereospecific nature of singlet photocycloadditions. Incomplete regiochemical or stereochemical integrity, sometimes attributed to a singlet biradical pathway, most likely results instead from leakage into the triplet manifold or from overirradiation: N. C. Yang, R. L. Yates, J. Masnovi, D. M. Shold, and W. Chiang, *Pure Appl. Chem.* **51,** 173 (1979); N. C. Yang, J. Masnovi, T. Wang, H. Shou, and D. H. Yang, *Tetrahedron* **37,** 3285 (1981); T.-Y. Wang, J.-D. Ni, J. Masnovi, and N. C. Yang, *Tetrahedron Lett.* **23,** 1231 (1982). The bottom of the pericyclic minimum probably corresponds to a funnel located at a geometry of symmetry lower than assumed in Figure 5.28 (critically heterosymmetric biradicaloid, cf. Section 5.1.2).

The derivation of orbital symmetry rules by the "interacting subunits method" in this text is patterned after J. Michl, *Mol. Photochem.* **4,** 287 (1972) and J. Michl, in *Chemical Reactivity and Reaction Paths* (G. Klopman, Ed.), Wiley, New York, 1974, p. 301.

Our treatment of **excimers and exciplexes** follows that given by H. Beens and A. Weller, in *Organic Molecular Photophysics,* Vol. 2 (J. B. Birks, Ed.), Wiley, London, 1975. For the original interpretations, see: T. Azumi, A. T. Armstrong, and S. P. McGlynn, *J. Chem. Phys.* **41,** 3839 (1964); J. N. Murrell and J. Tanaka, *Mol. Phys.* **7,** 363 (1964). See also: J. B. Birks, *Photophysics of Aromatic Molecules,* Wiley, New York, 1970; A Weller, *Pure Appl. Chem.* **54,** 1885 (1982); R. S. Davidson, in *Advances in Physical Organic Chemistry,* Vol. 19 (V. Gold and D. Bethell, Eds.), Academic Press, New York, 1983; N. Mataga and T. Kubota, *Molecular Interactions and Electronic Spectra,* Marcel Dekker, New York, 1970; M. Gordon and W. R. Ware, Eds., *The Exciplex,* Academic Press, New York, 1975; J. B. Birks, *Rep. Prog. Phys.* **38,** 903 (1975). The role of excimers and exciplexes in photocycloaddition of aromatics has been surveyed recently: J. J. McCullough, *Chem. Rev.* **87,** 811 (1987).

Triplet excimers are best detectable by T–T absorption: K. A. Zachariasse, R. Busse, U. Schrader, and W. Kühnle, *Chem. Phys. Lett.* **89,** 303 (1982); for a detailed recent discussion of the evidence, see E. C. Lim, *Acc. Chem. Res.* **20,** 8 (1987).

Sigmatropic and pseudosigmatropic shifts are formally related, in that a funnel in S_1 is expected at a critically heterosymmetric geometry, analogous to that of a diamond-shaped cyclobutadiene, with a finite energy difference between the two "nonbonding" orbitals ($\delta_{AB} > 0$, Section 4.6). Upon return to S_0, the migrating atom may initially find itself within the valence pull of each of the three carbons of the allylic chain.

CHAPTER 6

Linear Multicenter Reactions: One Active Orbital per Atom

There are many photochemical processes that can be viewed as proceeding by a more or less simultaneous bond-breaking and -making in a linear array of overlapping orbitals. Some of these, for example, stepwise photocycloadditions, are the nonconcerted analogs of the pericyclic reactions we have discussed in Chapter 5, while others, such as cis–trans isomerization in a diene, have no cyclic concerted analogs.

6.1 ELECTRONS IN A LINEAR ARRAY OF ORBITALS

Just like the cyclic interaction of four electrons in four orbitals, their linear interaction can be most simply modeled by a calculation on H_4. Two types of processes will be of foremost interest here: in one, the four H atoms will separate pairwise as in $H_2 + H_2$; in the other, they will separate into $H_3 + H$.

The $H_2 + H_2$ system. We shall consider the $H_2 + H_2$ case first. Since Figures 5.10–5.14 contain information on the energies of the G, S, and D singlet states for all trapezoidal geometries of H_4, they also display the properties of the system containing two H_2 molecules of equal but variable length approaching each other in a linear end-on fashion. Such a linear arrangement of H_4 is nothing but a limiting case of a trapezoid whose two parallel sides have zero separation. These geometries are represented by points in the bottom plane in Figures 5.10–5.14 (R = 0). The same results for symmetric linear geometries are shown once again schematically in Figures 6.1 and 6.2 as the more conventional plots of energy as a function of geometry, along with results for other geometries which correspond to the second step of the nonconcerted conversion, $H_2 + D_2 \rightarrow HD + H + D \rightarrow 2HD$.

On the left of Figure 6.1, two H_2 molecules H_AH_B and H_CH_D are brought together in a collinear fashion. Moving to the right in the front face, the two internal hydrogen atoms then form a new molecule, H_BH_C, and the external atoms H_A and H_D are simultaneously removed to infinity. The course of the energy curves is quite similar to that already seen for the rectangular case in that the G and D states again undergo an avoided touching. On the left, the prevailing singlet coupling into bonds in the sense of VB theory is A—B, C—D in the G state and A—D, B—C in the D state (cf. Figure 6.1A); on the right, the opposite is true. In the central region of the diagram, the G state is predominantly covalent, the D state is partly so, and the S state is strongly ionic. Their order is the same as found for intermediate and large squares: G, D, S. The behavior in the dissociation limit

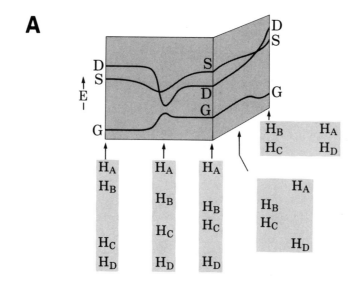
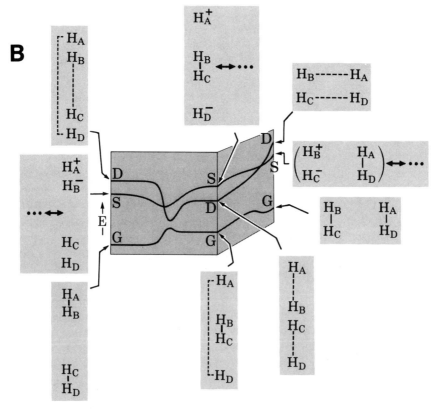

Figure 6.1 (A) Energies of the G, S, and D states of H_4 along the nonconcerted path for the $H_AH_B + H_CH_D \rightarrow H_BH_C + H_AH_D$ reaction. Front face: Symmetric linear geometries ($D_{\infty h}$). Side face: Trapezoidal geometries (C_{2v}). Schematic. (B) VB wave functions for the G, S, and D states of H_4 at three geometries.

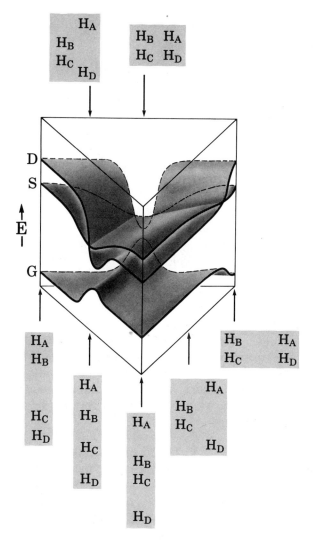

Figure 6.2 Energies of the G, S, and D states of H_4: A comparison of the concerted (back side) and nonconcerted (front two sides) reaction paths. Schematic.

also is similar to that found for rectangular geometries. The G state correlates with two ground-state H_2 molecules on the left and two singlet-coupled but infinitely separated H atoms plus a ground-state H_2 molecule on the right. The D state correlates with two triplet H_2 molecules on the left and two triplet-coupled but separated H atoms plus a triplet H_2 molecule on the right. The S state correlates with one ground-state and one singly excited H_2 molecule on the left and with one ground-state H_2 molecule and an ion pair on the right.

Essentially all that was said in Section 5.1.1 about the use of the simple four-state MO picture, of the VB picture, and of the subsystem configuration picture of rectangular H_4 applies here to symmetric linear H_4 also.

Since we are using the collinear approach of two H_2 molecules as simple model for the nonconcerted analogs of a $2_s + 2_s$ pericyclic reaction, it is of interest to display also information on those geometries that correspond to the second step of the nonconcerted path $H_AH_B + H_CH_D \xrightarrow{1} H_A + H_BH_C + H_D \xrightarrow{2} H_BH_C + H_AH_D$. In Figures 5.10–5.14, such geometric paths would involve an increase in R and a decrease in R_2, leading to the region in which the ground state enters the "vertical tube" of low energies.

In Figure 6.1, this path is represented by a line perpendicular to that of linear geometries. Along this path, the energies of the G, S, and D states develop in a fashion that is simply predictable because the H_BH_C molecule is now far removed from the atoms H_A and H_D. Since the latter are singlet-coupled in the G state, their approach will lower its energy as the second hydrogen molecule (H_AH_D) is formed. They are triplet-coupled in the D state, whose energy will therefore rise. In the S state, they represent an ion pair, so that the energy will decrease somewhat as the H_AH_D molecule is formed in its excited singlet state.

A comparison of the concerted and the nonconcerted paths is shown schematically in Figure 6.2. The front two faces correspond to those of Figure 6.1, and the back face corresponds to the rectangular path geometries of Figure 5.2. Although all the information contained in Figure 6.2 was already contained in Figures 5.10–5.14, the more usual type of presentation used in Figure 6.2 will permit a facile visualization of the difference between the concerted and two-step nonconcerted execution of the $H_AH_B + H_CH_D \rightleftarrows H_BH_C + H_AH_D$ process. In the ground state, the barrier that separates $H_AH_B + H_CH_D$ (left edge) from $H_AH_D + H_BH_C$ (right edge) forms a ridge whose continuous nature serves as a reminder that the forbidden nature of a ground-state $2_s + 2_s$ process is due to topology rather than symmetry and cannot be bypassed altogether by simply avoiding the most symmetric path. However, along the nonconcerted path the barrier is somewhat lower than along the concerted path. If the process were forced to proceed in the ground state after all, it would follow a nonconcerted path. In the case of H_4, of course, it is easier to dissociate one of the hydrogen molecules fully without involving the other.

In the S and D states, the minima occur near or at the midpoint of the concerted path, suggesting that under ordinary circumstances the return to the ground state will occur at a "concerted" geometry.

The H_4 results indicate that as a general rule the nonconcerted paths for ground-state forbidden pericyclic reactions are of relatively little interest for singlet photochemistry although they are normally followed in ground-state reactions of this type, if these occur at all. Indeed, there is no compelling experimental evidence that nonconcerted paths are followed in excited singlet reactions of this kind unless imposed by steric constraints. Rather, the prevalent retention of stereochemistry argues strongly that the concerted paths are followed.

The situation is different in the triplet state. As noted in Chapter 4, at biradicaloid geometries the S_0 and T_1 wave functions are quite similar (well correlated), and it is no surprise that both states prefer "loose" to "tight" biradicaloid geometries. Indeed, all examples in Section 6.2 deal with triplet photochemical processes.

The H_3 + H system. Turning attention to the H_3 + H separation next, we consider the effect of the removal of a terminal hydrogen from a linear geometry of H_4 represented by points in the bottom face of Figures 5.10–5.14 or by points in the front lines in Figures 6.1 and 6.2. Numerical calculations of the resulting changes in the energies of the G, D, and S states are not available, but the qualitative behavior is obvious from the VB representation of the wave functions (Figure 6.3).

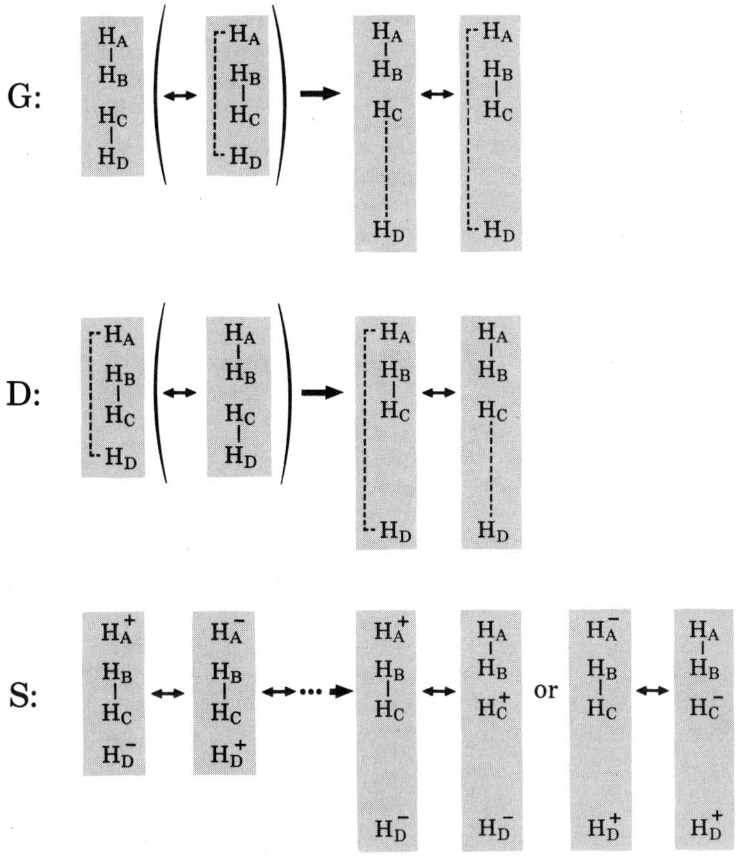

Figure 6.3 Correlation of VB wave functions of the G, S, and D states of H_4, starting at a symmetric linear geometry and removing a terminal hydrogen.

An alternative description in terms of MOs and configurations built from them is possible but very cumbersome and not particularly useful.

The removal of the hydrogen atom H_D will lead to the breaking of the H_C—H_D bond in the G state, which is predominantly described by a VB structure in which there is singlet coupling within the atom pairs $H_A H_B$ and $H_C H_D$, and will be energetically unfavorable. The product will be a pair of radicals, namely, an H atom and a linear H_3 in its ground electronic state (described as a combination of two resonance structures).

The same geometric change will have little effect on the energy of the D state, which contains only one H—H bond to start with, since it is predominantly described by a VB structure with a bond between H_B and H_C and singlet coupling but no effective bond between the distant atoms H_A and H_D. The product is again a pair of radicals, H˙ and H_3^*, with the latter now in an excited state, described by the other possible combination of the two resonance structures.

Finally, the S state, described by a mixture of a fair number of ionic VB structures, can be expected to increase in energy somewhat due to an increased distance at which charges are separated, as it dissociates into an ion pair.

An alternative description of the states of the $H_3 + H$ system can be provided in MO language, using the bonding $|1\rangle$, nonbonding $|0\rangle$, and antibonding $|-1\rangle$ MOs of linear H_3 and the isolated AO on the H atom. In the covalent G and D states, H_3 has three of the electrons and H has the fourth. In the G state, H_3 is in its ground electronic state, described well by the configuration $|1^2 0^1\rangle$. In the D state, H_3 is in its lowest excited state, described by a mixture of configurations $|1^1 0^2\rangle$ and $|1^2 - 1^1\rangle$. In the lowest ionic states, $H_3^+ + H^-$ and $H_3^- + H^+$, H_3^+ ($|1^2\rangle$) or H_3^- ($|1^2 0^2\rangle$) are in their respective ground states. The hole-pair structures $H_3^+ H^-$ and $H_3^- H^+$ correspond to the $|B^2\rangle$ and $|A^2\rangle$ configurations of the two-electron two-orbital model of biradicaloids (Sections 4.4 and 4.5). It is customary to refer to such configurations as $|Z_1\rangle$ and $|Z_2\rangle$, too. Whenever we adopt this usage, we define them explicitly in order to avoid confusion with the symbols defined in equations 4.46 and 4.47.

6.2 COVALENT BONDS

The principles of photochemical breaking and making of covalent, charged, and dative bonds have been discussed in Chapter 4 on the simplest examples of limited practical importance and are summarized in Figures 4.25 and 4.26. Presently, we shall consider more complicated cases of covalent bond-breaking in which one or both ends of the σ or π bond to be broken carry interacting substituents. These represent vehicles through which excitation with light of ordinary UV wavelengths can be accomplished by offering locally excited states at relatively low energies, and a study of the coupling through which energy flows into the bond to be broken represents an important aspect of photochemical theory. In the present formulation, poor coupling results in large barriers in potential energy surfaces, whose presence is qualitatively most easily detected by the stepwise method of producing correlation diagrams using the MOs or states of the isolated subunits as the zeroth order approximation (Section 3.3).

A catalog of important reaction types involving four electrons in a linear array of four orbitals is given in Figure 6.4 and indicates which section of Chapter 6 deals with those reaction types that have been selected as representative. Two of the active orbitals shown in the orbital interaction schemes in Figure 6.4 are always the two $2p_z$ carbon AOs of a C=C double bond; however, in the actual selection of examples, we use other conjugating substituents as well (in particular, phenyl).

6.2.1 Allylic and Benzylic Bond Cleavage

The simple model systems used in Chapter 4 to discuss the excited states of a σ bond are of little practical interest for organic photochemistry because they do not absorb in the most readily available part of the UV region. When the termini of the bond to be dissociated are provided with suitable substituents that serve as chromophores, more interesting systems result. However, most of them contain lone pairs, have low-energy $n\pi^*$ excitations, and therefore belong in Chapter 7.

Presently, we shall consider systems in which the light-absorbing auxiliary group is a benzene or a double-bond chromophore. We shall use this opportunity to consider the coupling of the singlet and triplet excitation energy from the chromophore to the reaction center in terms of surface diagrams. We shall find that σ bonds hyperconjugating with the π chromophore should be particularly prone to

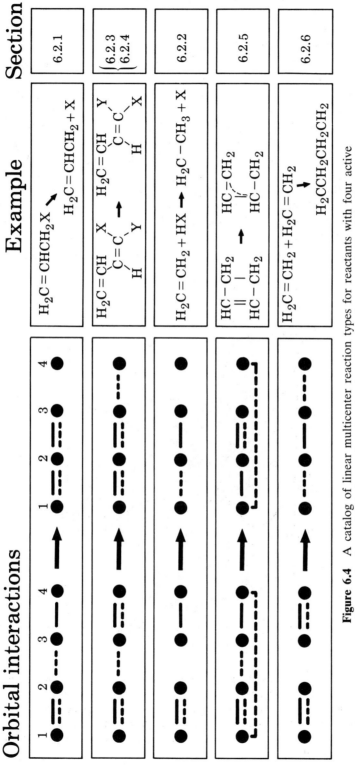

Figure 6.4 A catalog of linear multicenter reaction types for reactants with four active electrons in four orbitals in covalent bonds. Atomic orbitals are symbolized by dots, their primary interactions are indicated by solid lines, and inactive bonds are denoted by dashed lines.

294 LINEAR MULTICENTER REACTIONS: ONE ACTIVE ORBITAL PER ATOM

cleavage and shall consider the effect of their bond strength and bond polarity on the barrier encountered along the reaction path (Figure 6.5).

Description in terms of interacting subunits. Consider at first the zero-order states of a system containing a σ bond whose dissociation is to be investigated and a chromophore with locally excited states, as a function of the length of the bond of interest. At first, assume that there is no interaction between the two, in the spirit of Section 3.3. With an "insulating wall" in place, the S_0, S_1, and T_1 states of the bond will follow the shapes given by Figure 4.25 for the appropriate value of δ, the difference of the effective electronegativities of the orbitals used by the two centers to form the σ bond. The vertical excitation energies of the S_1 and T_1 states will generally increase with the increasing strength of the σ bond in the ground state with respect to heterolytic and homolytic dissociation, respectively.

The excitation energies of the locally excited states of the insulated chromophore will be clearly independent of the degree of stretching of the σ bond, and these states will thus produce curves that run parallel to the ground state (and others that run parallel to the other states of the σ bond, but these will be much too high

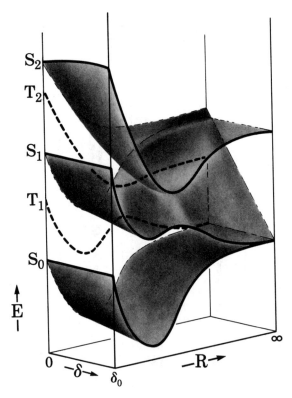

Figure 6.5 The energies of singlet states of a molecule containing a π chromophore with a locally excited state and an allylic (or benzylic) covalent σ bond as a function of the length of the latter (R) and of the difference between the effective electronegativities ($\delta = 0$; for a highly polar σ bond in a polar solvent, $\delta \simeq \delta_0$). The surfaces for the lowest triplet T_1 and for T_2 are assumed to be independent of δ and are shown as dashed curves in the $\delta = 0$ plane (back face). Schematic.

in energy to be of interest). Since at least the T_1 curve and, for a sufficiently polar σ bond, also the S_1 curve of the isolated σ bond descend rapidly as the bond is stretched, they will usually cut the curves corresponding to the zero-order states of the chromophore, yielding sharp breaks in the S_1 and T_1 surfaces (Figure 6.6). At this point, the S_1 and T_1 dissociation limits may be below or above the energy of the locally excited singlet and triplet states in energy, and they will be separated by barriers. If the energy of a locally excited state is below that needed for dissociation by more than a few kilocalories per mole, a photochemical process starting with a thermalized locally excited state and breaking the bond will have no chance to occur (such a process may still occur from a hot locally excited state if it manages to compete with vibrational relaxation). However, even if the energy of a locally excited state, singlet or triplet, is sufficient to reach the S_1 or T_1 dissociation limit, this may be prevented by the barrier that lies along the way if the locally excited state is indeed lower in energy than the vertical excited state of the σ bond (Figure 6.6).

Here, the degree of interaction that will result when the hypothetical "insulating wall" is removed will be essential. If the crossing of the locally excited state surface with the S_1 or T_1 surface of the σ bond is avoided strongly, the barrier may disappear altogether, making the desired reaction highly plausible. If it is not avoided at all, the reaction remains problematic (Figures 6.5 and 6.6C).

Factors that facilitate the reaction. The interaction matrix element vanishes if there is no overlap between the orbitals involved in the local excitation and the σ and σ^* orbitals of the bond to be dissociated. Thus, in practice it can be considered

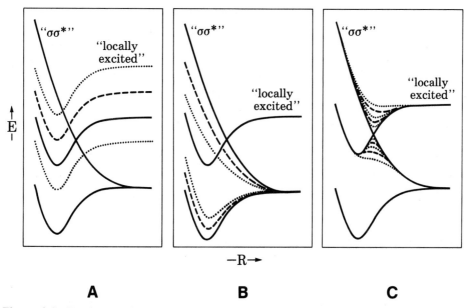

Figure 6.6 Barriers to dissociation in zero-order triplet states of a molecule containing a π chromophore with a locally excited state and an allylic (or benzylic) covalent σ bond as function of the length of R of the latter (schematic). (A) Effect of a variation in energy of the locally excited state; (B) effect of a variation in the σ bond strength; (C) effect of a variation in the degree to which the crossing is avoided.

negligible unless the bond to be dissociated is allylic (benzylic) with respect to the π chromophore. Its value will be optimized when the overlap is optimized, that is, when the bond to be dissociated is aligned for optimal hyperconjugation with the π system. Further optimization is achieved when the position of attachment on the π chromophore is such that the coefficients of the orbitals of the π system which are involved in its lowest singlet or triplet excitation are as large as possible (e.g., position 9 and not position 2 in anthracene).

The conditions that favor the dissociation of a σ bond in a molecule upon local excitation of a π chromophore by minimizing a potential barrier along the way can be summarized as illustrated in Figures 6.5 and 6.6:

(i) The local π excitation energy is large (Figure 6.6A).
(ii) The σ bond is weak (Figure 6.6B).
(iii) The excitation is into the triplet state, or intersystem crossing is efficient (Figure 6.5).
(iv) If the excitation is into the singlet state, the σ bond is polar (large δ_{AB}) and the solvent solvates ions well (Figure 6.5).
(v) The bond is allylic (benzylic) with respect to the excited π system (Figure 6.6C).
(vi) The bond is lined up as close to perpendicular to the plane of the π system as possible (Figure 6.6C).
(vii) The coefficients of the MOs involved in the π excitation are large in the position of attachment (Figure 6.6C).

Conditions (i)–(iv) optimize the position of the zero-order curve crossing, and conditions (v)–(vii) optimize the degree to which the crossing is avoided.

Two examples of favorable processes are shown on top of Figure 6.7. The C—Sn bond is weak, the electronegativity difference is large, and, besides, the presence of the heavy atom will promote intersystem crossing. The effective electronegativity of tris(p-aminophenyl)methyl is much lower than that of the cyano group, and a polar solvent will stabilize the resulting ion pair, providing a large δ_{AB}.

Two less favorable examples are shown at the bottom of Figure 6.7. The C—H bond is nonpolar, so that a triplet reaction will be required. It is also strong, making the process approximately thermoneutral. A significant barrier is likely to be present. The reaction does not proceed under ordinary conditions, but it can be performed when the triplet of the aromatic absorbs a second photon, providing a large energy excess.

6.2.2 Atom and Ion Transfer to Alkenes

The transfer of a neutral or charged atom or group X from the reagent X—Y to one of the termini of a C=C double bond involves the AO interaction scheme and the VB structure correlation diagrams depicted in Figures 6.8 and 6.9. Assuming that the reagent X—Y has no low-lying excited states, the familiar S_0, T_1, and S_2 states of the alkene appear on the left-hand side of the correlation diagram, plotted as a function of the double-bond twist angle. Depending on the choice of alkene, X, Y, and solvent, the relative energies of the product VB structures will vary. Two important cases are shown explicitly for the case of an uncharged bond

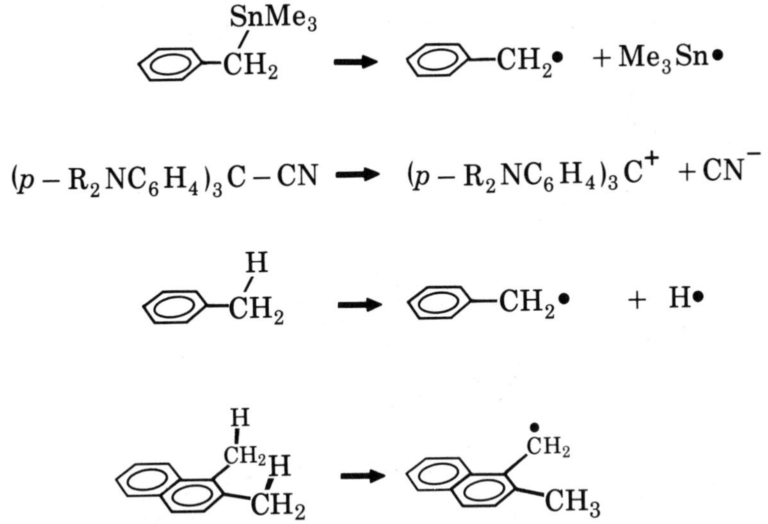

Figure 6.7 Examples of benzylic bond cleavage.

X—Y. The adjustments necessary for charged bonds such as X—Y$^+$ are obvious.

The nonpolar and the polar case. If the energy of the radical pair products, ·C—CX + Y·, lies below that of the ion pair products, C$^+$—CX + Y$^-$, Figure 6.8 applies and the ultimate photochemical products consist of a pair of radicals which then typically undergo further dark chemistry (the nonpolar case). If the energy of the radical pair products is higher than that of the ion pair products, Figure 6.9 applies and the ultimate photochemical products are a pair of ions which then typically also undergo further thermal transformations (the polar case). The net result in the nonpolar case is atom transfer to the double bond (e.g., H abstraction by norbornene), whereas in the polar case the net result is ion transfer to this bond (e.g., photohydration of styrene).

From the point of view of the group that is being transferred, the reaction corresponds to a net homolytic substitution in the former case and to a net nucleophilic or electrophilic substitution in the latter case, depending on the polarity of the ion pair produced. With ordinary alkenes, the polarity is C$^+$—CX + Y$^-$, but with very electron-poor alkenes and suitable Y, it could be C$^-$—CX + Y$^+$. From the point of view of the alkene, the nonpolar reaction is a radical addition and the polar reaction is an ionic addition. Best reaction partners XY for the former are those with X = H and a weak H—Y bond (hydrogen atom abstraction reactions). Best reaction partners for the latter are those with a positively charged X—Y bond, such as H$_3$O$^+$ (photohydration reactions); in such a case, the product actually is not an ion pair. As pointed out above, this requires only trivial adjustments in the theoretical description, since all the considerations embodied in Figures 6.8 and 6.9 still apply. Regardless of the net charges, the dot–dot and hole–pair VB structures still remain valid. Other electrophiles and nucleophiles X can be used; most investigations of this kind have been performed on arenes rather than on alkenes (aromatic photosubstitution).

298 LINEAR MULTICENTER REACTIONS: ONE ACTIVE ORBITAL PER ATOM

$$C=C + X-Y \rightarrow C-CX + Y$$

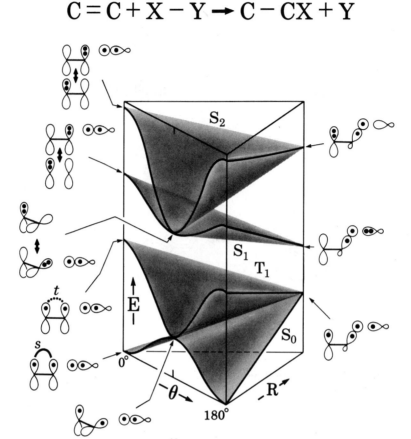

Figure 6.8 A state correlation diagram for an atom or ion transfer to an alkene, as a function of the twist angle θ and of the transfer reaction coordinate R. Depicted here is the nonpolar case in which the radical pair represents the product ground state.

The reaction paths. Under ideal conditions of rapidly reached infinite product separation and no outside intervention, the net outcome of the reaction is determined by the relative stability of the radical pair versus ion pair as stated above. In a solvent cage, matters become complicated. The correlation diagrams of Figures 6.8 and 6.9 suggest that the detailed course of the reaction will be a function of the multiplicity of the initial excitation.

In the S_0 state, the reaction will not occur except in special cases of extremely weak X—Y bonds or very strong electrophiles or nucleophiles X—Y, and we do not consider this trivial case. In the T_1 state, motion along the atom or ion transfer direction which represents the reaction coordinate R will produce a solvent-caged radical pair ·C—CX + Y· in its T_1 state in both the nonpolar case (Figure 6.8) and the polar case (Figure 6.9). Escape from the solvent cage will compete with intersystem crossing to the nearly isoenergetic singlet state (S_0 in the nonpolar case, S_1 in the polar case). This may be promoted by hyperfine interactions with magnetic nuclei present in the vicinity of the radical centers. The conversion to a singlet pair

$$C=C + X-Y \longrightarrow C-CX + Y$$

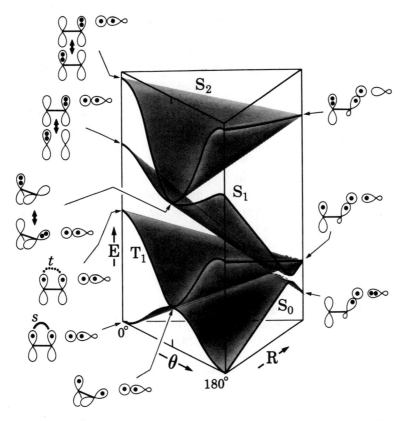

Figure 6.9 A state correlation diagram for an atom or ion transfer to an alkene, as a function of the twist angle θ and of the transfer reaction coordinate R. Depicted here is the polar case in which the ion pair represents the product ground state.

can be followed by disproportionation to yield the starting material or by recombination to yield a $_\pi 2 + {}_\sigma 2$ adduct, and the reader is referred to Section 6.5 for references to studies of chemically induced dynamic nuclear polarization (CIDNP) and of radical pair chemistry. In the polar case (Figure 6.9), another possibility exists, namely, the return of the radical pair from S_1 to S_0, that is, electron transfer between the two partners. If it occurs while the two radicals are still in a solvent cage, its rate may well be dictated by the solvent relaxation time unless it is exothermic already within a solvation shell adapted to the nonpolar radical pair, in which case the transfer may be practically instantaneous (subpicosecond) once S_1 is born.

In the S_1 state of the starting materials, motion along the atom or ion transfer coordinate R will first produce a solvent-cage ion pair $^+C-CX + Y^-$ (or $^-C-CX + Y^+$). In the nonpolar case, this should undergo an essentially instantaneous S_1 to S_0 internal conversion by electron transfer, forming a singlet radical pair, whose thermal chemistry will follow. In the polar case, a solvated ion pair will be produced and will typically undergo further thermal chemistry, such as nucleophilic solvent addition to the carbenium center.

It must be remembered that Figures 6.8 and 6.9 represent only correlation diagrams rather than actual potential energy surface plots. There may well be barriers in the surfaces that separate the starting from the final geometries. In particular, in the T_1 surface, one can expect a small barrier of the kind commonly encountered with radical abstraction reactions. In the S_1 surface, one should expect not only a barrier of the kind ordinarily encountered with proton or other electrophile transfer to an alkene, but also further complications related to the need to align the solvent properly to facilitate the polarization of the double bond by unbalancing the two hole–pair structures that contribute to the S wave function ("mixing S_1 and S_2"). This should be far easier at the twisted geometry, where S_1 and S_2 are nearly degenerate.

Otherwise, twisting of the alkene bond is detrimental to the reaction because it reduces the exothermicity or enhances the endothermicity of the process. The surfaces in Figure 6.8 have been drawn so as to illustrate this. The correlation from the T_1 state of the reactants (a small value of R) to the T_1 state of the products (a large value of R) runs downhill at planarity ($\theta = 0°$), but it runs uphill at orthogonality ($\theta = 90°$). Triplet states of olefins in which twist is prevented are particularly suitable for hydrogen abstractions, and indeed it has been observed for norbornene.

Finally, it needs to be recognized that the diagrams of Figures 6.8 and 6.9 present a very simplified view of the excited states of an alkene. In simple monoolefins the $\pi\pi^*$ excitation energy is so high that the presence of Rydberg states cannot be ignored, and these probably exhibit reactivity similar to that of a radical cation, changing the picture greatly. In more highly conjugated alkenes it may be quite acceptable to ignore the Rydberg states, but these will often have locally excited states whose energy increases along the transfer reaction coordinate R, so that barriers to the transfer will result, as indicated schematically in Figure 6.10.

Factors that facilitate the reaction. The conditions that favor a hydrogen atom abstraction by an alkene double bond and transfer of other atoms to an alkene double bond (C=C + X—Y → ·C—CX + Y·) are those that optimize exothermicity and minimize the barrier on the T_1 surface:

(i) triplet excitation or efficient intersystem crossing;
(ii) high excitation energy for any locally excited triplet state present;
(iii) weak X—Y bond (high stability for Y·);
(iv) strong C—X bond;
(v) high stability for the ·C—CX radical;
(vi) absence of double-bond twisting.

Finally, if electron transfer to yield an ion pair is to be prevented, a nonpolar solvent, a nonpolar X—Y bond, and a high energy for C$^+$—CX and C$^-$—CX are optimal.

The conditions that favor a transfer of a proton or another electrophile or of a nucleophile to an alkene double bond, C=C + X—Y → C$^+$—CX + Y$^-$ or

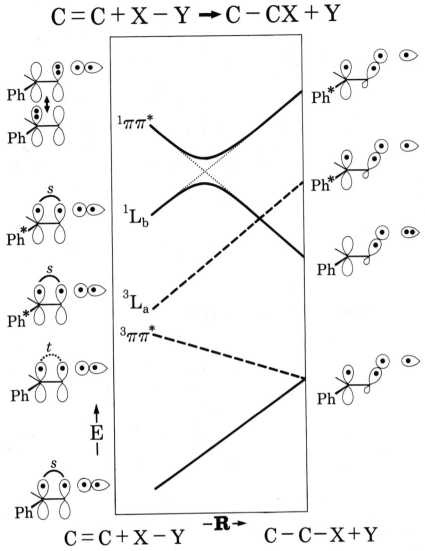

Figure 6.10 A state correlation diagram for an atom or ion transfer to styrene, before (dotted lines) and after (thick lines) interactions with the locally excited states are taken into account. Singlet (solid lines) and triplet (dashed lines) states.

C^-—$CX + Y^+$, are those that optimize the exothermicity and minimize the barrier on the S_1 surface:

(i) singlet excitation, no intersystem crossing (triplet excitation need not be fatal but is more likely to provide opportunities for side reactions);
(ii) high excitation energy for any possibly present locally excited lowest singlet state other than the alkene $\pi\pi^*$ state;
(iii) strongly polar X—Y bond (high stability for Y^- or Y^+);

(iv) strong C—X bond;

(v) high stability for the C^+—CX or C^-—CX ion in its ground state.

The net effect of double-bond twisting is not obvious.

It is likely that any photosubstitution reactions (and photoprotonation) of arenes follow a reaction path similar to that discussed here for alkenes. In arenes, however, "locally excited" states of fairly complicated nature are inevitably present and, at least for the S_1 surface, will frequently dictate the height of the resulting barriers (Figure 6.7). It is then not surprising that factors such as relative site reactivity are more readily correlated with excited state charge distributions than with the stabilities of the resulting ion or radical pairs as Figures 6.5 and 6.6 would otherwise suggest.

In conclusion, we note that most hydrogen abstraction and protonation reactions in photochemistry are not of the simple kind discussed here but, instead, involve $n\pi^*$ states of carbonyl compounds, imines, or azaaromatics. Because of the involvement of lone-pair electrons, such processes are discussed in Chapter 7 (Section 7.3).

6.2.3 Cis–Trans Isomerization: Butadiene

The cis–trans isomerization of 1,3-dienes around a double bond is a very common photochemical process and is believed to proceed by twisting. Upon singlet excitation, one bond can isomerize, whereas triplet excitation may lead to the isomerization of both bonds. This can be understood if the lifetime of the singly twisted species is very short in the S_1 state and longer in the T_1 state and/or if the barrier for the rotation of the second bond is high in the S_1 state and low in the T_1 state.

An investigation of a twisting reaction path starting at a planar geometry and leading to a singly twisted orthogonal geometry represents a good starting point. Figure 6.11 shows the AOs of the active region of the molecule at the start and the end of the path. It makes clear the four-electron four-orbital nature of the problem, as well as the linear character of the overlapping array of orbitals. The reaction path involves the partitioning of the four-orbital system into an allyl-like three-orbital system and an isolated orbital similar to that in a methyl radical.

Electronic states. We can now use our knowledge of the electronic states of the isoconjugate linear H_4 and of its $H_3 + H$ counterpart, although obviously there will be some differences because the interaction between the three-orbital and one-

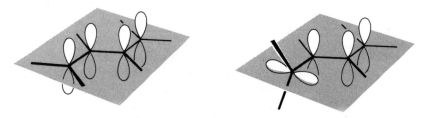

Figure 6.11 The atomic orbitals active in the cis–trans isomerization of 1,3-butadiene.

orbital components is not broken by stretching but by twisting. The former not only causes the overlap and resonance integrals to vanish but also causes the Coulomb and exchange integrals to vanish. In the latter case, only the former two are reduced to zero.

In the case of linear H_4, an understanding of the changes in the nature of the wave functions during a 3 + 1 separation is much more easily reached in VB terms than in MO terms, and the same is true in the case of butadiene. At the planar geometry, the G, S, and D states can be reasonably well described in either language (recall that the hole-pair structures are often called $|Z_1\rangle$ and $|Z_2\rangle$).

Figure 6.12 shows a representative one-dimensional cut through the potential energy hypersurfaces appropriate for the twisting of one of the double bonds in s-*trans*-1,3-butadiene. The motion involved is the twist of the bond accompanied by a simultaneous increase of the distance between the two carbon atoms as they move from planarity to orthogonal twist. All other geometrical variables are kept constant at values appropriate for butadiene in its equilibrium planar ground state. Rydberg states have been omitted. Otherwise, the calculations at intermediate twist angles are close to the 1982 state of the art. In spite of that, they are not very reliable because several states of the same symmetry need to be described

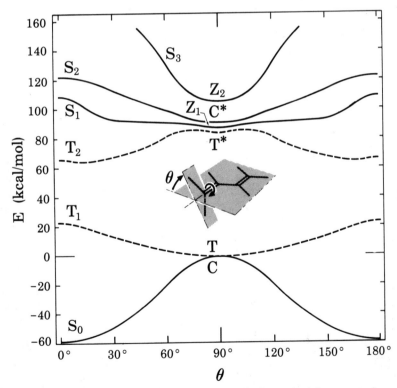

Figure 6.12 Potential energy curves for low-energy singlet and triplet states of s-*trans*-1,3-butadiene twisted at a terminal bond. The dominant VB-like structures are given at 90° twist (see Figure 6.14). Singlet (solid lines) and triplet (dashed lines) states. *Ab initio* large-scale CI calculation, reproduced by permission from V. Bonačić-Koutecký, M. Persico, D. Döhnert, and A. Sevin, *J. Am. Chem. Soc.* **104,** 6900 (1982).

simultaneously in a balanced manner. We shall return to this point below as we discuss the nature of the wave functions. At 0° and 90° twist, the symmetry is higher, so that better calculations are possible; these have been performed without changing the main features of the results.

The lowest two excited valence singlet states of the planar molecule are of A_g and B_u nature in the C_{2h} symmetry group. In Figure 6.12 the former lies significantly below the latter, but in a better calculation they are degenerate within a few kilocalories per mole; their actual order remains uncertain. At 90° twist angle, the point-group symmetry is C_s and the three low-energy singlet excited states lie within 20 kcal/mol of each other. The first and the third excited singlets are of A' symmetry, and the second state is of A'' symmetry.

The potential energy curve for the lowest triplet state shown in Figure 6.12 is more reliable than those for the other excited states. It is of symmetry B_u at planar geometry and A'' at 90° twisted geometry.

Planar geometry. The nature of the molecular wave functions obtained from these large-scale CI calculations at planar geometry is relatively easily described in MO terms (Figure 6.13) and agrees with expectations for the simple H_4 model. The ground state S_0 is well described by the ground configuration, and the ^1B_u excited singlet state is well described by the singly excited configuration $\pi_1 \rightarrow \pi_{-1}$. The 2^1A_g state is harder to describe, since it is represented by a mixture of approximately equal parts of three configurations. One of these results from the double excitation $\pi_1\pi_1 \rightarrow \pi_{-1}\pi_{-1}$. The other two are degenerate in the first approximation, and they result from the single excitations $\pi_1 \rightarrow \pi_{-2}$ and $\pi_2 \rightarrow \pi_{-1}$. The T_1 state is well represented by the triplet configuration corresponding to the single excitation $\pi_1 \rightarrow \pi_{-1}$, and the T_2 state is an approximately equal mixture of the triplet configurations obtained by the single excitations $\pi_1 \rightarrow \pi_{-2}$ and $\pi_2 \rightarrow \pi_{-1}$. All of these configurations involve only excitations from the top two occupied to the two lowest unoccupied π orbitals, so that the system can be viewed in terms of a four-electron four-orbital model. However, acceptable quantitative results are obtained from the CI calculations only if orbitals of σ symmetry are included in the excitations. The alternative VB description of the states also conforms to expectations based on linear H_4: S_0 corresponds to two ground-state ethylenes, the 2^1A_g state corresponds

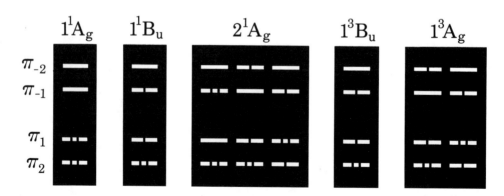

Figure 6.13 Leading configurations in the low-energy states of planar s-*trans*-butadiene.

to an overall singlet combination of two triplet ethylenes, and the ^1B_u state is a complicated mixture of hole-pair structures such as $C^+C^-C—C$.

Twisted geometry. As the twist angle increases from 0°, symmetry is lost, the B_u and A_g states begin to mix, and the reliability of the computed curve shapes decreases. These shapes depend quite sensitively on the details of the method of calculation and on the exact path chosen. A description of the nature of the wave

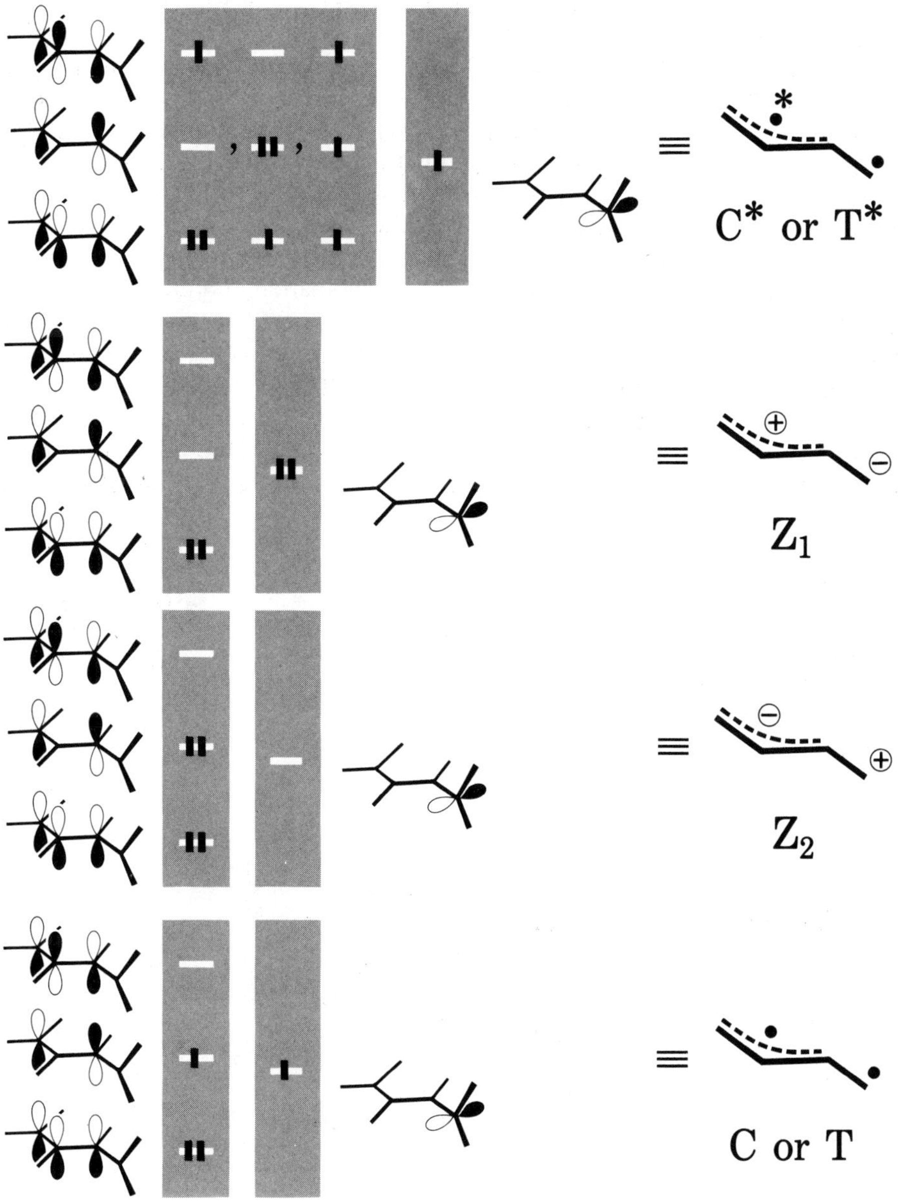

Figure 6.14 VB-like structures for 90° twisted 1,3-butadiene.

function in simple MO terms becomes difficult, but a VB description remains simple even as one approaches the limit of 90° twist (Figure 6.14). It can be presented at two levels. The first, referred to as a VB-like description, describes the states of the three-orbital allyl fragment in MO terms (i.e., occupancies of the bonding, nonbonding, and antibonding π orbitals) and uses the VB language only for the allyl–methyl coupling (i.e., singlet or triplet coupling). At exact 90° twist, it is then identical with the ordinary MO description. The second, a true VB description, describes also the states of the allyl fragment in VB terms.

At the 90° twisted geometry, two VB-like structures are far lower in energy than any other. These are the dot–dot structures $|C\rangle$ and $|T\rangle$ in which one electron is placed at the methylene group and three are placed into the π system of the allyl radical. Four further VB-like structures are of comparable but substantially higher energy and are responsible for the presence of the four close-lying excited states. Two of these are again of dot–dot nature, $|C^*\rangle$ and $|T^*\rangle$. Like $|C\rangle$ and $|T\rangle$, they have one electron on the methylene group and three in the π system of the allyl. Now, however, the allyl is in its lowest excited state, described in some detail below. The other two excited states are of the hole–pair type. The lower one is well approximated by $|Z_1\rangle$, a combination of a methyl anion with an allyl cation, and the upper one is approximated by $|Z_2\rangle$, a combination of a methyl cation with an allyl anion (Figures 6.14 and 6.15). The Z_1-dominated 90° twisted state can be further stabilized by pyramidalization of the terminal methylene group that carries the negative charge in the dominant $|Z_1\rangle$ configuration.

It should be noted that the "π-type" orbitals of each fragment, methyl and allyl, actually extend to some degree into the "σ-type" orbital of the other fragment by hyperconjugation. This should be particularly pronounced in the $|Z_1\rangle$ and $|Z_2\rangle$ configurations.

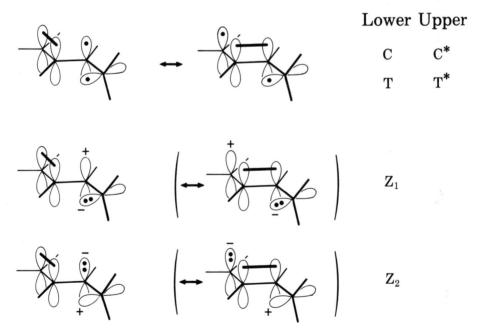

Figure 6.15 VB structures for 90° twisted 1,3-butadiene.

The nature of the excited state of the allyl radical, encountered in the VB-like structures $|C^*\rangle$ and $|T^*\rangle$, is described in MO terms in Figure 6.14 (a mixture of three configurations). In VB terms, it is the upper of the two states which result from the interaction of the two equivalent resonance structures of the allyl radical (Figure 6.15). The lower one is the ground state of allyl, which enters into $|C\rangle$ and $|T\rangle$.

As the twist angle decreases from 90°, the simplicity of the VB description of the S_0–S_3 states is gradually lost, as each state becomes a mixture of the four VB-like structures, $|C\rangle$, $|C^*\rangle$, $|Z_1\rangle$, and $|Z_2\rangle$. For each of the four singlet states, Figure 6.16 shows how its composition changes with the twist angle by showing the weights of the four VB-like structures. These are defined a little differently than in Figures 6.14 and 6.15 in order to keep them mutually orthogonal and are therefore enclosed in quotation marks. The procedure ceases to be applicable at small twist angles and does not permit an extrapolation to the planar geometry. Still, Figure 6.16 makes clear the presence of avoided crossings between the "C^*", "Z_1", and "Z_2" structures as the twist angle is varied.

Comparison with ethylene. We see that there are two substantial differences between the description of orthogonally twisted ethylene and orthogonally twisted butadiene. First, while the only two low-energy excited states of twisted ethylene were represented by hole–pair configurations, in twisted butadiene there are three low-energy excited singlets, one of which is represented by a dot–dot VB-like

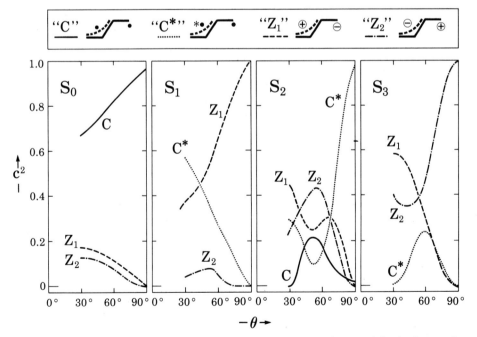

Figure 6.16 Weights of the four principal VB-like structures (see text) in the lowest four singlet states of s-*trans*-1,3-butadiene as a function of the twist angle θ. *Ab initio* large-scale CI calculations, reproduced by permission from V. Bonačić-Koutecký, M. Persico, D. Döhnert, and A. Sevin, *J. Am. Chem. Soc.* **104,** 6900 (1982).

structure $|C^*\rangle$. Second, unless the symmetry of the twisted ethylene is perturbed, the two zwitterionic structures that contribute to the lowest two excited singlet states mix equally so that the resultant states are nonpolar. In twisted butadiene the two hole–pair VB structures do not mix equally because the allyl moiety and methylene moiety have different effective electronegativities, so that the resulting states are polar ("sudden polarization").

Triplet reaction. The path for cis–trans isomerization in butadiene is relatively clear-cut when the initial excitation is into the T_1 state. When the T_1 energy is calculated as a function of the twist angle of a terminal methylene group along a more or less arbitrary reaction path, small barriers are likely to be encountered. Figure 6.17 illustrates this on an *ab initio* minimum basis set level in the case of rotation with all other geometric variables frozen. When all other geometric variables are optimized for every value of the bond angle, no barrier is encountered and the molecule is expected to relax smoothly to the 90° twisted form even at the lowest temperatures (Figure 6.17). This is a nice illustration of the dangers inherent in attempting to read too much significance into the fine details of potential energy curves calculated along unoptimized paths.

Simultaneous torsion of both terminal methylene groups is not energetically advantageous relative to torsion of a single end, since it causes the loss of the resonance stabilization of the allyl radical (~15 kcal/mol). Experimentally, isomerization at both ends following a single excitation event is observed and could be

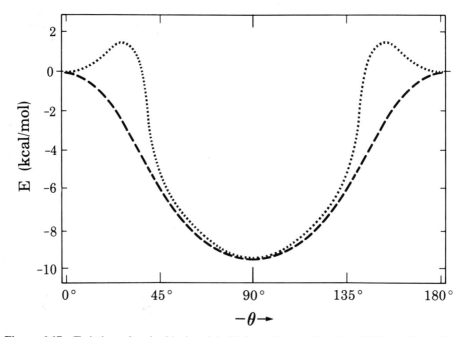

Figure 6.17 Twisting of a double bond in T_1 butadiene. *Ab initio* UHF, minimum basis set. Dotted line: All other geometric variables fixed. Dashed line: All other geometric variables optimized for each value of the twist angle θ. Adapted by permission from I. Ohmine and K. Morokuma, *J. Chem. Phys.* **73**, 1907 (1980).

due to thermal activation. It could also occur as a hot reaction in the T_1 state, since trajectory calculations on this surface suggest that the torsions at the two termini are very strongly coupled whereas coupling to other intramolecular modes is weaker. The hot T_1-state mechanism requires the vibrational energy transfer from one terminus to the other to compete successfully with vibrational cooling by transfer to the environment.

Subsequent return to the ground state should result in a partitioning between the 0° and 180° twisted geometries. As in ethylene (p. 214), spin–orbit coupling is small at the exactly 90° twisted geometries for which the S_0 state is quite accurately described by the dot–dot VB structure, so that the $T_1 \rightarrow S_0$ return is likely to occur during the thermal excursions of the triplet molecule into somewhat less twisted geometries, where S_0 acquires some hole–pair character and spin–orbit coupling increases, albeit at the cost of an increased S_0–T energy gap (cf. Section 4.6.4).

Singlet reaction. The situation is far more complicated in the singlet excited states. The S_1 surface is clearly very flat with respect to the twisting motion, and it is quite conceivable that more than one minimum can be found at different degrees of twist. For instance, as in ethylene (p. 214), CCH bending in twisted butadiene has been calculated to lead towards a funnel. Return from such minima to the ground state may then result in varying probabilities for partitioning between the two ground-state isomers.

Longer polyenes. Calculations of the quality shown here for ethylene and butadiene were not feasible for longer polyenes at the time of writing. For this reason we do not treat such polyenes here explicitly. Still, several trends are quite clear.

In the singlet manifold, as the length of the polyene chain is extended, the excited A_g state moves distinctly below the B_u state at planar geometry while the covalent structure $|C^*\rangle$ is favored relative to the zwitterionic structures $|Z_1\rangle$ and $|Z_2\rangle$ at the 90° twisted geometry, because the excitation energy of the polyenylmethyl radical decreases rapidly with the length of the chain from allyl to pentadienyl and beyond. The correlation of the A_g state of the planar geometry with the $|C^*\rangle$ state of the 90° twisted geometry is apparently not simple, since the S_1 surface shows a barrier. The existence of a local minimum at the planar geometry is demonstrated by the relatively long-lived fluorescence from the planar form, particularly at low temperatures.

An added complication is the availability of several isomerization sites. It is difficult to calculate accurately the relative size of the barriers that block the way from the starting planar geometry towards the various minima in S_1 that correspond to twisting along one double bond or another. Moreover, since the initial excitation often occurs at relatively short wavelengths, there may well be a contribution from hot excited-state processes, and their preferences for one or another path will be even harder to calculate.

In the triplet manifold, the problem posed by the presence of several minima in T_1, corresponding to the availability of various twist sites, is similar. In both cases, it is perhaps best to assume that the deepest minimum in the excited state is the one that is most readily reached, either because the downhill slope towards it is the steepest, or because the barrier on the way to it is the lowest. Such a parallel between thermodynamic stability and kinetic rate frequently works in ground-state processes. Clearly, one has to be prepared to abandon the assumption if better theory becomes available. Comparison with experiment is difficult at present because the relative energies of the various minima in S_1 or T_1 are not available.

6.2.4 Cis–Trans Isomerization: Styrene and Stilbene

Most cis–trans isomerizations of olefins that have been examined experimentally were performed on olefins carrying conjugating substituents at one or both carbon atoms, and butadiene was our first example. More complicated molecules of this class will usually have more than four active orbitals and high-quality calculations are not available for them.

Styrene. A still relatively simple prototype of this kind of olefin is styrene. This molecule typifies the usual complication encountered when a conjugated chromophore is added onto the C=C double bond: The chromophore has its own set of localized excited states which now need to be considered along with those of the ethylene moiety. Typically, the interaction of the two sets of zero-order states will cause avoided crossings in the reaction diagram. This kind of situation has been considered in general terms in Section 3.3, and it was already pointed out there that a frequent consequence will be the presence of more than one minimum on the S_1 or T_1 surface. Typically, there will be a "spectroscopic" minimum nearer the starting geometry in which the S_1 or T_1 states are primarily locally excited; then there will be a barrier, usually small, and then another, "nonspectroscopic" minimum at the biradicaloid geometry. In the case of olefin cis–trans isomerization, the latter will have its excitation concentrated in the twisted double bond.

A derivation of a schematic correlation diagram for the cis–trans isomerization of styrene using the concept of interacting subunits is shown in Figure 6.18. First, the energy of the molecule in its various states is drawn as a function of the twist angle θ as if an insulating wall were present between the phenyl and the vinyl subunits, permitting no interactions between them. Since the geometric changes are restricted to the double-bond subunit, the separation of the various states locally excited in the benzene ring, well known from experiment, is constant and independent of θ. The energies of the locally excited states of the double bond are those of Figure 4.35. The removal of the insulating wall will permit interactions between the zero-order states, indicated by arrows in Figure 6.18. For clarity, only the lowest resulting states are shown; these are the ones of photochemical interest.

At a planar geometry, the resulting excited states of styrene fit well the general picture of excited states of aromatic molecules. The lowest excited state is of ^1L_b character, almost unshifted from its position in benzene, and is followed by the ^1L_a singlet state, red-shifted considerably more. The lowest triplet is of ^3L_a nature and is also shifted noticeably.

At 90°, the states look very much like those of twisted ethylene. The two localized orbitals $|A\rangle$ and $|B\rangle$ now correspond to the nonbonding orbitals of a methyl group (CH_3) and of a benzyl group ($C_6H_5CH_2$). The ground state will be described by the configuration $^1|AB\rangle$, and the lowest triplet will be described by the configuration $^3|AB\rangle$. The lowest two excited singlet states are of zwitterionic (hole–pair) nature

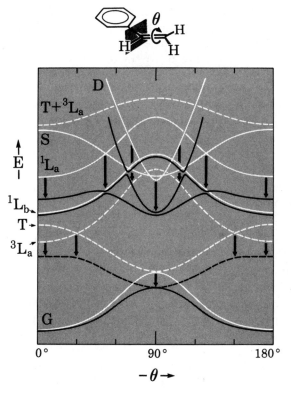

Figure 6.18 A stepwise construction of the state correlation diagram for the cis–trans isomerization of styrene (schematic). The white lines represent state energies of styrene in the presence of an insulating wall between subunits. The black lines represent the state energies after the removal of the wall. Singlet (solid lines) and triplet (dashed lines) states.

and contain $^1|A^2\rangle$, $^1|B^2\rangle$, and a dot–dot VB-like structure with locally excited $C_6H_5CH_2$ (not shown). Thus, the S_1 and S_2 states should be polarized (cf. propene).

In order to appreciate the course of the energy curves in the intermediate range of twist angles, we recognize that the characteristic configuration for this reaction path which corresponds to a localized $\pi \to \pi^*$ excitation in ethylene has the potential for interacting very strongly with the HOMO → LUMO configuration of the benzene moiety, since both of these orbitals have large coefficients at the point of attachment of the final substituent. This benzene configuration corresponds to the ^1L_a state and, thus, to the second excited singlet S_2 and the lowest excited triplet T_1. The interaction of the characteristic configuration with the ^1L_b locally excited state of the benzene chromophore should be weak, since it is described by a superposition of configurations obtained by the excitations $2 \to -1$ and $1 \to -2$, in which orbitals $|1\rangle$ and $|-1\rangle$ (i.e., HOMO and LUMO) have a large coefficient at the position of substituent attachment, but $|2\rangle$ and $|-2\rangle$ very nearly have a node there. If the vinyl group were contained in the C_1—C_4 axis of the phenyl ring, the $^{1,3}L_b$ states would be antisymmetric and the $^{1,3}L_a$ states as well as the ethylene π states would be symmetric with respect to reflection in a plane passing through the carbons C_1 and C_4 and perpendicular to the plane of the ring, so that the interaction would actually be symmetry-forbidden for L_b. As a net result of the interactions, one can thus expect the energies of S_2 and T_1 to rise only very mildly or not at all as the twisting angle is increased, whereas S_1 should increase in energy more distinctly (Figure 6.18). In the singlet manifold, the result is a barrier in the S_1 surface between the planar and orthogonal geometries. In the triplet manifold, the barrier in T_1 can be expected to be smaller or more likely absent altogether, because of the fairly strong interaction of the benzene ^3L_a state with the energetically close characteristic triplet configuration localized in the C=C subunit.

One can thus expect the S_1 singlet photoisomerization of styrene to be promoted by higher temperatures, whereas fluorescence and intersystem crossing should take over at low temperatures. The triplet cis–trans isomerization should proceed more readily even at lower temperatures.

Stilbene. Of all substituted olefins, stilbene is the one which has received the most attention from photochemists. Although its cis–trans isomerization is, in principle, similar to that in styrene, it differs in two important respects. First, in *trans*-stilbene there is a larger number of low-energy excited states, and the order of states at the planar geometry is ^1L_a below ^1L_b in energy. Second, stilbene has higher symmetry, and it is possible to keep a twofold rotation axis throughout the double-bond twisting path. We shall do so in the following example in order to simplify the description.

A qualitative sketch of a one-dimensional cut through the potential energy hypersurfaces appropriate for the isomerization of *trans*-stilbene is shown in Figure 6.19. It is based on an experimental investigation of *trans*-stilbene by polarized one-photon and two-photon absorption spectroscopy and a semiempirical calculation for the planar geometry. The parentage of the states of *trans*-stilbene in terms of the locally excited states of the benzene subunits is shown, with the plus and minus signs indicating the in-phase and the out-of-phase nature of the combination of the locally excited states. The S_1 state, $-L_a$, also has a considerable contribution from the ethylene $\pi \rightarrow \pi^*$ excitation. Because of the presence of two benzene chromophores, this interaction is even stronger here than it is in styrene, where it decreased the ^1L_a–^1L_b gap but did not reverse the order of the states (Figure 6.18). As in styrene, the ethylene $^1\pi\pi^*$ state hardly interacts with the ^1L_b locally excited states of the benzene rings at all. The T_1 state in the planar molecule is of mixed ethylene $^3\pi\pi^*$ and benzene ^3L_a character.

At 90° twisted geometry, the nature of the electronic wave functions can again be described in terms of the orbitals $|A\rangle$ and $|B\rangle$, localized on the two benzyl moieties present in the twisted molecule. The S_0 and T_1 states are described by the dot–dot $^1|AB\rangle$ and $^3|AB\rangle$ configurations, and the lowest excited singlets S_1 and S_2 are of hole–pair (zwitterionic) nature but also contain contributions from dot–dot configurations containing an excited benzyl.

Considering now geometries with an intermediate twist angle, we see that the $+L$ states will but that the $-L$ states will not interact with the characteristic $\pi\pi^*$ configuration as the latter descends in energy upon twisting from planarity. Crossings and avoided crossings result, as in the case of styrene. Along a less symmetric path than that chosen here, the $-L$ states will interact with the characteristic $\pi\pi^*$ configuration, but only weakly, and weakly avoided crossings result. The $^{1,3}L_a$ states

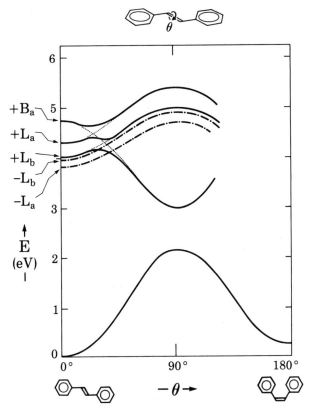

Figure 6.19 A state correlation diagram for the trans–cis photoisomerization of stilbene, suggested on the basis of experimental data and semiempirical calculations. Solid (dash–dot) lines correspond to symmetric (antisymmetric) states relative to the twofold rotation axis. Reproduced by permission from G. Hohlneicher and B. Dick, *J. Photochem.* **27**, 215 (1984).

T_1 and S_1 exhibit small barriers as a memory of these avoided crossings. The experimental value of the intrinsic S_1 barrier between *trans*-stilbene and the orthogonally twisted form is ~3.5 kcal/mol in the vapor phase and ~2.85 kcal/mol in alkane solvents.

One can also start the reaction with the cis isomer, which has moderate difficulties with steric hindrance. Its states generally lie higher in absolute energy, and the ^1L_b state lies again below ^1L_a as it did in styrene, in contrast to the trans isomer. Now, due to the increased exothermicity, the barriers in the states of L_a parentage, T_1 and S_2, should be wiped out altogether, and the others should at least be reduced. It appears likely that there actually is no local minimum at the cis geometry in the S_1 state, since the photoisomerization to the trans isomer seems to proceed without a barrier, presumably along a low-symmetry path.

The photochemical paths followed in the isomerization of *trans*-stilbene can be understood on the basis of Figure 6.19, although it represents only an idealization in that it is a particular one-dimensional cut along a path of particularly high symmetry. In the singlet process, a small minimum at the planar geometry in S_1

permits the molecules to fluoresce with some probability and, in a competing process, to escape to the twisted minimum. Return from the twisted minimum to the S_0 state leads to the production of the ground-state cis and trans isomers with comparable probabilities. In the triplet reaction, not shown in Figure 6.19, the excited trans isomer dwells for a short period of time in its planar spectroscopic minimum and can move over a small barrier to an orthogonally twisted minimum. At the latter, there is competition between rapid radiationless deactivation to S_0 and return to the planar geometry in the T_1 state, presumably again preferentially at geometries that are not exactly orthogonal. After return to S_0, the molecule may proceed with comparable likelihood to the cis and to the trans geometries.

6.2.5 Nonconcerted Ring Opening of Cyclobutene

The disrotatory and conrotatory concerted paths for the electrocyclic conversions of cyclobutene to butadiene were discussed in Section 5.2.1, and the disrotatory path was found to be favorable for the S_1 reaction. It was pointed out that the T_1 reaction was less likely to follow a concerted path in the absence of a special reason, such as a steric constraint or a charged perimeter, because of the tendency of the dot–dot VB structures of a T_1 (and S_0) biradicaloid to prefer those geometries at which the two nonbonding orbitals can avoid each other in space. This can be accomplished by performing the ring opening in a nonconcerted fashion, in which only one of the methylene groups is rotated to yield an s-*cis*-1,3-butadiene molecule twisted at one of its double bonds. One of the nonbonding orbitals is then localized on the allyl moiety and one on the methylene group. There is indeed a minimum in the T_1 surface at this geometry, which represents a joining point of the triplet surface paths of geometric isomerization of an s-cis diene and of the nonconcerted ring-opening of a cyclobutene. We have already discussed all of the low-energy electronic states at the twisted s-trans geometry thoroughly in Section 6.2.3, and those at the twisted s-cis geometry will be similar.

Figure 6.20 presents the VB structure and state correlation diagram for the nonconcerted ring-opening path, constructed using the method outlined in Section 3.2.2. An attempt was made to place the end points at correct relative positions in order to provide an idea of the reaction exothermicity for the various states. While S_0 clearly correlates uphill and is not favorable for the reaction, T_1 correlates smoothly downhill and provides a good driving force for the ring opening. The S_1 ($\pi\pi^*$) state of cyclobutene does not correlate with any one of the three lowest excited singlets of the twisted butadiene (C^*, Z_1, and Z_2) but, instead, correlates with an excited state in which the allyl moiety is excited into its second excited singlet state. This brings about an avoided crossing that may possibly cause the appearance of a barrier along the way. This is probably of little significance because the S_1 process is expected to follow the concerted ring-opening path.

A similar situation is expected for the opening of larger rings with an uncharged pericyclic perimeter, such as 1,3-cyclohexadiene. In keeping with the discussion in Section 5.1.3, however, in the case of charged perimeters and their heterocyclic analogs, concerted triplet and singlet mechanisms are likely (see Section 5.2.3).

6.2.6 Nonconcerted Cycloaddition and Cycloreversion

In Section 5.4 we considered the concerted reaction path for the 2 + 2 suprafacial cycloaddition of two double bonds. While "forbidden" in the S_0 state, the reaction

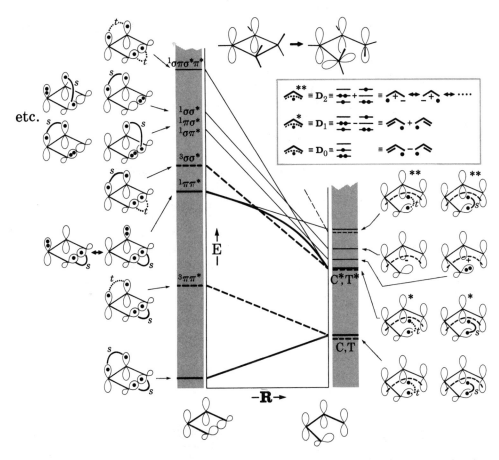

Figure 6.20 A VB structure (thin line) and state (thick line) correlation diagram for the nonconcerted ring-opening of cyclobutene. Singlet (solid lines) and triplet (dashed lines) states. D_0 stands for the ground state of the allyl radical, D_1 stands for its lowest covalent excited state, and D_2 stands for its lowest zwitterionic state.

is expected to be facile in the S_1 state unless barriers imposed by possibly present locally excited states intervene at the start of the reaction path. The driving force is provided by the presence of an excimer minimum and of a pericyclic minimum in the S_1 surface. The former serves as a feeder reservoir for the latter, which returns the molecules to the S_0 state. We have noted that the T_1 state correlated smoothly from reactants to products across the concerted reaction path and probably contained an apparent shallow triplet excimer minimum but, unlike S_1, contained no correlation-imposed minimum at the pericyclic geometry. The actual triplet excimer minimum does not lie along the concerted path, however (Section 5.4.2). The triplet reaction is likely to follow a nonconcerted path through the excimer minimum (Figure 5.31) and through similar biradicaloid geometries at which the nonbonding orbitals avoid each other in space, since this is energetically advantageous for the dot–dot VB structure that describes the T_1 wave function. It is likely that even in the S_1 state the cycloaddition reaction can be forced to follow a nonconcerted path in the presence of suitable steric constraints or substituents.

316 LINEAR MULTICENTER REACTIONS: ONE ACTIVE ORBITAL PER ATOM

Because of the linear, as opposed to cyclic, nature of the interaction between the two partners in a triplet excimer at its optimal geometry, triplet excimers properly belong in the present chapter. They were treated already in Chapter 5 only as a matter of convenience in the discussion of their wave function.

A generalization to $4_s + 4_s$ and other ground-state-forbidden reaction paths that involve an uncharged pericyclic perimeter is straightforward. As noted in Section 5.1.3, the situation is again quite different for systems with a charged perimeter and their heterocyclic analogs in that a concerted triplet path appears favorable, but much less is known experimentally.

Figure 6.21 shows a VB structure and state correlation diagram for the first step of the nonconcerted path, constructed using the methods of Section 3.2.2 and relying on the results of VB calculations for H_4 for further guidance (cf. Figure 6.2). Most of the VB structures are not shown, but they are implied by the labels of the states on both sides of the diagram (right, the states of the tetramethylene biradical; left, those of a pair of planar ethylene molecules). In the center, the states are labeled by the symbols used for the four-electron four-orbital system of linear H_4 in Section 6.1. The crossing of the G and D states is surely strongly avoided and probably leaves no vestige of a barrier in the resulting S_0 state. The

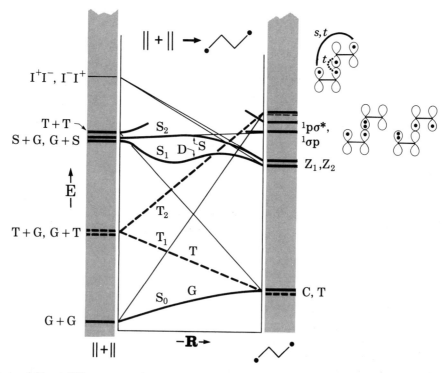

Figure 6.21 A VB structure (thin lines) and state (thick lines) correlation diagram for the nonconcerted addition of two ethylene molecules. Singlet (solid lines) and triplet (dashed lines) states. Labels on the right refer to states of a biradical, labels on the left refer to the states of ethylene ($I^\pm$ stands for radical cation and anion), and labels in the center refer to states of a four-electron four-orbital system.

correlation of the triplet states is particularly clear-cut, and the behavior of T_1 suggests the presence of a facile reaction path. In contrast, the excited singlet states do not correlate simply between the two limits, and the resulting avoided crossings provide ample opportunity for the presence of barriers. The concerted path towards an excimer and pericyclic minimum is more likely to be followed.

The second step of the nonconcerted reaction path (not shown) smoothly correlates the S_0 (C) state of the tetramethylene biradical with the S_0 state of cyclobutane and correlates the T_1 (T) state of the biradical with the T_1 state of cyclobutane. The correlation of the excited singlets is complicated by the high symmetry of cyclobutane, which introduces a large number of energetically similar configurations because all four C—C bonds are equivalent.

Although good-quality calculations are needed to obtain a more reliable picture of the potential energy surfaces, particularly the excited singlets, the correlation diagram is sufficient for discussing the photochemistry in the triplet state. From either triplet cyclobutane or an encounter of a triplet alkene with a ground-state alkene, we expect the formation of the tetramethylene biradical triplet. In the latter case we expect a small activation barrier similar in nature to those encountered in radical attack on alkenes.

Relatively good calculations are available for the various conformations of the tetramethylene biradical in its S_0 and T_1 states. They suggest that conformational transformations are very easy and that the S_0 and T_1 states are very close and cross repeatedly as a function of the nearly free rotation around the three C—C bonds. The T_1 biradical will therefore normally lose memory of its initial stereochemistry. The return to the S_0 surface occurs on the scale of about a hundred nanoseconds and is followed by very rapid formation of cyclobutane or fragmentation to two ethylene molecules. A detailed analysis requires a careful distinction between the gauche and the trans isomers and a more detailed knowledge of the rate of intersystem crossing as a function of molecular geometry than is presently available (Section 4.6.4). Only some qualitative guidelines to the types of distortions that are needed to secure large matrix elements of the spin–orbit coupling Hamiltonian have been published (See Section 6.5). Still, it can be claimed that the generally observed lack of stereochemical memory in the products of triplet cycloaddition and cycloreversion is in perfect agreement with theory.

6.3 CHARGED BONDS

We have already seen in Section 4.6 that the excited states of charged bonds are significantly different from those of ordinary neutral covalent bonds. This is summarized in a particularly simple way in Figures 4.25 and 4.26. The difference carries over to reaction paths involving linear arrays of four or more orbitals, which we shall discuss presently. These processes are important not only because of the parallelism between the behavior of S_1 and T_1 states of charged σ bonds, both of which should now lead to a photodissociative production of ions, but also because they result in charge translocation upon twisting in S_1 states of linear π-orbital arrays—in principle, even across quite long distances.

Although charged bonds can be both positively charged ("onium" cations) and negatively charged ("ate" anions), the former are far more common in practice, and we shall use them as our examples. Similar principles apply to the latter, except

318 LINEAR MULTICENTER REACTIONS: ONE ACTIVE ORBITAL PER ATOM

that electron photodetachment tends to compete with other photochemical processes in anions.

6.3.1 Bond Cleavage: Onium Salts

Excitation of a charged σ bond into its S_1 or T_1 state is expected to lead to a facile dissociation (Section 4.6). In practice, processes of this kind are normally studied in the presence of substituents that serve as chromophores, since the saturated parents absorb only at wavelengths that are inconveniently short. We shall consider a simple example of such a substituted charged σ bond whose dissociation can be considered as a four-orbital four-electron process, namely, the C—S$^+$ bond in an allyldialkylsulfonium ion. Examples of qualitative correlation diagrams for phenyl- and benzyl-substituted onium salts were given in Section 3.3 (Figures 3.14, 3.15, 3.17, and 3.18).

Figure 6.22 shows how the presence of the unsaturated substituent facilitates the C—S$^+$ bond cleavage. It is no longer necessary to excite into the high-lying $\sigma\sigma^*$ states; an excitation into the $\pi\pi^*$ states provides access to the dissociative part of the singlet and triplet energy curves, albeit over barriers whose size cannot be estimated without a calculation.

The construction of the schematic correlation diagram was performed by the

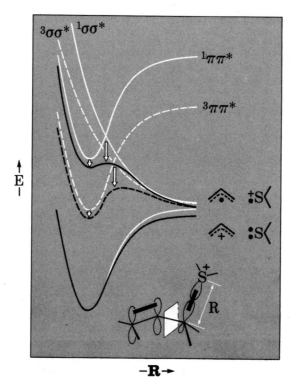

Figure 6.22 A stepwise construction of the state correlation diagram for the dissociation of the C—S$^+$ bond in an allylsulfonium cation (schematic). The white lines represent state energies in the presence of an insulating wall between subunits. The black lines represent the state energies after the removal of the wall.

now familiar stepwise method. Zero-order curves were obtained under the assumption of no interaction between the C=C and C—S$^+$ parts of the molecule, from the qualitative knowledge of the states of the subunits. The energies of the excitations localized on the C=C subunit do not depend on the length R of the C—S$^+$ bond, so that the energy curves for such states are parallel to the ground-state curve and are only displaced upward by the double-bond excitation energies.

When interaction between the subunits is introduced, state crossings are avoided, probably quite strongly, since the pertinent MOs have large coefficients at the critical carbon atoms previously separated by the insulating wall. The final state curves result.

The expected photodissociation products are the triplet and singlet allyl-radical–sulfide-radical cation pairs from triplet and singlet excitation, respectively. Of course, before the two radicals separate too far, the system may undergo intersystem crossing (promoted by spin–orbit coupling on the heavy sulfur atoms and possibly by hyperfine coupling) and internal conversion, respectively, into the ground state, represented by an allyl cation plus dialkylsulfide. Such conversion should be facilitated by the energetic proximity of the two states (the ionization potential of the allyl radical is close to that of dimethylsulfide).

It is worth noting that triplet-sensitized dissociation of benzylammonium salts leads exclusively to radical pair products, indicating that intersystem crossing is not competitive with escape from solvent cage. This is directly related to the bitopic nature of this bond dissociation process, which does not offer any facile spin-flipping mechanism such as would be available in the presence of a lone pair on one of the termini of the original bond (Section 7.2). In contrast, the dissociation of benzylammonium salts by direct irradiation yields a mixture of radical-pair- and ion-pair-derived products, since here the electron transfer between the two fragments is spin-allowed.

6.3.2 Atom and Ion Transfer to Charged Double Bonds

The discussion of atom and ion transfer to alkenes given in Section 6.2.2 applies to charged double bonds such as $R_2C=NR_2^+$, with three important differences.

First, the increased electron affinity of charged double bonds makes them much more effective as electron acceptors, and electron-transfer reactions are likely to dominate much of their photochemistry. Such donor–acceptor interactions are discussed in Section 6.4.

Second, what for transfer to alkenes used to be radical-pair and ion-pair products,

$$C=C + X-Y \longrightarrow \begin{cases} \cdot C-CX + Y \cdot \\ C^+-CX + Y^- \\ C^--CX + Y^+ \end{cases}$$

are now ion–neutral-pair products, and attacks by the two terminals of the double bond have quite different probabilities:

$$C=N^+ + X-Y \longrightarrow \begin{cases} \cdot C-N^+X + Y \cdot \\ N-CX + Y^+ \end{cases}$$

While the correlation diagrams discussed in Section 6.2.2 still apply in principle, the absence of charge separation for the N—CX + Y$^+$ products (as opposed to C$^-$—CX + Y$^+$) means that the energetic penalty associated with their formation in any but the most polar solvents has disappeared; this will be reflected in the state ordering on the right-hand side of the diagram in Figure 6.10. One expects nucleophilic attack in the S_1 state to be facile now, if electron-transfer reactions do not complicate matters. The products will be of the tetrahedral intermediate type and will often fragment to the starting materials, so that S_1 quenching will be the only net result.

The third aspect in which charged double bonds are different from uncharged ones is the near-degeneracy of S_0, S_1, and T_1 at twisted geometries (Section 4.8.2). This affects the region near $\Theta = 90°$ in the correlation diagrams in Figures 6.8 and 6.9. The twisted S_1 state now no longer enjoys the energy advantage it had over T_1 in transfer reactions to an uncharged double bond and undoubtedly undergoes rapid deactivation to S_0. It is then most likely quite essential to prevent the twisting by suitable steric constraints if an atom- or ion-transfer reaction is desired.

6.3.3 Charge Translocation: Acroleiniminium

The protonated Schiff base derived from acrolein provides an opportunity for investigation of the cis–trans isomerization around a C=C double bond and a C=N$^+$ double bond in the same molecule. Because of the presence of the positive charge in the delocalized π system, twisting along either of the double bonds must be viewed as a breaking of a charged π bond. This is true even in much larger systems of delocalized π electrons, as we shall see below on the example of rhodopsin.

C=C isomerization. Figure 6.23 shows a cut through the potential energy hypersurfaces pertinent to twisting around the C=C double bond. These results can be viewed as a result of a perturbation of the curves shown for butadiene in Figure 6.12. The geometries followed by the path chosen in Figure 6.23A correspond to a simultaneous twisting of the C=C double bond and its elongation. All other geometric parameters remain constant. Along the path chosen in Figure 6.23B, all geometric parameters are adjusted in a proportional fashion as the geometry changes from that of a vertically excited S_1 state to that of a relaxed orthogonally twisted S_1 state. The level of calculation is similar to that used for butadiene, but now there are fewer, if any, uncertainties in the lowest excited state since it is well removed from others. The differences between the two paths are minor.

At planar geometries the lowest two singlet excited states S_1 and S_2 can be viewed as correlating with the A_g and B_u states of butadiene, which are now heavily mixed. The most striking difference relative to butadiene, however, occurs at large twist angles; it is most easily described in terms of the VB-like structures introduced in Figure 6.14, which now take the form shown in Figure 6.23B. Because of the much larger electronegativity of the azaallyl moiety compared with the methylene group, the hole–pair VB-like structure Z_1 which places the positive charge onto the methylene carbon is of far lower energy than the other two singlet structures, C* and particularly Z_2. At orthogonality, where the lowest covalent structure C and the Z_1 structure are of different symmetry and cannot mix, the calculated S_0–

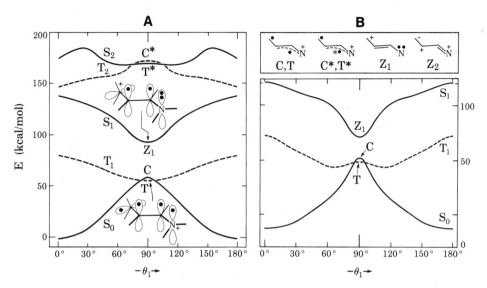

Figure 6.23 Potential energy curves for the low-energy singlet and triplet states of s-*trans*-acroleiniminium twisted at the C=C bond by angle θ_1. Simultaneously with the C=C twist, the C=C length was varied from 1.345 Å at planarity to 1.416 Å at perpendicular twist (A), or all bond lengths and angles were interpolated linearly from a geometry optimized for the planar S_0 state (vertical excitation) to a geometry optimized for the orthogonally twisted relaxed S_1 state (B). In the optimized S_0 state, the C=C distance is 1.334 Å; in the optimized S_1 state, it is 1.432 Å. The dominant VB-like structures are defined on the top right and assigned to states at 90° twist. *Ab initio* large-scale CI calculations. Adapted by permission from V. Bonačić-Koutecký, K. Schöffel and J. Michl, *Theor. Chim. Acta* **72**, 459 (1987).

S_1 state separation is therefore only about 20 kcal/mol, whereas in butadiene the S_0–S_1 gap was nearly 90 kcal/mol. The effects observed here are, of course, analogous to those already discussed in Chapter 4 for a single π bond. The dramatic difference between butadiene (Figure 6.12) and the protonated Schiff base (Figure 6.23) is limited to singlet states. The triplet-state energy curve is similar in both cases, and the state is well described by the covalent configuration in which one electron is placed on the methylene moiety and the other is placed on the azaallyl moiety. In the next higher triplet T_2 the structure is similar, but the azaallyl moiety is excited into its lowest $\pi\pi^*$ state (T^*).

Note the translocation of positive charge between the azaallyl and the methylene subunits upon going from S_0 to S_1. At 90° twist, this charge translocation is most clearly pronounced and approximately corresponds to a full unit of charge.

It is conceivable that a suitable variation of molecular geometry, and particularly, a suitable choice of polar solvent environment, will increase the electronegativity difference between the methyl and the azaallyl moieties of the 90° twisted molecule sufficiently for the S_0–S_1 gap to be reduced to zero so that a conical intersection will occur. According to the simple model of Chapter 4, this requires an equality of the energies of structures C and Z_1 and corresponds to an increase of the electronegativity difference δ sufficient for reaching the critical value δ_0 shown in Figures 4.24 and 4.26.

C=N isomerization. The next cis–trans isomerization to consider is that around the C=N$^+$ double bond. Figure 6.24 shows two cuts through the appropriate part of the nuclear configuration space. Along one path, the geometric change involves only the simultaneous twist around the C=N$^+$ double bond and its elongation. Along the other, the geometry is again varied in a linear fashion between that of the vertically excited S_1 state and that of the relaxed 90° twisted S_1 state. Now the electronegativity difference between the two moieties that are uncoupled at 90° twist is even larger than before. In fact, the difference δ is so large that along both paths chosen, the polar structure Z_1 is actually of lower energy at 90° than the covalent structure C. The difference is similar at the optimized S_1 geometry (Figure 6.24B) and at the more or less arbitrarily guessed twisted geometry of Figure 6.24A. The calculated S_0–S_1 differences at 90° twist are comparable with those observed for the C=C twist and are again much smaller than that in twisted butadiene. In terms of the discussion given in Chapter 4, this 90° twisted protonated Schiff base is represented by a δ value that is larger than δ_0 (Figures 4.24 and 4.26).

A conical intersection of S_0 and S_1 could most likely be reached by a suitable adjustment of molecular geometry and, particularly, by an adjustment of a polar solvent environment designed so as to decrease the electronegativity difference between the amino and the allyl moieties of the twisted molecule.

Once again, note the charge translocation between the allyl and the amino subunits upon going from S_0 to S_1. At 90° twist, this amounts approximately to a

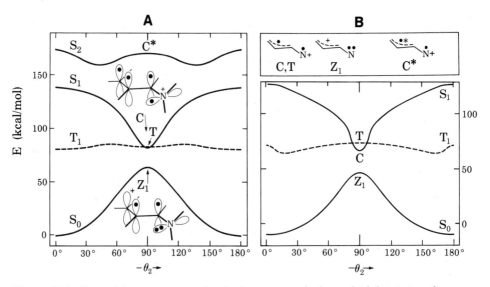

Figure 6.24 Potential energy curves for the low-energy singlet and triplet states of s-*trans*-acroleiniminium twisted at the C=N$^+$ bond by angle θ_2. Simultaneously with the C=N$^+$ twist, the C=N$^+$ length was varied from 1.30 Å at planarity to 1.37 Å at 90° twist (A), or all bond lengths and angles were interpolated linearly from a geometry optimized for the planar S_0 state (vertical excitation) to a geometry optimized for the 90° twisted relaxed S_1 state (B). In the optimized S_0 state, the C=N$^+$ distance is 1.288 Å; in the optimized S_1 state, it is 1.446 Å. The dominant VB-like structures are defined on the right and assigned to states at 90° twist. *Ab initio* large-scale CI calculations. Adapted by permission from V. Bonačić-Koutecký, K. Schöffel, and J. Michl, *Theor. Chim. Acta* **72**, 459 (1987).

transfer of a full unit charge. The S_0 and S_1 states of both the C=C twisted and the C=N$^+$ twisted forms of the acroleiniminium ion are quite close in energy, so that it is even conceivable that a suitable reorganization of the solvent molecules around the cation can switch the order of the C and Z_1 VB-like structures, making either one dominant in the ground state as long as orthogonal twist is preserved. In a suitable polar environment, the ion then could be bistable in its twisted form— once with a positive charge and suitable solvation on one end, and once on the other. Electronic excitation of either form would be likely to produce the other via return through the funnel (conical intersection) located at an intermediate geometric arrangement of the solvent molecules. The return from an excited state to S_0 would thus involve solvent motion. In order for the two isomeric ground-state forms to be truly long-lived, they would have to be prevented from leaving the orthogonally twisted geometry, perhaps by suitable geometric constraints. In Section 6.3.4 we shall consider the possibility that this type of charge translocation occurs upon electronic excitation of the visual pigment, rhodopsin.

The behavior of the triplet state differs from what we have seen for butadiene and for the C=C bond twist in acroleiniminium, in that it does not have a well-developed minimum at 90° twist. This is because at orthogonality the energy difference between the covalent configurations C and T is very small as always, but now the configuration C represents the excited state S_1 rather than the ground state S_0. Therefore at orthogonality the lowest triplet state is nearly degenerate with the first excited singlet state and not with the ground state of the molecule.

The comparison of Figures 6.23 and 6.24 suggests strongly that upon excitation into T_1 the protonated Schiff base will isomerize along the C=C double bond rather than along the C=N double bond. A statement of preference for twisting around one of the double bonds is much harder to make for the singlet state.

It is quite possible that in the protonated Schiff bases of longer polyenic aldehydes, the degree to which the electronegativity difference δ approaches δ_0 upon twisting depends more strongly upon which particular double bond is being twisted. This may lead to predictions of regioselectivity in the photochemical cis–trans isomerization. It should be noted that such charged delocalized π systems are quite common, for example, cyanine, diphenylmethane, and triphenylmethane dyes. In all of these, the same principles apply and a S_0–S_1 touching or near-touching will occur when the hole–pair and the dot–dot structures of the twisted form have similar energies. In cyanine dyes, the electronegativity difference parameter δ typically exceeds the critical value δ_0 so that the orthogonally twisted species has the hole–pair structure in the ground state. We believe that in these and in diphenylmethane and triphenylmethane dyes, protonated azastilbenes, and so on, the values of δ are close enough to δ_0 to produce a deep well in the S_1 surface, responsible for the often extremely rapid deactivation of the vertical S_1 states by twisting and return to the S_0 surface. Examples of a few structures of this type are **1-3**. Another one, rhodopsin, will be discussed in more detail as an archetypical case.

6.3.4 Charge Translocation: Rhodopsin

The chromophore of the visual pigment, rhodopsin, has been identified as the protonated form of the Schiff base of retinal, a higher vinylog of the acroleiniminium cation discussed in Section 6.3.3. The primary process is believed to involve a cis to trans isomerization around the C=C double bond between carbon atoms 11 and 12 (Figure 6.25), and possibly other presumably minor geometrical changes. The primary photoproduct, bathorhodopsin, is richer in energy by 35 kcal/mol and undergoes subsequently a series of further ground-state transformations that eventually trigger a nerve impulse and release retinal from the protein. Retinal is subsequently used for the reconstitution of rhodopsin.

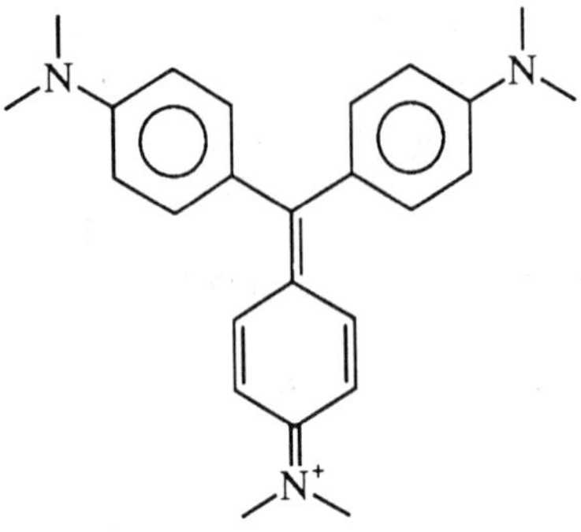

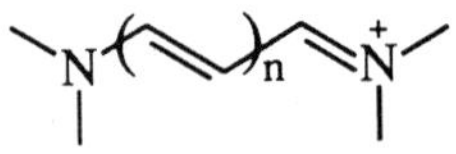

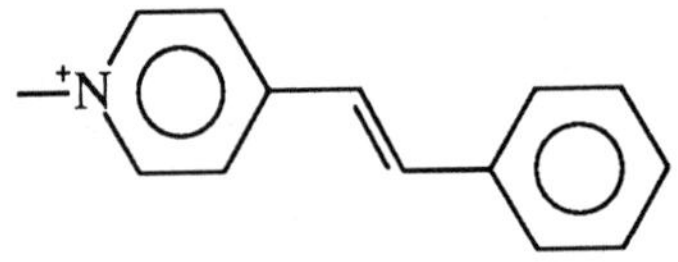

A large number of experimental investigations have established beyond much doubt that the primary photochemical step does not affect the proton on the nitrogen of the Schiff base. Yet, the 6-ps lifetime of the species initially produced by light absorption which decays to yield the primary photoproduct, bathorhodopsin, has been reported to increase by a factor of seven when all exchangeable hydrogens in rhodopsin are exchanged for deuterium. If this report is correct, and unless such deuteration is sufficient to change the protein structure enough to cause a large rate change, this coupling of the decay of the primary species to proton motion must involve one or more protons in the protein environment of the chromophore. This suggests that the charge translocation concepts discussed in Section 6.3.3 may be applicable and that the role of the proton motion is in the stabilization of the charge-translocated structure which converts the S_1 state into the ground S_0 state. While this clearly is nothing but a working hypothesis, it illustrates the principles alluded to in Section 6.3.3, and a more detailed description will be worthwhile to illustrate the use of these concepts even if it later turns out that the actual natural pigment does not follow this mechanism. An alternative explanation has been developed by Becker (Section 6.5).

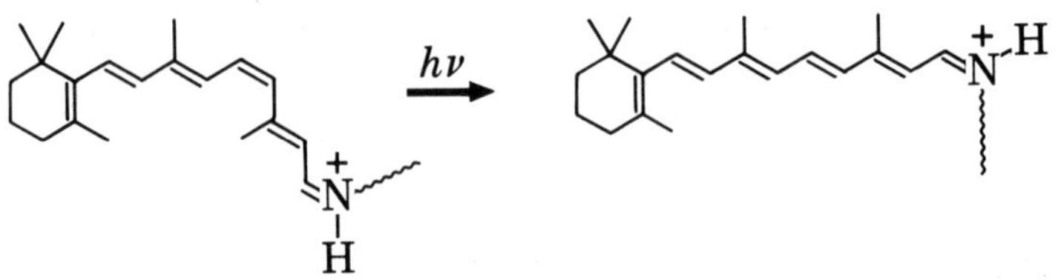

Figure 6.25 The conversion of the chromophore of rhodopsin to that of bathorhodopsin (schematic; other conformational changes may be involved as well).

Proton translocation. Protonation and deprotonation reactions and, in general, ion-transfer reactions, are a natural means for stabilizing and destabilizing charges on one or the other end of a photochemically charge-translocating molecule, thus producing two ground-state isomers. The acidity of groups located near the positively charged end of the chromophore will be decreased when the positive charge is removed to the other end, and their basicity will be increased. Functional groups located at the end of the molecule to which the positive charge migrates will respond to the charge translocation in the opposite fashion. Electronic excitation that leads to charge translocation by electron motion can thus effectively cause proton transfer from one end of the molecule to the other without actually moving a proton or any other nucleus more than a very short distance, simply by causing a proton to be picked up and covalently bound on one end and by causing another proton to be released from covalent bonding at the other end. This concept is quite general (Figure 6.26) but shall be illustrated here for the case of rhodopsin.

A variety of functional groups in the immediate vicinity of the chromophore could be envisaged to act in the desired capacity; one such arbitrary choice is shown in Figure 6.27, with a base shown hydrogen-bonded to the proton on the Schiff base nitrogen, with an acidic group close to the ionone ring, and with water molecules available at both locations. The corresponding shapes of the S_0 and S_1 surfaces are shown schematically in Figure 6.28. Initial vertical excitation of the cis isomer (cis-S_0) yields an excited state (cis-S_1) in which the positive charge has been partly removed from the nitrogen atom (step 1 in Figure 6.27). Upon twisting, the charge translocation towards the ionone ring becomes very pronounced, and the wave function can be well approximated by the hole–pair structure shown for twist-S_0, S_1. However, vibrational relaxation leading to the "twist-S_0, S_1" structure not only involves a twisting motion (step 2 in Figure 6.27) but simultaneously (step 3 in Figure 6.27) also involves a proton motion (i.e., motion to the right in Figure 6.28), since the charge translocation obtained in the twisted S_1 state greatly decreases the acidity of the proton bound to the Schiff base nitrogen and increases the acidity of the acidic group located near the ionone ring. Motion of the two

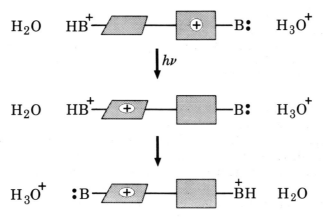

Figure 6.26 "Proton transfer" by photochemical charge translocation. The base B: on the right becomes more strongly basic and the acid BH⁺ on the left more strongly acidic upon excitation and resulting charge translocation, affecting the acid–base equilibria.

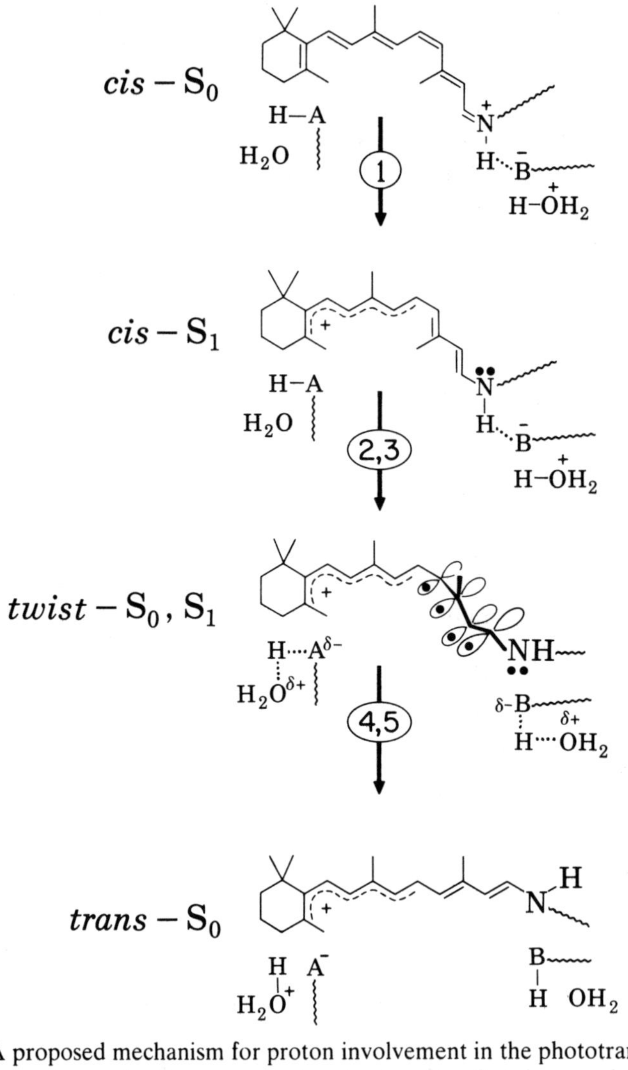

326 LINEAR MULTICENTER REACTIONS: ONE ACTIVE ORBITAL PER ATOM

Figure 6.27 A proposed mechanism for proton involvement in the phototransformation of rhodopsin to bathorhodopsin. Step 1: vertical excitation. Step 2: twist to orthogonality. Step 3: first part of proton motion. Step 4: second part of proton motion. Step 5: twist to planarity. Adapted by permission from V. Bonačić-Koutecký, J. Koutecký, and J. Michl, *Angew. Chem. Int. Ed. Engl.* **26**, 170 (1987).

protons and the concomitant decrease of the negative charge on the basic group B, along with the development of a partial negative charge on the acidic group AH, will occur until the dot–dot and hole–pair structures of the twisted chromophore become degenerate ($\delta_{AB} = \delta_0$), that is, until the bottom of the funnel shown in Figure 6.28 is reached. At this point, the S_0 and S_1 surfaces are degenerate or nearly so (twist-S_0, S_1 in Figure 6.27). One can expect a good fraction of the molecules, if not all of them, to continue to move to the right in Figure 6.28 as they proceed from the S_1 to the S_0 surface through the funnel, in a fashion that keeps dynamical memory of their immediate past geometry and the translocated charge distribution. On the S_0 surface, they continue the vibrational relaxation,

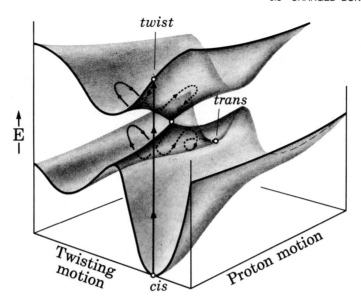

Figure 6.28 A schematic representation of conjectured energies of the S_0 and S_1 states of rhodopsin as a function of twist angle ($C_{11} - C_{12}$) and of proton motion in the environment.

completing the adjustment in the positions of the protons (step 4 in Figure 6.27) and returning from the twisted geometry towards planarity at the same time (step 5). They end up in the high-energy trans-S_0 minimum with an experimentally determined quantum yield of 0.7, or they return to the starting cis-S_0 minimum with a quantum yield of 0.3. There may be small barriers in the way of the proton motion involved in the cis-$S_1 \rightarrow$ twist-S_1, S_0 process, compatible with the experimental observation that at very low temperatures the decay process becomes temperature-independent and presumably proceeds only by tunneling.

The deuteration effect can now be easily understood and is not in contradiction with the observation that the Schiff base remains protonated throughout—it is other protons that move.

The effect of the "proton transfer" appears to be independent of the fact that a cis–trans isomerization occurred. As described, the above mechanism requires a twist to reach the funnel but not necessarily any further rotation. Yet, the occurrence of the cis–trans isomerization is most likely crucial if the "proton-transfer" product is not to return immediately back to the starting material once in its S_0 ground state. This is indicated in Figure 6.28 by the absence of barriers in the two valleys corresponding to proton motion in the cis and trans forms. If such return is not prevented, it will occur rapidly, and no further chain of nerve-impulse-producing events will be triggered.

Indeed, in order for the "proton-transfer" mechanism to yield a reasonably long-lived trans-S_0 product whose presence can trigger further action, one needs to postulate that the valley of trans geometries slopes to the right in Figure 6.28, that is, that the translocated charge distribution is the more stable. A structural reason for this might be found in the new hydrogen-bonding possibilities that the proton on the nitrogen of the Schiff base will have in the trans configuration. Of course, the charge-translocated product only needs to live for a very small fraction of a second in order to perform its work.

The relatively high energy content of the trans-S_0 form, responsible for triggering a complicated chain of further events, may be due to its poor fit into the pocket available in the protein but may, in part, also originate from the "proton transfer" effected in the process just described. This may cause an unfavorable charge distribution, substantial changes in the hydrogen bonding possibilities available to the various groups in the environment, and so on.

It remains to be seen whether events similar to those just outlined actually occur in the natural protein. Even if they do not, the above provides some intriguing guidelines towards the construction of artificial light-activated "proton-pumping" systems.

6.4 DATIVE BONDS AND OTHER DONOR–ACCEPTOR INTERACTIONS

Prototype systems with a dative bond in its various excited states have been discussed in Chapter 4, and the fundamental differences between the properties of covalent, charged, and dative bonds have been summarized in Figures 4.25 and 4.26. Presently, we shall consider the more common systems in which the participating "acceptor" and "donor" subunits, 𝔄 and 𝔅, respectively, do not just carry a single one-center orbital each, but are of more complicated nature. Most often, they are aromatic or other π-electron systems, and they complicate the picture by providing their own sets of locally excited electronic states. Donor–acceptor combinations of various kinds are of great practical interest in photochemistry because of the charge separation in some of their excited states and of the resulting occasional formation of reactive ions from nonionic starting materials.

6.4.1 Donor–Acceptor Pairs: CT Complexes, Exciplexes, Ion Pairs and Free Ions

In this section we consider donor–acceptor interactions characterized by sigma overlap between the components, in which the donor orbital, the acceptor orbital, or both are delocalized over more than a single center. Most often these orbitals are of the delocalized π type, for instance the donor orbitals of electron-rich olefins or aromatics and the acceptor orbitals of electron-poor aromatics, carbonyls, and so on. The delocalized nature of the acceptor (𝔄) and donor (𝔅) moieties complicate matters by providing locally excited states on one or both components at relatively low energy. At the same time, however, the presence of the readily accessible excited states provides facile access to photochemical processes.

Most donor–acceptor interactions of this type are relatively weak in the ground state. At times, one may be justified in talking about a dative bond analogous to that in the NH_3BH_3 prototype investigated in Chapter 4, but often the weak forces holding the components together in the ground state are primarily due to multipole–multipole and induced multipole interactions and not to the usually very small degree of actual charge transfer. Still, it is common to refer to such weakly held entities as charge-transfer (CT) complexes. In many cases, the ground state is not bound at all, at least not in solution, and such entities are referred to as exciplexes if they are of dipolar nature and if they are bound in their lowest excited S_1 or T_1

state. Typical distances between centers of mass of the components are of the order of 4 Å, and they can be considered as "tight" or "contact" ion pairs. Exciplexes are thus analogous to the excimers and particularly, heteroexcimers, which have been covered in Chapter 5. There actually is a continuous range of possibilities without a sharp delineation between heteroexcimers and exciplexes. Their theoretical description is based on identical principles, and we could have discussed them together rather than separately in Chapters 5 and 6, respectively. We have separated them because their typical photochemical behavior, if any, is quite different: Excimers and heteroexcimers normally occur as intermediates in pericyclic cycloadditions, whereas exciplexes also normally occur as intermediates in electron-transfer photochemistry. Because of their fundamental similarity, however, we shall need to refer the reader to Section 5.4.2 frequently.

Excited triple complexes are also known and can be viewed as exciplexes in which the donor is not an aromatic molecule but is, instead, a sandwich dimer of aromatic molecules, which has a lower oxidation potential. In the triple exciplex $\mathfrak{A}\mathfrak{B}\mathfrak{B}$, this part of the molecule has the structure of the well-known radical cation dimers $\mathfrak{B}\mathfrak{B}^+$, whereas the other part corresponds to a radical anion $\mathfrak{A}^-$.

In addition to the entities listed so far, we shall also need to refer to ion pairs and free ions. An ion pair consists of two ions of opposite charges, each with its solvation shell, located in near proximity (of the order of 7 Å between centers of mass). It is often referred to as a "loose" or "solvent-separated" ion pair. Solvated ions separated by larger distances are termed "free" ions.

Since it makes relatively little difference for the electronic structure of the various states whether an interacting donor–acceptor pair is or is not weakly bound in the ground state, we shall treat the photochemistry of charge-transfer complexes and exciplexes jointly. A practical difference is the accessibility of the charge-separated state of a CT complex by direct excitation into the weak charge-transfer absorption band, which is not readily available if the ground state is not bound (except as weak absorptions by collision complexes). In both cases, it is possible to access the excited state surfaces via excitation into one of the locally excited states.

Singlet exciplexes. When the components of an excited complex differ only a little, the description given for a singlet excimer in Section 5.4.2 applies with little change. We assume again that the usual situation applies, in which the partners $\mathfrak{A}$ and $\mathfrak{B}$ are uncharged closed-shell species. In the notation of Section 5.4.2, the ground state of the acceptor $\mathfrak{A}$ has an occupied orbital $|a\rangle$ and an empty orbital $|a'\rangle$, and the donor has an occupied orbital $|b\rangle$ and an empty orbital $|b'\rangle$. Their relative roles as an acceptor and donor, respectively, are dictated by the relative energies of their orbitals. For the best donor–acceptor interaction, $|a'\rangle$ lies below $|b'\rangle$ and $|b\rangle$ lies above $|a\rangle$.

When the properties of $\mathfrak{A}$ and $\mathfrak{B}$ differ only a little, the energies of $^1|\mathfrak{A}^*\mathfrak{B}\rangle$ and $^1|\mathfrak{A}\mathfrak{B}^*\rangle$ are still approximately equal and so are the energies of $^1|\mathfrak{A}^-\mathfrak{B}^+\rangle$ and $^1|\mathfrak{A}^+\mathfrak{B}^-\rangle$. They do not enter into the final wave functions with quite identical weights, and their contributions differ somewhat. The complex acquires polarity, with one of the partners negatively and the other positively charged. Still, all four configurations need to be considered even in a zero-order description. As mentioned above, such exciplexes are sometimes called heteroexcimers.

In typical exciplexes the partners differ more: The electron affinity of the acceptor

𝔄 is high and the ionization potential of the donor 𝔅 is low, so that the $^1|\mathfrak{A}^-\mathfrak{B}^+\rangle$ configuration is far lower in energy than the $^1|\mathfrak{A}^+\mathfrak{B}^-\rangle$ configuration. The latter then typically need not be considered at all in the discussion of low-energy states. In terms of an orbital description, the energy of the empty orbital $|a'\rangle$ is relatively low, the energy of the occupied orbital $|b\rangle$ is relatively high, and the orbitals $|a\rangle$ and $|b'\rangle$ do not need to be considered at all. In the simplest two-electron two-orbital description given in Chapter 4, one can then view the pair 𝔄𝔅 as a strongly heterosymmetric biradicaloid, with $|a'\rangle$ playing the role of the orbital $|A\rangle$ localized on 𝔄 and $|b\rangle$ playing the role of the orbital $|B\rangle$ localized on 𝔅. The configuration $^1|B^2\rangle$ lies far below $^1|A^2\rangle$ in energy ($\delta_{AB} > \delta_0$).

Of course, the two-electron two-orbital model described in Chapter 4 would not permit the description of locally excited states of 𝔄 and 𝔅, $^{1,3}|\mathfrak{A}^*\mathfrak{B}\rangle$ and $^{1,3}|\mathfrak{A}\mathfrak{B}^*\rangle$, and we shall therefore continue to use the notation $^1|\mathfrak{A}\mathfrak{B}\rangle$ in place of $^1|B^2\rangle$ and $^{1,3}|\mathfrak{A}^-\mathfrak{B}^+\rangle$ in place of $^{1,3}|AB\rangle$. However, we shall see below that it is useful to be aware of the equivalence.

The exciplex state will now be described by a wave function in which the contributions of the $^1|\mathfrak{A}^-\mathfrak{B}^+\rangle$ configuration and possibly of one or more locally excited configurations dominate, while that of the other charge-transfer configuration, $^1|\mathfrak{A}^+\mathfrak{B}^-\rangle$, is negligible. The state will be highly polar and in the limit will correspond to a contact ion pair, with the wave function $^1|\mathfrak{A}^-\mathfrak{B}^+\rangle$.

The zero-order description of such exciplexes becomes quite simple. Consider how the energies of the zero-order states vary for complexes of an acceptor 𝔄 with a series of donors 𝔅 (alternatively, for one donor with a series of acceptors). In the gas phase, this can be represented by the difference between the ionization potential of the donor $IP_\mathfrak{B}$ and the electron affinity of the acceptor $EA_\mathfrak{A}$. It has been found empirically that in solutions a better measure is the difference between the polarographic half-wave oxidation potential of the donor, $E_\mathfrak{B}^{ox}$, and the half-wave reduction potential of the acceptor, $E_\mathfrak{A}^{red}$ (measured against the same reference electrode). As $IP_\mathfrak{B} - EA_\mathfrak{A}$ or $E_\mathfrak{B}^{ox} - E_\mathfrak{A}^{red}$ increases, the energy of the charge-transfer configurations $^{1,3}|\mathfrak{A}^-\mathfrak{B}^+\rangle$ is raised with respect to the ground configuration $^1|\mathfrak{A}\mathfrak{B}\rangle$, since it is given by $IP_\mathfrak{B} - EA_\mathfrak{A} - C$ (see equation 5.16). In terms of the analysis given in Chapter 4 and summarized in Figure 4.25, $E_\mathfrak{B}^{ox} - E_\mathfrak{A}^{red}$ plays the role of δ_{AB}.

Experimentally, Rehm and Weller found

$$\Delta E = E_\mathfrak{B}^{ox} - E_\mathfrak{A}^{red} - \Delta \qquad (6.1)$$

In hexane, $\Delta = 0.15 \pm 0.10$ eV.

Figure 6.29 shows schematically a plot of the energy of the charge-separated dot-dot configuration, $^1|\mathfrak{A}^-\mathfrak{B}^+\rangle$, relative to the energy of the hole-pair ground configuration, $^1|\mathfrak{A}\mathfrak{B}\rangle$, against $E_\mathfrak{B}^{ox} - E_\mathfrak{A}^{red}$. The singlet and the triplet charge-separated configurations are degenerate to a good approximation, as we have already seen. This figure differs from Figure 4.24 only in that we now also show the energies of the lowest locally excited singlet and triplet configurations of the acceptor, $E(^1\mathfrak{A}^*\mathfrak{B})$ and $E(^3\mathfrak{A}^*\mathfrak{B})$. It is assumed for simplicity that the locally excited states of the donor, $^{1,3}|\mathfrak{A}\mathfrak{B}^*\rangle$, lie above these in energy, and they are not shown.

Now, when expressions 5.30–5.32 are used to estimate the interaction of the charge-separated configuration with the ground and locally excited configurations,

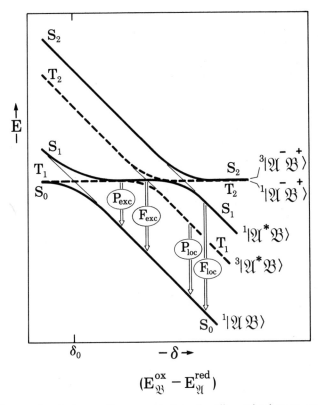

Figure 6.29 Interaction between the ground-state, locally excited state, and charge-separated configurations as a function of $E_{\mathfrak{B}}^{ox} - E_{\mathfrak{A}}^{red}$. Solid lines: singlets. Dashed lines: triplets. Thin lines: configuration energies. Schematic; note that the geometries of singlet and triplet excimers are generally not the same. Fluorescence (F) and phosphorescence (P) from locally excited (loc) and exciplex (exc) state are shown.

the thick lines shown in Figure 6.29 result. One would therefore expect that the exciplex emission energies for a series of donors will only be equal to their $E_{\mathfrak{B}}^{ox} - E_{\mathfrak{A}}^{red} - \Delta$ values, which describe the energy of the charge-separated configuration, within a limited range of $E_{\mathfrak{B}}^{ox} - E_{\mathfrak{A}}^{red}$ values, starting from the avoided touching of S_0 and S_1 on the left to the avoided touching of S_1 and S_2 on the right. The expected trend in the exciplex emission energies is shown in Figure 6.29 by a series of vertical arrows. As $E_{\mathfrak{B}}^{ox} - E_{\mathfrak{A}}^{red}$ grows and approaches the singlet excitation energy of the acceptor or the donor, whichever is lower, the singlet exciplex emission energy will be lower than $E_{\mathfrak{B}}^{ox} - E_{\mathfrak{A}}^{red} - \Delta$ because it asymptotically approaches the energy of the locally excited state. The mixing with the locally excited configuration then needs to be considered explicitly, and the two-configuration description tends to break down in this region of large $E_{\mathfrak{B}}^{ox} - E_{\mathfrak{A}}^{red}$ values; one is dealing with the already discussed heteroexcimers. An attempt to determine the value of Δ from the usual relation 6.1 for the exciplex emission energy ΔE will then yield values that are much too high, typically by 0.5–1.0 eV.

As $E_{\mathfrak{B}}^{ox} - E_{\mathfrak{A}}^{red}$ decreases below about 2 eV, the opposite deviations are found.

The charge-transfer configuration will then interact with the ground configuration and produce a minimum in the ground state, a stable donor–acceptor complex. The decrease in the ground-state energy will raise the emission energy of the exciplex. An attempt to determine Δ from equation 6.1 will then yield values that are much lower than the usual 0.15 ± 0.1 eV. Values of -0.1 and -0.2 eV have been observed for fluorescence of typical donor–acceptor complexes.

The expectations for state energies as a function of δ, reflecting the $E_\mathfrak{A}^{ox} - E_\mathfrak{A}^{red}$ difference, and of the separation R of the components, are summarized schematically in Figure 6.30, which has been obtained by including the locally excited states in Figure 4.25 and permitting them to interact with the charge-separated states (cf. the conversion of Figure 4.25 into Figure 6.5).

In the way of illustration, we show in Figure 6.31 a plot of the exciplex emission maxima of a series of complexes between derivatives of phthalic anhydride and cyano-substituted benzenes as a function of the energy difference $E_\mathfrak{A}^{ox} - E_\mathfrak{A}^{red}$. Note

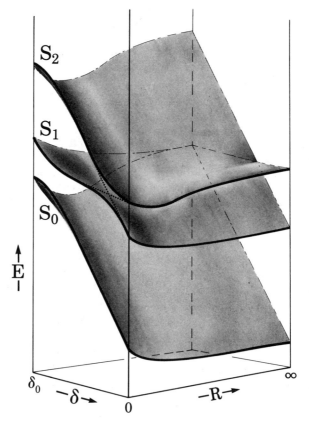

Figure 6.30 The energies of the singlet states of a donor–acceptor pair with the charge-separated configuration and a locally excited configuration as zero-order states (dotted lines) and after interaction. Plotted as a function of the separation R of the components and as a function of the difference δ between the effective electronegativities of the LUMO of the acceptor and the HOMO of the donor. For a very good donor and acceptor pair, charge separation at $R = \infty$ is thermoneutral and $\delta = \delta_0$. For more usual donor–acceptor pairs, charge separation at $R = \infty$ costs energy and $\delta > \delta_0$. Schematic.

6.4 DATIVE BONDS AND OTHER DONOR-ACCEPTOR INTERACTIONS

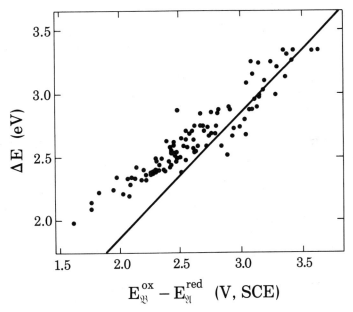

Figure 6.31 Energy ΔE of the fluorescence maxima for complexes of phthalic anhydride derivatives and of cyano-substituted benzenes with aromatics. The straight line represents equation 6.1. G. Rippen and K. Zachariasse, unpublished results; G. Rippen, Ph.D. dissertation, University of Göttingen, 1976 (reproduced by permission).

the gradual deviation from the straight line towards higher emission energies on the left, presumably due to ground-state stabilization (a part may be due to the wavelength dependence of the photomultiplier response).

As a specific example of a singlet exciplex, we take the complex of anthracene with diethylaniline. The pertinent experimental data for the two partners are $E(^1\mathfrak{A}^*_{L_a}\mathfrak{B}) = 3.28$ eV, $E(^1\mathfrak{A}\mathfrak{B}^*_{L_a}) = 4.45$ eV, $E(^3\mathfrak{A}^*_{L_a}\mathfrak{B}) = 1.82$ eV, $E(^3\mathfrak{A}\mathfrak{B}^*_{L_a}) = 3.33$ eV, $E(^1\mathfrak{A}^*_{L_b}\mathfrak{B}) \cong 3.5$ eV, $E(^1\mathfrak{A}\mathfrak{B}^*_{L_b}) = 3.84$ eV, $EA_\mathfrak{A} = 0.56$ eV and $IP_\mathfrak{B} = 7$ eV. The exciplex fluorescence maximum lies at 2.67 eV.

Assuming a geometry that maximizes the electrostatic attraction of the two ions (C—N bond along the anthracene long axis) and proceeding as in the case of pyrene excimer, one finds with the Coulomb integral (defined in equation 5.39) $J_\mathfrak{B} = 6$ eV,

$$H_4(S) = \langle ^1\mathfrak{A}^-\mathfrak{B}^+|\hat{H}|^1\mathfrak{A}\mathfrak{B}^*_{L_a}\rangle = -0.15 \text{ eV} \quad (6.2)$$

$$H_6(S) = \langle ^1\mathfrak{A}^-\mathfrak{B}^+|\hat{H}|^1\mathfrak{A}\mathfrak{B}\rangle = -0.17 \text{ eV} \quad (6.3)$$

while other interactions are negligible. Using perturbation theory, the exciplex wave function is

$$|S_1\rangle: 0.99|^1\mathfrak{A}^-\mathfrak{B}^+\rangle - 0.10|^1\mathfrak{A}\mathfrak{B}^*_{L_a}\rangle - 0.08|^1\mathfrak{A}\mathfrak{B}\rangle \quad (6.4)$$

and is essentially purely ionic (the calculated resonance stabilization is less than 0.01 eV).

The ground-state wave function is

$$|S_0\rangle: 0.99^1|\mathfrak{AB}\rangle + 0.08^1|\mathfrak{A}^-\mathfrak{B}^+\rangle \tag{6.5}$$

The ground state thus has a minimal degree of charge separation. Although these results cannot be considered quantitatively reliable, they provide some insight into the nature of the exciplex wave function.

Triplet exciplexes. Figure 6.29 also shows the energy of the triplet charge-separated dot-dot configuration $^3|\mathfrak{A}^-\mathfrak{B}^+\rangle$ and the energy of the lowest locally excited triplet state of the acceptor $^3|\mathfrak{A}^*\mathfrak{B}\rangle$. Experimentally, it is found that equation 6.1 is again valid, with $\Delta = 0.15 \pm 0.1$ eV. Because of the lower energy of the locally excited triplet, however, the linear region of the plot of the exciplex emission energies against $E_\mathfrak{B}^{ox} - E_\mathfrak{A}^{red}$ is even shorter. Moreover, the phosphorescence of triplet exciplexes that emit at 2 eV and below is extremely weak and hard to observe. It should be noted that there are significant structural differences between singlet and triplet exciplexes (Figure 5.31).

The gradual transition from an essentially pure charge-transfer triplet exciplex through the avoided crossing shown in Figure 6.29 to an essentially purely locally excited triplet state of the pair of partners can also be followed in the triplet ESR spectra. The average distance of the two unpaired electrons is approximately measured by the ESR parameter D^*. Its value increases by about a factor of four as one goes from the ion-pair triplet state at small $E_\mathfrak{B}^{ox} - E_\mathfrak{A}^{red}$ values to the locally excited triplet state at large $E_\mathfrak{B}^{ox} - E_\mathfrak{A}^{red}$ values. The change is half complete when the energies $E(^3\mathfrak{A}^-\mathfrak{B}^+)$ and $E(^3\mathfrak{A}^*\mathfrak{B})$ are equal; at that point the triplet exciplex wave function is half ionic and half locally excited. The difference between the singlet and triplet exciplex energy then measures directly the interaction element $H_3(T) = \langle^3\mathfrak{A}^-\mathfrak{B}^+|H|^3\mathfrak{A}^*\mathfrak{B}\rangle$. At this point, a very small change in $E_\mathfrak{B}^{ox} - E_\mathfrak{A}^{red}$ causes a large change in the D^* value.

If $E(^3\mathfrak{A}^-\mathfrak{B}^+)$ is less than $E(^3\mathfrak{A}^*\mathfrak{B})$, the exciplex has a predominantly charge-separated character. The singlet–triplet difference for the exciplex emission is very small, often less than 0.1 eV. If $E(^3\mathfrak{A}^-\mathfrak{B}^+)$ is larger than $E(^3\mathfrak{A}^*\mathfrak{B})$, the properties of the exciplex are dominated by the lowest locally excited triplet configuration. The singlet–triplet energy difference is large (over 0.2 eV), and the phosphorescence is hardly shifted (0–0.2 eV) relative to that emitted by an isolated acceptor molecule.

Figure 6.32 illustrates the situation on a series of exciplexes between 1,3,5-tricyanobenzene, with a triplet energy of 3.20 eV, and a series of donors, measured in a rigid glass. The gradual increase in the singlet–triplet exciplex energy difference and in the ESR D^* value with increasing $E_\mathfrak{B}^{ox} - E_\mathfrak{A}^{red}$ can be clearly seen. At the bottom, the weight of the charge-transfer contribution in the triplet exciplex state, calculated from the D^* value, is shown. The value of the interaction element between the charge transfer and the locally excited configuration, $H_3(T)$, can be estimated from the top plot as 0.2 ± 0.1 eV.

Solvent effects. Because of their highly dipolar nature, exciplexes can be strongly stabilized in polar solvents. When their wave function is approximated as

$$|S_1\rangle = C_{CT}^1|\mathfrak{A}^-\mathfrak{B}^+\rangle + C_{LOC}^1|\mathfrak{A}^*\mathfrak{B}\rangle \tag{6.6}$$

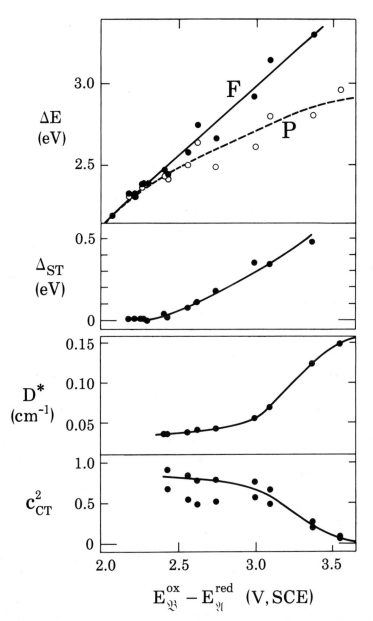

Figure 6.32 Complexes of 1,3,5-tricyanobenzene with a series of donors: Energies ΔE of the fluorescence (full circles, F) and phosphorescence (empty circles, P), singlet–triplet gaps Δ_{ST}, the zero-field splitting parameters D^*, and the weights C_{CT}^2 of the charge-separated structure in the triplet exciplex wave function (derived from D^*), plotted against $E_{\mathfrak{B}}^{ox} - E_{\mathfrak{A}}^{red}$. G. Rippen and K. Zachariasse, unpublished results (reproduced by permission).

their energy in the absence of solvent is obtained by diagonalizing the Hamiltonian matrix

$$\begin{matrix} ^1|\mathfrak{A}^-\mathfrak{B}^+\rangle: \\ ^1|\mathfrak{A}^*\mathfrak{B}\rangle: \end{matrix} \begin{pmatrix} E_{CT} & H_3 \\ H_3 & E_{LOC} \end{pmatrix} \qquad (6.7)$$

where E_{CT} and E_{LOC} are the energy of the configurations $^1|\mathfrak{A}^-\mathfrak{B}^+\rangle$ and $^1|\mathfrak{A}^*\mathfrak{B}\rangle$, respectively, and H_3 is defined in equation 5.20. If the dipole moment of the charge-separated configuration $^1|\mathfrak{A}^-\mathfrak{B}^+\rangle$ is labeled $\boldsymbol{\mu}_0$, the dipole moment of the exciplex will be $C_{CT}^2 \boldsymbol{\mu}_0$.

When the molecule is placed in a solvent, its dipole moment will cause the appearance of a reaction field **F** in the environment. In a simple continuum model, this will be

$$\mathbf{F} = \boldsymbol{\mu}_0 f_\varepsilon \qquad (6.8)$$

where

$$f_\varepsilon = \frac{2}{\rho^3} \frac{\varepsilon - 1}{2\varepsilon - 1} \qquad (6.9)$$

Here, ε is the dielectric constant of the solvent, and ρ is the equivalent sphere radius of the solvent cavity. The reaction field will affect the energy of the S_1 exciplex state by an amount equal to $C_{CT}^2 \mu_0^2 f_\varepsilon$, and it will have negligible effect on E_{LOC} and H_3. The S_1 energies and dipole moments are obtained by diagonalization. The polarization energy stored in the solvent,

$$E_{pol} = (1/2) C_{CT}^4 \mu_0^2 f_\varepsilon \qquad (6.10)$$

is added to obtain the free energy E_a of a rigid exciplex. This is shown in Figure 6.33 for a few typical values of the parameter H_3 and of the energy difference $E_{LOC} - E_{CT}$. Since the experimental values of μ_0^2/ρ^3 are about 1 eV at most, typical values of $\mu_0^2 f_\varepsilon$ for an exciplex range from $f = 0.19$ in hexane to $f = 0.48$ in acetonitrile. The exciplex is clearly stabilized in the more polar solvent. Its dipole moment approaches that of the pure charge-separated configuration as $\mu_0^2 f_\varepsilon$ and $E_{LOC} - E_{CT}$ increase and as H_3^2 decreases.

A similar treatment of an exciplex with the lowest-lying locally excited singlet state $^1|\mathfrak{A}\mathfrak{B}^*\rangle$ instead of $^1|\mathfrak{A}^*\mathfrak{B}\rangle$ yields an analogous result except that H_4 takes the place of H_3.

In order to obtain an expression for the energy of the emission maximum, which corresponds to a vertical transition, the destabilization of the Franck–Condon ground state relative to the relaxed ground state needs to be considered.

Ion pairs and free ions. Inspection of Figure 4.25 shows that the charge-separated state of a dative bond (small R) may well be approximately isoenergetic with an ion pair state (intermediate R) or a pair of separated ions (infinite R), when the δ value as determined by the donor and acceptor strengths of the neutral partners and the solvating power of the solvent are appropriate, and that all of these species

6.4 DATIVE BONDS AND OTHER DONOR–ACCEPTOR INTERACTIONS   337

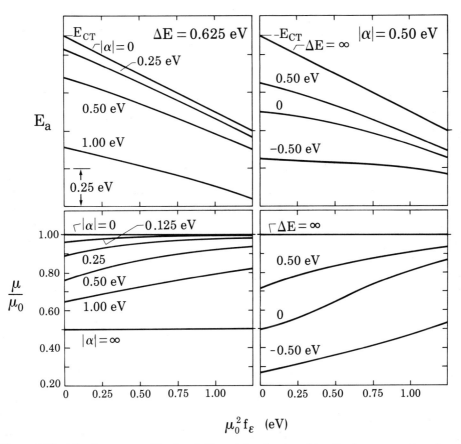

Figure 6.33 The free energy E_a of a rigid exciplex (top) and its dipole moment in units of μ_0 (see text) as a function of $\mu_0^2 f_\varepsilon$, with $\Delta E = E_{CT} - E_{LOC}$ and $\alpha = H_3$ (or H_4) as parameters. The curves on the left correspond to a series of values of $|\alpha|$, those on the right to a series of values of ΔE. Reproduced by permission from H. Beens and A. Weller, in *Organic Molecular Photophysics*, Vol. 2 (J. B. Birks, Ed.), Wiley, London, 1975, p. 159.

lie on the same energy surface. The finer details of the solvation effects are absent in this schematic drawing. In order to make the sketch applicable to more general donor–acceptor complexes under discussion presently, locally excited states have to be considered as well, and this has been done in Figure 6.30. In sufficiently polar solvents, exciplexes can thus dissociate into ion pairs and free ions in an adiabatic process requiring little thermal activation; conversely, oppositely charged ions can aggregate to yield ion pairs and exciplexes, producing electronically excited states in a process that can be considered chemiexcitation (and often chemiluminescence), since it starts from ground state ions. Of course, the ions are in their ground state only when considered one at a time, while the total system consisting of a free radical ion $\mathfrak{A}^-$ and a free radical ion $\mathfrak{B}^+$ is in its excited electronic state. Thus, no surface jumps are involved in this "chemiexcitation" process.

The orientation of the solvent molecules around $\mathfrak{A}^-$ and $\mathfrak{B}^+$ plays a crucial role in the description of their properties. In this sense, a variation of the R coordinate in Figures 4.25 and 6.30 does not stand merely for a change in the distance of the

338 LINEAR MULTICENTER REACTIONS: ONE ACTIVE ORBITAL PER ATOM

partners, but also a corresponding optimal adjustment of the solvent molecules. A preferable representation uses an effective solvation coordinate as an additional variable in the nuclear configuration space, in the fashion discussed below.

For instance, internal conversion from the S_1 surface to the S_0 surface, i.e., "back-transfer" of an electron producing the ground-state components $\mathfrak{A}$ and $\mathfrak{B}$, corresponds primarily to motion along such a solvation coordinate. This brings the system to an avoided crossing of the S_1 and S_0 surfaces, where the surface jump can occur with high probability.

6.4.2 Sigma-Bonded Pi-Donor–Pi-Acceptor Pairs: TICT States

We have seen in Chapter 4 that the interaction of a π donor $\mathfrak{B}$ with a π acceptor $\mathfrak{A}$, bonded to each other in a fashion permitting the π electrons of both subunits to interact, produced a dative bond in the ground state S_0, described well by a hole–pair structure $^1|\mathfrak{A}\mathfrak{B}\rangle = {}^1|B^2\rangle$ with a minor contribution from a charge-separated dot–dot structure $^1|\mathfrak{A}^-\mathfrak{B}^+\rangle = {}^1|AB\rangle$. Such systems then prefer a planar geometry in the S_0 state, unless steric constraints dictate otherwise. Twisting to orthogonality will break the dative bond by preventing the π electron donation from $\mathfrak{B}$ to $\mathfrak{A}$. Usually, however, the π dative bonds are relatively weak because bonding occurs at the expense of charge separation, and thermal cis–trans isomerization by twisting is facile, making it hard to isolate the isomers (cf. the partial C$\doteq$N double bonds of amides).

We have also seen that in the S_1 and T_1 excited states of a simple π dative bond, which are nearly degenerate ($K_{AB} \doteq 0$) and normally separated by a considerable energy gap from S_0, the orthogonally twisted geometry is strongly preferred, and we have discussed the reasons for this. At orthogonality, such excited species can be described as "charge-transfer biradicaloids"; they are described by the singlet or triplet dot–dot structures $^1|\mathfrak{A}^-\mathfrak{B}^+\rangle$ with a formal full negative charge on $\mathfrak{A}$ and a formal full positive charge on $\mathfrak{B}$.

In cases under consideration presently (e.g., those listed in Figure 6.34), the donor, the acceptor, or both will typically have locally excited states that need to be considered along with the two charge-separated excited states described above. Moreover, the radical anion of the acceptor, $\mathfrak{A}^-$, as well as the radical cation of the donor, $\mathfrak{B}^+$, will typically have very low energy excited states as well, so that a whole array of states needs to be considered. In obvious notation, some of these

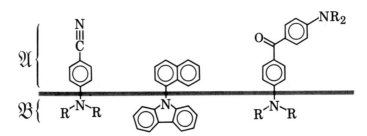

Figure 6.34 Some molecules for which fluorescence from a twisted internal charge-transfer (TICT) state has been observed.

zero-order states are $^1|\mathfrak{AB}\rangle$, $^1|\mathfrak{A}^-\mathfrak{B}^+\rangle$, $^1|\mathfrak{A}^*\mathfrak{B}\rangle$, $^1|\mathfrak{AB}^*\rangle$, $^1|\mathfrak{A}^{-*}\mathfrak{B}^+\rangle$, $^1|\mathfrak{A}^-\mathfrak{B}^{+*}\rangle$, and $^1|\mathfrak{A}^*\mathfrak{B}^*\rangle$. The relative ordering of these states normally depends on the twist angle. We have already noted that charge-separated states are favored at orthogonal geometries; many of the others behave like the ground state and are favored at planar geometries. Zero-order state crossings are thus likely to result, and these will be more or less strongly avoided when all interactions are considered. The Hamiltonian matrix elements needed for estimating the interactions between zero-order states are similar to those discussed for excimers in Section 5.4.2.

Figure 6.35 shows a few such cases in a schematic fashion. Depending on the relative positions of the charge-separated and locally excited zero-order states, there may be one or more minima in the lowest excited singlet state S_1 of the total molecule, and they may occur at the planar or at the 90° twisted geometry. In planar minima, the S_1 wave function could be expected to be primarily locally excited, but for an orthogonal minimum, it would be expected to be of the dot–dot type, $^1|\mathfrak{A}^-\mathfrak{B}^+\rangle = {}^1|\cdots AB\rangle$, characterized by charge separation. As is seen in Figure 6.35, the chances for an orthogonal minimum in S_1 increase as the energy of the charge-separated dot–dot structure $^1|\mathfrak{A}^-\mathfrak{B}^+\rangle$ decreases relative to that of the lowest locally excited state, and this depends on the electron affinity of the acceptor $\mathfrak{A}$ and the ionization potential of the donor $\mathfrak{B}$. As the former increases and the latter drops while the energy of local excitation is held constant, the orthogonal minimum in S_1 will develop and deepen. At first, it will be separated by sizable barriers from the planar minima; eventually, however, these will diminish and possibly disappear altogether.

Since the singlet–triplet splitting is normally large for the locally excited zero-order states and negligible for the charge-separated zero-order states, it will take a better acceptor and/or a better donor to obtain an orthogonal minimum in the

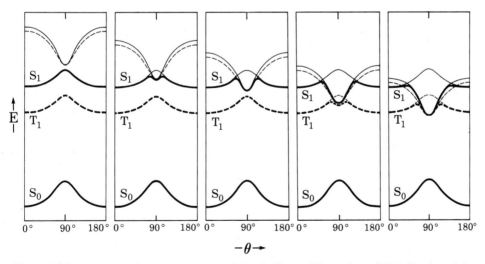

Figure 6.35 A schematic representation of the S_0, S_1, and T_1 surfaces (thick lines) and the zero-order locally excited and charge-separated configurations (thin lines). Singlets: solid lines. Triplets: dashed lines. The energy of the charge-separated configurations decreases from left to right.

T_1 state than in the S_1 state, and there will be a range of situations in which the S_1 already has an orthogonal minimum and T_1 not yet.

Because of the small mutual overlap density of the donor and acceptor orbitals, the dot–dot configuration contributes almost nothing to the absorption intensity, which stems predominantly from the locally excited zero-order states. Vertical excitation into a predominantly locally excited state, then, is the mechanism for accessing S_1 or T_1 in the region of planar geometries. Motion to the orthogonally twisted minimum in S_1 or T_1 would normally require thermal activation in order to overcome the barrier resulting from the crossing of zero-order states. Only for the best acceptor–donor combinations with relatively high-lying locally excited states would one expect barrierless access from planar geometries to the orthogonal minimum. This, then, is a good case for temperature-dependent dual fluorescence, and such fluorescence has been the standard means of detection of these orthogonally twisted charge-transfer states, dubbed TICT states for short. The fluorescence quantum yield will typically not be very large, since the radiative lifetime is expected to be very long: At 90° twist, the transition moment for going from the excited charge-separated state to the hole–pair ground state should be nearly zero, since the overlap density of the $|A\rangle$ and $|B\rangle$ orbitals is exceedingly small. Most of the observed fluorescent intensity is probably due to emission from molecules at somewhat less twisted geometries, permitting an admixture of locally excited states that can carry sizable oscillator strengths.

After radiative or radiationless return to the ground state from the orthogonal minimum in S_1 or T_1, one would expect vibrational relaxation to both the cis and trans forms. This is hard to detect because in these molecules the thermal cis–trans interconversion in S_0 is ordinarily quite fast.

Solvent effects. One of the most important factors that dictate the energy of the charge-separated dipolar dot–dot structure $^1|\mathfrak{A}^-\mathfrak{B}^+\rangle = {}^1|\cdots AB\rangle$ relative to the hole–pair structure $^1|\mathfrak{A}\mathfrak{B}\rangle = {}^1|\cdots B^2\rangle$ is the nature of the environment. Although emission from TICT states has been reported even in the gas phase, most of the reports deal with solutions in polar solvents, which lower the relative energy of the charge-separated structure. The wavelength of the TICT emission depends quite strongly on solvent polarity, and it is possible to go through some of the cases depicted in Figure 6.35 without changing the substrate, only the solvent.

A useful description of the behavior of molecules such as this in a polar solvent is obtained by adding a single additional geometrical variable to the nuclear configuration space of the substrate molecule, representing the collective effect of the orientation of the solvent molecules in the solvation shell. The range of interest for this "solvation variable" starts at a point representing the average solvent arrangement around the relatively nonpolar ground-state substrate and extends to a point representing the average solvent arrangement around the highly polar charge-separated state. Some of the solvent–solute interactions involved in the solvation may actually be quite specific, such as a solvent lone pair stabilizing a localized radical cation by the formation of a three-electron bond.

The change of the potential energy surfaces along the solvation coordinate is shown schematically in Figure 6.36 and is similar to that observed as one moves from the left to the right in Figure 6.35 (the full range shown there would probably not be covered even by going to the most polar solvent). We see that the devel-

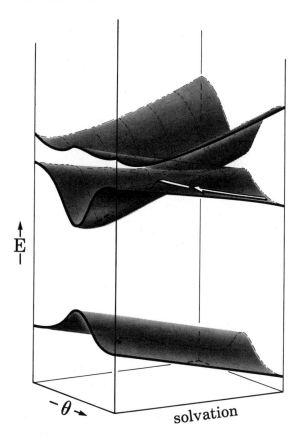

Figure 6.36 The ground and excited singlet surfaces resulting from the interaction of zero-order locally excited and charge-separated states as a function of twist angle θ and of solvent arrangement (schematic). The lowest energy path from a planar locally excited state to the twisted internal charge transfer (TICT) state is indicated by an arrow.

opment of the solute geometry towards the twisted minimum involves simultaneous twisting within the molecule and rotational and translational motion of the solvent, so that its rate should depend not only on factors such as the moment of inertia of the rotating group but also on factors describing the properties of the solvent such as its dielectric relaxation time.

This "averaged" or "effective" description of the structure of the surrounding solvent by a single coordinate is similar to the description of the position of acidic protons in the environment of the rhodopsin chromophore by a single coordinate in Figure 6.28.

An extreme case of a solvent effect results when 𝔄 and 𝔅 have the same structure, as in 9,9′-bianthryl (Figure 6.37). Its anthracene units have both a relatively low-lying acceptor orbital and thus a high electron affinity and a high-lying donor orbital and thus a low ionization potential, and either can act as a donor or as an acceptor. It is only the symmetry-breaking presence of a polar solvent such as alcohol that permits the two halves of the molecule to acquire sufficiently different properties for one to act as a donor and for the other to act as an acceptor in a charge-

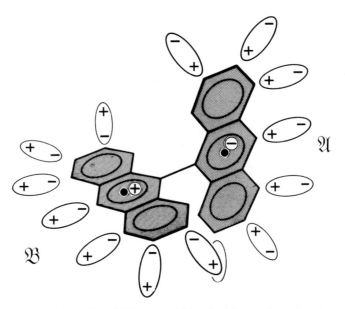

Figure 6.37 The TICT state of bianthryl in a polar solvent.

separated orthogonally twisted state, detected by its fluorescent emission. In this case, planar minima, which normally might be at lower energies even in highly polar solvents, are energetically handicapped by severe steric hindrance.

The theoretical description of this behavior can be couched in terms somewhat analogous to those used to describe the phenomenon of sudden polarization (Chapter 4). There, the Hamiltonian matrix element between two nearly isoenergetic dipolar hole–pair VB structures of a weakly heterosymmetric (singly pyramidalized) ethylene (C^+—C^- and C^-—C^+) became very small at nearly orthogonal angles of twist so that they became uncoupled, one dominating S_1, the other dominating S_2, making both highly polar. At other twist angles, the matrix element mixing the two structures become much larger than their energy separation and the structures were mixed nearly equally in the resulting states which therefore had very low polarity.

Presently, we again have two nearly isoenergetic VB-like dipolar structures, this time of the dot–dot type, $C_{14}H_9^{-}$—$C_{14}H_9^{+}$ and $C_{14}H_9^{+}$—$C_{14}H_9^{-}$. In a perfectly symmetric isolated molecule, they would actually be exactly degenerate, but geometric perturbations and solvent effects (Figure 6.37) can remove the degeneracy. At nearly orthogonal twist, enforced by steric hindrance, the matrix element mixing the two dot–dot configurations is very small, suggesting that already a very small perturbation of the geometry or by solvent will uncouple the two structures and induce charge separation in the resulting states.

However, the system differs from the case of simple sudden polarization in ethylene by the presence of the locally excited configurations $C_{14}H_9^*$—$C_{14}H_9$ and $C_{14}H_9$—$C_{14}H_9^*$ at comparable energies. Considering all four configurations, $^1|\mathfrak{A}\mathfrak{B}^*\rangle$, $^1|\mathfrak{A}\mathfrak{B}^*\rangle$, $^1|\mathfrak{A}^-\mathfrak{B}^+\rangle$, and $^1|\mathfrak{A}^+\mathfrak{B}^-\rangle$, a Hamiltonian matrix can be set up for the isolated system along the lines discussed in Section 5.4.2, using the notation introduced in

6.4 DATIVE BONDS AND OTHER DONOR–ACCEPTOR INTERACTIONS

equations 5.11 and 5.18–5.21 and taking advantage of the chemical identity of the $\mathfrak{A}$ and $\mathfrak{B}$ halves, each of which can equally well serve as a donor or as an acceptor:

$$\begin{array}{c} {}^1|\mathfrak{A}^*\mathfrak{B}\rangle: \\ {}^1|\mathfrak{A}^-\mathfrak{B}^+\rangle: \\ {}^1|\mathfrak{A}^+\mathfrak{B}^-\rangle: \\ {}^1|\mathfrak{A}\mathfrak{B}^*\rangle: \end{array} \begin{pmatrix} E_{LOC} & H_3 & H_4 & H_1 \\ H_3 & E_{CT} & H_2 & H_3 \\ H_4 & H_2 & E_{CT} & H_4 \\ H_1 & H_3 & H_4 & E_{LOC} \end{pmatrix} \quad (6.11)$$

The expressions for the off-diagonal elements are given in Section 5.4.2. H_1 can be written as the electrostatic interaction of the transition moments of $\mathfrak{A} \to \mathfrak{A}^*$ and $\mathfrak{B} \to \mathfrak{B}^*$ (equation 5.15) and therefore vanishes at the orthogonal geometry. Inspection of equations 5.19 for H_2, 5.20 for H_3, and 5.21 for H_4 shows that the only elements that are not necessarily negligible are H_3 and H_4 and that these can be approximated by the off-diagonal elements of the Fock operator between the donor orbitals on $\mathfrak{A}$ and $\mathfrak{B}$ in the case of H_3 and between the acceptor orbitals on $\mathfrak{A}$ and $\mathfrak{B}$ in the case of H_4. These elements are also quite small, since they can be written as a sum of contributions each of which is proportional to a resonance integral (and thus roughly to the overlap integral) between an AO on one subunit and an AO on the other subunit. At orthogonality, the AOs at carbons 9 and 9' have zero overlap, and only some very small contributions from non-nearest neighbors remain. Instead of attempting to calculate these small quantities, which occur as a sum in the resulting equations, one can treat their sum as an empirical parameter.

Now, the solvent can be introduced and its effect on the energies of the charge-separated configurations considered (the effect on the nonpolar configurations is negligible). This is done in a fashion already described for exciplexes in Section 6.4.1, and the dipole moment of the charge-separated structures is again labeled $\boldsymbol{\mu}_0$. If the molecule is in a polarized excited state, described well by a single such structure, the Hamiltonian matrix takes the form

$$\begin{pmatrix} E_{LOC} & H_3 & H_4 & 0 \\ H_3 & E_{CT} + \mu_0^2 f_\varepsilon & 0 & H_3 \\ H_4 & 0 & E_{CT} - \mu_0^2 f_\varepsilon & H_4 \\ 0 & H_3 & H_4 & E_{LOC} \end{pmatrix} \quad (6.12)$$

If the molecule is in a locally excited state, solvent effects are negligible and the $\mu_0^2 f_\varepsilon$ terms are absent in the Hamiltonian matrix element.

The energy of the S_1 state is taken to be the lower of two roots. One of these is obtained as the lowest root of the matrix 6.12, and the other is obtained as the lowest root of the matrix which is the same except that the $\mu_0^2 f_\varepsilon$ terms are absent. If the former is lower, the S_1 is of charge-separated character; if the latter is lower, it is of locally excited character. The energy of the S_1 state computed for a negligibly small value of the parameter $H_3 + H_4$, appropriate for orthogonally twisted 9,9'-bianthryl, is shown by the thick line in Figure 6.38 as a function of solvent polarity. The realistic value of 3000 cm^{-1} was chosen for the difference $E_{CT} - E_{LOC}$. At the bottom of Figure 6.38, the dipole moment of the S_1 state is shown in units of μ_0.

344 LINEAR MULTICENTER REACTIONS: ONE ACTIVE ORBITAL PER ATOM

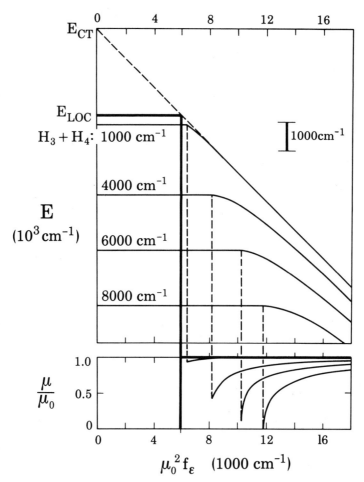

Figure 6.38 Top: solvent dependence of the energy of the S_1 molecule or complex including the solvent. Bottom: dipole moment of the S_1 state in units of μ_0 (see text). Thick lines: 9,9′-bianthryl. Thin lines: excimers. Reproduced by permission from H. Beens and A. Weller, *Chem. Phys. Lett.* **3**, 666 (1969).

In spite of its steric encumbrance, the 9,9′-bianthryl molecule is not totally restrained to the orthogonally twisted geometry. At somewhat smaller degrees of twist, $H_3 + H_4$ will begin to increase, but this is not likely to significantly affect the result shown in Figure 6.38: The S_1 state will be dipolar as long as the dipole solvation energy is sufficient to lower E_{CT} below E_{LOC}.

Because there is no intramolecular barrier to overcome in proceeding from an initial locally excited state to the dipolar state, the rate of the process is dictated by the relaxation properties of the solvent, and in ordinary low-viscosity solvents the observed rate constant is of the order of 10^{12} s^{-1}.

In this context it is of interest to inquire whether the nonpolar excimers discussed in Section 5.4.2 should not behave similarly to 9,9′-bianthryl and also develop a high degree of charge separation in a polar solvent. After all, they are also characterized by the same four configurations and the same matrix 6.11.

The fundamental difference between the cases is due to the π nature of the interaction between the two halves of 9,9'-bianthryl, which makes it easy to break their interaction almost completely by the simple expedient of orthogonal twisting without separating them apart, as opposed to the σ nature of the interaction between the partners in an excimer. The only way to reduce the interaction in the latter case is to pull the partners apart, thus destroying the excimer entity and producing two molecules or an ion pair.

Thus, while the off-diagonal elements in the matrix 6.11 were all very small for 9,9'-bianthryl, this will no longer be true for the anthracene excimer. The exciton coupling term H_1 may be substantial, and, more important, H_3 and H_4, which are proportional to the resonance integrals, and thus to overlap, are no longer negligible. This brings about a far more extensive mixing of the locally excited and the charge-separated configurations. The dipole moments of the resulting states are then no longer restricted to the values 0 and μ_0 as was the case for 9,9'-bianthryl, and their values will reflect the composition of the resulting wave functions. When these modifications are made in the above treatment of 9,9'-bianthryl, it becomes applicable to excimers as well. The results obtained for a series of values of the parameter $|H_3 + H_4|$ are shown as thin lines in Figure 6.38. For typical excimers, $|H_3 + H_4|$ has values on the order of 6000 cm^{-1}, and the largest $\mu_0^2 f_\varepsilon$ values available are of the order of 5000–8000 cm^{-1} ($\varepsilon = \infty$). Figure 6.38 makes it clear that for a typical excimer the S_1 state does not have a charge-separated nature even in the most polar solvent, as long as the components stay in contact. Indeed, experimentally, little dependence of the excimer fluorescence wavelength on solvent polarity is observed. A change in geometry which destroys the excimer by pulling its components apart will change the situation by reducing $|H_3 + H_4|$ as well as H_1. This will reduce the interaction of the locally excited configurations. The concomitant increase in the separation of the positive and negative charges raises the energy of the charge-separated configurations, but this will be partly compensated by the improved solvation of the individual ions. If the energy of the locally excited state of an isolated partner is sufficiently high and that of the solvated ions in the separated ion pairs sufficiently low, an ion pair might in principle indeed form from the excimer in a sufficiently polar solvent. Normally, of course, the process is exothermic in the opposite direction: Free ground-state radical anions $\mathfrak{A}^-$ and radical cations $\mathfrak{B}^+$ will diffuse together (on the S_1 surface of the total composite system) and produce excimers, detected by their fluorescence.

6.5 COMMENTS AND REFERENCES

An analysis of the states of **four electrons in a symmetric linear array of four orbitals** is given in W. Gerhartz, R. D. Poshusta, and J. Michl, *J. Am. Chem. Soc.* **98**, 6427 (1976). For calculations on the S_0 and T_1 states of the trimethylene biradical, see C. Doubleday, Jr., J. W. McIver, Jr., and M. Page, *J. Am. Chem. Soc.* **104**, 6533 (1982); for results for the tetramethylene biradical, see C. Doubleday, J. McIver, and M. Page, *J. Am. Chem. Soc.* **107**, 7904 (1985). A discussion of the effect of various geometric distortions on the magnitude of the spin–orbit coupling as a function of substitution on an assembly of two olefins is found in S. S. Shaik, *J. Am. Chem. Soc.* **101**, 3184 (1979), who, however, represents S_0 by a single closed-shell configuration and does not attempt to evaluate the relative energies of the seriously distorted geometric arrangements. Spin-orbit coupling in biradicals is discussed in L. Salem and C. Rowland, *Angew. Chem. Int. Ed. Engl.*, **11**, 92 (1972).

Cis–trans isomerization of dienes and higher polyenes is one of the best known photochemical processes and occurs both in the S_1 and in the T_1 states. For discussions of the isomerization of butadiene in the triplet state, see: I. Ohmine and K. Morokuma, *J. Chem. Phys.* **74**, 564 (1981); I. Ohmine and K. Morokuma, *J. Chem. Phys.* **73**, 1907 (1980); E. M. Evleth and R. A. Poirier, *J. Photochem.* **30**, 423 (1985). For calculations for isomerization in the singlet state, see: V. Bonačić-Koutecký, M. Persico, D. Döhnert, and A. Sevin, *J. Am. Chem. Soc.* **104**, 6900 (1982); M. Aoyagi, Y. Osamura, and S. Iwata, *J. Chem. Phys.* **83**, 1140 (1985); P. G. Szalay, A. Karpfen, and H. Lischka, *Chem. Phys.* **130**, 219 (1989); I. Ohmine, *J. Chem. Phys.* **83**, 2348 (1985), where CCH bending is considered. Excited singlet butadiene and its simple derivatives, as well as hexatriene, exhibit no fluorescence; this is compatible with twisting without a barrier, in agreement with the calculations quoted above. Higher polyenes, arylpolyenes, and α,ω-diarylpolyenes generally have a small barrier in the S_1 and T_1 surfaces along the twisting coordinate, or even two barriers, with separate minima at the cis, trans, and twisted geometries. Steric hindrance may remove some of these. For instance, in certain arylalkenes, the T_1 surface contains no minimum at the twisted and the cis geometries, so that a triplet *cis* species reacts adiabatically to yield a triplet trans product: T. Karatsu, T. Arai, H. Sakuragi, and K. Tokumaru, *Chem: Phys. Lett.* **115**, 9 (1985). Twisting around bonds that are single in S_0 is more difficult in both S_1 and T_1 and is normally not competitive in polyenes and arylethylenes. This is known as the NEER principle (nonequilibration of excited rotamers): H. J. C. Jacobs and E. Havinga, *Adv. Photochem.* **11**, 305 (1979).

An early *ab initio* calculation on the twisting of styrene was reported by G. L. Bendazzoli, G. Orlandi, P. Palmieri, and G. Poggi, *J. Am. Chem. Soc.* **100**, 392 (1978). For leading references to the photoisomerization of stilbene, see J. Saltiel and Y.-P. Sun, *J. Phys. Chem.* **93**, 8310 (1989). Several semiempirical calculations of the excited potential surfaces for the twisting motion have been reported and have generally overestimated the barrier in the S_1 surface; see, for example, G. Olbrich, *Ber. Bunsenges. Phys. Chem.* **86**, 209 (1982). A qualitatively correct account of the nature of the barrier was first given in G. Orlandi and W. Siebrand, *Chem. Phys. Lett.* **30**, 352 (1975) and was subsequently refined: J. B. Birks, *Chem. Phys. Lett.* **43**, 430 (1978); G. Hohlneicher and B. Dick, *J. Photochem.* **27**, 215 (1984).

The role of spin multiplicity in **cleavage of charged sigma bonds** carrying an absorbing substituent has been analyzed by J. R. Larson, N. D. Epiotis, L. E. McMurdie, and S. S. Shaik, *J. Org. Chem.* **45**, 1388 (1980). A general review of cation-forming photosolvolyses is found in S. J. Cristol and T. H. Bindel in *Organic Photochemistry* (A. Padwa, Ed.), Marcel Dekker, New York, 1983, Vol. 6, p. 327.

Atom and ion transfer to charged double bonds are likely, particularly when twisting is prevented. For a theoretical treatment of an intramolecular reaction of this type, see G. Trinquier, N. Paillous, A. Lattes, and J. P. Malrieu, *Nouv. J. Chim.* **1**, 403 (1977).

Cis–trans isomerization around double bonds carrying a positive charge, or conjugated with another double bond that does, is common. For leading references to the cis–trans isomerization of rhodopsin and related model compounds, see R. S. Becker, *Photochem. Photobiol.* **48**, 369 (1988); see also H. Shichi, *Biochemistry of Vision*, Academic Press, New York, 1983. The mechanism for the primary step proposed by Becker differs from the proposal described in Section 6.3.4, taken from V. Bonačić-Koutecký, J. Koutecký, and J. Michl, *Angew. Chem.* **26**, 170 (1987), which gives references to additional reviews; the deuterium isotope effect was originally reported in K. Peters, M. L. Applebury, and P. M. Rentzepis, *Proc. Natl. Acad. Sci. USA* **74**, 3119 (1977). Cis–trans isomerization of cyanine dyes is discussed in F. Dietz and S. K. Rentsch, *Chem. Phys.* **96**, 145 (1985). *Ab initio* calculations for the case of the protonated Schiff base of acrolein can be found in V. Bonačić-Koutecký, K. Schöffel, and J. Michl, *Theor. Chim. Acta* **72**, 459 (1987). Semiempirical calculations for the vinylogous case of the protonated Schiff base of 2,4-

pentadienal produced very similar results: G. J. M. Dormans, G. C. Groenenboom, W. C. A. van Dorst, and H. M. Buck, *J. Am. Chem. Soc.* **110,** 1406 (1988).

The analysis of **exciplexes** given in this chapter follows H. Beens and A. Weller, in *Organic Molecular Photophysics,* Vol. 2 (J. B. Birks, Ed.), Wiley, London, 1975, p. 159. Exciplexes are also discussed in M. Gordon and W. R. Ware, Eds., *The Exciplex,* Academic Press, New York, 1975. The **Rehm–Weller equation** (equation 6.1) was reported in D. Rehm and A. Weller, *Z. Phys. Chem. (NF)* **69,** 183 (1970). Additional references to excimers, exciplexes, charge-transfer complexes, ion pairs, and so on, are found in Section 5.5.

For the most recent reviews of **TICT states,** see: Z. R. Grabowski and J. Dobkowski, *Pure Appl. Chem.* **55,** 245 (1983); W. Rettig, *Angew. Chem. Int. Ed. Engl.* **25,** 971 (1986); E. Lippert, W. Rettig, V. Bonačić-Koutecký, F. Heisel, and A. Miehé, *Adv. Chem. Phys.* **68,** 1 (1987); E. M. Kosower, *Annu. Rev. Phys. Chem.* **37,** 127 (1986). The case of bianthryl was analyzed by H. Beens and A. Weller, *Chem. Phys. Lett.* **3,** 666 (1969); for recent results see T. J. Kang, M. A. Kahlow, D. Giser, S. Swallen, V. Nagarajan, W. Jarzeba, and P. F. Barbara, *J. Phys. Chem.,* **92,** 6800 (1988).

For a review of **CIDNP,** see: K. M. Salikhov, Yu. N. Molin, R. Z. Sagdeev, and A. L. Buchachenko, *Spin Polarization and Magnetic Effects in Radical Reactions* (Yu. N. Molin, Ed.), Elsevier, Amsterdam, 1984.

CHAPTER 7

Reactions with More Than One Active Orbital per Atom

In Chapter 4 we referred to the topicity of a reaction in which a bond between two atoms is cleaved as the total number of those valence orbitals on the two atoms that participate in the reaction. We can now more specifically define "participating" atom-centered orbitals as those that are not engaged in forming σ bonds in the products. In addition to the two, four, or six valence orbitals that were used to make the single, double, or triple bond present originally in the starting material but broken in the product, these can be lone-pair orbitals on one or both of the atoms involved, or they can be orbitals used to make π bonds terminating on one or both of these atoms. Although, in principle, one could include all valence lone-pair orbitals in the count, it is common to exclude those which are doubly occupied in all low-lying electronic states, such as the 2s lone pairs on oxygen or the halogens.

The simplest bond cleavage reaction is bitopic (two electrons, two orbitals), and this reaction type was treated thoroughly in Chapter 4. Chapters 5 and 6 represented an elaboration of the prototypes treated in Chapter 4, in that they discussed processes involving more than two centers; in all cases, however, each participating atom still participated only through one of its valence orbitals.

Presently, we shall treat processes of higher topicity. In these, at least one of the centers involved in the reaction process participates through two or more of its valence orbitals. Examples of bond cleavages of topicity three to six are given in Figure 7.1, in which the relevant valence orbitals are drawn explicitly.

The usefulness of the concept of topicity lies in the help it provides with the organization of the number and nature of low-lying electronic states of potential interest in the photochemical process, which is determined by the number of available orbitals and the number of electrons that occupy them. These numbers are listed for some of the important cases in Table 7.1 and in the right two columns of Figure 7.1.

It is often helpful to refer to the active orbitals, as well as the configurations derived from them, by labels. This is easy when at least a local plane of symmetry is available, and the labels σ, π are commonly used for the participating orbitals. When more than one configuration of a particular type is possible, these symbols can be used as subscripts to identify the orbitals that are singly occupied. Thus, $C_{\sigma,\sigma}$ stands for a dot–dot structure in which both singly occupied orbitals are of symmetry σ, and $C_{\sigma,\pi}$ stands for a structure in which one singly occupied orbital is of σ symmetry while the other is of π symmetry.

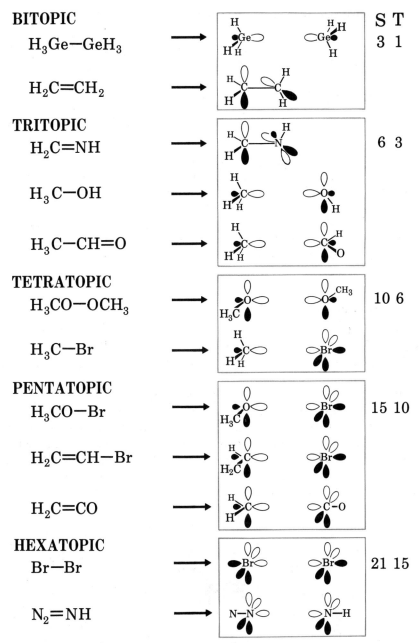

Figure 7.1 Topicity of bond cleavage. On the right, the number of singlet (S) and triplet (T) states available from the orbitals shown.

7.1 PI-BOND ISOMERIZATION

When one of the atoms of a double bond (or a conjugated system of such bonds) carries a lone pair, twisting to orthogonality produces a tritopic biradicaloid (Figure

TABLE 7.1 Bond Dissociation: Topicity and Configuration Count

		Number of Configurations[a]		
Topicity	Electrons	^{1}Dot–Dot	^{3}Dot–Dot	^{1}Hole–Pair
2	2	1	1	2
3	4	3	3	3
4	6	6	6	4
5	8	10	10	5
6	10	15	15	6

[a]Multiplicity is indicated by superscript. Except in bitopic dissociation, a dot-dot specification does not in itself identify charge distribution in the products, since both dots may but need not reside on the same fragment.

7.1). When two of the atoms carry lone pairs, such twisting produces a tetratopic biradicaloid, and so on. In the present section, we consider four examples of molecules with lone-pair atoms which can undergo a photochemical cis–trans isomerization by twisting.

7.1.1 Formaldimine

Formaldimine, $H_2C=NH$, represents a prototype of a Schiff base and can be viewed as a strongly perturbed ethylene. Unlike ethylene, and unlike its protonated form, formaldiminium ($H_2C=N^+H_2$), it has a lone pair whose presence is responsible for low-energy $^1n\pi^*$ and $^3n\pi^*$ states. At partially twisted geometries, symmetry no longer distinguishes σ and π orbitals, and this notation is then not really appropriate. Figure 7.2 shows the three critical orbitals and their interactions at three geometries.

The syn–anti isomerization around a C=N bond is important as an example of an S_0–S_1 touching (funnel, conical intersection). It is intrinsically an at least two-dimensional process, with the double-bond twist angle θ and the CNH valence angle ϕ being the important independent geometric variables. The variation of either one alone can bring about the conversion of the starting isomer into the final

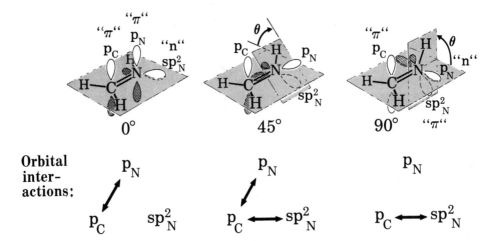

Figure 7.2 The valence orbitals involved in the twisting of the C=N bond in formaldimine and their interactions.

isomer, as can an infinite number of paths involving both geometric variables. Additional degrees of freedom, such as pyramidalization on the carbon atom, may be important.

In the following, we first consider two special paths through the two-dimensional space and subsequently discuss results for both dimensions.

One-dimensional cuts

The twist angle. The first path to be considered represents a twist around the C=N bond by an angle θ varying from 0° to 90° for a constant CNH valence angle ϕ held at a value in the vicinity of 120°. First, we construct a VB correlation diagram (Figure 7.3). In the planar imine, the lowest-energy VB structure is the covalent singlet. Its triplet counterpart is considerably higher in energy. Structures with

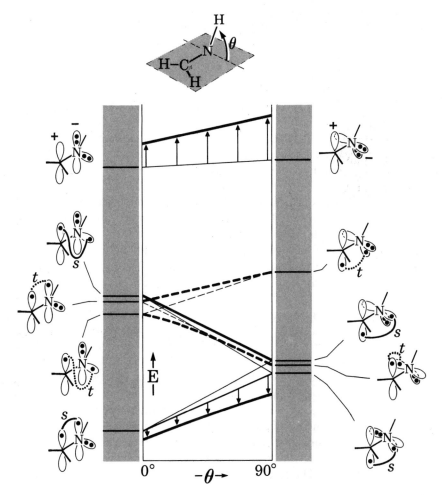

Figure 7.3 A VB structure (thin lines) and state (thick lines) correlation diagram for the twisting (θ) of formaldimine at CNH valence angle (ϕ) of about 120°, from $\theta = 0°$ to $\theta = 90°$. Singlets: solid lines. Triplets: dashed lines. Lines of intermediate thickness show an intermediate result in which only the interaction of the lowest two singlet VB structures has been considered.

separated π charges are of unequal energy. The one with the negative charge on nitrogen and positive on carbon is considerably lower and is the only one shown. It is seen in the top part of the left-hand side of the diagram in Figure 7.3. All three states mentioned so far are analogous to those in ethylene. In addition, however, there are two structures of the $n\pi^*$ type in which the lone-pair orbital is occupied by a single electron, and there are three electrons occupying orbitals of π symmetry. For both the singlet and triplet coupling case we only show in Figure 7.3 the VB structure with both of these electrons on the more electronegative nitrogen terminus, although the singlet and triplet structures with two electrons on the carbon, and therefore with separated charges, undoubtedly also will contribute to some degree. This mixing will actually also be responsible for the somewhat lower energy of the triplet relative to the singlet $n\pi^*$ state, since the singlet and triplet structures shown are actually almost exactly degenerate in energy.

On the right-hand side of Figure 7.3 the corresponding VB structures of the 90° twisted C=N double bond are shown. The lowest singlet structure is again of covalent nature but is at higher energy than it was before the twisting, since the interaction of the two "π"-symmetry orbitals on the carbon and the nitrogen is now less favorable. The orbital on carbon is still a pure p orbital that is well suited for π bonding. However, the orbital on nitrogen is a hybrid of the sp^2 type, tilted away from the axis of the carbon p orbital; only that part of it which is of p character can contribute to π bonding. Moreover, the lone pair is now in a pure p orbital, which is higher in energy than the sp^2 hybrid in which the lone pair was housed in the planar molecule. The corresponding triplet VB structure is again substantially higher in energy—although not as much as it was on the left-hand side, since the interaction of the two "π"-symmetry orbitals on the carbon and the nitrogen is now weaker. The charge-separated structure with both "π" electrons on the nitrogen atom is even less favorable. The remaining two structures are of the $n\pi^*$ type— one singlet and one triplet. It is not obvious just exactly what their energies are relative to the energy of the lowest covalent configuration, as will become clear when we discuss the effect of the CNH valence angle. The location shown on the right-hand side of Figure 7.3 is more or less arbitrary. It is assumed that the excitation of an electron from the lone-pair orbital of p character into the hybrid orbital on the nitrogen is favored by the partial s character of the latter, but that this is not sufficient to outweigh the loss of the stabilizing singlet coupling present in the lowest covalent structure. Just how this balance works out will obviously depend on the details of hybridization on the nitrogen atom.

Connecting the corresponding VB structures on both sides of the diagram produces the VB correlation diagram in the usual fashion. Since no symmetry is present along the reaction path except for the starting and ending point, curve crossings will be avoided; this is indicated for both the singlet and triplet structures. Moreover, in the case of the singlet structures, there will also be a mixing with the high-lying ionic structure. Its effect on the S_0 and S_2 states is indicated by arrows, the effect on S_1 is neglected. The final state correlation diagram is drawn in thick lines.

The valence angle. Next we shall consider the VB correlation diagram for a C=N double bond twisted to 90° as a function of the valence angle ϕ changing from 180° to approximately 90°. Figure 7.4 shows the relevant VB structures, already familiar

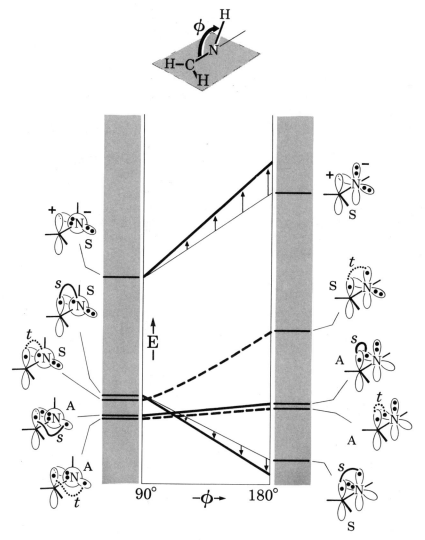

Figure 7.4 A VB structure (thin lines) and state (thick lines) correlation diagram for the folding of the CNH valence angle ϕ in orthogonally twisted formaldimine ($\theta = 90°$). Singlets: solid lines. Triplets: dashed lines.

from the right-hand side of Figure 7.3. Their energies depend on the valence angle, since it determines the state of hybridization of the "π" orbital on the nitrogen atom, varying from a pure p orbital on the right-hand side to an essentially s orbital on the left-hand side. During this process, the energy of this orbital decreases considerably, and this stabilizes those VB structures in which it is doubly occupied. On the other hand, the bonding interaction between this orbital and the p orbital on carbon gradually decreases, and this destabilizes those structures that derive their stability primarily from such a bonding interaction—in particular, the lowest covalent structure on the right-hand side. Consideration of the symmetry elements present indicates that the crossing that results in the diagram when the corresponding configurations on the left- and right-hand sides are connected will not be avoided. In the diagram we have indicated symmetries by the letters S and A corresponding to structures that are symmetric and antisymmetric, respectively, relative to a plane that contains the C=N bond and the axis of the p orbital on carbon. Only structures of like symmetry will interact when the state correlation diagram is deduced from the VB correlation diagram as indicated by arrows in Figure 7.4.

The crossing of the S and A symmetry singlet states in the correlation diagram of Figure 7.4 depends on the correctness of the estimated state order in the left-hand-side limit, A below S. If this indeed is the order, the crossing will imply an S_0–S_1 surface touching and a conical intersection, of considerable interest for photochemical processes. However, no experimental data are available for the state order assignment in the right-hand-side limit. This is a typical example of the kind of dilemma one frequently faces when using correlation diagrams in photochemistry: They are often suggestive but rarely conclusive. To settle the present issue, calculations are needed.

If the crossover between the A and S states at $\theta = 90°$ indeed occurs for some critical value of the valence angle $\phi = \phi_c$, the correlation diagram given in Figure 7.3 will only apply for $\phi \rangle \phi_c$. For $\phi \langle \phi_c$, the final state curves will still look the same but no avoided crossing will be involved.

The two-dimensional problem. The results of *ab initio* large-scale CI calculations are shown in Figures 7.5–7.10. We shall first consider results for the one-dimensional cuts for which we have just constructed correlation diagrams.

Figure 7.5 displays the state energy curves obtained upon varying the twist angle θ with simultaneous elongation of the C=N bond from 1.330 Å at planarity to 1.37 Å at 90° twist, keeping ϕ constant at 115°. They follow closely the state energy curves of the correlation diagram in Figure 7.3. Also, the nature of the wave functions is as expected.

In the singlet ground state S_0 the molecule has two electrons in an approximately sp^2-hybridized lone pair on the nitrogen and two in the bonding π orbital of the C—N bond. As the twist angle increases, the n and π orbitals begin to mix. At orthogonality, symmetry is restored and the two kinds of orbitals can again be clearly distinguished. Now the lone-pair orbital n is a 2p atomic orbital of the nitrogen, and the "π" and "π^*" combinations are formed from a 2p orbital on the carbon and an approximately sp^2 hybrid orbital on the nitrogen (Figure 7.2). Since at 90° twist a C=N double bond is still a double bond in the ground state, albeit a weakened one, it is not really analogous to an orthogonally twisted C=C double bond but, rather, to a C=C double bond pyramidalized at one end.

The larger electronegativity of nitrogen compared to carbon causes the π molecular orbital to be polarized towards nitrogen while π^* is polarized away from it, and the magnitude of this effect increases with the increasing twist angle. In terms of the VB structures shown in Figure 7.3, the covalent structure for the π bond dominates at planarity but the ionic structure dominates at 90° twist.

The lowest excited singlet and triplet states are of $n\pi^*$ nature. Thus at planarity the hybrid orbital on nitrogen contains only one electron and the π system contains three electrons, and at orthogonality the p orbital on nitrogen contains only one electron while the π system still contains three electrons.

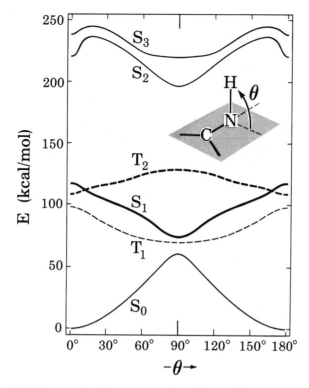

Figure 7.5 *Ab initio* large-scale CI state energies of formaldimine as a function of twist angle θ from 0° to 180° (CNH valence angle $\phi = 115°$). Reproduced by permission from V. Bonačić-Koutecký and M. Persico, *J. Am. Chem. Soc.* **105**, 3388 (1983).

The S_2 and T_2 states are of $n\pi^*$ nature (in a full calculation, Rydberg states would probably lie below S_2 at planar and nearly planar geometries). The S_0 and S_1 $n\pi^*$ states undergo an avoided touching between 0° and 90° of twist as shown in Figure 7.3 and then again between 90° and 180° of twist. This intended touching is avoided quite strongly in the regions of intermediate twist and does not noticeably affect the shapes of the potential energy curves.

Another avoided crossing in Figure 7.5 is that between the states of the molecule at $\theta = 0°$ and its states at $\theta = 180°$. Here, the S_0 state of the syn isomer ($\theta = 0°$) correlates with the S_1 $n\pi^*$ state of the anti isomer ($\theta = 180°$), and vice versa. This crossing should remain unavoided at the critical value of the valence angle ϕ_c for which the ground and $n\pi^*$ states of the orthogonal species, which differ in symmetry, are "accidentally" degenerate (Figure 7.4). For ϕ values on either side of ϕ_c, the crossing ought to be avoided; Figure 7.5 shows that this is indeed so, but the crossing is not avoided very strongly.

The next issue to address is the existence of the critical value of ϕ_c at which the A and S state cross in the orthogonally twisted molecule. Figure 7.6 agrees with the correlation diagram in Figure 7.4 and shows that the crossing indeed occurs at two different levels of calculation; for the particular choice of bond lengths used, $\phi_c \cong 90°$.

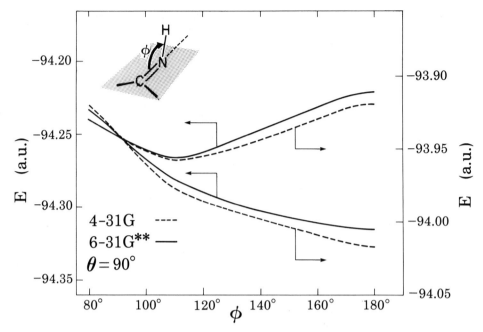

Figure 7.6 *Ab initio* large-scale CI state energies of orthogonally twisted formaldimine ($\theta = 90°$) as a function of the CNH valence angle ϕ, for two basis set choices. Reproduced by permission from V. Bonačić-Koutecký and J. Michl, *Theor. Chim. Acta* **68**, 45 (1985).

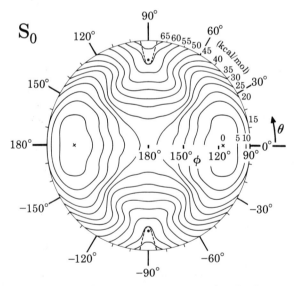

Figure 7.7 *Ab initio* large-scale CI energy contour map for the S_0 state of formaldimine, $CH_2{=}NH$. The radial variable is the CNH valence angle ϕ, and the angular variable is the twist angle θ; energies (in kcal/mol) are given relative to the two equilibrium geometries indicated by crosses. For details see text. Reproduced by permission from V. Bonačić-Koutecký and J. Michl, *Theor. Chim. Acta* **68**, 45 (1985).

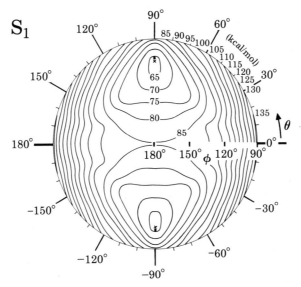

Figure 7.8 *Ab initio* large-scale CI energy contour map for the S_1 state of formaldimine, CH_2=NH. The radial variable is the CNH valence angle ϕ, and the angular variable is the twist angle θ; energies (in kcal/mol) are given relative to that of the S_0 state at its equilibrium geometry. For details see text. Reproduced by permission from V. Bonačić-Koutecký and J. Michl, *Theor. Chim. Acta* **68**, 45 (1985).

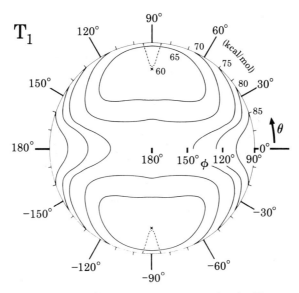

Figure 7.9 *Ab initio* large-scale CI energy contour map for the T_1 state of formaldimine, CH_2=NH. The radial variable is the CNH valence angle ϕ, and the angular variable is the twist angle θ; energies (in kcal/mol) are given relative to that of the S_0 state at its equilibrium geometry. For details see text. Reproduced by permission from V. Bonačić-Koutecký and J. Michl, *Theor. Chim. Acta* **68**, 45 (1985).

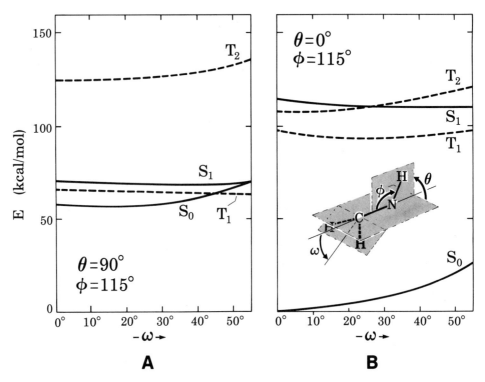

Figure 7.10 *Ab initio* large-scale CI calculation of the effect of pyramidalization on the carbon atom on state energies of formaldimine, CH_2=NH. Valence angle CNH, $\phi = 115°$. (A) Twist angle $\theta = 90°$; (B) twist angle $\theta = 0°$.

Figures 7.7–7.9 show the overall shape of the S_0, S_1, and T_1 surfaces as a function of the twist angle θ and the valence angle ϕ, with bond lengths optimized for the S_1 surface. The use of a polar plot, with θ as the angular variable and ϕ as the radial variable, maps the linear arrangement of the CNH atoms ($\phi = 180°$) at the origin and maps the right angle $\phi = 90°$ at the circumference of a circle. Planar geometries are located at the horizontal diameter ($\theta = 0°$ or $180°$), and orthogonally twisted geometries are located at the vertical diameter ($\theta = \pm 90°$). As the C=N bond is twisted at a fixed value of the valence angle ϕ, its geometries map on a circle with a center at the origin and radius dictated by the value of ϕ.

The two equilibrium geometries in the S_0 state ($\phi = 115°$, $\theta = 0°$ or $180°$) correspond to the syn and anti isomers (Figure 7.7). The transition state for their thermal interconversion lies at the origin (linear C=N—H). The dots at $\phi = 106.5°$ and $\theta = \pm 90°$ represent the S_0–S_1 touching points, and the dashed lines indicate the S_0–T_1 intersections.

In the S_1 state, orthogonally twisted geometries ($\theta = \pm 90°$) are strongly preferred. The two minima, at $\phi = 109°$, $\theta = \pm 90°$, indicated by crosses in Figure 7.8, are located very close to the S_0–S_1 touching points indicated by dots; these minima communicate through a transition state located at the linear C=N—H geometry in a fashion very similar to that already noted for the S_0 state.

The equilibrium geometry of the ground state is very unfavorable in the S_1 state and is located on a very steep slope. S_0–S_1 excitation should be followed by very rapid vibrational motion towards linear and orthogonal geometries. As soon as the S_0–S_1 touching area is reached, a jump to the S_0 state should follow with high probability, leaving little chance for fluorescence and intersystem crossing. This is in accord with the complete absence of fine structure in the observed absorption band and the failure to detect fluorescent emission from simple imines. It is conceivable that a dynamic memory effect will take the molecule preferably to one or the other minimum in the S_0 state, so that the quantum yield of the isomerization will deviate from the otherwise expected value of 0.5 (assuming no other photoprocesses); it could be as high as 1.0 or as low as 0.0. The latter is unlikely, since in substituted imines the photochemical syn–anti interconversion occurs readily.

The T_1 surface (Figure 7.9) is rather flat, but it, too, prefers orthogonally twisted geometries. Its two minima at $\phi = 112°$ and $\theta = \pm 90°$ are indicated by crosses. They lie very close to the S_0–S_1 intersection line; in the small enclosed regions beyond, T_1 is the ground state. Vertical S_0–T_1 excitation would be expected to send the molecule towards linear and orthogonal geometries. After reaching the vicinity of one or the other minimum, it should undergo fast intersystem crossing to the S_0 surface and then settle in either the syn or the anti minimum in the S_0 surface.

An additional geometric variable that needs to be considered is pyramidalization on the C atom. As shown in Figure 7.10, this type of distortion from the geometries considered so far is unfavorable in the S_0 state, but in the orthogonally twisted S_1 and T_1 states, it requires no energy until the tetrahedral geometry is reached, and probably beyond (Figure 7.10A). In the untwisted S_1 and T_1 states, this path actually leads downhill by a few kilocalories per mole.

A more detailed exploration of the three-dimensional nuclear configuration subspace spanned by twisting (θ), CNH bending (ϕ), and pyramidalization (ω), albeit at a much less sophisticated level of calculation, was used as one of the examples in Chapter 1 (Figure 1.5).

7.1.2 Acroleinimine

The electronic states of this molecule can be profitably related to those of both butadiene and formaldimine. The paths of interest are: (i) a geometric isomerization around the C=C double bond by twisting (θ_1); (ii) a geometric isomerization around the C=N double bond, either by twisting (θ_3) or by a valence angle (ϕ) change or a combination of both; and (iii) a geometric isomerization around the central single bond by twisting (θ_2).

C=C twist. We shall first consider the path corresponding to a twisting around the C=C double bond. Figure 7.11A shows the usual one-dimensional cut through the potential energy hypersurfaces in which the twist angle θ_1 varies from 0° to 180° while the C=C bond is simultaneously elongated as the geometry changes from planar to orthogonal. The electronic states of the planar molecule are related to those of butadiene except that the A_g and B_u excited states are mixed as a result of lower symmetry. As was the case in formaldimine, the presence of the lone pair on the nitrogen introduces a new low-lying state, which was absent in butadiene. Thus, the first excited singlet S_1 is the $n\pi^*$ state, in which one of the electrons in the lone pair is promoted into the lowest π^* orbital. Rydberg states have been omitted from the calculation with the usual justification. The T_1 state is of $\pi\pi^*$

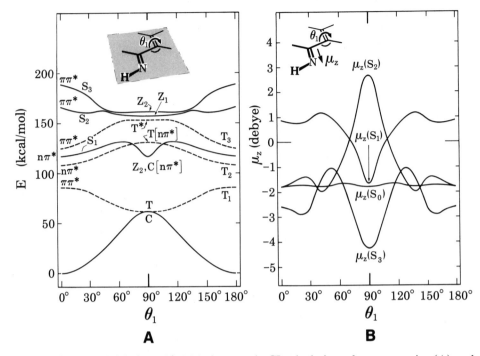

Figure 7.11 Acroleinimine. *Ab initio* large-scale CI calculation of state energies (A) and the z component of the molecular dipole moment (B) as a function of the angle of twist around the C=C bond (θ_1). Reproduced by permission from V. Bonačić-Koutecký and M. Persico, *J. Am. Chem. Soc.* **105**, 3388 (1983).

nature, quite analogous to that of butadiene. The triplet $n\pi^*$ state T_2 lies only a little below the singlet $n\pi^*$ state S_1 as expected.

The behavior of state energies upon twisting of the C=C double bond (Figure 7.11A) can be related to that observed for butadiene in Figures 6.12–6.17. As we did there, we shall refer to an allocation of electrons to fully delocalized orbitals as configurations (e.g., in the planar molecule) and to an allocation of electrons to partly localized orbitals as VB-like structures (e.g., in the orthogonal molecule). In butadiene, a barrier in S_0 and a very flat minimum in S_1 were found at 90° twist as a result of an avoided crossing between the ground configuration and the configuration obtained by the double excitation $\pi, \pi \rightarrow \pi^*, \pi^*$ as the molecule is twisted from 0° to 180°. As nearly orthogonal twist angles are reached, the $\pi, \pi \rightarrow \pi^*, \pi^*$ configuration is smoothly transformed into a form that can be well represented by the VB-like structure $^1|Z_2\rangle$. At 90° twist, S_0 is well represented by the dot–dot structure $^1|C\rangle$. Two of the next three closely spaced singlet states are well represented by the zwitterionic hole–pair structures $^1|Z_1\rangle$ and $^1|Z_2\rangle$ of opposite polarity, and the third is represented by the dot–dot structure $^1|C^*\rangle$. At intermediate twist angles, all three structures mix in a complicated manner, and avoided crossings are present.

In acroleinimine, the crossing of the ground and doubly excited $\pi, \pi \rightarrow \pi^*, \pi^*$ configurations that occurs in zero order is much less strongly avoided, for a reason already discussed in the case of protonated Schiff bases: At 90° twist, the high electronegativity of nitrogen makes one of the charge-separated (hole–pair) configurations ($^1|Z_2\rangle$) far more stable than the other ($^1|Z_1\rangle$), and indeed more stable than the $n\pi^*$ configuration. As before, a twist to an orthogonal geometry causes a gradual transformation of the $\pi, \pi \rightarrow \pi^*, \pi^*$ configuration into the $^1|Z_2\rangle$ structures. Simultaneously, the $n\pi^*$ configuration of the planar molecules is transformed into the $^1|C[n\pi^*]\rangle$ structure, which is a mixture of two VB-like structures that differ from the usual dot–dot structure $^1|C\rangle$ by $n \rightarrow \pi^*$ excitation in the azaallyl radical moiety (Figure 7.12). However, avoided crossing of the $\pi, \pi \rightarrow \pi^*, \pi^*$ and ground configurations at 90° twist is difficult to discern from the shapes shown in Figure 7.11A, since additional avoided crossings are present near 60° and 120°.

Their presence is caused by the above-mentioned change of the energy order of the $\pi, \pi \rightarrow \pi^*, \pi^*$ (or $^1|Z_2\rangle$) and the $n\pi^*$ (or $^1|C[n\pi^*]\rangle$) configurations as the twist proceeds to orthogonality. If it were not for the mutual interaction of the $\pi, \pi \rightarrow \pi^*, \pi^*$ (or $^1|Z_2\rangle$) and the $n\pi^*$ (or $^1|C[n\pi^*]\rangle$) configurations, the curve of the singlet $n\pi^*$ state and that of the triplet $n\pi^*$ state T_2 would be quite parallel to each other. As a result of the interaction, the crossing of the configurations is fairly strongly avoided and results in a barrier in the S_1 surface in the vicinity of 60° and 120°. The effects of the avoided crossing can be traced in the plot of the component

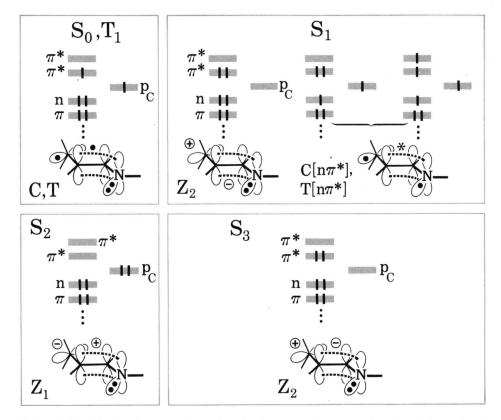

Figure 7.12 The dominant configurations in the electronic states of acroleinimine twisted around its C=C bond ($\theta_1 = 90°$). Adapted by permission from V. Bonačić-Koutecký and M. Persico, *J. Am. Chem. Soc.* **105**, 3388 (1983).

of the molecular dipole moment along the C=C bond (μ_z) as a function of the twist angle θ_1 (Figure 7.11B).

The small barrier in the T_2 triplet $n\pi^*$ state and the intended analogous barrier in the S_1 singlet $n\pi^*$ state at 90° twist have an easily understandable origin. The excitation from the n orbital into the π^* orbital becomes easier as the energy of the latter is decreased by twisting, but the ground-state energy increases even more rapidly because two electrons are present in the π orbital which is being strongly destabilized.

The shape and wave function of the lowest $\pi\pi^*$ triplet T_1 of acroleinimine are quite analogous to what they are in butadiene.

At 90° twist, the nature of the wave functions can be summarized as follows: In the S_0 and T_1 states, they are well represented by biradical dot–dot VB-like structures, C and T, respectively, as in butadiene. The wave function of the S_1 state is represented by a mixture containing (i) the hole–pair VB-like structure $^1|Z_2\rangle$ with a positive charge on the CH_2 terminus and a negative azaallyl moiety and (ii) two dot–dot structures of the $n\pi^*$ type in which the CH_2 moiety carries one electron and in which an electron has been excited from the lone pair to one of the π orbitals of the azaallyl moiety which therefore carries four π electrons. These are referred to jointly as $^1|C[n\pi^*]\rangle$. The T_2 state is represented by $^3|T[n\pi^*]\rangle$, composed of two dominant configurations (Figure 7.12).

C=N twist. The next photochemical path we shall discuss is the isomerization around the C=N double bond. It is defined by twisting (θ_3) and simultaneous elongation of the C=N double bond as the geometry changes from planar to orthogonal, keeping all other geometric parameters constant. The state energies along this path are shown in Figure 7.13A. The similarity to the case of twisting formaldimine is striking. Once again, the S_0 and S_1 ($^1n\pi^*$) states undergo a weakly avoided crossing at the orthogonal geometry. The nature of their electronic wave functions is similar as was the case for the twisting of formaldimine except that the CH_2 group has now been replaced by the CH_2=CH—CH group (Figure 7.14). The S_2 state, which is of mixed A_g–B_u character at planar geometries, gradually changes upon twisting into a state most easily characterized as a higher $n\pi^*$ state in which an electron has been promoted from the noninteracting lone pair on the nitrogen atom to one of the upper π^* orbitals. It is represented by a mixture of several configurations of this type (Figure 7.14). The S_3 state keeps its mixed A_g–B_u $\pi\pi^*$ character for all degrees of twist.

The T_1 state of the planar molecule is of $\pi\pi^*$ nature, and the T_2 state is of $n\pi^*$ nature. Upon twist, these states undergo an avoided crossing. At 90° twist, the lowest triplet T_1 is of $n\pi^*$ character, and its wave function is characterized by a structure containing one of the electrons of equal spin in the noninteracting lone-pair orbital and containing the other in the lowest antibonding orbital of the π system.

The nature of the wave functions of the singlet states is clearly reflected in μ_z (Figure 7.13B). Thus at orthogonal geometry ($\theta_3 = 90°$) the dipole moment is small for the states of $n\pi^*$ character, S_1 and S_2, but the negative end is still directed towards the nitrogen atom. In the ground state S_0 and in the $\pi\pi^*$ excited state S_3, the polarization of negative charge towards the nitrogen atom is much higher and the dipole moment is substantially larger. The avoided crossings that occur in the

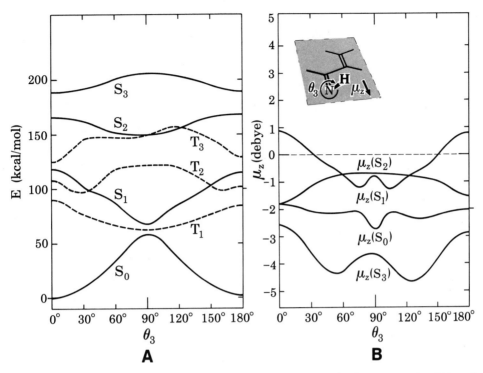

Figure 7.13 Acroleinimine. *Ab initio* large-scale CI calculation of state energies (A) and the z component of the molecular dipole moment (B) as a function of the angle of twist around the C=N bond (θ_3). Reproduced by permission from V. Bonačić-Koutecký and M. Persico, *J. Am. Chem. Soc.* **105**, 3388 (1983).

region between the planar and the fully twisted geometries are responsible for some of the striking variations of the calculated dipole moments as a function of the twist angle θ_3.

CNH inversion. As in formaldimine, there is a second linearly independent path for the syn–anti isomerization around the C=N double bond, namely, the change of the CNH valence angle ϕ. Figure 7.15 shows a cut along this reaction path as a function of ϕ, keeping all other geometric parameters constant. As in the case of formaldimine, the ground state S_0 shows a substantial barrier to the isomerization as a result of the rehybridization of the lone pair into an orbital of pure p nature which lies higher in energy. The barrier is again smaller than the barrier corresponding to the pure twisting motion. In the S_1 state, which is of $^1n\pi^*$ character, and in which the lone-pair orbital which is being rehybridized only contains one electron, the energy is nearly constant across most of the range of ϕ values.

In principle, the syn–anti isomerization around the C=N double bond requires the consideration of a two-dimensional diagram similar to the one shown for formaldimine. The one-dimensional cut displayed suggests that the two-dimensional picture will be qualitatively similar to that for formaldimine, at least for the S_0, S_1, and T_1 states. It has not been calculated.

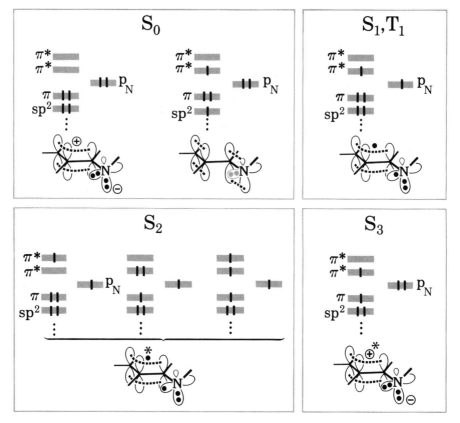

Figure 7.14 The dominant configurations in the electronic states of acroleinimine twisted around its C=N bond ($\theta_3 = 90°$). Reproduced by permission from V. Bonačić-Koutecký and M. Persico, *J. Am. Chem. Soc.* **105**, 3388 (1983).

C—C twist. The remaining geometric isomerization of interest is the twist around the single C_2—C_3 bond. Figure 7.16 shows a cut through the pertinent part of the potential energy surfaces in which the twist angle θ_2 around the C—C bond is varied and all other geometric parameters of the molecule are kept constant. As in butadiene, the barrier to this twist is very small in the ground state because only a little stabilization is derived from the conjugation of the two double bonds. In the $n\pi^*$ excited state S_1, the barrier is somewhat higher because the energy of the π^* orbital, into which the excitation occurs, is unfavorably affected by the twisting which destroys the conjugation of the two π bonds. Even in the highly excited states, this twisting around the C—C single bond is not an energetically favorable motion.

From the information available so far, it appears that photochemical isomerization of acroleinimine in its $n\pi^*$ S_1 state should preferably occur around the C=N double bond, whereas in the triplet $n\pi^*$ T_1 state the isomerization around the C=N and the C=C bonds both appear quite likely. Also, in the lowest $\pi\pi^*$ singlet state S_2 the two isomerizations could be competitive if internal conversion to the S_1 state does not intervene. The isomerization around the single C—C bond appears unlikely in any of the excited states but, of course, is facile in the ground state (cf. the NEER principle, p. 346).

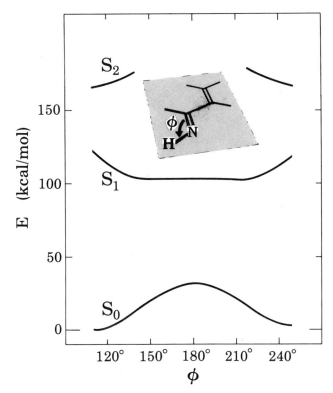

Figure 7.15 Acroleinimine. *Ab initio* large-scale CI energies of the S_0 and S_1 states as a function of the CNH valence angle ϕ. Reproduced by permission from V. Bonačić-Koutecký and M. Persico, *J. Am. Chem. Soc.* **105**, 3388 (1983).

7.1.3 Acrolein

Acrolein is the simplest representative of a group of compounds that has played an important role in the development of organic photochemistry, α, β-unsaturated carbonyl compounds. The reaction path to be discussed here is the geometric isomerization around the $C_3\!\!=\!\!C_4$ double bond (twisting angle θ). Its importance reaches well beyond acyclic enone chemistry, in that it now appears that this is the initial step even in some cycloalkenone 2 + 2 cycloadditions as well as rearrangements, not discussed in this book. By analogy to butadiene and acroleinimine, twisting around the $C_2\!\!-\!\!C_3$ single bond should be difficult in the excited state, facile in the ground state, and therefore of limited interest in photochemistry except as a hot ground-state reaction (cf. the NEER principle, p. 346).

Figure 7.17 shows a one-dimensional cut through the potential energy hypersurfaces in which the twist angle θ varies from 0° to 180° while the C═C bond is simultaneously elongated as the geometry changes from planar to orthogonal. Both the nature of the wave functions and the changes of the wave functions along the twisting coordinate are very similar to those already described for C═C twisting in acroleinimine (Section 7.1.2).

In the planar molecule, S_1 is the $n\pi^*$ state and S_2 is a $\pi\pi^*$ state that corresponds to a mixture of the butadiene A_g and B_u states. The $n\pi^*$ and $\pi\pi^*$ triplets represent the T_1 and T_2 states. They are nearly degenerate, and their exact order is not

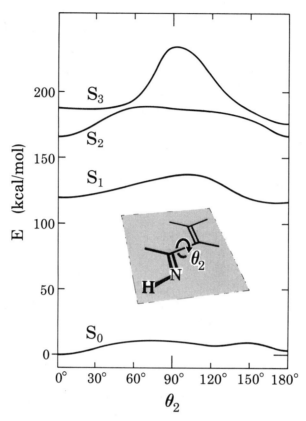

Figure 7.16 Acroleinimine. *Ab initio* large-scale CI singlet state energies as a function of the twist angle θ_2 around the C—C bond. Reproduced by permission from V. Bonačić-Koutecký and M. Persico, *J. Am. Chem. Soc.* **105**, 3388 (1983).

reliably determined by the calculation. The $n\pi^*$ excitation energies are distinctly lower in acrolein than in acroleinimine. Although the lone-pair orbitals are held more firmly, the energy of the π^* orbital is much lower in the former and provides the dominant effect.

In the 90° twisted molecule, the S_0 state is represented by the dot–dot structure $^1|C\rangle$ consisting of the —$CH_2^{\cdot}$ radical attached to the oxaallyl O=CH—CH$^{\cdot}$— radical, in perfect analogy to acroleinimine and butadiene. In the S_1 state, the hole–pair structure $^1|Z_2\rangle$ (analogous to that shown in Figure 7.12) dominates. Since oxygen is even more electronegative than nitrogen, the S_0–S_1 separation is now even smaller than in C=C twisted acroleinimine and much smaller than in twisted butadiene.

As in twisted acroleinimine, another structure makes an important contribution to the S_1 state as well. This is the $^1|C[n\pi^*]\rangle$ structure, obtained from the dot–dot structure $^1|C\rangle$ by promotion of a lone-pair electron to the "nonbonding" orbital of the oxaallyl moiety. The other state resulting from the $^1|Z_2\rangle - {}^1|C[n\pi^*]\rangle$ mixing is the S_2 state. The $^1|Z_2\rangle - {}^1|C[n\pi^*]\rangle$ crossing is strongly avoided, in a manner already described for C=C twisting in acroleinimine. Quite flat S_1 and S_2 surfaces result. Whether the S_1 surface actually slopes downhill as the twist angle approaches 90° along some path cannot be determined in the absence of geometry optimization.

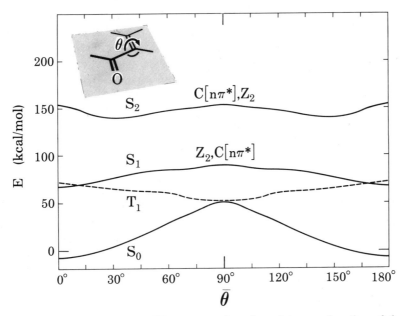

Figure 7.17 *Ab initio* large-scale CI state energies of acrolein as a function of the C=C twist angle θ.

The T_1 and T_2 states of twisted acrolein, described by the $^3|T\rangle$ and $^3|T[n\pi^*]\rangle$ structures, respectively, are quite analogous to the twisted states of acroleinimine. Both are again relatively flat and featureless.

In summary, it appears that both S_1 and T_1 acrolein can twist its C_3=C_4 bond quite easily, with only small, if any, barriers along the way but without a pronounced driving force for the twist.

7.1.4 Diazene

Diazene continues the series of increasing complexity which started with ethylene and proceeded to formaldimine. In this molecule, each of the termini of the double bond carries a lone pair. As a result, a minimal analysis of the electronic states involves six electrons in four orbitals. This is sufficiently complicated that the general case has not been considered in all its complexity in Chapter 4. Fortunately, only a few of the states will be of photochemical interest.

Cis–trans isomerization around the N=N double bond can be accomplished along three independent paths: a twisting motion and changes of the NNH valence angle at either one of the two nitrogens. These motions can be combined so that the relevant cut for the isomerization path is three-dimensional.

Instead of exploring the three-dimensional pertinent part of the nuclear configuration hyperspace, we shall limit our attention to two one-dimensional cuts through it. The first one corresponds to the twisting motion, and the other corresponds to a valence angle change at one of the nitrogen atoms.

The planar states. Because the cis and trans forms of diazene differ in their properties, the diagrams showing state energies as a function of the geometric variable will not be symmetric about their midpoint. However, the differences between the electronic states of the two isomers are relatively minor, and the electronic structure of both isomers can be discussed together.

The relevant molecular orbitals are of two types, namely, π and n. The π and π^* orbitals are completely analogous to those of ethylene. The lone-pair orbitals are delocalized over both nitrogen atoms. The hybrid on one of the nitrogens is combined in a bonding or an antibonding fashion with the hybrid on the other nitrogen as shown in Figure 7.18. In the trans isomer, the symmetric combination n_S is of higher energy, lies above the π orbital, and represents the HOMO of the molecule, while the antisymmetric combination n_A is of lower energy and lies below the π orbital. In the cis isomer, the order of the symmetric n_S and antisymmetric n_A combinations is reversed.

The lowest excited states of diazene correspond to triplet and singlet $n\pi^*$ excitations. Their symmetry is B_g for the trans compound (point symmetry group C_{2h}) and B_2 for the cis isomer (C_{2v} symmetry). At considerably higher energies, a group of states lie close together. These are: the $\pi\pi^*$ state, B_u (trans) or B_1 (cis); another $n\pi^*$ state, A_u (trans) or A_2 (cis) involving excitation from the lower-energy lone-pair combination; and an A_g (trans) or A_1 (cis) double n, $n \rightarrow \pi^*$, π^* excited state in which both electrons have been removed from the upper-energy lone-pair orbital and moved to the π^* orbital. Two of the three states have triplet analogs, and of these the $\pi\pi^*$ triplet B_u (trans) or B_1 (cis) lies significantly lower in energy. Most of the excitation energies are slightly higher in the cis than in the trans isomer.

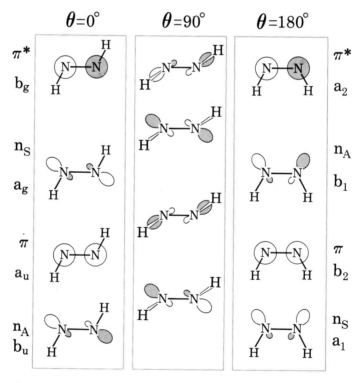

Figure 7.18 Molecular orbitals of diazene for three values of the N=N twist angle θ.

N=N twist. Among the various photochemical paths, we shall first consider the twisting motion. Figure 7.19 displays a cut through the energy surfaces for a twisting path between the cis and trans isomers, calculated by an *ab initio* large-scale CI procedure, including Rydberg states. The NN distance was 1.252 Å for the planar geometries and was gradually increased to 1.30 Å at orthogonality. All other geometric parameters were kept constant.

The MO correlation diagram which corresponds to this one-dimensional cut is shown in Figure 7.20. Avoided crossings between π and lone-pair orbitals are indicated by a dotted line, and these crossings make it clear why the diagram differs from that for ethylene (Figure 3.6) as a result of the presence of the lone pairs. In diazene, unlike ethylene, it is the lone-pair orbital of a starting isomer which becomes the antibonding π^* orbital of the product isomer. According to the dis-

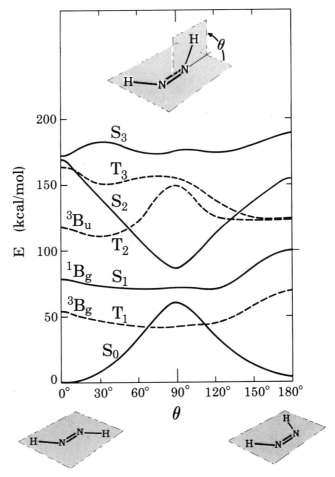

Figure 7.19 *Ab initio* large-scale CI state energies of diazene HN=NH as a function of the twist angle θ. Rydberg states are included.

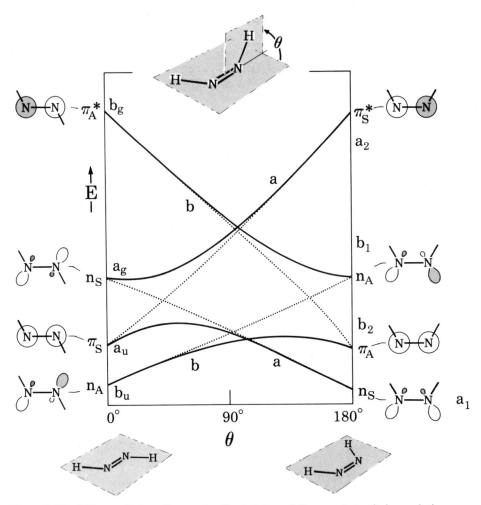

Figure 7.20 MO correlation diagram for the twisting of diazene. Intended correlations are indicated by dotted lines; actual correlations are indicated by solid lines.

cussion given in Chapter 3, the ground state should encounter a barrier along the twisting motion path, the singly excited configuration $n\pi^*$ should not change its energy very much, and the doubly excited configuration n, n → π^*, π^* should decrease in energy as it correlates with the ground configuration of the product. These considerations permit an intuitive understanding of the shape of the energy curves in Figure 7.19: The energy of the singlet and triplet $n\pi^*$ states remains almost constant across the diagram, and there is an avoided crossing between the ground state and a state corresponding to the doubly excited n, n → π^*, π^* configuration.

A striking difference between the results for diazene shown in Figure 7.19 and the analogous diagram for ethylene itself (Figure 4.35) is the small degree to which the crossing between the ground and doubly excited states is avoided in diazene and the fact that the doubly excited state does not lie below the singly excited state at 90° twist. A related difference between the two diagrams is the singlet–triplet splitting at 90°. In diazene the T_1 state lies significantly below S_0, whereas in ethylene the order is the opposite. The proximity of the three singlets and one triplet that result from the simplest 3×3 CI description of a biradical, which is encountered in the case of diazene, has not come up in any of the other molecules that we have considered so far.

The molecular orbitals at 90° twist are shown in Figure 7.18 and are pairwise approximately degenerate. Exact degeneracy is reached at the twist angles defined by the crossing points evident in Figure 7.20. It does not occur exactly at 90° as it did in ethylene because of the lower symmetry of the diazene molecule. At orthogonality, the symmetry-adapted delocalized molecular orbitals (Figure 7.18) are represented by linear combinations of hybrid orbitals on each nitrogen. The exact details of the hybridization depend on the NNH valence angle, but it can be thought of as approximately trigonal. All configurations that contribute to the S_0, S_1, S_2, and T_1 states have the set of the two most stable among these orbitals doubly occupied. They differ in the occupancy of the top two orbitals which are seen in Figure 7.20 to correlate with the upper lone pair and the π^* orbitals of the planar species. At the crossing point, which occurs at a little less than 90° twist, the S_1 and the T_1 states have both the HOMO and the LUMO orbital singly occupied while the S_0 and S_2 states are represented by linear combinations of configurations in which one of these orbitals, either HOMO or LUMO, is occupied twice. In S_0 the combination is out-of-phase; in S_2 it is in-phase, as discussed in Chapter 4.

As usual at orthogonal twist geometry, it is instructive to transform the MO picture into the VB description. The orthogonal localized orbitals $|A\rangle$ and $|B\rangle$ shown in Figure 7.21 are no longer restricted essentially to a single key orbital on one or the other terminus of the double bond as was the case in ethylene, since this p orbital is lined up appropriately for overlap and interaction with the hybrid orbital on the other atom of the double bond. We have already seen this type of interaction on the example of formaldimine and emphasized that it depends strongly

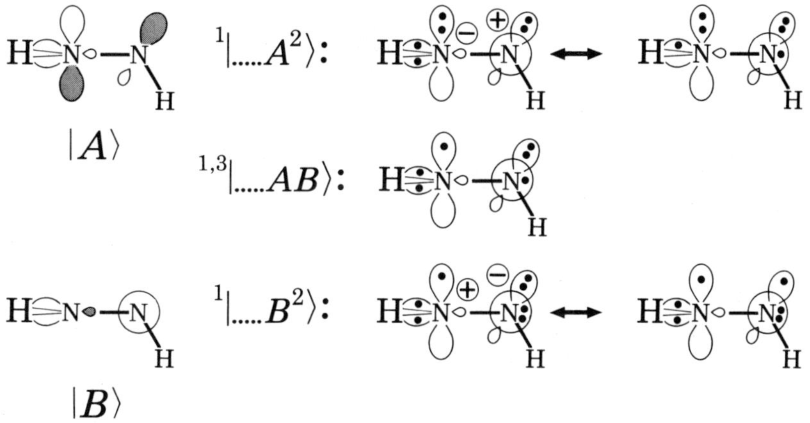

Figure 7.21 The "nonbonding" orbitals of orthogonally twisted diazenes and the four configurations resulting from their occupancy by two electrons.

on the valence angle and hybridization on the N atom. At first, we assume approximately sp^2 hybridization.

The partial delocalization of each of the orbitals $|A\rangle$ and $|B\rangle$ is clear upon inspection of Figure 7.21. The interaction between the p orbital and the hybrid orbital on the neighboring nitrogen is of the antibonding character both in $|A\rangle$ and in $|B\rangle$. Note that both of the corresponding bonding combinations are doubly occupied in all low-energy states and correspond to lower-energy MOs in the MO diagram of Figure 7.18. As we shall see in the following, the partial delocalization of $|A\rangle$ and $|B\rangle$ onto the other nitrogen atom has profound consequences for state energies.

The substantially lower energy of two of the MOs which hold four electrons in all low-lying states permits us to treat the low-lying states again in terms of a two-electron two-orbital model. As a result, the transformation of the CI wave functions based on molecular orbitals into VB structures is completely analogous to that which we have already seen in Chapter 4.

The S_0 state is described by the dot–dot configuration $^1|AB\rangle$, and the T_1 state is described by the configuration $^3|AB\rangle$, neither of which involves charge separation. In the simplest approximation, the triplet should lie below the singlet by twice the exchange integral K_{AB}. In ethylene, where $|A\rangle$ and $|B\rangle$ were each well localized on a single center, this integral is very small because it contains only two-center contributions. In the present case, its magnitude is considerably larger since both $|A\rangle$ and $|B\rangle$ are partially delocalized on both nitrogen atoms. As a result, the triplet is stabilized significantly below the singlet and is expected to be the ground state at the twisted geometry, whereas in the twisted ethylene this was not so, due to the dynamic spin polarization effect. This effect is undoubtedly also operative in the present case but is not sufficient to overcome the greatly increased magnitude of K_{AB}.

The S_1 and S_2 states are represented by the out-of-phase and in-phase mixture of the configurations $|A^2\rangle$ and $|B^2\rangle$, also shown in Figure 7.21. In ethylene, both of these configurations had very high energies because they involved charge separation. In the present case, as a result of the partial delocalization of both orbitals, the charge separation is far smaller and the energies of both states are not much above S_0. Because in the simplest model the singly excited S_1 state should lie $2K_{AB}$ below the doubly excited S_2 state, and because we have already seen that the exchange integral is much larger than was the case for ethylene, it does not come as a surprise that even at the orthogonal geometry, the ordering of the excited states is singly excited below doubly excited.

NNH inversion. The next one-dimensional reaction path we consider is the change of the NNH valence angle on one of the nitrogens in the planar molecule. Figure 7.22 shows a one-dimensional cut through the energy hypersurfaces resulting upon such a valence angle change, keeping all the other geometric parameters of the molecule constant.

As in the analogous reaction path in formaldimine, the most obvious change occurs for the lone pair on the nitrogen atom which changes from an approximately sp^2-hybridized to a pure p orbital at 180° NH valence angle. The resulting increase in the energy of the lone-pair electrons is reflected in the barrier observed in the S_0 state. Although relatively difficult, this ground-state reaction path is still favored over the twisting path discussed earlier.

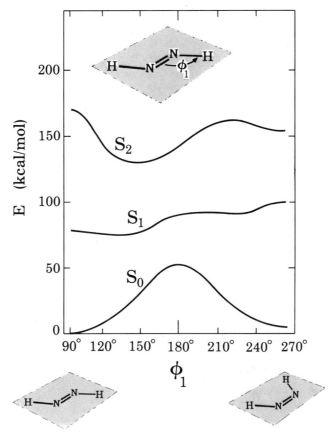

Figure 7.22 *Ab initio* large-scale CI singlet state energies of planar diazene HN=NH as a function of one of the NNH valence angles ϕ. Rydberg states are included.

In the $n\pi^*$ excited singlet and triplet states, the occupancy of the lone-pair orbital is reduced to one and the barrier essentially disappears. One of the higher states, the doubly excited singlet $|n, n \rightarrow \pi^*, \pi^*\rangle$, actually decreases in energy considerably as the linearizing motion is followed.

Although the three-dimensional pertinent part of the hyperspace defined by the twisting coordinate Θ and the NNH valence angles ϕ_1 and ϕ_2 has not yet been explored fully by reliable calculations, it appears quite likely that the photochemical paths followed will be primarily based on the twisting motion both in the singlet and triplet states, probably with some adjustment of the NNH valence angles, as was the case in formaldimine.

7.2 BOND DISSOCIATION

As shown in the examples given in Figure 7.1, when one of the two atoms linked by a single bond carries a lone pair, dissociation of the bond by stretching produces a tritopic biradical. When one of the two atoms carries two lone pairs, or when each of the atoms carries one, such dissociations produce a tetratopic biradical. In the presence of three lone pairs, a pentatopic biradical results; in the presence of four, a hexatopic biradical is produced. Biradicals of topicity higher than two result also from the stretching of vinylic bonds. Dissociation of a double bond by stretching produces a tetratopic product in the absence of lone pairs or further double bonds; in their presence, pentatopic or hexatopic products result.

An increase of the topicity number beyond two has important consequences for the shape of the potential energy curve of the S_1 state. For a bitopic bond rupture, this curve is not dissociative unless the two centers differ vastly in electronegativity, as in a charged bond (Chapter 4), whereas for a dissociation of higher topicity the S_1 curve typically is strongly dissociative, since two or more different radical-pair wave functions are available at the dissociation limit (e.g., $C_{\sigma,\sigma}$ and $C_{\sigma,\pi}$). Then, not only the S_0 state but also the S_1 state and possibly other singlet states correlate with this low-energy limit. In this sense, lone pairs and double bonds can be said to promote singlet photochemical cleavage of adjacent bonds.

Atom- and ion-transfer reactions are complex in that they involve more than a mere dissociation by stretching of one bond; however, they have also been included in this section.

7.2.1 Tritopic. Amines, Alcohols, Norrish I

In Section 7.1 we have seen how the introduction of a lone pair on a doubly bonded atom introduced new low-lying electronic states and provided new paths for geometric isomerization. Presently, we shall consider the effect of the increased topicity on the dissociation of a single bond.

We analyze first the simplest case, in which one of the atoms forming the bond also carries a lone pair. We use the dissociation of the C—N bond in methylamine to model this tritopic process. The second case is the rupture of the C—O bond in methanol, where the oxygen atoms carries two lone pairs of considerably different energy. If only the more available electrons from the higher-energy lone pair are viewed as participating in the reaction, the process is still tritopic, but if both lone pairs are included, the process is tetratopic. This example illustrates some of the arbitrariness associated with the concept of topicity.

Both the CH_3NH_2 and CH_3OH examples are somewhat artificial from the standpoint of ordinary organic photochemistry, since they require vacuum UV excitation. However, they provide simple model cases on which the basic principles are readily demonstrated. Also, the relatively high excitation energies involved force us to face up to the issue of the treatment of Rydberg states. We find that even in these highly unfavorable cases, a simple treatment based on valence orbitals alone still produces a picture that is correct in many respects. In particular, the essential difference between bitopic dissociation, in which the T_1 state is directly dissociative while the S_1 state is bound, and tritopic dissociation, in which S_1 and T_1 are both dissociative, is reproduced correctly.

A third example of a tritopic process, more directly interesting from the viewpoint of the organic photochemist, is the dissociation of a C—C single bond attached to a carbonyl group, known as Norrish I cleavage.

C—N bond dissociation

A correlation diagram. Figure 7.23 shows the VB structure correlation diagram for the dissociation of CH_3NH_2 in which the amino group is kept planar and a

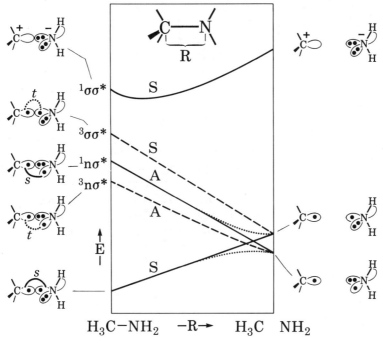

Figure 7.23 Dissociation of the C—N bond in methylamine in C_s (solid lines) and C_1 (dotted lines) symmetry. A VB structure and state correlation diagram.

plane of symmetry is present; dotted lines are used to indicate the changes that occur when the symmetry is lowered.

The dot–dot structures of the ground state and the $^3\sigma\sigma^*$ state of the starting molecule correlate with the singlet and triplet states of the radical pair $CH_3\cdot$ + $\cdot NH_2$, respectively. Both wave functions are symmetric (S) with respect to the CNH_2 plane, however, and the correlation therefore is with radical pair states in which the NH_2 radical is excited into its n → p state.

The radical pair states in which $\cdot NH_2$ is in its ground state are antisymmetric (A) with respect to the plane of symmetry and correlate with the singlet and triplet $n\sigma^*$ states of the starting molecule.

When symmetry is lowered, the A and S wave functions interact and the dotted curves result, providing a smooth endothermic dissociation of the ground state and giving both the S_1 and T_1 excited states a directly dissociative shape. Clearly, the $^1\sigma\sigma^*$ state of the starting molecule correlates with an ion-pair state of the product, as it did in the bitopic case, but the presence of low-energy $n\sigma^*$ states on the left-hand side of Figure 7.23, and, above all, the presence of a low-lying excited singlet state in the radical-pair product, provide a purely dissociative course for S_1, which is absent in the bitopic case. This similarity in the S_1 and T_1 surfaces is the main characteristic that distinguishes processes of higher topicity from bitopic ones. In the former, singlet and triplet excitations should both be expected to induce bond-breaking; in the latter, triplet excitation should be far more effective.

Rydberg states. In the real CH_3NH_2 molecule at its equilibrium geometry, the valence shell $n\sigma^*$ configurations are high in energy and Rydberg states represent the lowest excited states. Since the latter are not shown in the correlation diagram in Figure 7.23, its value must be questioned.

Indeed, the representation of vertical excited states of CH_3NH_2 on the left-hand side of Figure 7.23 is not correct. However, the energy of the Rydberg configurations does not change much along the reaction coordinate, since this energy is largely dictated by the chemical nature of the atoms involved; these configurations will have essentially no effect on the low-energy states on the right-hand side of Figure 7.23. The conclusions concerning the dissociative nature of S_1 and T_1 will therefore remain unchanged, but the detailed shapes of the potential energy surfaces near the initial CH_3NH_2 geometry will be affected by crossings and avoided crossings with the Rydberg configurations and cannot be obtained from the correlation diagram of Figure 7.23. The process of gradual transformation of the vertical singlet and triplet excited states from Rydberg to valence character is sometimes referred to as "de-Rydbergization." It needs to be understood in detail if the photochemical dynamics of gas-phase processes induced by vacuum UV light are to be unraveled, but it has little consequence for routine organic photochemistry in solution. In the particular case of aliphatic amines, which fluoresce strongly in solution, the presence of the Rydberg states clearly causes a presence of local minima in the S_1 surface near the ground-state equilibrium geometry.

C—O bond dissociation

Correlation diagrams. The VB structure correlation diagram for the dissociation of the C—O bond in methanol is shown in Figure 7.24A. It differs from the correlation diagrams for methylamine (Figure 7.23) in only one important respect: Excitations from the second lone pair on the oxygen atom, which has considerable 2s character and lies fairly low in energy (n_s), into the σ^* orbital introduce a pair of $^3n_s\pi^*$ and $^1n_s\pi^*$ configurations in addition to the already familiar $^3n_p\pi^*$ and $^1n_p\pi^*$ configurations. In the dissociation limit, all four correlate with radical-pair states $CH_3\cdot$ + $\cdot OH$, as do the ground and $^3\sigma\sigma^*$ configurations. The combination of the $CH_3\cdot$ radical with the ground state of the $\cdot OH$ radical gives rise to two pairs of singlet and triplets, since $\cdot OH$ has a doubly degenerate ground state of $^2\Pi$ symmetry; the unpaired electron is found in either of the 2p oxygen orbitals perpendicular to the OH bond, and a lone pair is present in the other. The combination of the $CH_3\cdot$ radical with the np excited state of the $\cdot OH$ radical yields only one singlet–triplet pair, since this state is not degenerate ($^2\Sigma^+$); the unpaired electron is found in the hybrid orbital parallel to the OH bond, and two lone pairs are present in the 2p orbitals perpendicular to the bond.

Which VB structure of CH_3OH correlates with which VB structure of the radical pair depends on the path chosen for the dissociation. In the path chosen in Figure 7.24A, the departing OH radical gradually aligns so as to make C—O—H collinear. Then, a crossing results in the VB structure correlation diagram, but it is avoided

---

Figure 7.24 Dissociation of the C—O bond in methanol. A VB structure (dotted lines) and state (thick lines) correlation diagram (A). *Ab initio* large-scale CI state energies without (B) and with (C) Rydberg orbitals in the AO basis. Part C is reproduced by permission from R. J. Buenker, G. Olbrich, H.-P. Schuchmann, B. L. Schürmann, and C. von Sonntag, *J. Am. Chem. Soc.* **106**, 4362 (1984).

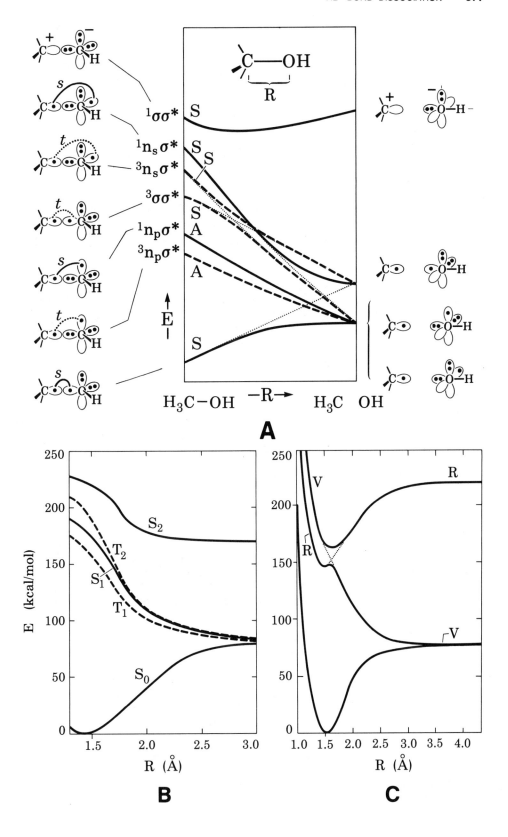

as indicated in the figure when resonance among the structures is introduced and a state correlation diagram thus produced. If the departing OH radical is instead allowed to align so as to make C—O—H a right angle, there is no crossing at the VB structure correlation level. Obviously, once the two fragments are sufficiently separated, their relative orientation has no effect on the energy of the systems, and the final state correlation is independent of the path chosen, even if the intermediate VB structure correlation diagram is not.

The number of radical-pair states in the dissociation limit in Figure 7.24A agrees with Table 7.1. When the 2s orbital is not allowed to participate, there are two low-energy radical-pair singlets and two triplets corresponding to the $CH_3 \cdot + \cdot OH$ ($^2\Pi$) combination. The third dot-dot state pair involves charge separation ($CH_3^- + OH^+$). If the 2s orbital is allowed to participate, there is a third singlet and a third triplet with no charge separation, resulting from the $CH_3 \cdot + \cdot OH$ ($^2\Sigma^+$) combination. This state is considerably higher in energy, and at times it may not be necessary to consider it, in which case the dissociation of the C—O bond can be viewed as a tritopic process. Of course, in a small saturated molecule such as CH_3OH, vertical excited states are even much higher in energy, and for most purposes one will wish to include the $CH_3 \cdot + \cdot OH$ ($^2\Sigma^+$) radical-pair state and to treat the C—O bond dissociation as a tetratopic process. Either way, it is clear that the correlation diagram differs dramatically from that for a bitopic dissociation in that both S_1 and T_1 are dissociative.

Figure 7.24B shows the potential energy curves for the first two excited singlets and triplets calculated without the inclusion of Rydberg orbitals for a CH_3OH dissociative path in which the COH angle is kept constant at its equilibrium value in methanol. The similarity to the qualitative correlation diagram of Figure 7.24A is striking.

Rydberg states. As already discussed in the case of the C—N bond dissociation, the introduction of Rydberg configurations is essential for the understanding of the potential energy curves in the region of vertical geometries but is expected to have little effect for the lower states in the dissociation limit. This is illustrated in Figure 7.24C, which shows the results of calculations for the three lowest singlet states for the C—O dissociation of methanol with inclusion of Rydberg orbitals.

At geometries close to ground-state equilibrium, the lowest excited state is of Rydberg (R) nature. Along the dissociative path, its energy increases and roughly follows that of the ground state. As a result, the steeply descending curve of the valence excited state (V) undergoes an avoided crossing with the Rydberg curve ("de-Rydbergization").

C—C bond dissociation: Norrish I. The photochemical cleavage of a single bond in position α to a carbonyl group has been known for a long time as the Norrish reaction of type I. A VB structure and state correlation diagram for this process is shown in Figure 7.25A for the C—C bond cleavage in acetaldehyde to yield CH_3 and CHO. In the dissociation limit, a linear geometry is assumed for the CHO radical. Since the ground state of linear CHO is the doubly degenerate $^2\Pi$ state, two singlet and two triplet radical pairs with CH_3 are possible, as expected for a tritopic dissociation process. The upper states in the dissociation limit correspond to the $^1\pi\pi^*$ and $^3\pi\pi^*$ locally excited states of the carbonyl function. At the acetaldehyde geometry, the usual collection of $n\pi^*$ and $\pi\pi^*$ singlets and triplets is

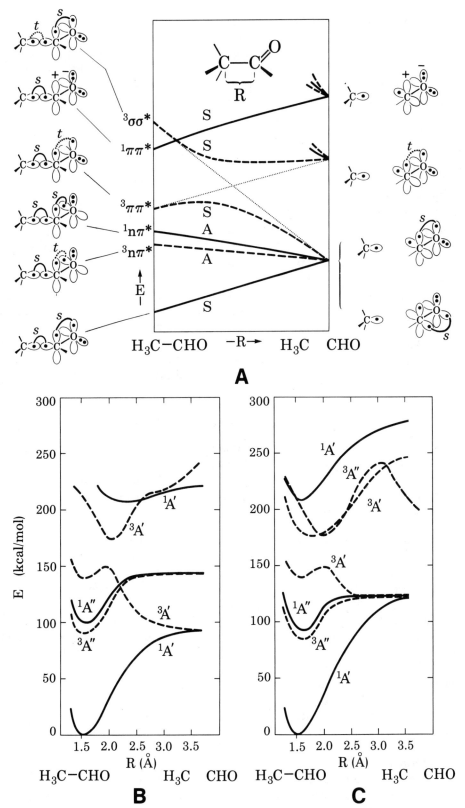

Figure 7.25 Norrish I C—C bond cleavage. A VB structure and state correlation diagram (A), and *ab initio* large-scale CI calculations of state energies, in which only the C—C distance R is varied, and the formyl group remains bent (B), or the formyl group is gradually linearized as R increases (C).

shown, followed by the $^3\sigma\sigma^*$ triplet of the C—C bond, the characteristic configuration for this dissociation reaction path. The symmetry labels in Figure 7.25 indicate states that are symmetric (S,A') or antisymmetric (A,A") with respect to reflection in the CCHO symmetry plane. Upon going from VB structures to states, the crossing of the lines representing the energies of the $^3\pi\pi^*$ and $^3\sigma\sigma^*$ structures is avoided, so that in the end it is the $^3\pi\pi^*$ state rather than the much higher $^3\sigma\sigma^*$ state that correlates with the ground state of products. A memory of the crossing is likely to remain and result in a barrier in the $^3\pi\pi^*$ curve, separating the starting molecule from the product radical pair.

The $n\pi^*$ state curves can also be expected to have barriers, for a different reason. In the initial stages of the dissociation, the HCO moiety is surely severely bent, as it is in the starting material. If it remained bent as the dissociation proceeds, the curves for the A symmetry $^1n\pi^*$ and $^3n\pi^*$ states would reach a higher energy limit than those for the S symmetry ground and $^3\pi\pi^*$ states. The energy difference would be just the $n\pi^*$ excitation energy of bent CHO radical, which is zero only at the linear geometry and is calculated to be fairly substantial at bent geometries (~50 kcal/mol at 120°). It is this higher limit that is aimed for at the initial stages of the dissociation process, so that a slight barrier can be anticipated.

The results of calculations (Figure 7.25B,C) are in qualitative agreement with the conclusions drawn from the correlation diagram (Figure 7.25A). In Figure 7.25B, the dissociation proceeds to a bent HCO radical; in Figure 7.25C, it proceeds to a linear HCO radical.

7.2.2 Tetratopic. Alkyl Halides, Peroxides, Azo Compounds

An example of tetratopic bond dissociation has in effect already been discussed in Section 7.2.1, namely, the C—O bond cleavage in methanol with inclusion of the oxygen 2s AO. Three additional examples will be briefly mentioned here: the cleavage of a C—Br bond, the cleavage of an O—O bond, and α-cleavage in an azo compound.

Carbon–halogen bond dissociation. The dissociation of the C—Br bond in methyl bromide involves a rupture of a bond, one of whose terminal atoms carries two lone pairs, not counting the low-energy 4s electron pair, and therefore corresponds to a tetratopic process. If excitations from the 4s electron pair are included, the process will have to be considered pentatopic.

Now, both the singlet and the triplet radical pairs obtained in the dissociation limit exhibit a triple degeneracy in the first approximation, since the unpaired electron on the bromine atom can be in any one of the three 4p orbitals. This is apparent in Figure 7.26, which shows the result of an *ab initio* large-scale CI calculation using the pseudopotential approximation for bromine inner-shell electrons. As the $CH_3\cdot$ and $Br\cdot$ radicals approach, the 4p orbital that is directed along the C—Br line becomes distinct from the other two; thus a nondegenerate singlet ground state (^1A) and a doubly degenerate $^1n\sigma^*$ state (^1E), as well as their triplet analogs (^3A, ^3E), result. Clearly, both S_1 and T_1 are again dissociative. Note that the description of the excited electronic states at the short C—Br distances is inadequate because Rydberg configurations have been ignored. As discussed previously, this has little effect on the conclusions of interest here, but it might introduce local minima in some of the excited states and thus affect dissociation quantum yields.

In a more realistic approximation, spin–orbit coupling must be considered because Br is a relatively heavy atom. Then, the triple degeneracy of the Br ground state is lifted and $^2P_{3/2}$ and $^2P_{1/2}$ states results, each of which can now combine with the ^2A state of the methyl radical. The insert in Figure 7.26 shows that the ground state, all three components of the $^3\sigma\sigma^*$ state, and four of the six components of the $^3n\sigma^*$ state correlate with the lower dissociation limit $^2P_{3/2}$.

Therefore, excitation into S_1 or T would be expected to cause dissociation of the carbon–halogen bond, but motion on the S_1 potential energy curve produces an excited halogen atom. This is the basis of operation of the iodine atom laser.

Intersystem crossing in the product radical pair is facile because of the presence of the lone pairs on the halogen atom, and, in general, similar mixtures of radical-pair- and ion-pair-derived products result from direct and from triplet-sensitized excitation of benzyl halides in polar solvents. This contrasts with the situation in benzylammonium salts (Section 6.3.1).

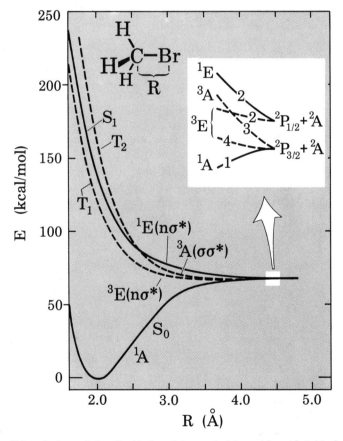

Figure 7.26 Dissociation of the C—Br bond in methyl bromide. *Ab initio* large-scale CI state energies (pseudopotential for Br).

O—O Bond dissociation. The dissociation of the O—O bond in methyl peroxide represents a case in which each of the singly bonded atoms carries a lone pair, and it again corresponds to a tetratopic process. If we wished to consider excitations for the 2s lone pairs or the oxygens as well, the process would have to be viewed as hexatopic, but we shall not do so. It is useful to combine the 2p lone pairs on the two oxygen atoms into the in-phase combination n_+ and the higher-energy out-of-phase combination n_-. Once again, Rydberg configurations will be ignored, with a disastrous effect on the description of vertical excited states but no effect on the conclusions of primary interest here.

Both dissociation products, $CH_3O\cdot$ and $HO\cdot$, have doubly degenerate ground states, ^2E and $^2\Pi$, respectively, so that they can be combined into four singlet and four triplet radical pairs. The symmetries and energies of the ground, $^3\sigma\sigma^*$, $^{1,3}n_+\sigma^*$, and $^{1,3}n_-\sigma^*$ states that originate in this dissociation limit depend on the mutual orientation of the two fragments. In Figure 7.27, the resulting potential energy curves are shown for the simple case in which the C—O—O—H moiety is coplanar, so that electronic states can be classified as symmetric (A') or antisymmetric (A'') with respect to reflection in this plane. Furthermore, it is simplest to assume that the C—O—O and O—O—H valence angles are both 90°, so that the oxygen 2s orbitals are not included in primary bonding interactions.

It is then seen that the dissociation limit generates a ground-state singlet, a strongly antibonding $^3\sigma\sigma^*$ triplet, and similarly strongly antibonding doubly excited $^{1,3}n_+n_- \to \sigma^*$, σ^* singlets and triplets of A' symmetry, and $^{1,3}n_+ \to \sigma^*$ and $^{1,3}n_- \to \sigma^*$ state pairs of A'' symmetry. Obviously, both S_1 and T_1 are again purely dissociative states.

When spin–orbit coupling is introduced, the very high degeneracy at the dissociation limit is again partially lifted, since the $J = 1/2$ and $J = 3/2$ states of both $CH_3O\cdot$ and $\cdot OH$ are no longer degenerate. A detailed description of these effects is of little interest in the present context because they do not affect the basic conclusions from Figure 7.27: Both singlet and triplet excitation of the peroxide linkage are likely to produce radical pairs with very high efficiency. Since the COOH dihedral angle in the ground-state equilibrium geometry of acyclic peroxides is not zero but close to 90°, and since Rydberg configurations have not been considered in Figure 7.27, it is not useful for the discussion of vertical excited states and of the possible presence of shallow minima in their potential energy curves whose existence might reduce the dissociation quantum yields below unity.

α-Cleavage in azo compounds. The dissociation of one or both α bonds in an azo compound has been investigated on the model case of *cis*-diazene using a minimum basis set. Single-bond dissociation was found to be much favored and shall now be discussed briefly.

In contrast to the other tetratopic processes described in this section, α-cleavage in the lowest singlet and triplet excited states of an azo compound is not purely dissociative, and there are local minima in the S_1 and T_1 surfaces at geometries close to that of the ground state. This is suggested from the correlation diagram for the N—H bond scission in diazene, whose lowest excited states, $^3n_- \to \pi^*$ and $^1n_- \to \pi^*$, correlate with the triplet and singlet states of the radical pair $H\cdot$ +

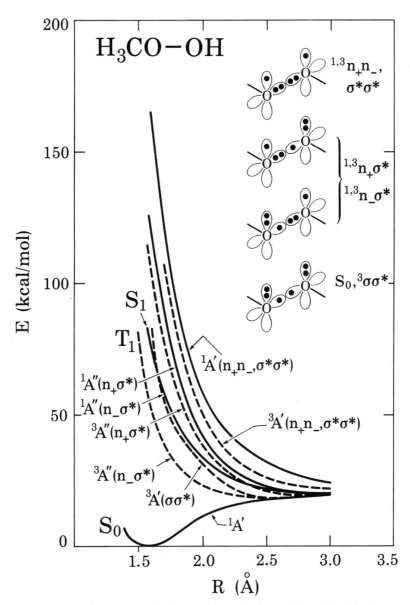

Figure 7.27 Dissociation of the O—O bond in methyl peroxide. *Ab initio* large-scale CI state energies.

·NNH*, where the asterisk indicates excitation of the unpaired electron into the π^* orbital (**1**). The ground state of the ·NNH radical contains only two π electrons, and the unpaired electron is in an orbital of σ symmetry (**2**). The singlet combination H· + ·NNH correlates with the ground state and the triplet combination H· + ·NNH with the $^3\sigma \to \sigma^*$ excited state ($^3n_- \to \sigma_+^*$) of diazene. The published calculations are in agreement with expectations and suggest that α-cleavage in the $^3n_- \to \pi^*$ and $^1n_- \to \pi^*$ states of diazene requires activation energy.

384 REACTIONS WITH MORE THAN ONE ACTIVE ORBITAL PER ATOM

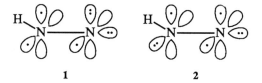

7.2.3 Pentatopic. Vinyl Halides, Ketenes, Diazoalkanes

In this section we shall consider three examples of pentatopic bond dissociation processes: C—Br bond cleavage in vinyl bromide, C=C bond cleavage in ketene, and C—N bond cleavage in diazomethane.

Vinylic carbon–halogen bond dissociation. Figure 7.28A shows the state correlation diagram and Figure 7.28B shows the computed potential energy curves

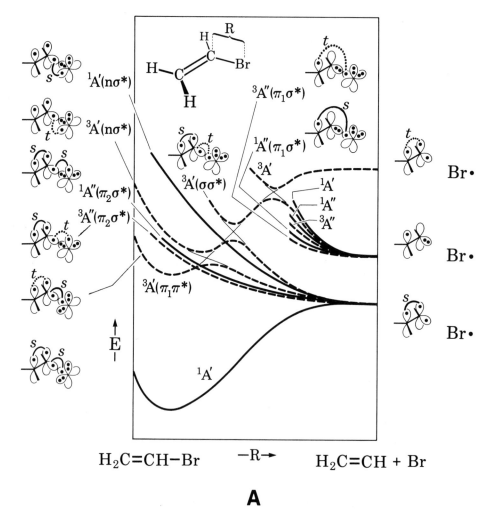

Figure 7.28 Dissociation of the C—Br bond in vinyl bromide. (A) A VB structure and state correlation diagram. (B) *Ab initio* large-scale CI state energies.

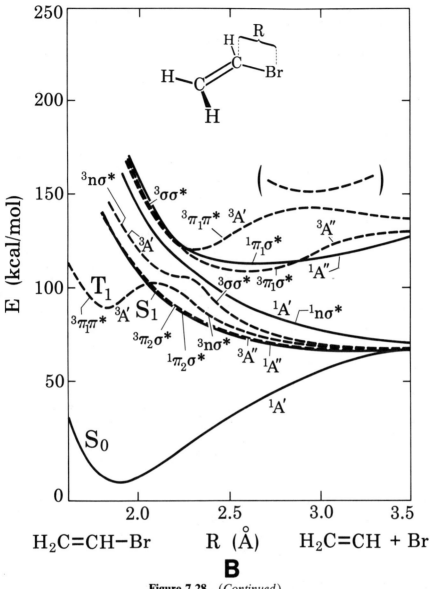

Figure 7.28 (*Continued*)

for the dissociation of the C—Br bond in vinyl bromide, assuming that all valence angles remain constant and that a plane of symmetry is kept throughout. The states are either symmetric (A′) or antisymmetric (A″) relative to mirroring in this plane. Rydberg configurations were not included. This probably does not affect the low-lying $^3\pi\pi^*$ triplet of vinyl bromide but would affect higher-energy states, which are not shown. Only the correlation limits of the "characteristic" $^{1,3}n\sigma^*$ configurations for the bond-breaking process are shown schematically.

These dissociation limits are shown without regard for spin–orbit coupling. The lowest one corresponds to the combination of a ground-state vinyl radical and a bromine atom. The energy for this limit is somewhat overestimated because the vinyl radical has not been allowed to relax to its equilibrium geometry (linear C=C—H), but this is of no consequence at present. The states ending in this limit are just those ending in the lowest dissociation limit for the tetratopic CH_3—Br bond cleavage in Figure 7.26, except that there the T_1 and S_1 states were degenerate ($^{1,3}E$) and presently they are split because of the lower symmetry. Introduction of spin–orbit coupling would again convert this dissociation limit into two, corresponding to the $^2P_{1/2}$ and $^2P_{3/2}$ states of the bromine atom (cf. Figure 7.26).

The pentatopic nature of the C—Br dissociation process in vinyl bromide is due to the availability of the $2p_z$ orbital on the substituted carbon atom. This greatly increases the number of low-energy states in Figure 7.28 relative to Figure 7.26. First, a new low-energy dissociation limit is available in which a $\pi \rightarrow sp^2$ excited state of the vinyl radical is combined with a Br atom, and three pairs of singlet and triplet states originate in this limit.

Second, and more important, a locally excited $^3\pi_1\pi^*$ configuration of the double bond has been introduced at low energy (π_1 is localized primarily on the C=C double bond, and π_2 is localized primarily on the p orbital of the bromine atom). Its course parallels that of the ground state as the C—Br bond is stretched. The crossing between the characteristic triplet configuration, $^3n\sigma^*$, and the locally excited $^3\pi_1\pi^*$ configuration will be weakly avoided (their interaction element will be small because the n, σ^*, π_1, and π^* orbitals are all distinct). This ought to result in a local triplet minimum near the vertical geometry separated by a barrier from the dissociation limit. A lowering of the symmetry will cause the $n\sigma^*$–$\pi_1\pi^*$ crossing to be more strongly avoided, so that the favored escape path from the locally excited $^3\pi_1\pi^*$ minimum can be expected to follow a path through out-of-plane bent geometries.

A locally excited $^1\pi_1\pi^*$ configuration is also present, but it lies much higher in energy and is not shown in Figure 7.28. The $^1n\sigma^*$ surface is likely to cut it much earlier, or perhaps to lie below it altogether, so that a prediction of a locally excited-state minimum at vertical geometries cannot be made for the singlet state. Besides, the energies involved in the singlet excitation are so high that Rydberg configurations cannot be ignored.

Our expectation, then, is that S_1 excitation has the potential for effectively cleaving the C—Br bond whereas T_1 excitation may require some thermal activation before the bond ruptures, with the difference in the expected behavior being ultimately due to the much higher $^1\pi_1\pi^*$ excitation energy relative to $^3\pi_1\pi^*$.

Ketene photofragmentation. The photodissociation of ketenes is an important source of carbenes. We shall consider the photodissociation of ketene itself as a prototype process of this kind. It belongs into the category of two-center double-bond dissociations and is pentatopic in nature.

Two of the relevant orbitals are the nonbonding orbitals of CH_2 and three are located on the carbon monoxide fragment, one being the lone pair on the carbon atom and two being the π^* orbitals of the triple bond.

The MO energies of the starting material and products are shown schematically in Figure 7.29. They are labeled by point-group symbols corresponding to the C_{2v} point-group symmetry of ketene itself. The orbitals relevant for the reaction are the three π orbitals of ketene, analogous to the three π orbitals of allyl, of b_1 symmetry, the totally symmetric orbitals of a_1 symmetry, and the "in-plane π'

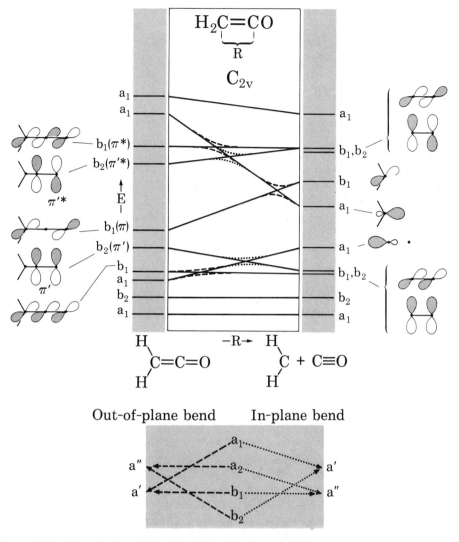

Figure 7.29 Dissociation of the C=C bond in ketene. MO correlation diagram for C_{2v} symmetry. See text.

orbitals" or "antisymmetric σ orbitals," analogous to the orbitals of ethylene, of b_2 symmetry.

Linear path. First, we shall consider a molecular correlation diagram appropriate for a linear dissociation path in which the C_{2v} symmetry is kept throughout. Connecting orbitals of like symmetries produces the picture shown in Figure 7.29. It is seen that the HOMO of ketene becomes the LUMO of the products (namely, the p orbital on the carbon in CH_2), whereas the HOMO of the products (namely, the approximately sp^2-hybridized orbital on CH_2) correlates with one of the higher-lying unoccupied orbitals of ketene and not with its LUMO or even its second

388   REACTIONS WITH MORE THAN ONE ACTIVE ORBITAL PER ATOM

LUMO. The orbital crossing is thus of the abnormal type discussed in Chapter 3, and the presence of barriers in the configuration and state correlation diagrams is immediately expected. The configuration correlation diagram for the C_{2v} symmetry path is shown in Figure 7.30 by thin lines, and the expected state correlation diagram resulting after the crossing of configurations of like symmetry is avoided is shown there by solid lines. The lowest-energy excited triplet and singlet states correspond to a mixture of configurations. In the dominant configuration, an electron is promoted from the HOMO, which is a b_1 π orbital, similar to the nonbonding orbital of allyl, to the LUMO, which is an in-plane oriented b_2 π'^* orbital. The next higher pair of singlet and triplet states is of A_1 symmetry, and each is again represented by a mixture of configurations. One of them corresponds to a $\pi \rightarrow \pi^*$ excitation, and the other corresponds to a $\pi' \rightarrow \pi'^*$ promotion.

On the product side, the lowest-energy excitations are localized on the carbene moiety. In the triplet manifold, the only low-lying state corresponds to the $^3B_1$ triplet ground state of CH_2 with one electron in each of the nonbonding orbitals on the carbon atom. There are two low-lying singlet states. One of them is the lowest $^1A_1$ singlet of carbene with both nonbonding electrons in its hybridized

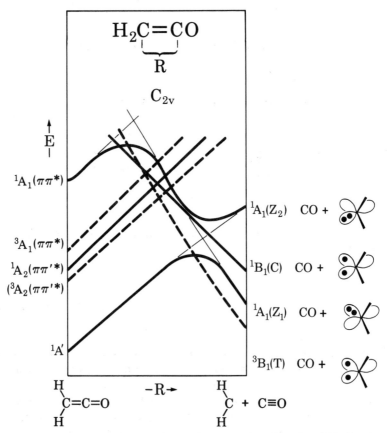

Figure 7.30 Dissociation of the C=C bond in ketene. Configuration (thin lines) and state (thick lines) correlation diagram for C_{2v} symmetry.

orbital, and the other is the $^1B_1$ singlet excited carbene state with one electron in each of the nonbonding orbitals. The next higher singlet is the $^1A_1$ doubly excited state of carbene, with the 2p orbital doubly occupied.

The correlation displays the pattern expected for a ground-state correlation-forbidden path with an abnormal orbital crossing. Not only the S_0 state but also the several lowest excited states, both singlet and triplet, contain large barriers. Thus, this reaction is forbidden not only in a ground state but also in the lowest singlet and triplet excited states.

In-plane bend. Clearly, a deviation from the C_{2v} symmetry is required if the reaction is at all to proceed easily, whether photochemically or thermally. The obvious way of searching for a path in which the correlation-imposed barriers are lowered is to permit the unavoided crossings of Figure 7.30 to become avoided by symmetry lowering using a suitable perturbation. An in-plane bend of the CCO moiety will remove one of the two planes of symmetry of the C_{2v} group so that both the A_2 and B_1 states of this group will become A'' states of the C_s group. This will permit the crossing between the $^{3,1}A_2$ and $^{3,1}B_1$ states to become avoided, lowering the barrier in the lower surfaces, S_1 and T_1. Upon this distortion, the A_1 and B_2 representations both become A'. The resulting effect on the MO and configuration correlation diagram is shown in Figures 7.29 and 7.31. At the MO correlation level the reaction remains formally ground-state-forbidden because the crossing of the ascending bonding with the descending antibonding orbital has not been removed (Figure 7.29). At the configuration and state levels, a lower barrier is expected for reaction from the S_1 and T_1 states (Figure 7.31).

Out-of-plane bend. An alternative symmetry distortion is that which will cause the crossing to be avoided at the MO level. This is an out-of-plane distortion of the carbonyl group which will cause the representations A_1 and B_1 of the C_{2v} group to become A' of the resulting C_s group while A_2 and B_2 become A''. Then the MO correlation diagram formally becomes of the symmetry-allowed type in that the HOMO of the ketene can be connected directly with the HOMO of the products. However, a memory of the crossing is likely to persist even for relatively large angles of bending out-of-plane, and this is indicated in the MO diagram shown in Figure 7.29. The lowest excited state of A_1 symmetry is now free to mix with the B_1 state, which descends steeply towards the products side. The correlation-imposed barrier in the A_1 state can be expected to be reduced, as indicated in the configuration and state correlation diagrams shown in Figure 7.32.

Double bend. Removal of the remaining symmetry by twisting of the CH_2 group around the C—C bond will promote further mixing of configurations that are of A' and A'' symmetry in the C_s group and is likely to lead to a further decrease of the barriers in the excited surfaces. This analysis of the fragmentation of ketene thus provides a nice illustration of the tendency of excited-state molecules to follow reaction paths of low or no symmetry.

Singlet and triplet reaction paths. No calculated surfaces of adequate quality are available at present for display. It is clear, however, that the S_1 state of ketene is not conducive to fragmentation, since it correlates with a product state in which

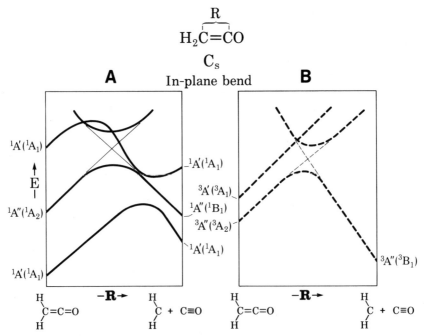

Figure 7.31 Dissociation of the C=C bond in ketene. State correlation diagram for C_{2v} (thin lines) and in-plane bent C_s (thick lines) symmetry. (A) Singlets. (B) Triplets.

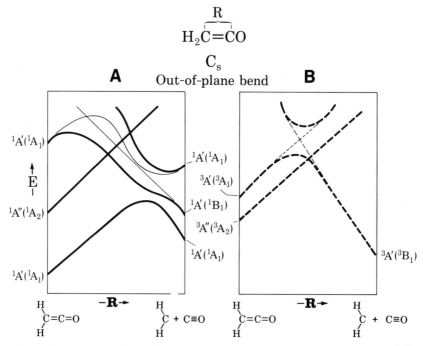

Figure 7.32 Dissociation of the C=C bond in ketene. State correlation diagram for C_{2v} (thin lines) and out-of-plane bent C_s (thick lines) symmetry. (A) Singlets. (B) Triplets. Adapted by permission from S. Yamabe and K. Morokuma, *J. Am. Chem. Soc.* **100,** 7551 (1978).

CH$_2$ is in its lowest excited state, so that the process would be very strongly endothermic. The known singlet photodissociation of ketene from the low-lying vibrational levels of its singlet $^1A_2$ or singlet $^1A_1$ excited state undoubtedly proceeds by internal conversion to the vibrationally excited $^1A_1$ ground state followed by a hot reaction of the ground state.

The dissociation of triplet ketene is much less endothermic. At the minimum basis set level, barriers have been calculated for both the $^3A_2$ and $^3A_1$ states, which are nearly degenerate in ketene itself, at least along the symmetric paths investigated so far. As we have indicated, it is likely that further loss of symmetry will reduce these barriers and possibly remove them altogether. As a result, triplet ketene does have a reasonable chance to dissociate to triplet products during its lifetime at room temperature or if it contains excess vibrational energy.

Diazomethane photofragmentation. Diazoalkanes are another important photochemical as well as thermal source of carbenes. They are isoelectronic with ketenes, whose photodissociation we have just discussed in considerable detail, and we can therefore outline the nature of the pentatopic dissociation of the C=N bond of diazomethane in much less detail.

The linear dissociation path possesses C_{2v} symmetry throughout as it produces the N$_2$ and CH$_2$ fragments. The MO correlation diagram is identical to that for ketene (Figure 7.29) and is characterized by an abnormal MO crossing. The configuration and state correlation diagram is also the same as for ketene (Figure 7.30), except for minor differences in state energies, and shows the anticipated barriers in the S_0, S_1, and T_1 states. The reaction is thus "forbidden" in all three states; however, it is now strongly exothermic in the T_1 state and roughly thermoneutral in the S_1 state, whereas the dissociation of ketene was very roughly thermoneutral in the T_1 state but strongly endothermic in the S_1 state.

Figure 7.33A shows the results of an *ab initio* large-scale CI calculation for the C_{2v} dissociation of diazomethane. The similarity to the correlation diagram of Figure 7.30 is striking, and the size of the correlation-imposed barriers is seen to decrease in the order S_0, S_1, and T_1. A feature that was not apparent in the correlation diagram is a shallow minimum in the T_1 curve, corresponding to a weakly bound complex between triplet carbene and an N$_2$ molecule.

As in ketene, in-plane bending will lower the barriers in S_1 and T_1 by causing the A$_2$–B$_1$ crossing to be avoided, since both states will be of A" symmetry in the C_s group. This avoidance of a conical intersection is shown in Figure 7.33B. Out-of-plane bending removes the MO crossing (Figure 7.29) but not the A$_2$–B$_1$ state crossings, and its effect on the state energies is shown in Figure 7.33C. Finally, Figure 7.33D shows the state energies along a path of no symmetry, in which both in-plane and out-of-plane bending are introduced. A path of this kind is most likely to represent the low-energy escape from the local minimum in the S_1 or T_1 state towards the dissociated products.

Large in-plane bending angles connect the geometry of diazomethane with that of diazirine. Their photochemical interconversion will be discussed in Section 7.5.3.

7.2.4 Hexatopic. Azides, Bromine

We have selected two examples of hexatopic bond dissociation processes: expulsion of a nitrogen molecule from an azide and dissociation of a halogen molecule.

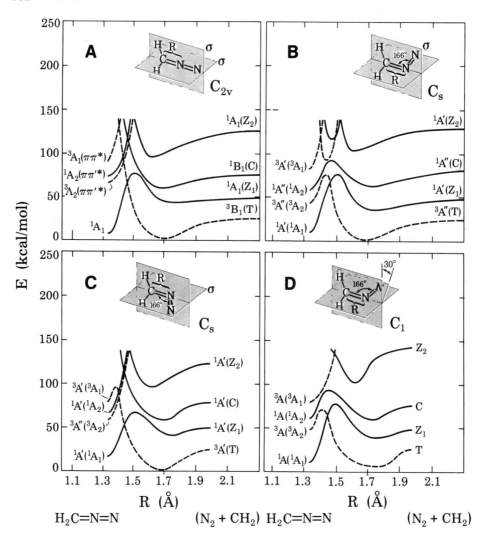

Figure 7.33 Dissociation of the C=N bond in diazomethane. *Ab initio* large-scale CI state energies. Singlets (solid lines) and triplets (dashed lines). (A) C_{2v} symmetry. (B) C_s symmetry (in-plane bent). (C) C_s symmetry (out-of-plane bent). (D) C_1 symmetry (in-plane and out-of-plane bent).

Hydrazoic acid photofragmentation. The fragmentation of azides to molecular nitrogen and a nitrene or its rearrangement product is another important organic photochemical process. We shall model it here by the dissociation of hydrazoic acid to HN and N_2. The process is related to the photodissociation of ketene into CH_2 and CO and of diazomethane to CH_2 and N_2 except that now symmetry is lower because one of the CH bonds has, in effect, been replaced by the nitrogen lone pair. Therefore the correlation diagrams for HN_3 will have the appearance of those for the in-plane bent dissociation path of ketene or diazomethane which also had no more than C_s symmetry, as long as we assume the simplest dissociation path for HN$_3$ in which the symmetry plane is kept throughout. All MO and state crossings will again be avoided if this assumption is removed.

The MO correlation diagram is shown in Figure 7.34, and its great similarity to the MO correlation diagram for an in-plane C$_s$ ketene (or diazomethane) dissociation, shown in Figure 7.29 by dotted lines, is readily apparent. The most obvious difference is due to the higher symmetry of the NH product compared with CH$_2$. In CH$_2$, one of the nonbonding orbitals was represented approximately by an sp^2 hybrid and thus was lower in energy than the other, which was a p orbital. In nitrene, however, both of these orbitals are 2p atomic orbitals on nitrogen and are degenerate at infinite separation of the two dissociation products. The allowed a$_1$–b$_1$ MO crossing for the C$_{2v}$ symmetry path in Figure 7.29 is now replaced by a situation in which the HOMO and the LUMO of the starting materials have simply both become nonbonding and degenerate on the product side. Either way, the reaction is of the ground-state forbidden kind, since a biradical-like geometry is reached either close to the product geometry (ketene, diazomethane) or at the product geometry (hydrazoic acid).

The construction of the configuration and state correlation diagrams (Figure 7.35) is facilitated by the similarity to the case of in-plane C$_s$ dissociation of ketene or diazomethane (Figure 7.31). Particularly the barriers in the S$_1$ (^{1}A″) and T$_1$ (^{3}A″) state curves are analogous to those that were a memory of the A$_2$–B$_1$ crossing, unavoided in C$_{2v}$ symmetry, in the C$_s$ dissociation of in-plane bent ketene or dia-

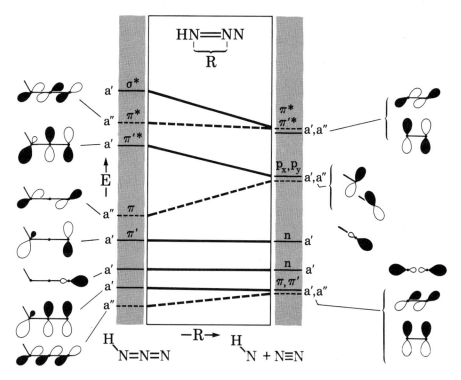

Figure 7.34 Dissociation of the HN=NN bond in hydrazoic acid. MO correlation diagram for C$_s$ symmetry.

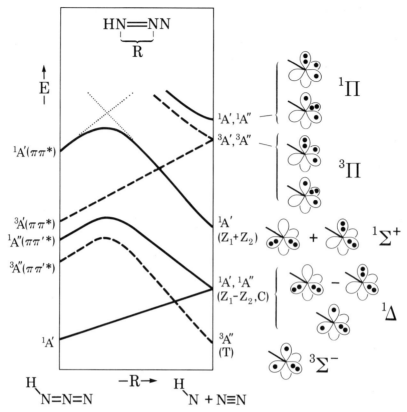

Figure 7.35 Dissociation of the HN=NN bond in hydrazoic acid. Configuration (dotted lines) and state (thick lines) correlation diagram for singlets (solid lines) and triplets (dashed lines). C_s symmetry.

zomethane. There are two main differences between Figure 7.35 and Figure 7.30. First, the decomposition of hydrazoic acid is more favorable thermodynamically. This means that all of the states on the right-hand side of the diagram are pulled down in energy. The difference is readily traced to the special stability of the N_2 molecule. This converts paths that went far uphill in the case of ketene into more nearly flat curves for the decomposition of hydrazoic acid. Second, the higher symmetry of the nitrene product relative to CH_2 introduces degeneracies on the right-hand side of the diagram. Figure 7.36 shows the result of an *ab initio* large-scale CI calculation for a one-dimensional cross-cut through the potential energy hypersurface pertinent for the dissociation of hydrazoic acid to nitrene and molecular nitrogen. The geometry for this calculation was optimized for hydrazoic acid at an STO-3G level, and the two fragments were then separated in a linear manner without changing any other geometric parameters such as their bond lengths. The results show all the qualitative features expected from the correlation diagram of Figure 7.35—in particular, the memory of the crossings that were not avoided in the isoelectronic ketene and diazomethane in-plane bend dissociation paths.

The barriers in both the S_1 and T_1 states are quite low relative to those en-

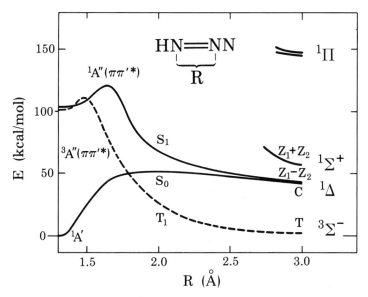

Figure 7.36 Dissociation of the HN=NN bond in hydrazoic acid. *Ab initio* large-scale CI state energies for singlets (solid lines) and triplets (dashed line). C_s symmetry.

countered for ketene. Similarly as was the case there, it is probable that they will be reduced further when symmetry is lowered by an out-of-plane bend that will permit the interaction between states of A' and A" symmetries. Thus, all indications are that both the S_1 and T_1 states will be photoactive with respect to the fragmentation, and this agrees well with known experimental facts.

Dissociation of molecular bromine. Perhaps the simplest example of a hexatopic two-center dissociation is the dissociation of halogen molecules to atoms. Strictly speaking, this is not organic photochemistry, but its consideration is helpful to the understanding of processes that are. The orbitals involved are the doubly occupied σ, π, and π' orbitals of the bromine molecule and their antibonding σ^*, π^*, and π'^* counterparts on the left-hand side of a correlation diagram, as well as the three p orbitals on each of the bromine atoms on the right-hand side of the diagram (Figure 7.37A). If excitations from the valence s orbitals are considered as well, the process is actually octatopic.

When it comes to configuration and state correlation diagrams, the bromine molecule provides an extreme example of a system that has a very high degeneracy at the lowest dissociation limit. It is possible to produce nine different covalently dissociated singlet states and the same number of covalently dissociated triplet states for the two ground-state bromine atoms. Thus, 18 curves altogether merge to a common limit for infinite internuclear separation of the two bromine atoms. This, of course, is only true if spin–orbit coupling is neglected. When it is recognized that the bromine atom can be either in its $^2P_{1/2}$ or $^2P_{3/2}$ state, three closely spaced levels are actually present at the dissociation limit. Of the total 36 substates of the system consisting of two bromine atoms, nine singlets, and 3×9 triplet components, 16 correlate with the lowest limit in which both bromine atoms are in the

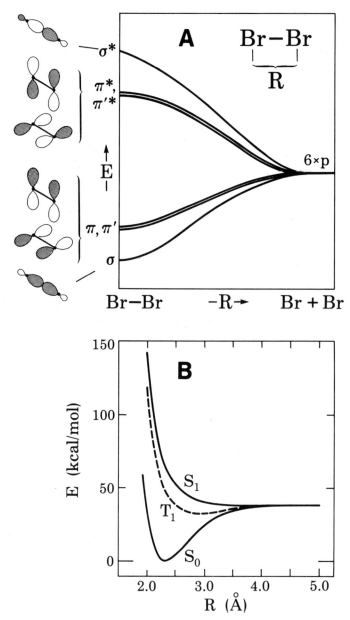

Figure 7.37 Dissociation of the Br—Br bond. (A) MO correlation diagram. (B) *Ab initio* large-scale CI state energies (pseudopotential) for the lowest two singlets (solid lines) and the lowest triplet (dashed line).

$^2P_{3/2}$ state, 16 with the next higher one in which one of the bromine atoms is in the $^2P_{1/2}$ state, and 4 with the highest one in which both bromine atoms are in the $^2P_{1/2}$ state. In the absence of magnetic field, of course, some of these substates are degenerate. Details of the course of the resulting multitude of dissociation curves are not of interest in the present context. It is worthwhile to note the extremes to which the increase in topicity leads when it is remembered that, in the bitopic case, there were only two covalent dissociated states, one singlet, and one triplet, and already the lowest excited singlet state had to be of zwitterionic (or Rydberg) nature.

A calculation of the potential energy curves for the S_0, S_1, and T_1 states of the bromine molecule are shown in Figure 7.37B. It is of the *ab initio* kind with large CI and a large basis set, but using a pseudopotential description of the inner shells and neglecting spin–orbit coupling. The very high degeneracy at the dissociation limit is not shown. Clearly, S_1 and T_1 are both dissociative.

A noteworthy feature of the potential energy diagram is the bound nature of the lowest triplet state, in which one of the π electrons is promoted into the σ^* orbital. In this state, the molecule has a three-electron σ bond and a three-electron π bond and can be described qualitatively as a loose complex of two bromine atoms. Not surprisingly, it is very highly reactive; for instance, it attacks readily organic multiple bonds. All other halogens and interhalogen compounds such as iodine chloride behave similarly.

7.3 CARBONYL ADDITION TO OLEFINS

The attack of an excited carbonyl compound on a multiple bond eventually leading to the formation of a four-membered ring is a very common mode of reaction for ketones and aldehydes. The most common example is the addition of a carbonyl compound to an olefin producing an oxetane, known as the Paterno–Büchi reaction.

Figure 7.38 shows the potential energy diagram for an addition of formaldehyde to ethylene, reproduced from an early simple *ab initio* calculation using a minimum basis set and 3 × 3 CI. The one-dimensional cross-section through the potential energy hypersurfaces shown corresponds to a nonconcerted attack by the lone-pair

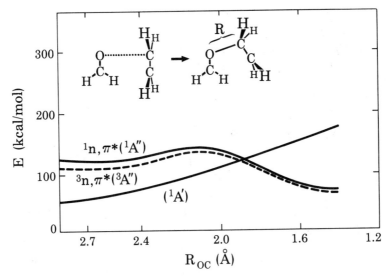

Figure 7.38 Carbonyl addition to olefin. Minimum basis set *ab initio* 3 × 3 CI calculation of state energies. Reproduced by permission from L. Salem, *J. Am. Chem. Soc.* **96**, 3486 (1974).

orbital of the oxygen atom on one of the termini of the double bond. This reaction path is believed to be typical of additions to electron-rich olefins. The geometry chosen is a particularly simple one preserving a plane of symmetry throughout.

The T_1 and S_1 states of the carbonyl compound correspond to the triplet and singlet $n\pi^*$ excitations. The states of the resulting biradical are those expected for a typical biradical with almost degenerate S_0 and T_1 states and a zwitterionic S_1 state lying considerably higher in energy. An activation barrier is present in both the S_1 and T_1 surfaces, but in the absence of geometry optimization, its height is difficult to discuss. Its physical origin is undoubtedly related to that of the barrier observed for an addition of a radical to an olefin.

The reaction on the triplet surface is expected to proceed in a straightforward manner to the biradical product at its lowest-energy geometry. The triplet biradical then undergoes rotations around single bonds and loses the initially present stereochemical information.

The S_0 and S_1 surfaces touch at a point near the middle of the reaction path. This touching will be avoided as soon as the path taken has less symmetry, as will be the case in most realistic situations. If the excited S_1 molecule overcomes the initial barrier, it is very likely to return to the S_0 state at or near the touching point of the two surfaces. Subsequently, it can either continue towards the biradical geometry or return to the ground state of the starting material. Thus one can expect this initial step of the addition to be less efficient for a singlet than for a triplet excited molecule.

The triplet or singlet biradical, of course, is not the product finally isolated or most often even observed, and its further fate is dictated by intersystem crossing and ordinary thermal processes. In polar solvents, these may include dissociation to the same pair of radical ions that would have been formed originally by a simple electron transfer from the electron-rich olefin to the excited ketone. It should be noted that the triplet reaction normally does not proceed when the T_1 of the olefin lies below the T_1 of the carbonyl compound, and triplet energy transfer occurs instead.

The stereospecific carbonyl addition to an electron-poor double bond in the $n\pi^*$ singlet state is believed to proceed along a concerted path in a manner somewhat analogous to the processes discussed in Section 5.4. It presumably involves a nucleophilic attack by the electron-rich π system of the excited carbonyl compound on the π system of the olefin.

7.4 ATOM AND ION TRANSFER

We have already discussed two types of proton-transfer equilibria (or, more generally, atom and ion transfer equilibria).

Transfer to a lone pair. The first case (Figure 7.39A) involves a linear system of three interacting orbitals containing four electrons, where in the starting material one terminal atom, and in the products the other terminal atom, holds two of the electrons as a lone pair, while the other two are used to form a σ bond between the remaining two atoms. When the central atom is a proton, it can hydrogen-bond to the lone pair in an encounter complex or in an intramolecular fashion.

A. Proton transfer to lone pair

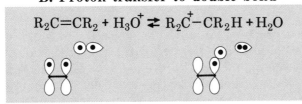

B. Proton transfer to double bond

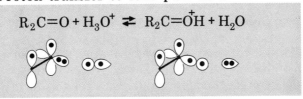

C. Proton transfer to lone pair on a double bond

$$R_2C=O + H_3O^+ \rightleftarrows R_2C=\overset{+}{O}H + H_2O$$

Figure 7.39 A catalog of proton-transfer reactions.

The reaction then represents a conversion of a hydrogen bond to a true covalent bond and vice versa, and involves minimal electronic rearrangement. The activation energy barrier separating the starting materials from the products is typically very small in the exothermic direction and barely exceeds the endothermicity in the opposite direction. In this first case, the three-orbital four-electron system remains unperturbed in low-energy excited states of the molecule, in which electronic excitation is localized elsewhere, typically in a π chromophore. Insofar as the excitation into S_1 or T_1 changes the relative stability of the reactant and the product, it will shift the position of the proton-transfer equilibrium in the S_1 or T_1 state relative to its position in the S_0 state and may cause a proton shift to occur. Since along the proton-transfer reaction coordinate the S_0, S_1, or T_1 surfaces were approximately parallel to one another, with no allowed or avoided touching, this type of ion-transfer process is normally adiabatic. Although photochemical in nature, it bears considerable resemblance to photophysical processes and was discussed in Section 2.6.

Transfer to a double bond. In addition, we have also considered proton and atom transfer to an alkene double bond (Figure 7.39B; see also Section 6.3.2). These processes involve a much more profound change in chemical bonding and electronic structure. They involve a linear array of four interacting orbitals containing four electrons, two of which initially form the π bond and two a σ bond. After a proton transfer, two form a new σ bond between the central atoms, and two form a lone pair on one of the terminal atoms, leaving a positively charged carbenium ion center on the terminal carbon. After a hydrogen atom transfer, two of the four electrons still form a new σ bond, but each of the remaining two now occupies singly one of the terminal atoms. If the transfer is intermolecular, there is no question as to whether a proton transfer or a hydrogen atom transfer occurred, since this can be determined by a mere inspection of total charges on the two partners. The distinction is less clear-cut for intramolecular transfer; the one case that is totally ambiguous is treated separately, as a sigmatropic reaction, and has been dealt with in Section 5.3.

Transfer to a lone pair on a double bond. Presently, we are ready to consider the third, most complicated case (Figure 7.39C), in which the ion transfer occurs to a lone pair located on a doubly bonded atom, so that a total of at least five orbitals and six electrons have the potential of being directly involved (Figure 7.39C). This provides a rich variety of possible electronic states at low energies in the primary reaction system. Whether all of these five orbitals and six electrons indeed are involved is a function of the reaction path considered. For certain symmetric reaction paths the situation is no more complicated than case A (Section 2.6).

Once again, the question arises as to whether an ion transfer or an atom transfer will take place. If the transfer is intermolecular, the outcome can be readily discerned by inspection of the total charges in the reaction partners before and after reaction. We shall see below that the actual outcome may be affected by factors such as the choice of the initial excited state and polarity of the medium. If the transfer is intramolecular, the distinction between ion transfer and atom transfer may be much less clear; this case will be considered separately.

7.4.1 Intermolecular Transfer

Hydrogen atom abstraction by an isolated C=O or C=N group. We are now ready to consider one of the most characteristic photochemical reactions, namely, intermolecular hydrogen atom abstraction by excited carbonyl compounds. We shall use this opportunity to illustrate the application of VB correlation diagrams of Chapter 3 to a relatively complicated case, in which it is no longer very useful to think in terms of all the individual VB structures contributing to the various states involved, because of their excessive number. Instead, one uses the leading VB structures to determine the overall symmetry of the state wave functions. This approach is possible if a sufficiently symmetric model path can be selected for consideration.

As a model system, we chose the abstraction of a hydrogen atom from methane by a carbonyl group. First, we represent the states on both sides of the correlation diagram by their dominant VB structures, and then we shall use their symmetry to produce the desired correlation (Figure 7.40).

Since methane will be in its ground state in all low-energy situations, on the left-hand side we need information on the low-energy excited states of the carbonyl group. Their construction is patterned after the states of ethylene discussed in Chapter 3 and takes into account both the electronegativity difference between the two termini of the C=O double bond and the presence of one of the lone pairs on oxygen. The other pair is in a predominantly 2s orbital at a much lower energy and is of no interest in the present context.

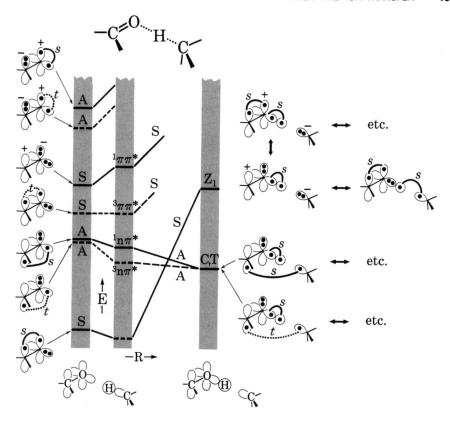

Figure 7.40 Hydrogen or proton abstraction by a carbonyl. VB structures and a state correlation diagram for singlets (solid lines) and triplets (dashed lines). C_s symmetry.

The left-hand side of Figure 7.40 represents schematically the energies of the VB structures. The bonding singlet-coupled covalent structure lies considerably below its antibonding triplet counterpart, and both are symmetric (S) with respect to reflection in the plane of the carbonyl goup and its attached atoms (the COHC plane). One of the charge-separated structures, C^+—O^-, is also shown at a somewhat higher energy. The other charge-separated symmetric structure, which would place a negative charge on the carbon and a positive one on oxygen, is much higher in energy and is not shown.

Structures in which the lone-pair orbital is occupied only singly are antisymmetric (A). In the lowest two such structures the electron is taken out of the lone pair and placed into the oxygen p orbital of π symmetry, which then contains an electron pair. The two unpaired electrons can be either singlet- or triplet-coupled; however, this makes very little difference for their energy because these two orbitals are orthogonal and well separated in space, so that their exchange integral is comparable to that in twisted ethylene and very small. It is also possible to place the electron that has been taken out of the lone-pair orbital onto the carbon atom. This will lead to VB structures with negative charge on the carbon orbital and positive charge on the oxygen. These are higher in energy and again occur as a singlet- and triplet-coupled pair. In this case the two unpaired electrons are in atomic orbitals that are orthogonal but located on the same atom, leading to a much larger exchange integral. This will cause a substantially larger energy splitting between the singlet and triplet VB structure.

Further to the right in Figure 7.40 is shown the effect of the mixing of the VB structures of like symmetries and of like spin multiplicity to produce molecular states. Since the VB structure basis is not orthogonal, the stabilization of the lower member of an interacting pair will be smaller than the destabilization of the upper one. The covalent ground-state structure mixes somewhat with the higher-energy ionic structure and is stabilized to produce the ground state of the carbonyl compound, whereas the latter is destabilized to produce its $\pi\pi^*$ excited singlet state. The triplet $\pi\pi^*$ state is described by a single VB structure in this approximation.

The situation is different for the singlet and triplet $n\pi^*$ states. These originate predominantly from the VB structures with single occupancy in the lone pair and no charge separation, but mixing with the higher-lying charge-separated VB structures of A symmetry needs to be considered. It is stronger in the triplet case because the triplet charge-separated VB structure is lower in energy, and this is responsible for the more significant separation between the singlet and triplet components of the resulting $n\pi^*$ state. Still, the separation is not very large relative to the separation of the two components of the $\pi\pi^*$ state. Clearly, in the VB picture the $^1n\pi^*$–$^3n\pi^*$ splitting originates primarily in the differential admixture of charge-separated structures into the two nearly degenerate low-energy antisymmetric VB structures $^{1,3}[\dot{C}-\dot{O}\cdot]$. Therefore, it will be affected by the relative energies of the neutral and charge-separated structures with a single electron in the lone-pair orbital and, thus, by the electronegativity difference between the two termini of the double bond.

Next, we consider the right-hand side of the diagram in Figure 7.40. Two radical centers are present in the low-lying states of the products. One is the p orbital on the carbonyl carbon, and the other is an sp^3 hybrid on the methane carbon atom. The two covalent VB structures shown, a singlet and a triplet, will be the lowest in energy and will be degenerate if the products are infinitely separated from each other. Both have a three-electron π bond and contain an odd number of electrons in AOs antisymmetric with respect to the COHC plane, and the wave functions are of A symmetry. In higher energy structures, one or the other nonbonding AO can be occupied doubly, producing charge-separated VB structures, or a three-electron bond can be present between one or another pair of orbitals in the σ framework. These high-energy structures are of S symmetry, and their exact location on the energy scale is immaterial for the following argument and is not shown in Figure 7.40. Instead, we show only the energy of the lowest energy state that results from this mixing (Z_1).

Finally, we note that the lowest state on the right-hand side must lie above the lowest state on the left-hand side by about one bond energy. This fixes the relative positions of the two scales. The resulting diagram contains an unavoided singlet state crossing along the reaction path. This crossing would be avoided if symmetry were lowered, but as long as the deviation from the plane of symmetry is only small, it will not be avoided very much.

Figure 7.41 shows the potential energy curves calculated by a very simple *ab initio* method (minimum basis set, 3×3 CI) for a prototype system, formaldehyde, abstracting hydrogen from methane. The path for the one-dimensional cut shown was chosen so as to preserve a plane of symmetry throughout and is indicated

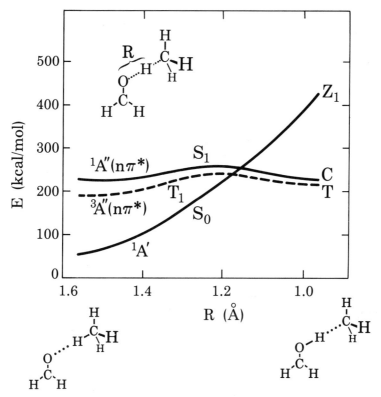

Figure 7.41 Hydrogen or proton abstraction from methane by formaldehyde. *Ab initio* 3 × 3 CI state energies (minimum basis set) for singlets (solid lines) and triplets (dashed line). C_s symmetry. Reproduced by permission from L. Salem, *J. Am. Chem. Soc.* **96,** 3486 (1974).

schematically. The course of the calculated curves follows faithfully the expectations based on the simple correlation diagram of Figure 7.40. The calculated triplet $n\pi^*$ state of the aldehyde (T_1) lies a little below its singlet $n\pi^*$ excited state (S_1). The product radical pair has its S_0 and T_1 states essentially degenerate; at infinite separation they would be exactly degenerate. Its S_1 state is of zwitterionic nature and is considerably higher in energy. The triplet T_1 state correlates essentially straight across the diagram and exhibits only a minor barrier, whose origin can be understood in terms of the "natural" correlation diagram constructed using the characteristic configuration concept (Chapter 3). At the symmetric planar geometries chosen, the S_1 state touches the S_0 state near the middle of the reaction path. This touching corresponds to a tight geometry for the biradical, whereas the completely dissociated geometry is its loose geometry. As paths of lower symmetry are considered, the touching becomes avoided, but even then it represents a likely point of return of the excited system to the S_0 surface. In the triplet state the photochemical abstraction reaction is likely to proceed completely to the right-hand side of the diagram; reformation of the starting materials is not possible unless spin flip occurs, and the system then returns on the S_0 surface. On the other hand, in the excited singlet reaction, return to S_0 is likely to occur at the tight geometry, and subsequent return along the S_0 surface to the starting materials can compete with the further progress towards the dissociated radical pair.

Although we have used the specific example of a carbonyl group in the above discussion, it is clear that identical principles apply to the isolated imino group.

Hydrogen atom abstraction by more complex functional groups.

An entirely analogous analysis can be applied to hydrogen atom abstractions by conjugated carbonyl or imino groups. A particularly important class of disguised imines are the azaaromatics, in which the C=N bond forms a part of the aromatic ring system.

The correlation diagrams for the hydrogen atom abstraction by these more complex functionalities are essentially identical to that of Figure 7.40 except for one important aspect. While for simple carbonyl and imino compounds the $^3\pi\pi^*$ state lies well above the $^3n\pi^*$ state, and the $^1\pi\pi^*$ state much higher than the $^1n\pi^*$ state, this need no longer be the case if the chromophore is more complex. In many highly conjugated carbonyl compounds, imines, and azaaromatics, the lowest $^3\pi\pi^*$ state actually lies below the $^3n\pi^*$ state, and even $^1\pi\pi^*$ often lies below $^1n\pi^*$. The correlation diagram acquires one of the forms shown in Figure 7.42 accordingly, and it is obvious that a correlation-imposed barrier in the T_1 (S_1) surface results unless the $^3n\pi^*$ ($^1n\pi^*$) state is lowest. The size of the barrier will increase as the $\pi\pi^*$ state dips further below $n\pi^*$ in energy. The $n\pi^*$-$\pi\pi^*$ crossing can be avoided by choosing an unsymmetric path that permits $n\pi^*$-$\pi\pi^*$ mixing, thus reducing the barrier somewhat as shown by dotted lines in Figure 7.42. Still, it is clear that for systems whose $\pi\pi^*$ state lies sufficiently far below $n\pi^*$, the reaction will no longer be possible. In view of the shorter lifetime of the excited singlet relative to the

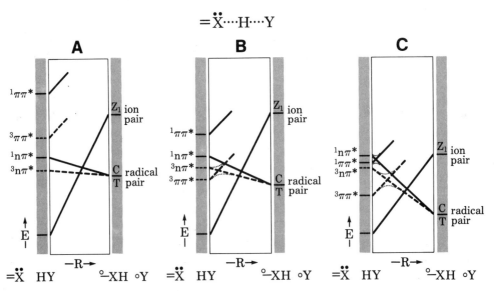

Figure 7.42 Hydrogen or proton abstraction by a carbonyl. State correlation diagram for three relative positions of the $n\pi^*$ and $\pi\pi^*$ states for singlets (solid lines) and triplets (dashed lines) at C_s symmetry. Dotted lines indicate the effects of symmetry lowering. The circles in the formulas indicate either dots (radicals) or charges (ions).

triplet, for a given size of the $n\pi^*$–$\pi\pi^*$ gap, this will happen in the S_1 state before it does in the T_1 state. Just really when it happens will depend on the overall exothermicity of the S_1 and T_1 processes and therefore on the choice of the reaction partners. As the strength of the bond that originally holds the hydrogen atom is reduced, the slopes of all the correlation lines in Figure 7.42 become more favorable for the reaction and the barriers are reduced.

It is sometimes said that the reacting molecule needs to be excited to its $^3n\pi^*$ (or $^1n\pi^*$) state for the abstraction to proceed. This is not accurate, as Figure 7.42 clearly shows: The energy needed is defined by the top of the activation barrier, and this may well be less than the vertical $n\pi^*$ excitation energy.

Hydrogen atom versus proton transfer. The relative stability of the radical-pair and ion-pair states on the right-hand side of the correlation diagrams in Figure 7.42 depends strongly on the nature of the atoms X and Y and on solvent polarity. When the lone pair on X: is strongly basic, the proton on YH strongly acidic, and the medium polar, the ion-pair state may well lie below the radical-pair state. Instead of being strongly endothermic, the $=$X: $+$ HY $\rightarrow$ $=$X$^+$H $+$ Y$^-$ correlation shown in Figure 7.42 may actually be exothermic, so that proton transfer occurs measurably already in the S_0 state, for instance, in the system pyridine–hydrogen iodide in water. The system then reverts to the case A of Figure 7.39, and the pK_a values in the S_0, S_1, and T_1 states are related by the Förster cycle as discussed in Section 2.6.

An interesting situation results when the ion-pair and the radical-pair product energies are nearly degenerate (cf. Section 6.2.2). Then, the instantaneous arrangement of solvent molecules in the vicinity may control the state order and determine which product pair represent S_0 and which S_1. An initially formed radical pair formed in a solvent environment optimized for the T_1 state may then transform by electron transfer to an ion pair essentially irreversibly at a rate dictated by the solvent relaxation time. The situation is thus similar to that involved in the ion-pair formation from exciplexes (Section 6.4.1), but it involves partners that are each in their ground state.

7.4.2 Intramolecular Transfer. Phototautomerization, Norrish II

In the case of the intramolecular transfer, the distinction between an atom transfer and an ion transfer is much less clear-cut. There is no difficulty if the molecular framework that separates the atom- or ion-donor portion of the molecule from the atom- or ion-acceptor portion provides effective electronic insulation and does not participate in the primary bonding changes in the reaction (e.g., a saturated chain or ring). In the opposite case, electron density flow through the framework between the donor and acceptor portions may occur freely, accompanied by appropriate bond switching, making the decision as to whether an atom or an ion transfer is occurring quite arbitrary and dependent on the choice of the VB structures that are felt to best represent the ground and excited states.

An illustration is provided in Figure 7.43 on the example of *o*-hydroxyacetophenone, with the two alternative representations of the same set of reactions shown as A and B. In scheme A, the view is taken that vertical electronic excitation, which redistributes electron density in the π system from the phenolic oxygen to the carbonyl oxygen, corresponds to a transfer of a full negative charge from the

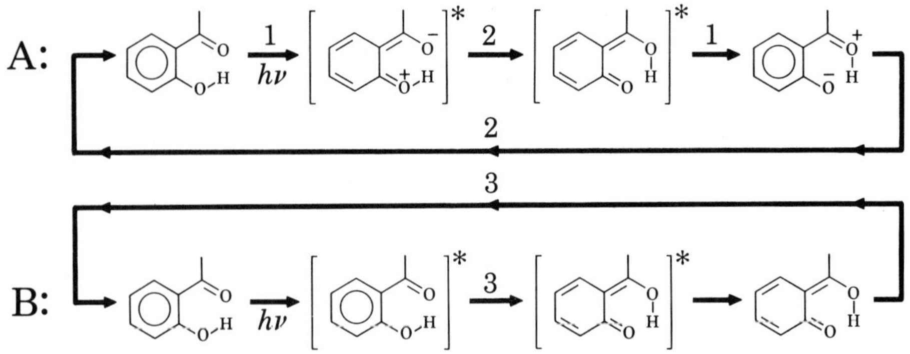

Figure 7.43 Alternative views of phototautomerization. (1) Charge transfer. (2) Proton transfer. (3) Atom transfer.

acidic center to the basic center in the molecule. This makes the former much more acidic and the latter much more basic than they were in the ground state, and it changes the position of the acid–base equilibrium. The atomic motion involved in the reaction is then logically called proton transfer from an acidic to a basic functionality. Return to the ground state is then viewed as another electron transfer, and re-formation of the starting material is regarded as another acid–base reaction step.

In scheme B, the view is taken that the charge distribution upon vertical electronic excitation is nowhere near as extreme as implied by scheme A, and the best description of the structure still involves no formal charge separation. The actual reaction step in which nuclei move is then logically called an atom transfer.

Clearly, whether nomenclature A or B is more appropriate will not only depend on the choice of substrate, solvent, and a particular electronic excited state but will also depend on the subjective judgment as to whether the charge transferred from the acidic group to the basic group in the vertical electronic excitation step is close enough to being a full negative charge. We shall not attempt to resolve this nomenclature problem and shall avoid it by referring to this type of transfer process as phototautomerization.

Reaction paths for phototautomerization: Planar versus pericyclic.

Another source of ambiguity in the description and nomenclature of intramolecular transfer reactions of type C in Figure 7.39 is the possible overlap of this category with the category of sigmatropic reactions (Section 5.3). As illustrated on the example of o-methylacetophenone in Figure 7.44, two paths with different orbital interaction schemes are possible for phototautomerization. The planar path (A) is characterized by a σ–π distinction, with the primary atomic motion modifying the interactions within an allyl-like system of three linearly interacting orbitals, which may contain three or four electrons depending on the $n\pi^*$ or $\pi\pi^*$ nature of the excited state. The pericyclic path (B) is characterized by an out-of-plane motion of the atom which is being transferred and an uninterrupted cycle of primary orbital interactions. The lone-pair orbital interacts with one of the members of this cyclic orbital arrangement, and complicates the situation relative to the ordinary sigmatropic

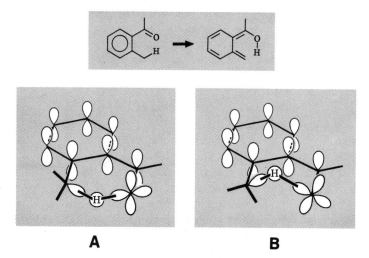

Figure 7.44 Alternative geometric paths for phototautomerization. (A) Planar. (B) Pericyclic.

path considered in Section 5.3. Because of the presence of this lone-pair orbital, the usual distinction between the suprafacial and antarafacial sigmatropic paths has been lost in this case.

Norrish II reactions. Carbonyl compounds containing an aliphatic chain can undergo not only intermolecular but also intramolecular hydrogen atom abstraction reactions if steric conditions permit. The most readily abstracted hydrogen atoms are those located in the γ position to the carbonyl group. Their transfer follows the concepts developed in Section 4.3.1 for intermolecular reactions except that it leads to intramolecular radical pairs (1,4-biradicals). When formed on the T_1 surface, these have a lifetime on the order of dozens of nanoseconds because spin flip is relatively slow. They decay by return to S_0, followed either by reversal to starting materials, ring closure to a cyclobutane, or fragmentation to an enol and an olefin. The return to starting materials corresponds to an abstraction of a hydrogen on a C·—OH group by an alkyl radical. It is slowed down in hydrogen-bonding solvents in which the C·—OH group is stabilized. When formed on the S_1 surface, the singlet biradicals—so far unobserved—presumably have a very short lifetime and also decay in the ways mentioned above. The fragmentation to an olefin and an enol is known as the Norrish II process.

7.5 THREE-MEMBERED RING HETEROCYCLES: RING-OPENING AND FRAGMENTATION

The photochemistry of heterocycles is extremely rich and varied. Of the few processes that are sufficiently simple for inclusion on our list of elementary photochemical steps, we have selected the ring-opening and fragmentation of four three-membered ring systems: oxirane, aziridine, oxaziridine, and diazirine.

7.5.1 Oxirane and Aziridine

C—O and C—N bond-breaking. Our first examples, the C—O and C—N bond cleavages in oxirane and aziridine, serve as prototypes for a significant fraction of the photochemistry of substituted oxiranes and aziridines. Another large fraction of their photochemistry has to do with cleavage of the C—C bond which is likely to occur in a concerted electrocyclic fashion in both the S_1 and T_1 states and has already been discussed in Section 5.2.3.

The elementary step to be considered here produces a 1,3-biradical from the initial three-membered ring heterocycle.

Figure 7.45 represents one-dimensional cuts through the potential energy hypersurfaces of oxirane and aziridine appropriate for the cleavage of one of the bonds to the heteroatom. The calculations are of the simple *ab initio* type and use a minimum basis set with limited configuration interaction. Because of this, they are only of semiquantitative significance but should contain all the important qualitative features. The reaction path displayed involves a gradual increase of the CCO or CCN valence angle ϕ from its equilibrium value of about 60° in the initial three-membered ring to the tetrahedral value of 109°. Simultaneously, other bond angles and lengths are changed in a linear fashion towards values considered reasonable for the biradical geometry on the basis of a separate optimization. In this structure the plane of the terminal CH_2 group remains perpendicular to the CCO plane. It was ascertained that the rotation of this CH_2 group around the C—C bond has little effect on the energy of the ground state.

The potential energy curves for both oxirane and aziridine look very much like those shown for a tritopic two-center bond dissociation process discussed in Section 7.2.1 (Figure 7.23). This is hardly surprising, considering the weak interaction of the two radical centers through the central CH_2 group. The substantial difference between the process discussed in Section 7.2.1 and the present case is the presence of a large ring strain in the three-membered ring which is relieved in the bond dissociation process. This makes the ground-state energy increase between the starting heterocycle and the biradical structure considerably smaller than was the case for a simple two-center dissociation.

The results for oxirane and aziridine are very similar. The only notable difference is the splitting of the $C_{\sigma\sigma}$, $T_{\sigma\sigma}$ and $C_{\sigma\pi}$, $T_{\sigma\pi}$ pairs of biradical states. These differ in the distribution of the three electrons present on the heteroatom in the covalent structures for the biradical. In the $\sigma\sigma$ form, the AO that originally formed the bond that was broken (σ) is occupied once, and the lone-pair orbital on the heteroatom (π) is occupied twice. In the $\sigma\pi$ structure, the opposite is true. In the case of oxirane cleavage, the energies of the two orbitals, σ and π, differ very little because both are essentially pure p atomic orbitals on the oxygen atom. For this reason, the structures of the states $C_{\sigma\sigma}$ and $C_{\sigma\pi}$ are nearly degenerate, and so are those of the states $T_{\sigma\sigma}$ and $T_{\sigma\pi}$. In the case of aziridine, the two orbitals on nitrogen differ considerably in their electronegativity. One of them is a p orbital, whereas the other is a hybrid orbital of approximately sp^2 character. Their orientation depends on the orientation of the N—H bond relative to the CCN plane. For the geometry used in the construction of Figure 7.45, this N—H bond was oriented out of the CCN plane so that the p orbital of nitrogen was of σ symmetry. In the lower-energy configurations $C_{\sigma\sigma}$ and $T_{\sigma\sigma}$, this p orbital is occupied once and the hybrid orbital

7.5 THREE-MEMBERED RING HETEROCYCLES: RING-OPENING AND FRAGMENTATION

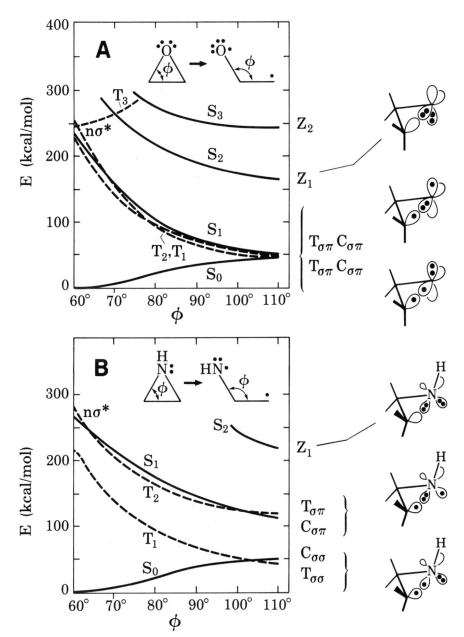

Figure 7.45 Dissociation of the carbon–heteroatom bond in oxirane (A) and aziridine (B). *Ab initio* limited CI state energies (minimum basis set) for singlets (solid lines) and triplets (dashed lines) as a function of the CCO or CCN valence angle ϕ. C_s symmetry for oxirane, N—H bond out of CCN plane in aziridine. Reproduced by permission from B. Bigot, A. Devaquet, and A. Sevin, *J. Org. Chem.* **45**, 97 (1980).

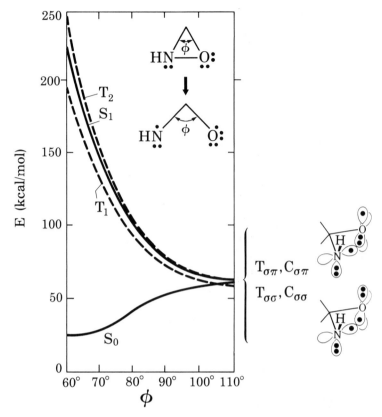

Figure 7.46 Dissociation of the O—N bond in oxaziridine. *Ab initio* limited CI state energies (minimum basis set) for singlets (solid lines) and triplets (dashed lines) as a function of the NCO valence angle ϕ. C_s symmetry. Reproduced by permission from B. Bigot, D. Roux, A. Sevin, and A. Devaquet, *J. Am. Chem. Soc.* **101**, 2560 (1979).

twice. In the upper dot–dot configurations $C_{\sigma\pi}$ and $T_{\sigma\pi}$, the opposite is true and their energy is therefore considerably higher.

The state diagrams for oxiranes and aziridines of actual photochemical interest are likely to be more complicated than the model just presented. In the parent compounds and their derivatives with saturated substituents, Rydberg states will represent the lowest excited states at the initial geometry. Since they do not decrease significantly in energy as the reaction path is followed, sooner or later they will undergo a crossing or an avoided crossing with the decreasing valence states. Thus they may represent a local minimum acting as a reservoir for molecules that can escape towards the dissociation path. In oxiranes and aziridines containing unsaturated substituents, that is, those with chromophores that make the usual solution photochemical studies possible, there will typically be locally excited states at the initial geometries which will usually increase in energy just as the ground state does when the C—O or C—N bond is extended and will undergo a crossing or an avoided crossing with the descending characteristic configurations of $n\sigma^*$ type. Once again, local minima at the initial geometry are likely to result and to serve a similar function as discussed above for the Rydberg states.

7.5 THREE-MEMBERED RING HETEROCYCLES: RING-OPENING AND FRAGMENTATION 411

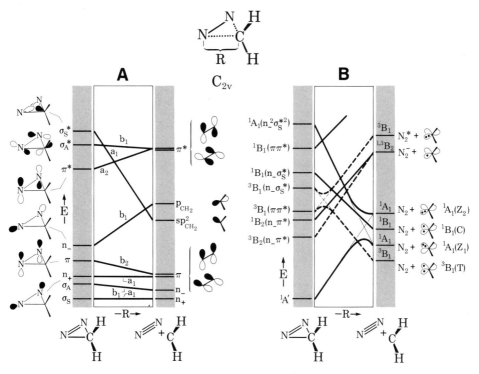

Figure 7.47 Fragmentation of diazirine into $CH_2 + N_2$, linear path (C_{2v} geometry). (A) MO correlation diagram. (B) Configuration (thin lines) and state (thick lines) correlation diagram for singlets (solid lines) and triplets (dashed lines).

C—C bond-breaking. It is not clear to what degree a nonconcerted breaking of the C—C bond by its extension and simultaneous rotation of one of the CH_2 groups is competitive in the triplet state with the favored disrotatory concerted reaction path. The experimental evidence suggests strongly that it competes poorly and that the reaction is quite stereospecific even in the triplet state.

7.5.2 Oxaziridine

In oxaziridine, the ruptures of the O—N and O—C bonds are both of photochemical interest. The photodissociation of the O—N bond produces a biradical that undergoes a subsequent thermal reaction and rearranges to formamide. The step of interest in the present context is the conversion of oxaziridine into the biradical. Because the biradical is of the 1,3 type with limited interactions through the saturated CH_2 group, the energies of its electronic states are not very sensitive to the rotation of the N—H bond around the C—N bond. Because of the limited interaction between the two radical centers, the energy diagram for this ring-opening is essentially identical with that of the simple two-center tetratopic dissociation of an acyclic N—O bond (cf. Section 7.2.2). In this sense it resembles the C—O and C—N bond dissociations in oxirane and aziridine, which were also essentially two-center processes. Calculations at the same relatively crude level as those just dis-

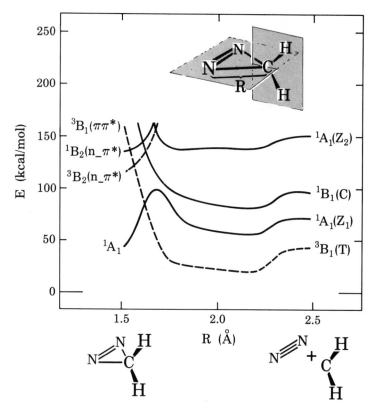

Figure 7.48 Fragmentation of diazirine into $CH_2 + N_2$, linear path (C_{2v} symmetry). *Ab initio* large-scale CI state energies for singlets (solid lines) and triplets (dashed lines).

cussed for oxirane and aziridine have been performed for a one-dimensional cut through the potential energy hypersurfaces in which only the NCO valence angle is increased and other geometric parameters are varied linearly towards final values obtained by choosing the OCN angle to be 110° and optimizing the rest of the geometry for the ground state (Figure 7.46). The energies of the S_1 and T_1 states drop precipitously as the ring-opening reaction proceeds, whereas the energy of the S_0 state increases. The presence of ring strain in the starting structure is reflected in the relatively small ground-state energy difference between the biradical product and the starting heterocycle.

The breaking of the C—O bond of oxaziridine is involved in the photochemical interconversion of oxaziridines with nitrones. Here at least two dimensions of the potential energy hypersurface need to be considered, corresponding to the increase of the ONC angle and to the rotation of the CH_2 group relative to the ONC plane. Another geometric parameter that ought to be considered as well is the degree of pyramidalization on the nitrogen atom. In the following we shall only briefly consider the first two motions in the two-dimensional space and note the effect of the degree of pyramidalization of the nitrogen on some of the state energies.

The first motion is the breaking of the three-membered ring of oxaziridine by extending the C—O distance without twisting the CH_2 group. *Ab initio* calculations at the crude level described above have been performed for this path, keeping the nitrogen pyramidalized. The behavior of the S_0, S_1, and T_1 states during this motion was found to be just that expected for a simple two-center tritopic dissociation of a C—O bond. The S_0 state increases in energy, less so than usual because of the presence of ring strain at the starting geometry, whereas the S_1 and T_1 states descend abruptly in energy as the bond is broken. The resulting curves look very similar to those for the N—O bond dissociation shown in Figure 7.46. The exact situation at the biradical geometry is sensitive to the degree of pyramidalization on the nitrogen, since the $C_{\sigma\sigma}$, $T_{\sigma\sigma}$ and $C_{\sigma\pi}$, $T_{\sigma\pi}$ configuration pairs are affected by the pyramidalization in different ways. In general, however, the biradical has two singlet and two triplet states, all nearly degenerate at low energy as expected for a tritopic case. Note that the biradical geometry can also be described as a twisted nitrone, possibly with a pyramidalized nitrogen. Calculations at a similar level indicate that the S_0 energy of the nitrone increases with the twist while those of its T_1, S_1, and T_2 states decrease, which is not surprising for the twisting of a double bond.

Photochemical processes involving the rupture or establishment of the C—O bond then can be viewed as follows. The excitation of oxaziridine into its S_1 or T_1 valence state leads to the breaking of the C—O bond and production of the same biradical which also results from excitation of a nitrone to its T_1 or S_1 state. Return to the S_0 state can then be followed by ring closure to produce the oxaziridine or untwisting to produce the planar nitrone. The details of the processes may be much more complicated than this simple discussion indicates because the presence of several low-lying electronic states for the biradical form and the availability of the CH_2 twist degree of freedom may cause the presence of several minima from which return to S_0 is possible, and different minima may be populated with different probabilities depending on the starting point. More detailed calculations are required before a meaningful discussion of these details becomes possible.

7.5.3 Diazirine

The photofragmentation of diazirines represents an important source of carbenes. The linear dissociation path that possesses C_{2v} symmetry is ground-state forbidden as is seen from the MO correlation diagram in Figure 7.47A. The orbital crossing is of the "abnormal" type. Indeed, a configuration correlation diagram (Figure 7.47B) suggests the presence of barriers in the S_1 and T_1 surfaces as well, since the dominant configurations in these states are of the $^{1,3}n_-\pi^*$ ($^{1,3}B_2$) type and correlate with very high lying configurations of the product, whereas the S_1 and T_1 states of the product are of $^{1,3}B_1$ symmetry. An *ab initio* large-scale CI calculation yields potential energy curves (Figure 7.48) that agree with expectations based on the correlation diagram.

A B_1–B_2 state crossing can be avoided when the planes of symmetry are removed. Figure 7.49 shows the potential energy curves calculated along one such low-symmetry fragmentation path. First, the ring is partially opened by increasing one of the CNN valence angles to $\phi = 135°$ while simultaneously rotating the CH_2 group by 20° about the HCH angle bisector. Then, the shorter of the C—N bonds is extended to R = 2.5 Å. Comparison of Figures 7.48 and 7.49 shows that the first step (increase in ϕ) greatly reduces the S_1 and T_1 energies and that the original B_1–B_2 crossing is avoided in the second step (increase in R). It is likely that some

414 REACTIONS WITH MORE THAN ONE ACTIVE ORBITAL PER ATOM

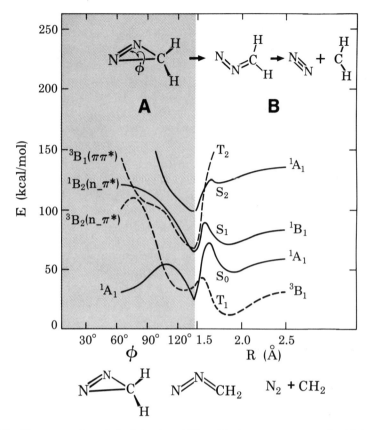

Figure 7.49 Fragmentation of diazirine into $CH_2 + N_2$, low-symmetry path. *Ab initio* large-scale CI state energies (double zeta) for singlets (solid lines) and triplets (dashed lines). (A) Increase in the CNN valence angle ϕ with a constant rotation angle for the CH_2 group. (B) Subsequent increase in the C—N separation R.

low-symmetry path such as this is the favored path for the fragmentation of diazirine in its S_0, S_1, and T_1 states.

The striking drop in the S_1 and T_1 energies along the unsymmetric path (increasing ϕ) suggests the possible existence of a minimum in one or both of these surfaces at a partially opened ring geometry. Return to S_0 through such a minimum would then be responsible for the known photochemical interconversion of diazirines and diazo compounds.

Figure 7.50 shows the results of *ab initio* large-scale CI calculations for three such ring-opening paths. Along these, an NNC valence angle ϕ was increased gradually from 60° to 180° while bond lengths and over valence angles were simultaneously changed to those characteristic of the ground state of diazomethane. In part A, the CH_2 group was gradually rotated about the axis that bisects HCH angle until at $\phi = 180°$ it was turned by a full 90°. In part B, it was rotated more slowly so that at $\phi = 180°$ it was turned by 60°. In part C, it was not rotated at all so that the NNC plane remained a plane of symmetry throughout.

In all cases, the S_1 and T_1 surfaces go through a minimum near $\phi = 150°$ and

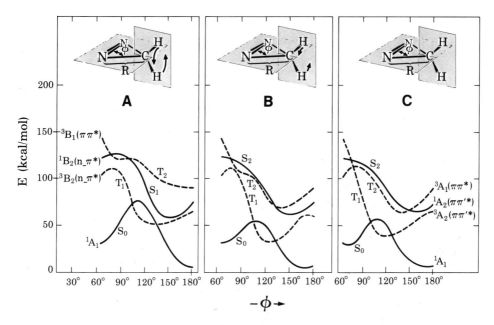

Figure 7.50 Ring-opening of diazirine (left) to diazomethane (right). *Ab initio* large-scale CI state energies (double zeta) for singlets (solid lines) and triplets (dashed lines), as a function of the CNN valence angle ϕ, with simultaneous rotation and gradual planarization of the CH$_2$ group. The rate of rotation would bring the final rotation angle to 90° (A), 60° (B), or 0° (no CH$_2$ rotation) (C), if ϕ were allowed to increase to 180°.

$\phi = 120°$, respectively. In the S$_1$ surface, these minima can be reached without any barrier; in the T$_1$ surface, they can be reached by going over a small barrier. The symmetry-lowering rotation of the CH$_2$ group lowers the T$_1$ barrier. Considerable further effort would be required to establish whether bound minima are indeed present in S$_1$ and T$_1$ near the geometries suggested by Figure 7.50 and whether they can be reached from the vertical excited diazirine geometry without going over a barrier.

This final example illustrates the difficulties encountered in the search of photochemical pathways, even for simple molecules. Clearly, the results are suggestive rather than definitive.

7.6 COMMENTS AND REFERENCES

For the development of the concept of **topicity** and its application to numerous classes of reactions, see W. G. Dauben, L. Salem, and N. J. Turro, *Acc. Chem. Res.* **8,** 41 (1975).

Cis–trans isomerizations in linear conjugated systems containing both C=C and C=N bonds are common; see, for instance, R. S. Becker and K. Freedman, *J. Am. Chem. Soc.* **107,** 1477 (1985). Calculations for formaldimine can be found in: P. Russegger, *Chem. Phys.* **34,** 329 (1978); V. Bonačić-Koutecký and M. Persico, *J. Am. Chem. Soc.* **105,** 3388 (1983); V. Bonačić-Koutecký and J. Michl, *Theor. Chim. Acta* **68,** 45 (1985). The second of these also deals with acroleinimine, which can isomerize around the C=C or the C=N bond. For recent experimental evidence implicating ground-state *trans*-cyclohexenone intermediates in processes previously believed to proceed in the T_1 state, see D. I. Schuster, D. B. Brown, L. J. Capponi, C. A. Rhodes, J. C. Scaiano, P. C. Tucker, and D. Weir, *J. Am. Chem. Soc.* **109,** 2533 (1987).

Dissociation reactions of topicity higher than two are also very common [e.g., photosolvolysis of organic halides: S. J. Cristol and T. H. Bindel, *Organic Photochemistry,* Vol. 6 (A. Padwa, Ed.), Marcel Dekker, New York, 1983; P. J. Kropp, *Accounts Chem. Res.* **17,** 131 (1984)]. They have been the subject of numerous theoretical investigations. A verification of the topicity rules at a semiempirical level was reported by E. M. Evleth and E. Kassab, *J. Am. Chem. Soc.* **100,** 7859 (1978). A discussion of the rate of intersystem crossing in the radical-pair products was given by J. R. Larson, N. D. Epiotis, L. E. McMurdie, and S. S. Shaik, *J. Org. Chem.* **45,** 1388 (1980).

Examples of *ab initio* **calculations** are: **methanol** [R. J. Buenker, G. Olbrich, H.-P. Schuchmann, B. L. Schürmann and C. von Sonntag, *J. Am. Chem. Soc.* **106,** 4362 (1984)]; **Norrish I** [J. Schüle and M. Klessinger, *Theor. Chim. Acta* **72,** 35 (1987)]; **azirine** [B. Bigot, A. Sevin and A. Devaquet, *J. Am. Chem. Soc.* **100,** 6924 (1978)]; **ketene** [S. Yamabe and K. Morokuma, *J. Am. Chem. Soc.* **100,** 7551 (1978)]; **diazomethane and diazirine** [B. Bigot, R. Ponec, A. Sevin, and A. Devaquet, *J. Am. Chem. Soc.* **100,** 6575 (1978)]; **hydrazoic acid** [A. Sevin, J. P. LeRoux, B. Bigot, and A. Devaquet, *Chem. Phys.* **45,** 305 (1980)]; **oxirane** [B. Bigot, A. Sevin, and A. Devaquet, *J. Am. Chem. Soc.* **101,** 1095, 1101 (1979); R. de Maré, in *Computational Theoretical Chemistry* (I. G. Czismadia and R. Daudel, Eds.), Reidel, Dordrecht, 1981, p. 335]; **aziridine** [B. Bigot, A. Devaquet, and A. Sevin, *J. Org. Chem.* **45,** 97 (1980)]; **oxaziridine** [B. Bigot, D. Roux, A. Sevin, and A. Devaquet, *J. Am. Chem. Soc.* **101,** 2560 (1979)]; α-**cleavage of azo compounds** [B. Bigot, A. Sevin and A. Devaquet, *J. Am. Chem. Soc.* **100,** 2639 (1978)].

Carbonyl additions to olefins (Paterno–Büchi) have been treated at the PMO level, in which only a relative estimate of the initial slope of the excited state surface is obtained [W. C. Herndon, *Topics Curr. Chem.* **46,** 141 (1974)]. For a treatment at a simple *ab initio* level, and a treatment of **hydrogen abstraction by carbonyls** and α-cleavage of carbonyl compounds (**Norrish I**), see B. Bigot, A. Devaquet, and N. J. Turro, *J. Am. Chem. Soc.* **103,** 6 (1981). Semiempirical computations of the **Norrish II** reaction path can be found in M. J. S. Dewar and C. Doubleday, *J. Am. Chem. Soc.* **100,** 4935 (1978).

A review of phototautomerism is found in W. Klöpffer, *Adv. Photochem.* **10,** 311 (1977).

CHAPTER 8

Epilogue

We have now completed our survey of the role of electronic structure in elementary photochemical reaction steps. In the introductory Part A, we provided a review of the basic theoretical tools (Chapters 1 and 3) and a brief phenomenological survey of photochemical processes (Chapter 2). A more mathematical overview of the theoretical tools is given in Appendixes I–III. Of these, Appendix I is quite elementary and contains introductory textbook material. Appendix II contains mathematical details that we did not wish to include in Chapter 4. Appendix III provides a quick reference to the most important current *ab initio* methods.

In Part B, we first considered, in a fair amount of detail, the electronic aspects of simple two-center reactions, that is, the variation of the electronic nature of a simple two-electron bond in its various states of excitation as a function of the degree of stretching (σ bonds) or twisting (π bonds). The attention that we have paid to this subject (Chapter 4) is far out of proportion with the actual number of organic photochemical reactions that are so simple in the laboratory: Most known elementary processes involve cyclic (Chapter 5) or linear (Chapter 6) arrays of simple bonds or, even more likely, involve bonds that we do not consider simple (one or both participating atoms carry lone pairs). The amount of space devoted to analyzing the breaking of a single σ or π bond reflects our belief that this provides the basis for understanding other elementary processes, in which such breaking is assisted by interaction with additional bonds. This also is in line with our frequent use of the "interacting subunits" approach in the analysis of reaction paths.

We realize that our selection of elementary photochemical steps for discussion in Part B is rather arbitrary. We have not included some simple steps important primarily in vacuum–UV photochemistry, nor have we included many that are of great importance in organic solution photochemistry but appeared too complex for a reasonably reliable *ab initio* treatment. Aromatic photosubstitution, photoisomerization, and photocycloaddition, as well as di-π-methane, enone, and dienone rearrangements, are just a few examples of important and classical photoprocesses that have not been treated. Also, we believe that many of the concepts are useful in inorganic photochemistry as well, particularly that of main group elements. For example, the extrusion of a silylene from an oligosilane chain is isolectronic with the disrotatory electrocyclic closure of 1,3-butadiene. In this sense, mastery of the subject matter presented in this text will permit applications well outside the immediate area discussed here.

We emphasize again that our goal has not been to offer the reader a comprehensive theoretical treatment of all organic photochemical reactions or an exhaustive review of the literature dealing with the subject but, rather, to offer a flavor of what can be done in getting a first-principles understanding of the relation between electronic structure and photochemical reactivity. We believe that such understanding not only satisfies intellectual curiosity but also provides connection between superficially unrelated processes and suggests new experiments. We apologize to authors whose theoretical papers have not been selected for illustration or whose papers have not been listed in the quite incomplete summaries at the end of each chapter.

While the particular *ab initio* calculations used throughout the text will soon be obsolete, if they are not already, we hope that the underlying physical models will remain useful for some time to come. However, we wish to emphasize that in our opinion the theoretical understanding of organic photochemistry—like its experimental counterpart—is at its mere beginning, and that many exciting fundamental developments await us in the future. Surely, by concentrating mostly on the present state of the understanding of the electronic aspects of photochemistry, this book avoids the larger theoretical issue of dynamics, where much exciting progress is occurring right now. In a sense, however, an understanding of the mere "static" issue of electronic structure and potential energy surfaces is a prerequisite for a reliable application of dynamical theories. If we have succeeded in offering some small measure of such understanding and if this contributes to the growth of a new generation of organic photochemists, we shall be richly rewarded for our efforts.

APPENDIXES

APPENDIX I

The MO and VB Methods Illustrated on the H$_2$ Molecule

In the minimum basis set approximation, the two electrons of H$_2$ are permitted to move in a space spanned by a 1s orbital on each center. We shall denote these AOs as $|A(\mathbf{r})\rangle$ and $|B(\mathbf{r})\rangle$, where $\mathbf{r}$ is the position vector of an electron, and we shall assume their form to be the same as in the H atom:

$$|A(\mathbf{r})\rangle = \frac{1}{\sqrt{\pi}} e^{-r_A} \tag{I.1}$$

$$|B(\mathbf{r})\rangle = \frac{1}{\sqrt{\pi}} e^{-r_B} \tag{I.2}$$

We use r to represent the length of the vector $\mathbf{r}$; $r_A = |\mathbf{r} - \mathbf{R}/2|$ is the distance of an electron from nucleus A, $r_B = |\mathbf{r} + \mathbf{R}/2|$ is the distance of an electron from nucleus B, and $\mathbf{R}$ is the vector from A to B. The origin of the coordinate system is in the center of the molecule. All distances will be measured in atomic units (1 a.u. = 0.5297 Å). The first electron's spatial coordinates are collected in the vector $\mathbf{r}_1$; its spin coordinate is ξ_1. The coordinates of the second electron are $\mathbf{r}_2$ and ξ_2.

We shall now proceed to find exact solutions for the electronic wave functions in this basis set, known as the full configuration interaction solution or, less commonly, full VB solution. The objective is to illustrate the VB and MO approaches and some fundamental properties of the electronic states of a system of two electrons on two centers. If the objective were to obtain accurate results for H$_2$, a far better basis set or a numerical method would have to be used.

The VB approach

VB structures. In the VB procedure, we first construct simple many-electron wave functions (VB structures) by occupying the AOs by the two electrons in all possible ways. There are four dot–dot structures, that is, those without net charge separation. Their spin-adapted form is obtained by making suitable linear combinations of the simple product functions, $|A(\mathbf{r}_1)\rangle|B(\mathbf{r}_2)\rangle$, $|A(\mathbf{r}_1)\rangle|\overline{B}(\mathbf{r}_2)\rangle$, $|\overline{A}(\mathbf{r}_1)\rangle|B(\mathbf{r}_2)\rangle$, and $|\overline{A}(\mathbf{r}_1)\rangle|\overline{B}(\mathbf{r}_2)\rangle$. Spin functions $\alpha(\xi_1)$ and $\alpha(\xi_2)$ are not shown explicitly, and the spin functions $\beta(\xi_1)$ and $\beta(\xi_2)$ are shown by placing a bar above the AO symbol. Spin adaptation is required by the Pauli principle, which demands that the total electronic wave function be antisymmetric with respect to the interchange of the coordinates of electrons 1 and 2 and thus with respect to the interchange of the subscripts 1 and 2.

The spin-adapted dot–dot VB structures $^1|AB\rangle$, $^{3(1)}|AB\rangle$, $^{3(0)}|AB\rangle$, and $^{3(-1)}|AB\rangle$ are most easily written in the form of Slater determinants. Explicit detail is shown for the singlet covalent structure $^1|AB\rangle$, whose space part is symmetric and spin part antisymmetric relative to the interchange of the two hydrogen atoms in the molecule, and for the triplet structure $^{3(0)}|AB\rangle$, whose space part is antisymmetric and spin part is symmetric. In progressively simplified notation, we have

$$^1|AB\rangle = \frac{1}{\sqrt{2+2S_{AB}^2}}[|A(\mathbf{r}_1)\rangle|B(\mathbf{r}_2)\rangle + |B(\mathbf{r}_1)\rangle|A(\mathbf{r}_2)\rangle]$$

$$\times \frac{1}{\sqrt{2}}[|\alpha(\xi_1)\rangle|\beta(\xi_2)\rangle - |\beta(\xi_1)\rangle|\alpha(\xi_2)\rangle]$$

$$= \frac{1}{2\sqrt{1+S_{AB}^2}}\left\{\begin{vmatrix} A(1) & \overline{B}(1) \\ A(2) & \overline{B}(2) \end{vmatrix} - \begin{vmatrix} \overline{A}(1) & B(1) \\ \overline{A}(2) & B(2) \end{vmatrix}\right\}$$

$$= \frac{1}{2\sqrt{1+S_{AB}^2}}(|A\overline{B}| - |\overline{A}B|) \tag{I.3}$$

$$^{3(0)}|AB\rangle = \frac{1}{\sqrt{2-2S_{AB}^2}}[|A(\mathbf{r}_1)\rangle|B(\mathbf{r}_2)\rangle - |B(\mathbf{r}_1)\rangle|A(\mathbf{r}_2)\rangle]$$

$$\times \frac{1}{\sqrt{2}}[|\alpha(\xi_1)\rangle|\beta(\xi_2)\rangle + |\beta(\xi_1)\rangle|\alpha(\xi_2)\rangle]$$

$$= \frac{1}{2\sqrt{1-S_{AB}^2}}(|A\overline{B}| + |\overline{A}B|) \tag{I.4}$$

In the simplified notation, we have first replaced the argument $\mathbf{r}_i$, the position vector of the i-th electron, by the label i of that electron. We have then introduced further simplification in which only the diagonal of the determinant is written and the arguments are understood. The fraction in front is chosen so as to guarantee that the VB structures are normalized, and S_{AB} is the overlap integral of orbitals $|A\rangle$ and $|B\rangle$.

There are also two hole–pair (ionic) structures in which both electrons are placed on the same hydrogen atom which then carries a negative charge while the other atom carries a positive one. This sounds like extreme charge separation, and indeed it is extreme if the internuclear distance R is large. However, at close distances of approach the AOs $|A\rangle$ and $|B\rangle$ overlap significantly so that the charge separation is small. The expressions for all six VB structures have been given in Chapter 4 (equations 4.10–4.15).

The two hole–pair structures are sometimes also called zwitterionic to emphasize the presence of charges of both signs. We shall use the expression "hole–pair VB structure" for the simple functions given in expressions 4.10 and 4.12. We shall reserve the expression "zwitterionic VB structures" for the symmetry-adapted combinations

$$^1|Z_2\rangle = \frac{1}{\sqrt{2(1 + S_{AB}^2)}} [^1|A^2\rangle + {}^1|B^2\rangle] \tag{I.5}$$

$$^1|Z_1\rangle = \frac{1}{\sqrt{2(1 - S_{AB}^2)}} [^1|A^2\rangle - {}^1|B^2\rangle] \tag{I.6}$$

which we shall encounter in the following (cf. equations 4.46 and 4.47). The usual chemical notation for either $^1|Z_1\rangle$ or $^1|Z_2\rangle$ is $H^-H^+ \leftrightarrow H^+H^-$. If the two need to be distinguished, a plus or a minus sign can be placed above the double-headed arrow.

The mixing of VB structures (resonance). In order to obtain state wave functions as linear combinations of VB structures and to evaluate their energies, we need to diagonalize the Hamiltonian matrix. The resulting energies were given without a derivation in Table 4.1. The Hamiltonian matrix has the form

$$\begin{pmatrix} \langle^1AB|\hat{H}|^1AB\rangle & \langle^1AB|\hat{H}|^1A^2\rangle & \langle^1AB|\hat{H}|^1B^2\rangle & 0 & 0 & 0 \\ \langle^1A^2|\hat{H}|^1AB\rangle & \langle^1A^2|\hat{H}|^1A^2\rangle & \langle^1A^2|\hat{H}|^1B^2\rangle & 0 & 0 & 0 \\ \langle^1B^2|\hat{H}|^1AB\rangle & \langle^1B^2|\hat{H}|^1A^2\rangle & \langle^1B^2|\hat{H}|^1B^2\rangle & 0 & 0 & 0 \\ 0 & 0 & 0 & \langle^{3(1)}AB|\hat{H}|^{3(1)}AB\rangle & 0 & 0 \\ 0 & 0 & 0 & 0 & \langle^{3(0)}AB|\hat{H}|^{3(1)}AB\rangle & 0 \\ 0 & 0 & 0 & 0 & 0 & \langle^{3(-1)}AB|\hat{H}|^{3(-1)}AB\rangle \end{pmatrix} \tag{I.7}$$

The Hamiltonian operator $\hat{H}$ for the H_2 molecule can be written as the sum of the operator $\hat{H}_1(1)$ for the first hydrogen atom, the operator $\hat{H}_2(2)$ for the second hydrogen atom, and the operator $\hat{H}'(1,2)$ representing the interatomic interactions:

$$\hat{H} = \hat{H}_1(1) + \hat{H}_2(2) + \hat{H}'(1,2) \tag{I.8}$$

$$\hat{H}_1(1) = -\hbar^2\nabla_1^2/2m - \frac{e^2}{r_{1A}} \tag{I.9}$$

$$\hat{H}_2(2) = -\hbar^2\nabla_2^2/2m - \frac{e^2}{r_{2B}} \tag{I.10}$$

$$\hat{H}'(1,2) = -\frac{e^2}{r_{1B}} - \frac{e^2}{r_{2A}} + \frac{e^2}{r_{12}} + \frac{e^2}{R} \tag{I.11}$$

Here, m is electron mass, $\nabla_i^2 = (\partial^2/\partial x_i^2 + \partial^2/\partial y_i^2 + \partial^2/\partial z_i^2)$, and $r_{12} = \sqrt{(x_1 - x_2)^2 + (y_1 - y_2)^2 + (z_1 - z_2)^2}$ is the distance of the two electrons.

The AOs $|A\rangle$ and $|B\rangle$ are eigenfunctions of the atomic Hamiltonians $\hat{H}_1$ and $\hat{H}_2$, with eigenvalues E_H equal to the energy of an isolated H atom, -13.6 eV:

$$\hat{H}_1|A(\mathbf{r}_{1A})\rangle = E_H|A(\mathbf{r}_{1A})\rangle \tag{I.12}$$

$$\hat{H}_2|B(\mathbf{r}_{2B})\rangle = E_H|B(\mathbf{r}_{2B})\rangle \tag{I.13}$$

Since we are now ignoring spin–orbit coupling and other spin-related corrections, the matrix elements between the singlet and triplet VB structures as well as those between the different triplet structures vanish by integration over spin. For a spin-free Hamiltonian, the matrix element separates into a product of integrals and $\int \beta(\xi)\alpha(\xi)\,d\xi$ vanishes as a result of the orthogonality of the spin functions α and β.

The block-diagonal form of the Hamiltonian matrix I.7 means that the three triplet VB structures undergo no further mixing and already represent the wave functions of the three components of the triplet state $|T_1\rangle, |T_0\rangle, |T_{-1}\rangle$. Their energy $E(T_1) = E(T_0) = E(T_{-1})$ can be evaluated by substituting the explicit expressions for $\hat{H}$ and $^3|AB\rangle$:

$$E(T) = \langle ^3AB|\hat{H}|^3AB\rangle = \frac{J'_{AB} - K'_{AB}}{1 - S^2_{AB}} \tag{I.14}$$

where

$$J'_{AB} = \int A(\mathbf{r}_1)B(\mathbf{r}_2)\,\hat{H}\,A(\mathbf{r}_1)B(\mathbf{r}_2)\,d\mathbf{r}_1 d\mathbf{r}_2 \tag{I.15}$$

$$K'_{AB} = \int A(\mathbf{r}_1)B(\mathbf{r}_2)\,\hat{H}\,A(\mathbf{r}_2)B(\mathbf{r}_1)\,d\mathbf{r}_1 d\mathbf{r}_2 \tag{I.16}$$

The first of these integrals, J'_{AB}, is of the "Coulomb" type, with each electron assigned to the same orbital on both sides of the $\hat{H}$ operator. The second, K'_{AB}, is of the "exchange" type, with the electrons interchanged. These two types of integrals appear quite generally in the evaluation of matrix elements over spin-adapted wave functions.

Substitution of the explicit form of the Hamiltonian shown in equations I.8–I.11 and of the explicit form of the AOs given in equations I.1 and I.2, along with the use of equations I.12 and I.13, yields

$$J'_{AB} = \int A(\mathbf{r}_1)B(\mathbf{r}_2)(\hat{H}_1 + \hat{H}_2 + \hat{H}')A(\mathbf{r}_1)B(\mathbf{r}_2)\,d\mathbf{r}_1 d\mathbf{r}_2$$

$$= 2E_H + 2J''_{AB} + J_{AB} + \frac{e^2}{R} = 2E_H + J^{vb}_{AB} \tag{I.17}$$

where

$$J''_{AB} = \int A(\mathbf{r}_1)\frac{-e^2}{r_{1B}}A(\mathbf{r}_1)\,d\tau_1 = \int B(\mathbf{r}_2)\frac{-e^2}{r_{2A}}B(\mathbf{r}_2)\,d\tau_2 \tag{I.18}$$

$$J_{AB} = \int [A(\mathbf{r}_1)B(\mathbf{r}_2)]^2\frac{e^2}{r_{12}}\,d\mathbf{r}_1 d\mathbf{r}_2 \tag{I.19}$$

The "Coulomb integral" of VB theory, J^{vb}_{AB}, is defined by

$$J^{vb}_{AB} = 2J''_{AB} + J_{AB} + \frac{e^2}{R} \tag{I.20}$$

and is usually labeled J or Q.

In a similar fashion,

$$K'_{AB} = \int A(\mathbf{r}_1)B(\mathbf{r}_2)(\hat{H}_1 + \hat{H}_2 + \hat{H}')A(\mathbf{r}_2)B(\mathbf{r}_1)\,d\mathbf{r}_1 d\mathbf{r}_2$$
$$= 2E_H S^2_{AB} + 2S_{AB}K''_{AB} + K_{AB} + e^2 S^2_{AB}/R = 2E_H S^2_{AB} + K^{vb}_{AB} \quad (I.21)$$

where

$$K''_{AB} = \int A(\mathbf{r}_1) \frac{-e^2}{r_{1B}} B(\mathbf{r}_1)\,d\mathbf{r}_1 = \int A(\mathbf{r}_2) \frac{-e^2}{r_{2A}} B(\mathbf{r}_B)\,d\mathbf{r}_2 \quad (I.22)$$

and

$$K_{AB} = \int A(\mathbf{r}_1)B(\mathbf{r}_2) \frac{e^2}{r_{12}} A(\mathbf{r}_2)B(\mathbf{r}_1)\,d\mathbf{r}_1 d\mathbf{r}_2 \quad (I.23)$$

The "exchange integral" of VB theory, K^{vb}_{AB}, is defined by

$$K^{vb}_{AB} = 2K''_{AB}S_{AB} + K_{AB} + \frac{e^2}{R} S^2_{AB} \quad (I.24)$$

and is usually labeled K or J.

The final expression for the energy of the triplet state, then, is

$$E(T_1) = 2E_H + \frac{e^2}{R} + \frac{2J''_{AB} - 2S_{AB}K''_{AB} + J_{AB} - K_{AB}}{1 - S^2_{AB}}$$
$$= 2E_H + \frac{J^{vb}_{AB} - K^{vb}_{AB}}{1 - S^2_{AB}} \quad (I.25)$$

Simple physical significance can be associated in a pictorial way with the individual terms in this expression. The energy deviates from that of two isolated hydrogen atoms in the following ways: (i) the repulsion of the two nuclei (e^2/R); (ii) the repulsion of the two electrons in their respective AOs, given by $J_{AB} - K_{AB}$, since they have equal spins (cf. Section 1.3.2); (iii) the attraction of each electron in its respective AO to the nucleus of the other atom, J''_{AB}; and (iv) an amount related to the attraction of the nonclassical overlap electron density to the two nuclei, $S_{AB}K''$; this density has moved out of the region between the nuclei. Formally, this is as if a positively charged overlap density repelled the nuclei. The denominator corrects for the fact that the integrals in the nominator are computed for nonorthogonal orbitals.

The one-center nuclear attraction integral has the value

$$J''_{AB} = e^2[-1/R + e^{-2R}(1 + 1/R)] \quad (I.26)$$

and the two-center nuclear attraction integral is given by

$$K''_{AB} = -e^2 e^{-R}(1 + R) \tag{I.27}$$

The overlap integral is given by

$$S_{AB} = e^{-R}(1 + R + R^2/3) \tag{I.28}$$

The two-electron Coulomb and exchange repulsion integrals are the hardest to evaluate. The results are

$$J_{AB} = e^2[1/R - e^{-2R}(1/R + 11/8 + 3R/4 + R^2/6)] \tag{I.29}$$

$$K_{AB} = e^2/5 \{-e^{-2R}(-25/8 + 23R/4 + 3R^2 + R^3/3)$$
$$+ (6/R)[S^2_{AB}(\gamma + \log R) + S'^2_{AB} \operatorname{Ei}(-4R)$$
$$- 2S_{AB}S'_{AB} \operatorname{Ei}(-2R)]\} \tag{I.30}$$

where

$$S'_{AB} = e^R(1 - R + R^2/3) \tag{I.31}$$

the function Ei is the integral logarithm, and Euler's constant γ equals 0.5772....

Now we can turn our attention to the wave functions of the three singlet states, to be obtained by the diagonalization of the 3 × 3 matrix in the upper left corner of matrix I.7. For its diagonal matrix elements we obtain

$$E(^1AB) = \langle ^1AB|\hat{H}|^1AB\rangle = 2E_H + \frac{e^2}{R} + \frac{2J''_{AB} + 2S_{AB}K''_{AB} + J_{AB} + K_{AB}}{1 + S^2_{AB}}$$
$$= 2E_H + \frac{J^{vb}_{AB} + K^{vb}_{AB}}{1 + S^2_{AB}} \tag{I.32}$$

$$E(^1A^2) = \langle ^1A^2|\hat{H}|^1A^2\rangle = E(^1B^2) = \langle ^1B^2|\hat{H}|^1B^2\rangle = E_{H^-} + 2J''_{AB} + \frac{e^2}{R}$$
$$= 2E_H + E^{CT}_{AB} \tag{I.33}$$

The charge-transfer energy is defined by

$$E^{CT}_{AB} = E_{H^-} - 2E_H + 2J''_{AB} + \frac{e^2}{R}$$
$$= -E_H - (E_H - E_{H^-}) + 2J''_{AB} + \frac{e^2}{R} \tag{I.34}$$

where $-E_H$ is the ionization potential of the hydrogen atom, $E_H - E_{H^-}$ is its electron affinity, and $2J''_{AB} + e^2/R$ is the energy of electrostatic attraction between H^- and H^+.

Simple physical significance can again be associated with these results. The expression for the energy of the dot–dot structure resembles that for the energy of the triplet structure, with some important differences: The repulsion of the two electrons is $J_{AB} + K_{AB}$, as it should be, since they are now singlet-coupled (cf.

Section 1.3.2), and the term related to the attraction of the overlap density to the nuclei, $2S_{AB}K''_{AB}$, enters with the opposite sign. This is again sensible since now the overlap electron density is negatively charged; i.e., the interference term from the overlap of the two AOs enhances electron density in the internuclear region.

The expression I.33 for the energy of the ionic VB structure, H^-H^+, is even easier to understand. It is given by the energy of the H^- ion and the energy of the electrostatic attraction between H^- and H^+. In the present approximation, $E_{H^-} = -3/8$ a.u. $= -10.2$ eV, which is higher than the energy of an H atom with the electron removed to infinity, -13.6 eV. In reality, H^- is bound and its energy is $E_{H^-} = -0.53$ a.u. $= -14.36$ eV. For the present purposes, it does not matter what the actual absolute value of E_{H^-} is, since we are primarily interested in its dependence on the internuclear distance R.

The off-diagonal elements are

$$\langle ^1AB|\hat{H}|^1A^2\rangle = \langle ^1AB|\hat{H}|^1B^2\rangle$$

$$= \frac{2}{\sqrt{1+S_{AB}^2}}\left[\left(2E_H + J''_{AB} + \frac{e^2}{R}\right)S_{AB} + K''_{AB} + (AA|AB)\right] \quad (I.35)$$

$$\langle ^1A^2|\hat{H}|^1B^2\rangle = \left(2E_H + \frac{e^2}{R}\right)S_{AB}^2 + 2K''_{AB}S_{AB} + K_{AB}$$

$$= 2E_H S_{AB}^2 + K_{AB}^{vb} \quad (I.36)$$

where the hybrid electron repulsion integral $(AA|AB)$ is defined by

$$(AA|AB) = \langle A(1)A(2)\left|\frac{e^2}{r_{12}}\right|A(1)B(2)\rangle \quad (I.37)$$

and represents the repulsion of the overlap density with the density due to an electron in orbital $|A\rangle$. Its evaluation yields

$$(AA|AB) = e^{-R}\left(R + \frac{1}{8} + \frac{5}{16R}\right) - e^{-3R}\left(\frac{1}{8} + \frac{5}{16R}\right) \quad (I.38)$$

These off-diagonal elements consist of terms that depend on the existence of overlap density; it appears to the first power in each of the terms in equation I.35 and to the second power in each of the terms in equation I.36. This means that the off-diagonal elements vanish as the internuclear distance R becomes infinitely large and the VB structures then become eigenstates, as is reasonable on physical grounds.

We can now block-diagonalize the 3×3 matrix of singlets by transforming the $^1|A^2\rangle$, $^1|B^2\rangle$ VB structures into the zwitterionic wave functions $^1|Z_1\rangle$ and $^1|Z_2\rangle$ (equations I.5 and I.6). It acquires the form:

$$\begin{array}{c} ^1|AB\rangle: \\ ^1|Z_2\rangle: \\ ^1|Z_1\rangle: \end{array} \begin{pmatrix} \langle ^1AB|\hat{H}|^1AB\rangle & \langle ^1AB|\hat{H}|^1Z_2\rangle & 0 \\ \langle ^1AB|\hat{H}|^1Z_2\rangle & \langle ^1Z_2|\hat{H}|^1Z_2\rangle & 0 \\ 0 & 0 & \langle ^1Z_1|\hat{H}|^1Z_1\rangle \end{pmatrix} \quad (I.39)$$

The energy of the eigenstate $^1|Z_1\rangle$, which we shall refer to as the S state, is then given by

$$E(^1Z_1) = E(S) = \langle ^1Z_1|\hat{H}|^1Z_1\rangle = \left\{ E_{H^-} + \frac{e^2}{R} + 2J''_{AB} \right.$$
$$\left. - \left[\left(2E_H + \frac{e^2}{R}\right) S^2_{AB} + 2S_{AB}K''_{AB} + K_{AB} \right] \right\} \bigg/ (1 + S^2_{AB}) \quad (I.40)$$

The diagonal element corresponding to the symmetric combination of the ionic configurations is

$$E(^1Z_2) = \langle ^1Z_2|\hat{H}|^1Z_2\rangle = \left\{ E_{H^-} + \frac{e^2}{R} + 2J''_{AB} \right.$$
$$\left. + \left(2E_H + \frac{e^2}{R}\right) S^2_{AB} + 2S_{AB}K''_{AB} + K_{AB} \right\} \bigg/ (1 + S^2_{AB}) \quad (I.41)$$

The solution of a quadratic equation for the 2 × 2 block of matrix I.39 is

$$\left.\begin{array}{l} E(G) \\ E(D) \end{array}\right\} = \{E(^1AB) + E(^1Z_2) - 2\Delta_{AB}\langle ^1AB|\hat{H}|^1Z_2\rangle \mp [2(1 - \Delta^2_{AB})]^{-1}$$
$$\times \sqrt{\begin{array}{l} [E(^1AB) + E(^1Z_2) - 2\Delta_{AB}\langle ^1AB|\hat{H}|^1Z_2\rangle]^2 \\ -4(1 - \Delta^2_{AB})[E(^1AB)E(^1Z_2) - \langle ^1AB|\hat{H}|^1Z_2\rangle^2] \end{array}} \quad (I.42)$$

where

$$\Delta_{AB} = 2S_{AB}/(1 + S_{AB}) \quad (I.43)$$

The lower energy state, $|G\rangle$, represents the ground state S_0 at all values of R. The upper state, $|D\rangle$, represents the S_2 state of the H_2 molecule at all values of R in the present approximation. Figure 4.6 shows the dependence of the energies of the VB structures, of the off-diagonal matrix elements between them, and of the final state energies on the internuclear distance R calculated for AOs similar to those in equations I.1 and I.2 but using an exponential coefficient of 1.4. All three singlet states are bound, and the triplet is purely repulsive. In the picture employed here, the principal reason for the repulsive nature of the triplet is the removal of overlap density from the internuclear region where it would have been attracted to both nuclei (the "exchange term" $-S_{AB}K''_{AB}$). The reason for the presence of the minimum in the E(G) curve is the addition of this overlap density into the internuclear region (the "exchange term" $+S_{AB}K''_{AB}$). The two hole–pair configurations, $^1|A^2\rangle$ and $^1|B^2\rangle$, are bound primarily because of the mutual attraction of the ions. The interaction between $^1|A^2\rangle$ and $^1|B^2\rangle$, which yields $^1|Z_1\rangle$ and $^1|Z_2\rangle$, is relatively weak, and thus the latter are still bound. At short internuclear distances, all the curves rise rapidly because the point-charge repulsion of the nuclei is then quite inadequately shielded by the electron density, only a small fraction of which is then between them. The interaction (resonance) between the dot–dot structure $^1|AB\rangle$

and the zwitterionic structure $^1|Z_2\rangle$ does not change the overall picture very much but is still significant. It reflects the physical fact that electron repulsion cannot keep the electron motions completely correlated. As long as there is some overlap, there is some probability that both electrons will be found in the vicinity of one of the nuclei. Understandably, the importance of the mixing decreases as R increases. The effect of the $^1|AB\rangle$, $^1|Z_2\rangle$ interaction on the ground-state curve $|G\rangle$ is to make the minimum somewhat deeper and to shift it to a smaller R value. Its effect on the $|D\rangle$ state is to make its minimum somewhat shallower and to shift it to a larger R value.

The MO approach

Molecular orbitals. In the MO procedure, we first construct one-electron wave functions that extend over the whole molecule. These MOs will be obtained as linear combinations of the same two AOs which we have used in the VB procedure. In general, finding optimal coefficients in linear combinations is a tedious iterative process, but our case is simple enough that they are determined by symmetry:

$$|1(\mathbf{r}_1)\rangle = \frac{1}{\sqrt{2 + 2S_{AB}}} [|A(\mathbf{r}_1)\rangle + |B(\mathbf{r}_1)\rangle] \qquad (\text{I}.44)$$

$$|-1(\mathbf{r}_1)\rangle = \frac{1}{\sqrt{2 - 2S_{AB}}} [|A(\mathbf{r}_1)\rangle - |B(\mathbf{r}_1)\rangle] \qquad (\text{I}.45)$$

We shall now see that the energy of an electron in the symmetric combination $|1\rangle$ is lower than that of an electron in the antisymmetric combination $|-1\rangle$. The former is a bonding MO, whereas the latter is an antibonding MO. If only one electron were present in the system, the species in question actually would be H_2^+. The Hamiltonian operator for H_2^+ is $\hat{h}(1) + e^2/R$, where

$$\hat{h}(1) = -\hbar^2 \nabla_1^2/2m - e^2/r_{1A} - e^2/r_{1B} \qquad (\text{I}.46)$$

and its expectation values are

$$\begin{aligned} h_{\pm 1} &= \langle \pm 1(1)|\hat{h}(1)|\pm 1(1)\rangle \\ &= \frac{1}{1 \pm S_{AB}} [\langle A(1)|\hat{h}(1)|A(1)\rangle \pm \langle A(1)|\hat{h}(1)|B(1)\rangle] \\ &= E_H + \frac{J''_{AB} \pm K''_{AB}}{1 \pm S_{AB}} \end{aligned} \qquad (\text{I}.47)$$

where the upper signs apply to $|1\rangle$ and the lower signs to $|-1\rangle$. The physical significance of the individual terms in the expression I.47, E_H, J''_{AB}, and K''_{AB}, is already familiar from the discussion of the VB model. Since K''_{AB} is negative, $|1\rangle$ is of lower energy than $|-1\rangle$. It can be noted that $|1\rangle$ is less bonding than $|-1\rangle$ is antibonding. The total energy of H_2^+ as a function of R is equal to $h_{\pm 1} + e^2/R$.

Configurations and their mixing. Next, we turn our attention to H_2 and construct many-electron wave functions (MO configurations) by occupying the MOs by the two electrons in all possible ways. These are obtained in their spin-adapted form by taking suitable linear combinations of the simple product functions and have been listed in Chapter 4: equations 4.1–4.6, with $a = |-1\rangle$ and $b = |1\rangle$. In order to obtain state wave functions as linear combinations of configurations and to obtain their energies, we need to diagonalize a Hamiltonian matrix. The results were given without proof in Table 4.1.

As in the VB case, the triplet part is already diagonal and the singlet part separates into a 2 × 2 and a 1 × 1 part. The Hamiltonian operator $\hat{H}$ for the H_2 molecule is now written as the sum of the one-electron operators $\hat{h}(1)$ for the first electron and $\hat{h}(2)$ for the second electron, the two-electron operator $\hat{g}(1,2) = e^2/r_{12}$ for the electron–electron repulsion, and the nuclear–nuclear repulsion term e^2/R:

$$\hat{H} = \hat{h}(1) + \hat{h}(2) + \hat{g}(1,2) + e^2/R \tag{I.48}$$

The three triplet configurations represent the wave functions of the three components of the triplet state, namely, $|T_1\rangle$, $|T_0\rangle$, and $|T_{-1}\rangle$. Their energy $E(T_1) = E(T_0) = E(T_{-1})$ is evaluated by substituting the explicit forms of $\hat{H}$ and $^3|1-1\rangle$ (cf. Table 4.1):

$$E(T) = \langle ^3(1-1)|\hat{H}|^3(1-1)\rangle = h_1 + h_{-1} + J_{1,-1} - K_{1,-1} + e^2/R \tag{I.49}$$

The Coulomb (J) and exchange (K) integrals are already familiar (equations 1.4 and 1.5) and can be expressed in terms of AOs by substituting from equations I.44 and I.45:

$$J_{1,-1} = \langle 1(\mathbf{r}_1)|\langle -1(\mathbf{r}_2)|\frac{e^2}{r_{12}}|1(\mathbf{r}_1)\rangle|-1(\mathbf{r}_2)\rangle$$

$$= \frac{1}{4(1-S_{AB}^2)}(J_{AA} + J_{BB} + 2J_{AB} - 4K_{AB}) \tag{I.50}$$

$$K_{1,-1} = \langle 1(\mathbf{r}_1)|\langle -1(\mathbf{r}_2)|\frac{e^2}{r_{12}}|-1(\mathbf{r}_1)\rangle|1(\mathbf{r}_2)\rangle$$

$$= \frac{1}{4(1-S_{AB}^2)}(J_{AA} + J_{BB} - 2J_{AB}) \tag{I.51}$$

For the one-electron energies $h_{\pm 1}$, substitution from equations I.44 and I.45 yields

$$h_{\pm 1} = E_H + \frac{J''_{AB} \pm K''_{AB}}{1 \pm S_{AB}} \tag{I.52}$$

Substitution of equations I.50, I.51, and I.52 into the MO expression for the triplet energy, equation I.49, yields the result already obtained from the VB procedure, equation I.25.

The physical interpretation of the result I.49 is simple: The energy of the triplet state is the sum of the energies h_1 and h_{-1} of the two electrons, one in orbital $|1\rangle$, the other in $|-1\rangle$, in the absence of electron–electron repulsions, plus the average energy of their repulsion, given by $J_{1,-1} - K_{1,-1}$ since their spins are the same, plus the energy due to nuclear repulsion.

The triplet energy can also be cast in terms of the orbital energies $\varepsilon_{\pm 1}$ which already contain corrections for electron–electron repulsions (Table 4.1). The energy ε_1 is that of an electron in the orbital $|1\rangle$ in the ground singlet configuration $^1|1^2\rangle$, so that it is equal to minus the ionization potential (Koopmans' theorem). It is given by the sum of the one-electron energy for an electron in $|1\rangle$ and the repulsion with the other electron:

$$\varepsilon_1 = h_1 + J_{1,1} \qquad (I.53)$$

The energy ε_{-1} is that of an electron in the orbital $|-1\rangle$ in the ground state of the radical anion H_2^-, so that it is equal to minus the electron affinity of H_2 (Koopmans' theorem). It equals the sum of the one-electron energy for an electron in $|-1\rangle$ and the repulsion with the other two electrons in $|1\rangle$: $J_{1,-1}$ with that of the opposite spin and $J_{1,-1} - K_{1,-1}$ with that of the same spin:

$$\varepsilon_{-1} = h_{-1} + 2J_{1,-1} - K_{1,-1} \qquad (I.54)$$

The energy of the ground configuration $^1|1^2\rangle$ is not equal simply to the sum of the nuclear repulsion energy and of the two-electron orbital energies, $2\varepsilon_1 + e^2/R$, since their repulsion would then be counted twice, and is evaluated by direct substitution as

$$E(^1|1^2\rangle) = \langle 1^2|\hat{H}|1^2\rangle = 2h_1 + J_{1,1} + e^2/R = 2\varepsilon_1 - J_{1,1} + e^2/R \qquad (I.55)$$

where

$$\begin{aligned}
J_{1,1} &= \langle 1(\mathbf{r}_1)|\langle 1(\mathbf{r}_2)|\frac{e^2}{r_{12}}|1(\mathbf{r}_1)\rangle|1(\mathbf{r}_2)\rangle \\
&= \frac{1}{4(1+S_{AB})^2}[J_{AA} + J_{BB} + 2J_{AB} + 4K_{AB} \\
&\quad + 4(AA|AB) + 4(AB|BB)] \qquad (I.56)
\end{aligned}$$

In keeping with the discussion in Section 1.3.2, the energy of the $^3|1\,-1\rangle$ triplet exceeds that of the $^1|1^2\rangle$ configuration by the orbital energy difference $\varepsilon_{-1} - \varepsilon_1$ reduced by the electron repulsion term $J_{1,-1}$:

$$\begin{aligned}
E(T) = E(^3|1\,-1\rangle) &= E(^1|1^2\rangle) + \varepsilon_{-1} - \varepsilon_1 - J_{1,-1} \\
&= \varepsilon_1 + \varepsilon_{-1} - J_{1,1} - J_{1,-1} + e^2/R \qquad (I.57)
\end{aligned}$$

The singlet configuration $^1|1\,-1\rangle$ also represents an eigenstate and does not mix with others. It is referred to as the S state. Its energy is obtained by substituting the explicit forms of $\hat{H}$ (equations I.46 and I.48) and of the orbitals $|1\rangle$ (equation I.44) and $|-1\rangle$ (equation I.45):

$$\begin{aligned}
E(S) = E(^1|1\ -1\rangle) &= \langle ^1(1\ -1)|\hat{H}|^1(1\ -1)\rangle \\
&= E(^1|1^2\rangle) + \varepsilon_{-1} - \varepsilon_1 - J_{1,-1} + 2K_{1,-1} \\
&= h_1 + h_{-1} + J_{1,-1} + K_{1,-1} + e^2/R \\
&= \varepsilon_1 + \varepsilon_{-1} - J_{1,1} - J_{1,-1} + 2K_{1,-1} + e^2/R
\end{aligned} \qquad (I.58)$$

The physical interpretation of this expression is straightforward, whether it is cast in terms of the one-electron energies $h_{\pm 1}$ or orbital energies $\varepsilon_{\pm 1}$: the singlet configuration energy exceeds that of the corresponding triplet by $2K_{1,-1}$ for reasons discussed at length in Section 1.3.2.

Finally, we need to diagonalize the 2×2 block of the Hamiltonian matrix spanned by the singlet configurations $^1|1^2\rangle$ and $^1|-1^2\rangle$. The matrix elements are again evaluated by direct substitution. The result for $^1|1^2\rangle$ has already been given in equation I.55, and the result for $|-1^2\rangle$ is

$$\begin{aligned}
E(^1|-1^2\rangle) = \langle -1^2|\hat{H}|-1^2\rangle &= 2h_{-1} + J_{-1,-1} + e^2/R \\
&= 2\varepsilon_{-1} + J_{-1,-1} - 4J_{1,-1} + 2K_{1,-1} + e^2/R
\end{aligned} \qquad (I.59)$$

For the off-diagonal matrix element we obtain

$$\langle 1^2|\hat{H}|-1^2\rangle = K_{1,-1} \qquad (I.60)$$

Diagonalization then yields the energies of the ground (G) and the doubly excited (D) states:

$$\left.\begin{array}{c} E(G) \\ E(D) \end{array}\right\} = h_1 + h_{-1} + (J_{1,1} + J_{-1,-1})/2 + e^2/R$$
$$\mp \sqrt{[h_1 - h_{-1} + (J_{1,1} - J_{-1,-1})/2]^2 + K_{1,-1}^2} \qquad (I.61)$$

The state energies are thus given by the average energy of the $^1|1^2\rangle$ and $^1|-1^2\rangle$ configurations, minus or plus a contribution whose magnitude exceeds half the difference in the $^1|1^2\rangle$ and $^1|-1^2\rangle$ energies by an amount determined by the size of the off-diagonal mixing element $K_{1,-1}$. Note that the splitting about the average energy $[E(^1|1^2\rangle) + E(^1|-1^2\rangle)]/2$ is symmetric because the configurations $^1|1^2\rangle$ and $^1|-1^2\rangle$ are orthogonal. The analogous VB result (equation I.42) was more complicated and yielded an unsymmetric splitting because the structures $^1|AB\rangle$ and $^1|Z_1\rangle$ were nonorthogonal. However, the average energies of $^1|1^2\rangle$ and $^1|-1^2\rangle$ and of $^1|AB\rangle$ and $^1|Z_1\rangle$ and the respective splittings are such that the final results for $E(G)$ and $E(D)$ are identical. This is seen in Figures 4.5 and 4.6, which also show the weights of the individual VB structures and MO configurations, respectively, in the final states.

This exposition of the MO and VB methods follows J. C. Slater, *Quantum Theory of Molecules and Solids, Vol. 1: Electronic Structure of Molecules*, McGraw-Hill, New York, 1963.

APPENDIX II

The Two-Electron Two-Orbital Model of Biradicals

The two-electron two-orbital model provides an approximation to the electronic structure of biradicals which considers explicitly only two electrons in two approximately nonbonding orbitals ("active space") and represents the others by a fixed core. The model is so simple that it is amenable to an exact algebraic solution, yet sufficiently realistic that it provides a useful way of looking at biradicals, although they typically have many more than two electrons, and these have more than two orbitals available to them.

Fundamentals. Let the two active orbitals $|\mathcal{A}\rangle$ and $|\mathcal{B}\rangle$ be chosen orthogonal. The important one-electron quantities are the overlap and Hückel resonance integrals,

$$S_{\mathcal{AB}} = \langle \mathcal{A}(1)|\mathcal{B}(1)\rangle = 0 \tag{II.1}$$

$$h_{\mathcal{AB}} = \langle \mathcal{A}(1)|\hat{h}(1)|\mathcal{B}(1)\rangle \tag{II.2}$$

where $\hat{h}(1)$ is the one-electron part of the Hamiltonian. The important two-electron quantities are the Coulomb, exchange, and hybrid repulsion integrals already defined in Appendix I:

$$J_{\mathcal{AB}} = \langle \mathcal{A}(1)\mathcal{B}(2)|\frac{e^2}{r_{12}}|\mathcal{A}(1)\mathcal{B}(2)\rangle \tag{II.3}$$

$$K_{\mathcal{AB}} = \langle \mathcal{A}(1)\mathcal{A}(2)|\frac{e^2}{r_{12}}|\mathcal{B}(1)\mathcal{B}(2)\rangle \tag{II.4}$$

$$\langle \mathcal{AA}|\mathcal{AB}\rangle = \langle \mathcal{A}(1)\mathcal{A}(2)|\frac{e^2}{r_{12}}|\mathcal{A}(1)\mathcal{B}(2)\rangle \tag{II.5}$$

It is useful to define several secondary quantities:

$$\gamma_{\mathcal{AB}} = 2h_{\mathcal{AB}} + (\mathcal{AA}|\mathcal{AB})^* + (\mathcal{BB}|\mathcal{BA}) \tag{II.6}$$

$$\gamma^-_{\mathcal{AB}} = (\mathcal{AA}|\mathcal{AB})^* - (\mathcal{BB}|\mathcal{BA}) \tag{II.7}$$

$$\delta_{\mathcal{AB}} = h_{\mathcal{AA}} - h_{\mathcal{BB}} + (J_{\mathcal{AA}} - J_{\mathcal{BB}})/2 \tag{II.8}$$

$$K'_{\mathcal{AB}} = [(J_{\mathcal{AA}} + J_{\mathcal{BB}})/2 - J_{\mathcal{AB}}]/2 \tag{II.9}$$

$$E_0 = h_{\mathcal{A}\mathcal{A}} + h_{\mathcal{B}\mathcal{B}} + (J_{\mathcal{A}\mathcal{A}} + J_{\mathcal{B}\mathcal{B}})/4 + J_{\mathcal{A}\mathcal{B}}/2 \tag{II.10}$$

$$E(T) = E_0 - K_{\mathcal{A}\mathcal{B}} - K_{\mathcal{A}\mathcal{B}} \tag{II.11}$$

Here, $\gamma_{\mathcal{A}\mathcal{B}}$ corresponds roughly to twice the Hückel resonance integral between the orbitals $|\mathcal{A}\rangle$ and $|\mathcal{B}\rangle$, whereas $\delta_{\mathcal{A}\mathcal{B}}$ reflects the relative electronegativities of the orbitals $|\mathcal{A}\rangle$ and $|\mathcal{B}\rangle$ and corresponds approximately to the difference of their Hückel "Coulomb" integrals. We always choose $|\mathcal{B}\rangle$ to be the more electronegative orbital, so that $\delta_{\mathcal{A}\mathcal{B}} \geq 0$.

A complete basis set for the electronic states of the model system is provided by the six two-electron functions

$$^{3(u)}|\mathcal{A}\mathcal{B}\rangle = (1/\sqrt{2})[|\mathcal{A}(1)\rangle|\mathcal{B}(2)\rangle - |\mathcal{B}(1)\rangle|\mathcal{A}(2)\rangle]|\Theta_u\rangle,$$

$$u = x,y,z \text{ or } -1,0,1$$

$$^1|\mathcal{A}\mathcal{B}\rangle = (1/\sqrt{2})[|\mathcal{A}(1)\rangle|\mathcal{B}(2)\rangle + |\mathcal{B}(1)\rangle|\mathcal{A}(2)\rangle]|\Sigma\rangle \tag{II.12}$$

$$^1|\mathcal{A}^2 - \mathcal{B}^2\rangle = (1/\sqrt{2})[|\mathcal{A}(1)\rangle|\mathcal{A}(2)\rangle - |\mathcal{B}(1)\rangle|\mathcal{B}(2)\rangle]|\Sigma\rangle$$

$$^1|\mathcal{A}^2 + \mathcal{B}^2\rangle = (1/\sqrt{2})[|\mathcal{A}(1)\rangle|\mathcal{A}(2)\rangle + |\mathcal{B}(1)\rangle|\mathcal{B}(2)\rangle]|\Sigma\rangle$$

where the singlet spin function Σ and the three triplet spin functions Θ_{-1}, Θ_0, and Θ_1 are those used in Chapters 1 and 4, and the triplet spin functions Θ_x, Θ_y, and Θ_z are defined in the usual manner:

$$\Theta_x = -(1/\sqrt{2})[\alpha(1)\alpha(2) - \beta(1)\beta(2)]$$

$$\Theta_y = (i/\sqrt{2})[\alpha(1)\alpha(2) + \beta(1)\beta(2)] \tag{II.13}$$

$$\Theta_z = (1/\sqrt{2})[\alpha(1)\beta(2) + \beta(1)\alpha(2)]$$

In this basis, the Hamiltonian matrix is

$$\begin{array}{r} ^1|\mathcal{A}^2 - \mathcal{B}^2\rangle: \\ ^1|\mathcal{A}^2 + \mathcal{B}^2\rangle: \\ ^1|\mathcal{A}\mathcal{B}\rangle \quad : \\ ^3|\mathcal{A}\mathcal{B}\rangle \quad : \end{array} \begin{pmatrix} E(T) + 2K'_{\mathcal{A}\mathcal{B}} & \delta_{\mathcal{A}\mathcal{B}} & \gamma_{\mathcal{A}\mathcal{B}} & 0 \\ \delta_{\mathcal{A}\mathcal{B}} & E(T) + 2(K'_{\mathcal{A}\mathcal{B}} + K_{\mathcal{A}\mathcal{B}}) & \gamma_{\mathcal{A}\mathcal{B}} & 0 \\ \gamma_{\mathcal{A}\mathcal{B}} & \gamma_{\mathcal{A}\mathcal{B}} & E(T) + 2K_{\mathcal{A}\mathcal{B}} & 0 \\ 0 & 0 & 0 & E(T) \end{pmatrix}$$

$$\tag{II.14}$$

in the absence of outside fields.

Orbital transformations. There is an infinite number of equivalent possible choices of the orthogonal basis orbitals $|\mathcal{A}\rangle$ and $|\mathcal{B}\rangle$. Starting with an arbitrary choice $|\mathcal{A}\rangle$ and $|\mathcal{B}\rangle$, any other choice can be reached by a unitary transformation characterized by parameters ω and ϕ:

$$(|\mathcal{A}_{\omega\phi}\rangle, |\mathcal{B}_{\omega\phi}\rangle) = (|\mathcal{A}\rangle, |\mathcal{B}\rangle) \mathsf{U}(\omega, \phi) \tag{II.15}$$

where

$$U(\omega, \phi) = \begin{pmatrix} \cos \omega & -e^{-i\phi} \sin \omega \\ e^{i\phi} \sin \omega & \cos \omega \end{pmatrix} \quad (\text{II}.16)$$

The two-electron functions in the transformed basis are

$$\begin{pmatrix} {}^1|\mathscr{A}_{\omega\phi}^2 - \mathscr{B}_{\omega\phi}^2\rangle \\ {}^1|\mathscr{A}_{\omega\phi}^2 + \mathscr{B}_{\omega\phi}^2\rangle \\ {}^1|\mathscr{A}_{\omega\phi} \mathscr{B}_{\omega\phi}\rangle \\ {}^3|\mathscr{A}_{\omega\phi} \mathscr{B}_{\omega\phi}\rangle \end{pmatrix} = \begin{pmatrix} \cos 2\omega + 2\sin^2\phi \sin^2\omega & i\sin 2\phi \sin^2\omega & \cos\phi \sin 2\omega & 0 \\ -i\sin 2\phi \sin^2\omega & 1 - 2\sin^2\phi \sin^2\omega & i\sin\phi \sin 2\omega & 0 \\ -\cos\phi \sin 2\omega & i\sin\phi \sin 2\omega & \cos 2\omega & 0 \\ 0 & 0 & 0 & 1 \end{pmatrix} \begin{pmatrix} {}^1|\mathscr{A}^2 - \mathscr{B}^2\rangle \\ {}^1|\mathscr{A}^2 + \mathscr{B}^2\rangle \\ {}^1|\mathscr{A}\mathscr{B}\rangle \\ {}^3|\mathscr{A}\mathscr{B}\rangle \end{pmatrix} \quad (\text{II}.17)$$

Of all the choices of the orbital pair $|\mathscr{A}\rangle, |\mathscr{B}\rangle$, three are of special importance; all of them make $\gamma^-_{\mathscr{A}\mathscr{B}}$ vanish. These are the most localized real orbital pair $|A\rangle, |B\rangle$, the most delocalized real orbital pair $|a\rangle, |b\rangle$, and the pair of most delocalized orbitals related by complex conjugation, $|c\rangle, |c^*\rangle$. They are mutually related by

$$|a\rangle = (|A\rangle - |B\rangle)/\sqrt{2} \quad (\text{II}.18)$$

$$|b\rangle = (|A\rangle + |B\rangle)/\sqrt{2} \quad (\text{II}.19)$$

$$|c\rangle = (|A\rangle + i|B\rangle)/\sqrt{2} \quad (\text{II}.20)$$

$$|c^*\rangle = (|A\rangle - i|B\rangle)/\sqrt{2} \quad (\text{II}.21)$$

The orbital set $|a\rangle, |b\rangle$ is obtained from $|A\rangle, |B\rangle$ by the transformation $U(\pi/4, 0)$ ($|A_{\pi/4,0}\rangle = |a\rangle$, $|B_{\pi/4,0}\rangle = |b\rangle$), and the set $|c\rangle, |c^*\rangle$ is obtained by the transformation $U(\pi/4, \pi/2)$ ($|A_{\pi/4,\pi/2}\rangle = |c^*\rangle$, $|B_{\pi/4,\pi/2}\rangle = i|c\rangle$). The most localized real orbital pair $|A\rangle, |B\rangle$ is defined by the condition that the repulsion integrals J_{AB} and K_{AB} be as small as possible.

The orbitals $|\mathscr{A}\rangle, |\mathscr{B}\rangle$ are most often chosen to be real and then, only orthogonal transformations $U(\omega, 0)$ need to be considered. These convert an orbital pair $|\mathscr{A}\rangle, |\mathscr{B}\rangle$, to the pair $|\mathscr{A}_\omega\rangle, |\mathscr{B}_\omega\rangle$. The effect on the two-electron basis functions is given by

$$\begin{pmatrix} {}^1|\mathscr{A}_\omega^2 - \mathscr{B}_\omega^2\rangle \\ {}^1|\mathscr{A}_\omega^2 + \mathscr{B}_\omega^2\rangle \\ {}^1|\mathscr{A}_\omega \mathscr{B}_\omega\rangle \\ {}^3|\mathscr{A}_\omega \mathscr{B}_\omega\rangle \end{pmatrix} = \begin{pmatrix} \cos 2\omega & 0 & \sin 2\omega & 0 \\ 0 & 1 & 0 & 0 \\ -\sin 2\omega & 0 & \cos 2\omega & 0 \\ 0 & 0 & 1 & 1 \end{pmatrix} \begin{pmatrix} {}^1|\mathscr{A}^2 - \mathscr{B}^2\rangle \\ {}^1|\mathscr{A}^2 + \mathscr{B}^2\rangle \\ {}^1|\mathscr{A}\mathscr{B}\rangle \\ {}^3|\mathscr{A}\mathscr{B}\rangle \end{pmatrix} \quad (\text{II}.22)$$

In order to describe the effect of the transformation from $|\mathscr{A}\rangle, |\mathscr{B}\rangle$ to $|\mathscr{A}_\omega\rangle$, $|\mathscr{B}_\omega\rangle$ on the various other quantities occurring in the model, we use subscripted quantities such as J_ω, K_ω, γ_ω^-, etc., as shorthand notation for $J_{\mathscr{A}_\omega \mathscr{B}_\omega}$, $K_{\mathscr{A}_\omega \mathscr{B}_\omega}$, $\gamma^-_{\mathscr{A}_\omega \mathscr{B}_\omega}$, etc., respectively. Quantities without subscripts refer to $\omega = 0$, that is, $J_{\mathscr{A}\mathscr{B}}$, $K_{\mathscr{A}\mathscr{B}}$, $\gamma^-_{\mathscr{A}\mathscr{B}}$, etc. The quantities E, E(T), $(K + K')$, $(J - K)$, $(\gamma^2 + \delta^2)$, and $[(K' - K)^2 + (\gamma^-)^2]$ are invariant, and

$$(\delta_\omega, \gamma_\omega) = (\delta, \gamma) U(2\omega, 0) \quad (\text{II}.23)$$

$$[(K'_\omega - K_\omega), \gamma_\omega^-] = [(K' - K), \gamma^-] U(4\omega, 0) \quad (\text{II}.24)$$

The important derivatives with respect to ω are

$$\frac{dK'_\omega}{d\omega} = -\frac{dK_\omega}{d\omega} = -\frac{dJ_\omega}{d\omega} = 2\gamma_\omega^-$$

$$\frac{d\gamma_\omega^-}{d\omega} = 4(K_\omega - K'_\omega) \tag{II.25}$$

$$\frac{d\gamma_\omega}{d\omega} = -2\delta_\omega$$

$$\frac{d\delta_\omega}{d\omega} = -2\gamma_\omega$$

The rotation angle ω needed to produce either $|A\rangle,|B\rangle$, or $|a\rangle,|b\rangle$ from an arbitrarily chosen initial set $|\mathscr{A}\rangle,|\mathscr{B}\rangle$ is given by

$$\omega = (1/4)\tan^{-1}[\gamma^-/(K' - K)] \tag{II.26}$$

This equation has a positive root ω_+ and a negative root $\omega_- = \omega_+ - \pi/4$ in the range of interest, $-\pi/4 \leq \omega \leq \pi/4$. Considering that the sign of $[dK'_\omega/d\omega]_{\omega=0}$ is equal to the sign of γ^- and that the second derivative of K'_ω is

$$\frac{d^2 K'_\omega}{d\omega^2} = 8(K'_\omega - K_\omega) \tag{II.27}$$

we see that

$$\begin{aligned}(|A\rangle,|B\rangle) &= (|\mathscr{A}\rangle,|\mathscr{B}\rangle)\ \mathsf{U}\ (\omega_-,0) \\ (|a\rangle,|b\rangle) &= (|\mathscr{A}\rangle,|\mathscr{B}\rangle)\ \mathsf{U}\ (\omega_+,0)\end{aligned} \quad \text{if } \gamma^- > 0 \tag{II.28}$$

$$\begin{aligned}(|A\rangle,|B\rangle) &= (|\mathscr{A}\rangle,|\mathscr{B}\rangle)\ \mathsf{U}\ (\omega_+,0) \\ (|a\rangle,|b\rangle) &= (|\mathscr{A}\rangle,|\mathscr{B}\rangle)\ \mathsf{U}\ (\omega_-,0)\end{aligned} \quad \text{if } \gamma^- < 0 \tag{II.29}$$

If $\gamma^- = 0$ and $K' > K$, then $|\mathscr{A}\rangle,|\mathscr{B}\rangle$ are equal to $|A\rangle,|B\rangle$ and $(|a\rangle,|b\rangle) = (|\mathscr{A}\rangle,|\mathscr{B}\rangle)\ \mathsf{U}\ (\pm \pi/4, 0)$. Finally, if $\gamma^- = K' - K = 0$, all orbital choices are localized to the same degree and K, K', J, and γ^- are each invariant with respect to ω.

It is useful to note the identities

$$\begin{aligned} K_{AB} &= K'_{ab} \\ K'_{AB} &= K_{ab} \end{aligned} \tag{II.30}$$

Perfect biradicals. In these, $\gamma_{\mathscr{AB}} = \delta_{\mathscr{AB}} = 0$, and the Hamiltonian matrix will be diagonal in the basis chosen above if the orbital pair $|\mathscr{A}\rangle,|\mathscr{B}\rangle$ is chosen so as to make $\gamma^-_{\mathscr{AB}}$ vanish. For real orbitals, this permits the choices $|A\rangle,|B\rangle$ and $|a\rangle,|b\rangle$; the energies and wave functions of the triplet and three singlet states relative to E_0, then, are:

Wave function	Energy
S_2: $(1/\sqrt{2})^1\|A^2 + B^2\rangle = (1/\sqrt{2})^1\|a^2 + b^2\rangle$	$K'_{AB} + K_{AB} = K'_{ab} + K_{ab}$
S_1: $(1/\sqrt{2})^1\|A^2 - B^2\rangle = {}^1\|ab\rangle$	$K'_{AB} - K_{AB} = -K'_{ab} + K_{ab}$
S_0: ${}^1\|AB\rangle = (1/\sqrt{2})^1\|a^2 - b^2\rangle$	$-K'_{AB} + K_{AB} = K'_{ab} - K_{ab}$
T_1: ${}^3\|AB\rangle$	$-K'_{AB} - K_{AB} = -K'_{ab} - K_{ab}$

(II.31)

For other choices of real orbitals $|\mathscr{A}\rangle$ and $|\mathscr{B}\rangle$, $\gamma^-_{\mathscr{AB}}$ does not vanish and the Hamiltonian matrix is only block-diagonal. The expression for the wave function of the S_2 state, then, is $(1/\sqrt{2})^1|\mathscr{A}^2 + \mathscr{B}^2\rangle$; the expression for its energy is $K'_{\mathscr{AB}} + K_{\mathscr{AB}}$, and those for the T_1 state are ${}^3|\mathscr{AB}\rangle$ and $-K'_{\mathscr{AB}} - K_{\mathscr{AB}}$, respectively. The wave functions of the S_1 and S_0 states have the form of linear combinations of ${}^1|\mathscr{A}^2 - \mathscr{B}^2\rangle$ and ${}^1|\mathscr{AB}\rangle$, and their energies can be written as $\pm[(K'_{\mathscr{AB}} - K_{\mathscr{AB}})^2 + (\gamma^-)^2]^{1/2}$. Of course, the arbitrary decision concerning the choice of the orbital basis set has no real effect on the state wave functions nor energies, and it affects only their apparent form.

Perfect biradicals for which $K_{AB} = 0$ are called pair biradicals, since K_{AB} can vanish exactly only for biradicals whose localized orbitals A and B are infinitely far apart and which therefore are a pair of noninteracting radicals. In pair biradicals, the S_0 and T_1 states are degenerate within the present approximation, as are S_1 and S_2. The separation of the two pairs of states is $2K'_{AB}$. In these biradicals, the components of the degenerate S_1, S_2 state can also be written as ${}^1|A^2\rangle$ and ${}^1|B^2\rangle$, one with a positive and the other with a negative dipole moment along the line joining the centers of the two localized orbitals. Application of an outside electric field along this direction will split these two states by the Stark effect [the Stark perturbation takes the form of δ_{AB} when the Hamiltonian matrix II.14 is written in the $|A\rangle, |B\rangle$ basis]. In practice, K_{AB} cannot vanish exactly, and only "nearly" pair biradicals are possible.

Perfect biradicals for which $K_{AB} = K'_{AB}$ are called axial biradicals, since this equality is normally imposed by the presence of a threefold or higher axis of symmetry. In axial biradicals, the S_0 and S_1 states are degenerate; T_1 lies $2K_{AB}$ lower, and S_2 lies $2K_{AB}$ higher. In these biradicals, the projection of orbital angular momentum into the symmetry axis is a useful approximate quantum number; the properly symmetry-adapted choice of orbitals $|\mathscr{A}\rangle, |\mathscr{B}\rangle$ are the complex orbitals $|c\rangle, |c^*\rangle$ (in one of these, the electron circulates clockwise and in the other, counterclockwise, around the symmetry axis). In this basis, the four wave functions are $T_1: i^3|cc^*\rangle$, $S_0: (i/\sqrt{2})^1|c^{*2} - c^2\rangle$, $S_1: (1/\sqrt{2})^1|c^{*2} + c^2\rangle$, $S_2: {}^1|cc^*\rangle$. In the T_1 and S_2 states the orbital angular momenta of the electrons in the $|c\rangle$ and $|c^*\rangle$ orbitals cancel. In the degenerate S_0, S_1 state, the two component states can be written as ${}^1|c^2\rangle$ and ${}^1|c^{*2}\rangle$, one with a positive and the other with a negative angular momentum component along the symmetry axis. Application of an outside magnetic field along this axis will split these two states by the Zeeman effect. The Zeeman perturbation matrix element takes the form δ_{cc^*} when the Hamiltonian matrix II.14 is written in the c,c* orbital basis.

Perfect biradicals of a general kind exhibit no state degeneracies in the model. T_1 is the lowest state, and S_0 lies $2K_{AB}$ higher. The S_1 state lies $2K'_{AB}$ above T_1, and the S_2 state lies $2K_{AB}$ above S_1.

Any choice of orthogonal orbitals $|\mathcal{A}\rangle, |\mathcal{B}\rangle$ represents acceptable natural orbitals of a perfect biradical. The occupancy of both natural orbitals is unity in each of the four electronic states, demonstrating that they are all of equally "open shell" nature.

Biradicaloids. In these "imperfect" biradicals, we have either $\gamma_{\mathcal{AB}} \neq 0$ or $\delta_{\mathcal{AB}} \neq 0$, or both, when $\gamma'_{\mathcal{AB}} = 0$. The Hamiltonian matrix II.14 now consists of a 3 × 3 block for singlet states and a 1 × 1 block for T_1. The structural perturbations that introduce nonzero $\gamma_{\mathcal{AB}}$ and/or $\delta_{\mathcal{AB}}$ typically also affect E_0 and $E(T)$, but as long as we are interested only in energy splitting, this need not be considered. In the following, we express singlet energies relative to $E(T)$.

Diagonalization of the 3 × 3 block produces wave functions in the form

$$|S\rangle = C_{i,-} {}^1|\mathcal{A}^2 - \mathcal{B}^2\rangle + C_{i,+} {}^1|\mathcal{A}^2 + \mathcal{B}^2\rangle + C_{i,0} {}^1|\mathcal{AB}\rangle \quad \text{(II.32)}$$

for the three singlets i = 0, 1, and 2. Their density matrices can be shown to be

$$\rho(S_i) = \begin{pmatrix} 1 + 2C_{i,+}C_{i,-} & 2C_{i,0}C_{i,+} \\ 2C_{i,0}C_{i,+} & 1 - 2C_{i,+}C_{i,-} \end{pmatrix} \quad \text{(II.33)}$$

and the occupation numbers $n_{I,II}$ for the two natural orbitals are

$$n_{I,II} = 1 \pm 2C_{i,+}\sqrt{1 - C_{i,+}^2} \quad \text{(II.34)}$$

While in the perfect biradical we had $n_I = n_{II} = 1$ for all states, in the limit of a very strong perturbation $\gamma^2_{\mathcal{AB}} + \delta^2_{\mathcal{AB}}$, they now approach 0 and 2 for the S_0 state. In this limit, the "biradicaloid" then no longer has any biradical-like character but represents an ordinary closed-shell species. The arbitrary nature of the borderline between the two limiting regions is obvious.

Physical significance of the perturbation parameters $\gamma_{\mathcal{AB}}$ and $\delta_{\mathcal{AB}}$. The sum $\gamma^2_{\mathcal{AB}} + \delta^2_{\mathcal{AB}}$ is independent of the choice of the orbital basis $|\mathcal{A}\rangle, |\mathcal{B}\rangle$. However, for any perturbation there is an orbital choice that makes $\delta_{\mathcal{AB}}$ vanish and another that makes $\gamma_{\mathcal{AB}}$ vanish.

We have already pointed out two examples in which the nature of the perturbation dictates the most convenient choice of the orbitals $|\mathcal{A}\rangle, |\mathcal{B}\rangle$, which makes $\gamma_{\mathcal{AB}}$ and $\gamma'_{\mathcal{AB}}$ vanish and maximizes $\delta_{\mathcal{AB}}$, thus causing a decoupling of the "balanced" wave functions ${}^1|\mathcal{A}^2 + \mathcal{B}^2\rangle$ and ${}^1|\mathcal{A}^2 - \mathcal{B}^2\rangle$ into their constituents ${}^1|\mathcal{A}^2\rangle$ and ${}^1|\mathcal{B}^2\rangle$. The examples were the electric (Stark) and magnetic (Zeeman) perturbations by outside fields. They have not been included in the Hamiltonian matrix II.14 but can be shown to enter as δ_{AB} and δ_{cc^*}, respectively. The former therefore calls for the choice $|\mathscr{A}\rangle, |\mathscr{B}\rangle = |A\rangle, |B\rangle$, and the latter calls for the choice $|\mathscr{A}\rangle, |\mathscr{B}\rangle = |c\rangle, |c^*\rangle$ (in both instances, $\gamma_{AB}^- = \gamma_{cc^*}^- = 0$). The former removes the (near) degeneracy of the S_1, S_2 state in (nearly) pair biradicals and converts the nonpolar $^1|A^2 \pm B^2\rangle$ wave functions into the polar functions $^1|A^2\rangle$ and $^1|B^2\rangle$ of the perturbed species. The latter removes the degeneracy of the S_0, S_1 state in axial biradicals and converts $^1|c^2 \pm c^{*2}\rangle$, which carries no magnetic moment, into the magnetized functions $^1|c^2\rangle$ and $^1|c^{*2}\rangle$ of the perturbed species.

When the initial degeneracy is imperfect, either because the biradical is only approximately of the pair kind (K_{AB} small but nonzero, as in twisted ethylene) or of the axial kind (e.g., slightly Jahn–Teller distorted), or because $\gamma_{\mathscr{AB}}^2 + \delta_{\mathscr{AB}}^2$ is small but nonzero due to structural perturbations, or both, this uncoupling of the $^1|\mathscr{A}^2 \pm \mathscr{B}^2\rangle$ wave functions into the $^1|\mathscr{A}^2\rangle$, $^1|\mathscr{B}^2\rangle$ functions by the Stark or Zeeman effects becomes progressively more difficult—now, one is dealing with a weakly avoided surface touching. As long as the degree of avoidance is small, the change from the "coupled" to the "uncoupled" type of wave function will occur quite abruptly as either the strength of the outside field or the strength of the degeneracy-removing structural perturbation is varied through the critical region. This abrupt uncoupling is known as "sudden polarization" in the case of the electric field effect. In practice, an internal rather than an outside electric field is usually imposed in that δ_{AB} is varied by a suitable structural change. The uncoupling is known as "angular momentum quenching" in the case of the magnetic field effect. Structural perturbations are not capable of mimicking the magnetic field and making δ_{cc^*} different from zero.

A similar uncoupling of $^1|a^2 \pm b^2\rangle$ into $^1|a^2\rangle$ and $^1|b^2\rangle$ by an outside field is not possible, since no available field produces a perturbation that behaves like δ_{ab} in the Hamiltonian matrix II.14. However, just as it was possible to effectively replace an outside electric field by a structural perturbation δ_{AB} that represented an introduction of an energy difference between the localized orbitals $|A\rangle$ and $|B\rangle$ (biradical "polarization"), it is possible to effect the decoupling of $^1|a^2 \pm b^2\rangle$ into $^1|a^2\rangle$ and $^1|b^2\rangle$ by a structural perturbation δ_{ab} that represents an introduction of an energy difference between the delocalized orbitals a and b, that is, by bond formation. However, since $^1|a^2 + b^2\rangle$ and $^1|a^2 - b^2\rangle$ cannot be even approximately degenerate to start with, because one represents S_2 and the other S_0, this decoupling cannot be very abrupt. Therefore, while we have sudden polarization and sudden angular momentum quenching, we do *not* have "sudden bond formation."

Structural perturbations of perfect biradicals that can be described by δ_{AB} alone produce "heterosymmetric biradicaloids." Those that can be described by δ_{ab} alone produce "homosymmetric biradicaloids." There are no structural perturbations that can be described by δ_{cc^*} alone.

Structural perturbation whose description requires both δ_{AB} and δ_{ab} lead to "nonsymmetric" biradicaloids. This is the most common case in practice, in which choice of the orbital pair $|\mathscr{A}\rangle, |\mathscr{B}\rangle$ that causes $\gamma_{\mathscr{AB}}$ to vanish and allows the perturbation to be expressed through $\delta_{\mathscr{AB}}$ alone (note that $\gamma_{\mathscr{AB}}^2 + \delta_{\mathscr{AB}}^2$ is invariant) is none of the special cases $|A\rangle, |B\rangle$; $|c\rangle, |c^*\rangle$; $|a\rangle, |b\rangle$ but, instead, some general orbital pair. Then, $\gamma_{\mathscr{AB}}^-$ no longer vanishes, and two off-diagonal elements in the Hamiltonian matrix are now nonzero.

It is then probably best to choose arbitrarily one of the three special orbital choices for which $\gamma_{\mathscr{AB}}^- = 0$, say, the localized orbitals $|A\rangle, |B\rangle$, and to accept the existence of two nonzero matrix elements describing the perturbation, δ_{AB} and γ_{AB}.

Since $\delta_{AB} = \gamma_{ab}$ and $\delta_{ab} = \gamma_{AB}$, it is clear that structural perturbations can be equally well described by their effect in the localized orbital basis $|A\rangle, |B\rangle$ (δ_{AB}, γ_{AB}) and the delocalized orbital basis $|a\rangle, |b\rangle$ (δ_{ab}, γ_{ab}). In the following, we choose to work with the localized orbital $|A\rangle, |B\rangle$ throughout. In analogy to VB theory, we refer to $^1|A^2\rangle$ and $^1|B^2\rangle$ as hole–pair configurations (a pair of electrons on one of the biradical centers, none on the other) and to $^1|AB\rangle$ and $^3|AB\rangle$ as dot–dot configurations (one electron on each center).

Homosymmetric biradicaloids. In these, $\gamma_{AB} = \delta_{ab} \neq 0$ and $\delta_{AB} = \gamma_{ab} = 0$; that is, the localized orbitals $|A\rangle, |B\rangle$ have equal energies but interact, and the delocalized orbitals $|a\rangle, |b\rangle$ have different energies and do not interact. In order for the two orthogonal orbitals $|A\rangle$ and $|B\rangle$ to interact, they cannot be perfectly localized apart. Typical examples are two hybrid orbitals on the same atom or a pair of Löwdin-orthogonalized orbitals on neighboring atoms. The state wave functions and their energies relative to E(T) are:

	Wave function	Energy	
S_2:	$\cos\alpha\,^1|A^2 + B^2\rangle + \sin\alpha\,^1|AB\rangle$	$K'_{AB} + 2K_{AB} + \sqrt{K'^2_{AB} + \gamma^2_{AB}}$	
S_1:	$^1|A^2 - B^2\rangle$	$2K'_{AB}$	(II.35)
S_0:	$-\sin\alpha\,^1|A^2 + B^2\rangle + \cos\alpha\,^1|AB\rangle$	$K'_{AB} + 2K_{AB} - \sqrt{K'^2_{AB} + \gamma^2_{AB}}$	

$$\alpha = (1/2)\tan^{-1}(\gamma_{AB}/K'_{AB}) \tag{II.36}$$

As $|\gamma_{AB}|$ increases, S_0 is stabilized relative to T. The two states are degenerate when $|\gamma_{AB}| = 2[K_{AB}(K_{AB} + K'_{AB})]^{1/2}$. For larger $|\gamma_{AB}|$, S_0 lies below T.

The natural orbitals are $|a\rangle, |b\rangle$. Their occupation numbers are obtained from equation II.34. For the usual case $\gamma_{AB} < 0$, we have

$$\left.\begin{array}{l} n_a(S_0) = n_b(S_2) \\ n_b(S_0) = n_a(S_2) \end{array}\right\} = 1 \pm \sin 2\alpha = 1 \pm [1 + (K'_{AB}/\gamma_{AB})^2]^{-1/2}$$
$$n_a(S_1) = n_b(S_1) = n_a(T) = n_b(T) = 1 \tag{II.37}$$

With increasing strength of the perturbation, γ_{AB}/K'_{AB}, S_0, and S_2 become "closed-shell" states, while S_1 and T_1 preserve their "open-shell" (biradicaloid) nature.

Heterosymmetric biradicaloids. In these, $\delta_{AB} = \gamma_{ab} \neq 0$, $\gamma_{AB} = \delta_{ab} = 0$; that is, the localized orbitals $|A\rangle, |B\rangle$ have different energies but do not interact, and the delocalized orbitals $|a\rangle, |b\rangle$ have equal energies and do interact. The state wave functions and their energies relative to E(T) are:

		Energy	
S_2:	$|2\rangle = \cos\beta\,^1|A^2 + B^2\rangle + \sin\beta\,^1|A^2 - B^2\rangle$	$2K'_{AB} + K_{AB} + \sqrt{K^2_{AB} + \delta^2_{AB}}$	
S_1 or S_0:	$|1\rangle = -\sin\beta\,^1|A^2 + B^2\rangle + \cos\beta\,^1|A^2 - B^2\rangle$	$2K'_{AB} + K_{AB} - \sqrt{K^2_{AB} + \delta^2_{AB}}$	(II.38)
S_0 or S_1:	$^1|AB\rangle$	$2K_{AB}$	

$$\beta = (1/2)\tan^{-1}(\delta_{AB}/K_{AB}) \tag{II.39}$$

APPENDIX II: THE TWO-ELECTRON TWO-ORBITAL MODEL OF BIRADICALS   **441**

The order of the wave functions that represent S_0 and S_1 depends on the magnitude of δ_{AB} (which is positive by definition). For $\delta_{AB} = \delta_0 = 2[K'_{AB}(K'_{AB} - K_{AB})]^{1/2}$, their energies are equal. For $\delta_{AB} < \delta_0$, $^1|AB\rangle$ is the lower (S_0) of the two lowest singlet states; for $\delta_{AB} > \delta_0$, it is the higher (S_1) of these two states.

The natural orbitals are $|A\rangle, |B\rangle$. Their occupation numbers obtained from equation II.34 are

$$\left.\begin{array}{l}n_B(|1\rangle) = n_A(|2\rangle) \\ n_A(|1\rangle) = n_B(|2\rangle)\end{array}\right\} = 1 \pm \sin^2\beta = 1 \pm [1 + (K_{AB}/\delta_{AB})^2]^{-1/2}$$

$$n_A(^1|AB\rangle) = n_B(^1|AB\rangle) = n_A(T) = n_B(T) = 1 \tag{II.40}$$

With increasing strength of the perturbation δ_{AB}/K_{AB}, $|1\rangle$ and $|2\rangle$ become "closed-shell" states, while $^1|AB\rangle$ and T_1 keep their "open-shell" (biradicaloid) nature.

Nonsymmetric biradicaloids. In these, both $\delta_{AB} = \gamma_{ab} \neq 0$ and $\gamma_{AB} = \delta_{ab} \neq 0$; that is, the localized orbitals have different energies and interact as well. The general expressions for state wave functions and energies can be written explicitly, but they are cumbersome because the solution of an equation of the third degree is involved. The crossing of the S_0 and S_1 states at δ_0 is now avoided.

For the special case of a perturbed axial biradical ($K_{AB} = K'_{AB}$), however, explicit solutions for wave functions and energies relative to $E(T)$ are simple:

$$\begin{pmatrix}|S_2\rangle \\ |S_1\rangle \\ |S_0\rangle\end{pmatrix} = \begin{pmatrix}\cos\alpha\sin\beta & \cos\beta & \sin\alpha\sin\beta \\ -\sin\alpha & 0 & \cos\alpha \\ \cos\alpha\cos\beta & -\sin\beta & \sin\alpha\cos\beta\end{pmatrix}\begin{pmatrix}^1|A^2 - B^2\rangle \\ ^1|A^2 + B^2\rangle \\ ^1|AB\rangle\end{pmatrix}$$

Energy:
- $3K_{AB} + \sqrt{K^2_{AB} + \gamma^2_{AB} + \delta^2_{AB}}$
- $2K_{AB}$
- $3K_{AB} - \sqrt{K^2_{AB} + \gamma^2_{AB} + \delta^2_{AB}}$ \tag{II.41}

$$\alpha = \tan^{-1}(\gamma_{AB}/\delta_{AB})$$

$$\beta = (1/2)\tan^{-1}(\sqrt{\gamma^2_{AB} + \delta^2_{AB}}/K_{AB}) \tag{II.42}$$

Additional information on the two-electron two-orbital model and leading references can be found in: V. Bonačić-Koutecký, J. Koutecký, and J. Michl, *Angew. Chem. Internat. Ed. Engl.* **26**, 170 (1987); J. Michl and V. Bonačić-Koutecký, *Tetrahedron* **44**, 7559 (1988).

Appendix III

Computational Methods: SCF-CI and GVB-CI

Basis sets. The use of AO basis sets is prevalent in modern calculations of molecular electronic structure described in this appendix, and we provide a brief introduction to this subject first.

Slater-type AOs (STOs) were the first to be used. The form of the Slater 1s, 2s, and 2p orbitals is

$$|1s\rangle = (\zeta^3/\pi)^{1/2}\exp(-\zeta r)$$

$$|2s\rangle = (\zeta^5/3\pi)^{1/2} r \exp(-\zeta r)$$

$$|2p\rangle = (\zeta^5/\pi)^{1/2} \begin{Bmatrix} x \\ y \\ z \end{Bmatrix} \exp(-\zeta r)$$

where x,y,z are the Cartesian coordinates of the electron, $r = (x^2 + y^2 + z^2)^{1/2}$, and ζ is known as the orbital exponent. Similar equations can be written for 3s, 3p, 3d, and higher Slater orbitals. The actual values of the orbital exponents are fixed by the so-called Slater rules.

A minimum AO basis set of Slater orbitals contains one AO for each orbital of each subshell that is at least partially occupied in the ground state of the atom (1s for H, He; 1s 2s for Li, Be; 1s 2s 2p for C through Ne; etc.).

Slater AOs are not used much in practice because they lead to difficult integrals in actual calculations. It is computationally easier to approximate each Slater AO as a linear combination of several Gaussian-type orbitals (GTOs) with fixed coefficients (contracted GTOs):

$$|1s'\rangle = \sum_{t=1}^{T} d_{1s,t} \exp(-\alpha_{1s,t} r^2)$$

$$|2s'\rangle = \sum_{t=1}^{T} d_{2s,t} \exp(-\alpha_{2s,t} r^2)$$

$$|2p'\rangle = \sum_{t=1}^{T} d_{2p,t} \exp(-\alpha_{2p,t} r^2)$$

In the popular basis sets introduced by Pople, the contraction coefficients $d_{1s,t}$,

$d_{2s,t}$, etc., have been chosen so as to minimize the difference between $|1s\rangle$ and $|1s'\rangle$, $|2s\rangle$ and $|2s'\rangle$, etc., but the exponents of ns and np orbitals have been chosen equal. The standard notation for minimal basis sets of this type is STO-TG. Of these, STO-3G has seen most use.

Minimum basis sets with fixed exponents are quite inadequate for molecular calculations. It is normally necessary to introduce additional flexibility into the wave function by introducing additional AOs. An improvement is the replacement of each basis function by two independent functions—one with a smaller and one with a larger exponent ("inner" and "outer" part). Such basis sets are said to be of the "double-zeta" type. In "split-valence" basis sets, the doubling of the basis is performed only for the valence shell, assuming that the splitting of the inner shell is less important for the description of chemical bonds. A tripling of each basis function produces a triple-zeta basis set, etc.

Examples of the standard notation for Pople's basis sets are 4-31G (1s is a contraction of four Gaussians; 2s and 2p are each represented by a pair of functions—the inner one is a contraction of three Gaussians, the outer one is a single Gaussian) and 6-311G (1s: contraction of six Gaussians; 2s, 2p: tripled, with one contraction of three Gaussians and two single Gaussians).

In order to describe the deformation (polarization) of AOs in a molecule by the presence of other atoms, functions of higher angular quantum number are added (p for an s orbital, etc.). The presence of such polarization functions on second-row atoms (d orbital on C, etc.) is indicated by an asterisk, and their additional presence on hydrogen atoms is denoted by two asterisks—hence 6-31G*, 6-31G**. Finally, in anions and Rydberg excited states it is important to include diffuse AOs with quite small exponents, and their presence in Pople-type bases is indicated by one or two plus signs in an analogous manner, for example, 3-21+G and 3-21++G.

Many other AO basis sets in addition to Pople's are in use. An example of a common double-zeta basis set introduced by Dunning is [4s 2p] contracted from (9s 5p). A leading reference is provided at the end of Appendix III.

The self-consistent-field (SCF) method. In this method the electronic wave function is assumed to have the form of an antisymmetrized product of spinorbitals, known as the *Slater determinant* $|\phi_0\rangle$:

$$|\phi_0\rangle = \hat{\mathscr{A}} | 1(1)\rangle | 2(2)\rangle \cdots |i(i)\rangle|j(j)\rangle \cdots |n(N)\rangle$$

$$= \frac{1}{\sqrt{N!}} \sum_{r=1}^{N!} (-1)^{P_r} \hat{P}_r | 1(1)\rangle | 2(2)\rangle \cdots |i(i)\rangle|j(j)\rangle \cdots |n(N)\rangle \quad \text{(III.1)}$$

The number of electrons is N, the operator $\hat{P}_r$ permutes the electrons labeled by the numbers in parentheses, the label r identifies the permutations, whose number is N!, and i labels the spinorbitals. The operator $\hat{\mathscr{A}}$ is known as the antisymmetrizer. A spinorbital is a product of a function of the spatial coordinates of an electron (an orbital) and a function of its spin coordinate:

$$|i(k)\rangle = |i(k)\rangle|\eta_i(k)\rangle \quad \text{(III.2)}$$

The spin functions are

$$\eta_i(k) = \begin{cases} |\alpha(k)\rangle \\ |\beta(k)\rangle \end{cases} \tag{III.3}$$

The orbitals in $|\phi_0\rangle$ are chosen orthonormal:

$$S_{ij} = \langle i(1)|j(1)\rangle = \delta_{ij} \tag{III.4}$$

The expectation value of energy is

$$E = \langle \phi_0|\hat{H}|\phi_0\rangle = \langle \phi_0|\sum_{k=1}^{N} \hat{h}(k) + \frac{1}{2}\sum_{k\neq l=1}^{N}\frac{e^2}{r_{kl}}|\phi_0\rangle$$

$$= \frac{1}{N!}\sum_{r,r'=1}^{N!}(-1)^{P_r+P_{r'}}\hat{P}_r\hat{P}_{r'}\langle l(1)|\cdots\langle i(i)|\cdots\langle j(j)|\cdots\langle n(N)|$$

$$\times \left\{\sum_{k=1}^{N}\hat{h}(k) + \frac{1}{2}\sum_{k\neq l=1}^{N}\frac{e^2}{r_{kl}}\right\}|l(1)\rangle\cdots|i(i)\rangle\cdots|j(j)\rangle\cdots|n(N)\rangle \tag{III.5}$$

where $\hat{h}(1)$ is the one-electron part of the Hamiltonian. The operator $\hat{P}_r$ permutes electrons in the bra, and $\hat{P}_{r'}$ permutes electrons in the ket. Because of the orthogonality, most of the terms in the sums over r and r' vanish. Nonvanishing contributions to E are obtained for the one-electron Hamiltonian $\hat{h}(k)$ only if $P_r = P_{r'}$, and for the two-electron part e^2/r_{kl} only if $P_r = P_{r'}$, or if P_r and $P_{r'}$ differ only in two indices, that is, in one transposition:

$$E = \sum_{i=1}^{N}\langle i(1)|\hat{h}(1)|i(1)\rangle + \frac{1}{2}\sum_{ij}\left\{\langle i(1)|\langle j(2)|\frac{e^2}{r_{12}}|i(1)\rangle|j(2)\rangle\right.$$

$$\left. - \langle i(1)|\langle j(2)|\frac{e^2}{r_{12}}|i(2)\rangle|j(1)\rangle\right\}$$

$$= \sum_{i=1}^{N}\langle i(1)|\left\{\hat{h}(1) + \frac{1}{2}\sum_{j=1}^{N}\langle j(2)|\frac{e^2}{r_{12}}(1-\hat{P}_{12})|j(2)\rangle\right\}|i(1)\rangle \tag{III.6}$$

where i and j are summation indices. It is useful to define the one-electron operators $\hat{J}_j$ and $\hat{K}_j$:

$$\hat{J}_j(1)|i(1)\rangle = \langle j(2)|\frac{e^2}{r_{12}}|j(2)\rangle|i(1)\rangle \tag{III.7}$$

$$\hat{K}_j(1)|i(1)\rangle = \langle j(2)|\frac{e^2}{r_{12}}|i(2)\rangle|j(1)\rangle \tag{III.8}$$

Here, the integration is carried out over coordinates of the second electron. For real orbitals,

$$\langle i(1)|\hat{J}_j(1)|i(1)\rangle = \langle j(1)|\hat{J}_i(1)|j(1)\rangle = J_{ij} = J_{ji} \tag{III.9}$$

$$\langle i(1)|\hat{K}_j(1)|i(1)\rangle = \langle j(1)|\hat{K}_i(1)|j(1)\rangle = K_{ij} = K_{ji} \qquad (III.10)$$

K_{ij} vanishes unless the spins of electrons in the spinorbitals i and j are parallel.

$$E = \sum_{i=1}^{N}\left\{h_{ii} + \frac{1}{2}\sum_{j=1}^{N}(J_{ij} - K_{ij})\right\} \qquad (III.11)$$

where $h_{ii} = \langle i(1)|\hat{h}(1)|i(1)\rangle$.

The closed-shell case. All orbitals are either doubly occupied or empty. Then, the indices in the sum come in pairs. For example, in the first sum, $i = \mathfrak{s}$ and $i = \mathfrak{t}$, so that

$$|\mathfrak{s}\rangle = |p\rangle|\alpha\rangle$$
$$|\mathfrak{t}\rangle = |p\rangle|\beta\rangle \qquad (III.12)$$

and in the second sum, $j = |\mathfrak{u}\rangle$ and $j = |\mathfrak{v}\rangle$, so that

$$|\mathfrak{u}\rangle = |r\rangle|\alpha\rangle$$
$$|\mathfrak{v}\rangle = |r\rangle|\beta\rangle \qquad (III.13)$$

The energy expression then simplifies:

$$E = \sum_{i=1}^{N/2}\left\{2h_{ii} + \sum_{j=1}^{N/2}(2J_{ij} - K_{ij})\right\}, \qquad 1 \le i \le N/2 \qquad (III.14)$$

with

$$h_{ii} = \langle i|\hat{h}|i\rangle \qquad (III.15)$$

$$J_{ij} = \langle i(1)|\langle j(2)|\frac{e^2}{r_{12}}|i(1)\rangle|j(2)\rangle = J_{ji} \qquad (III.16)$$

$$K_{ij} = \langle i(1)|\langle j(2)|\frac{e^2}{r_{12}}|j(1)\rangle|i(2)\rangle = K_{ji} \qquad (III.17)$$

The reason J_{ij} appears four times is that all four combinations of $\mathfrak{s},\mathfrak{t}$ with $\mathfrak{u},\mathfrak{v}$ are possible, as the integration over i is separated from the integration over j. On the other hand, K_{ij} appears only twice because the spinorbitals i and j must have the same spin part if K_{ij} is not to vanish.

The closed-shell plus open-shell case, with parallel spins in the open shell. Some orbitals are doubly occupied (closed-shell part, *cl*), others are singly occupied with all spins parallel (either $|i\rangle = |p\rangle|\alpha\rangle$ or $|i\rangle = |p\rangle|\beta\rangle$, open-shell part, *op*), and the rest are empty. The contributions to the energy from the case i, j ∈ *cl* are the same as those just given for closed shell. The results for i ∈ *cl*, j ∈ *op*, i ∈ *op*, j ∈ *cl* and i ∈ *op* and j ∈ *op* follow.

The sum in the energy expression in spinorbitals can be partitioned:

$$\sum_{i,j} = \sum_{i,j \in cl} + \sum_{i,j \in op} + \sum_{i \in cl}\sum_{j \in op} + \sum_{i \in op}\sum_{j \in cl}$$

$$= \sum_{i,j \in cl} + \sum_{i,j \in op} + 2\sum_{i \in cl}\sum_{j \in op} \qquad (III.18)$$

In the energy expression in terms of MOs, the indices i and j are used for the doubly occupied MOs and k,l are used for singly occupied MOs:

$$E = \sum_{i \in cl}\left[2h_{ii} + \sum_{j \in cl}(2J_{ij} - K_{ij})\right] + \sum_{k \in op} h_{kk}$$

$$+ \sum_{i \in cl}\sum_{k \in op}(2J_{ik} - K_{ik}) + \sum_{l \in op}\sum_{l \langle k \in op}(J_{kl} - K_{kl})$$

$$= 2\sum_{i \in cl} h_{ii} + \sum_{k \in op} h_{kk} + \sum_{i \in cl} J_{ii} + \sum_{i \langle j \in cl}(4J_{ij} - 2K_{ij})$$

$$+ \sum_{i \in cl}\sum_{k \in op}(2J_{ik} - K_{ik}) + \sum_{l \in op}\sum_{i \langle k \in op}(J_{kl} - K_{kl}) \qquad (III.19)$$

There is only one contribution from Coulombic repulsion of the two electrons placed in the doubly occupied orbital i.

For i and j $\in$ cl, we have two electrons in orbital i and two electrons in orbital j, therefore we have four times the Coulomb integral corresponding to the interaction of one electron in one orbital with another electron in the other orbital. The exchange integral is contributed twice, by the two parallel electron spin arrangements.

For i $\in$ cl and k $\in$ op, we obtain J_{ik} twice since two electrons occupy i and one occupies k. Only one of two electrons occupying orbital i has its spin parallel with that of the electron in orbital k, and K_{ik} therefore only appears once.

For l $\in$ op and k $\in$ op, orbitals k and l each contain only one electron and their spins are parallel, so that J_{kl} and K_{kl} are each contributed once. These considerations make it easy to write down the expression for the energy, equation III.19.

The closed-shell plus open-shell case, with both α and β spins in the open shell. The open shell is now divided in two, labeled *opα* and *opβ*:

$$E = \sum_{i \in cl}\left[2h_{ii} + \sum_{j \in cl}(2J_{ij} - K_{ij})\right] + \left(\sum_{k \in op\alpha} + \sum_{k \in op\beta}\right) h_{kk}$$

$$+ \sum_{i \in cl}\left(\sum_{k \in op\alpha} + \sum_{k \in op\beta}\right)(2J_{ik} - K_{ik}) + \sum_{(l \langle k) \in op\alpha}(J_{kl} - K_{kl})$$

$$+ \sum_{(l \langle k) \in op\beta}(J_{kl} - K_{kl}) + \sum_{l \in op\alpha}\sum_{k \in op\beta} J_{kl} \qquad (III.20)$$

A Slater determinant with two open shells—one containing electrons with α, and the other containing β spins—is not an eigenstate of the total spin operator.

It is therefore necessary to introduce a proper linear combination of Slater determinants. These share a common closed shell and differ in the assignment of spin in the open-shell MOs. To find the linear combination of these Slater determinants that is an eigenstate of the total spin operator, off-diagonal matrix elements between the individual Slater determinants are needed.

If two Slater determinants differ only in one spinorbital, $|\mathfrak{r}\rangle \to |\mathfrak{s}\rangle$, we obtain

$$\langle\phi|\hat{H}|\phi_{\mathfrak{r}\to\mathfrak{s}}\rangle = \frac{1}{N!} \sum_{l,l'=1}^{N!} (-1)^{P_l+P_{l'}} \hat{P}_l \hat{P}_{l'} \langle \mathfrak{l}(1)|\cdots\langle \mathfrak{r}(r)|\cdots\langle \mathfrak{n}(N)|$$

$$\times \left[\sum_{k=1}^{N} \hat{h}(k) + \frac{1}{2} \sum_{k\neq m=1}^{N} \frac{e^2}{r_{km}}\right] |\mathfrak{l}(1)\rangle\cdots|\mathfrak{s}(r)\rangle\cdots|\mathfrak{n}(N)\rangle \quad \text{(III.21)}$$

In the one-electron part, only the case $\hat{P}_l = \hat{P}_{l'}$ contributes, whereas in the two-electron part, both $\hat{P}_l = \hat{P}_{l'}$ and $\hat{P}_l \neq \hat{P}_{l'}$ contribute; however, in the latter case, l and l' may differ at most in a single transposition.

$$\langle\phi|\hat{H}|\phi_{\mathfrak{r}\to\mathfrak{s}}\rangle = \langle \mathfrak{r}(1)|\hat{h}(1)|\mathfrak{s}(1)\rangle$$

$$+ \frac{1}{2} \sum_i \left[\langle \mathfrak{r}(1)|\langle \mathfrak{i}(2)|\frac{e^2}{r_{12}} (1 - \hat{P}_{12})|\mathfrak{i}(2)\rangle|\mathfrak{s}(1)\rangle \right.$$

$$+ \left. \langle \mathfrak{i}(1)|\langle \mathfrak{r}(2)|\frac{e^2}{r_{12}} (1 - \hat{P}_{12})|\mathfrak{s}(2)\rangle|\mathfrak{i}(1)\rangle \right] \quad \text{(III.22)}$$

Since the integration variables 1 and 2 can be exchanged, one obtains

$$\langle\phi|\hat{H}|\phi_{\mathfrak{r}\to\mathfrak{s}}\rangle = \langle \mathfrak{r}(1)| \left[\hat{h}(1) + \sum_i (\hat{J}_i - \hat{K}_i) \right] |\mathfrak{s}(1)\rangle \quad \text{(III.23)}$$

If two Slater determinants differ in two spin orbitals, $|\mathfrak{r}\rangle \to |\mathfrak{s}\rangle$ and $|\mathfrak{u}\rangle \to |\mathfrak{v}\rangle$, the matrix element between them is

$$\langle\phi|\hat{H}|\phi_{\mathfrak{r}\to\mathfrak{s},\,\mathfrak{u}\to\mathfrak{v}}\rangle = \langle \mathfrak{r}(1)|\langle \mathfrak{u}(2)|\frac{e^2}{r_{12}}|\mathfrak{s}(1)\rangle|\mathfrak{v}(2)\rangle$$

$$- \langle \mathfrak{r}(1)|\langle \mathfrak{u}(2)|\frac{e^2}{r_{12}}|\mathfrak{s}(2)\rangle|\mathfrak{v}(1)\rangle \quad \text{(III.24)}$$

Two special cases are particularly important. In the first, the Slater determinants differ by the spin-flipping of two electrons:

$$|\mathfrak{r}\rangle = |a\rangle|\alpha\rangle, \quad |\mathfrak{s}\rangle = |a\rangle|\beta\rangle$$
$$|\mathfrak{u}\rangle = |b\rangle|\beta\rangle, \quad |\mathfrak{v}\rangle = |b\rangle|\alpha\rangle \quad \text{(III.25)}$$

We then obtain

$$\langle\phi|\hat{H}|\phi_{a\to\bar{a},\bar{b}\to b}\rangle = -\langle a(1)|\langle b(2)|\frac{e^2}{r_{12}}|a(2)\rangle|b(1)\rangle = -K_{ab} \qquad (III.26)$$

In the other case, the determinants differ by a double excitation from a doubly occupied orbital $|a\rangle$ to a virtual orbital $|b\rangle$:

$$\begin{aligned}|\mathfrak{r}\rangle &= |a\rangle|\alpha\rangle & |\mathfrak{s}\rangle &= |b\rangle|\alpha\rangle \\ |\mathfrak{u}\rangle &= |a\rangle|\beta\rangle & |\mathfrak{v}\rangle &= |b\rangle|\beta\rangle\end{aligned} \qquad (III.27)$$

We obtain

$$\langle\phi|\hat{H}|\phi_{a^2\to b^2}\rangle = \langle a(1)|\langle a(2)|\frac{e^2}{r_{12}}|b(1)\rangle|b(2)\rangle = K_{ab} \qquad (III.28)$$

In the general case the spin-adapted wave function can be written as

$$\Psi = \sum_K C_K \phi_K \qquad (III.29)$$

where ϕ_K are Slater determinants. The expectation value for the Hamiltonian is

$$E = \langle\Psi|\hat{H}|\Psi\rangle = \sum_K \sum_L C_K C_L \langle\phi_K|\hat{H}|\phi_L\rangle \qquad (III.30)$$

Taking into account the expressions for the off-diagonal matrix elements, the general expression for E in spatial orbitals takes the form

$$E = \sum_{i,j}\left\{\omega_{ij} h_{ij} + \sum_{k,l} \alpha_{ijkl} \langle i(1)|\langle j(2)|\frac{e^2}{r_{12}}|k(2)\rangle|l(1)\rangle\right\} \qquad (III.31)$$

Only in special cases, E takes a form in which one of the summations is restricted to the set of occupied orbitals M,

$$\begin{aligned}E &= \sum_{i \in M} \langle i(1)|\omega_i \hat{h} + \sum_{j \in M}(\alpha_{ij}\hat{J}_j - \beta_{ij}\hat{K}_j)|i(1)\rangle \\ &= \sum_{i \in M}\left\{\omega_i h_{ii} + \sum_{j \in M}(\alpha_{ij}J_{ij} - \beta_{ij}K_{ij})\right\}\end{aligned} \qquad (III.32)$$

and which is usually used in the SCF procedure. In the Hartree–Fock SCF procedure, ω_i, α_{ij}, and β_{ij} are fixed numbers. In the multiconfigurational SCF (MC SCF) procedure, a variation of these quantities is included in the proper way in the optimization.

Euler equations. In the variational procedure, the orthonormality of the spatial parts of the MOs must be respected:

$$\langle i|j\rangle - \delta_{ij} = 0 \qquad (III.33)$$

In order to do so, one constructs the following expression:

$$L = E - 2 \sum_i \sum_j \varepsilon_{ij} (\langle i|j\rangle - \delta_{ij}) \tag{III.34}$$

where ε_{ij} are Lagrange multipliers. The factor -2 is included merely for convenience, and electron labels are not shown explicitly. For arbitrary variations δi and δj, we have

$$\delta L = \delta E - 2 \sum_{i,j} \varepsilon_{ij} [\langle \delta i|j\rangle + \langle i|\delta j\rangle] = 0 \tag{III.35}$$

and

$$\delta E = \sum_i \omega_i \{\langle \delta i|\hat{h}|i\rangle + \langle i|\hat{h}|\delta i\rangle\} + \sum_{i,j} \{\alpha_{ij} [\langle \delta i|\hat{J}_j|i\rangle + \langle i|\hat{J}_j|\delta i\rangle + \langle \delta j|\hat{J}_i|j\rangle + \langle j|\hat{J}_i|\delta j\rangle] - \beta_{ij}[\langle \delta i|\hat{K}_j|i\rangle + \langle i|\hat{K}_j|\delta i\rangle] + \langle \delta j|\hat{K}_i|j\rangle + \langle j|\hat{K}_i|\delta j\rangle]\} \tag{III.36}$$

where α_{ij} and β_{ij} are symmetric.

Substituting equation III.35 into equation III.36, one obtains the Euler equations

$$\left\{\omega_i \hat{h} + 2 \sum_{j \in M} (\alpha_{ij} \hat{J}_j - \beta_{ij} \hat{K}_j)\right\} |i\rangle = 2 \sum_{j \in M} \varepsilon_{ij}|j\rangle$$

$$\langle i| \left\{\omega_i \hat{h} + 2 \sum_{j \in M} (\alpha_{ij}\hat{J}_j - \beta_{ij} \hat{K}_j)\right\} = 2 \sum_{j \in M} \langle j|\varepsilon_{ji} \tag{III.37}$$

The Hartree–Fock equations are

$$\hat{F}_i|i\rangle = \sum_{j \in M} \varepsilon_{ij}|j\rangle \quad \text{for all } i \in M \tag{III.38}$$

where $\hat{F}_i$ is the Hartree–Fock operator

$$\hat{F}_i = \frac{1}{2} \omega_i \hat{h} + \sum_{j \in M} (\alpha_{ij} \hat{J}_j - \beta_{ij} \hat{K}_j) \tag{III.39}$$

The second Euler equation can be written as

$$\langle i|\hat{F}_i = \sum_{j \in M} \langle j|\varepsilon_{ji} \tag{III.40}$$

Because of orthogonality of the MOs, one obtains

$$\langle j|\hat{F}_i|i\rangle = \varepsilon_{ij}$$
$$\langle i|\hat{F}_i|j\rangle = \varepsilon_{ji} \tag{III.41}$$

For a Hermitian Hartree–Fock operator $\hat{F}_i = \hat{F}_i^\dagger$,

$$\varepsilon_{ij} = \varepsilon_{ji}^\dagger \tag{III.42}$$

and for real orbitals,

$$\langle i|\hat{F}_i|j\rangle = \langle i|\hat{F}_j|j\rangle \tag{III.43}$$

Now, it is possible to specify which values ω_i, α_{ij}, and β_{ij} should have by comparing the general Hartree–Fock equation with the energy expressions derived earlier. We show the result for three cases: (i) closed shell, (ii) closed shell plus open shell with parallel spins, and (iii) closed shell plus open shell with opposite spins:

(i) $\omega_i = 2$, $\quad \alpha_{ij} = 2$, $\quad \beta_{ij} = 1 \quad$ for all $i \in M$

$$\hat{F}_i = \hat{h} + \sum_j (2\hat{J}_j - \hat{K}_j)$$

(ii) $\omega_i = 2$, $\quad \alpha_{ij} = 2$, $\quad \beta_{ij} = 1 \quad i \in M, i,j \in cl \tag{III.44}$

$\quad \omega_i = 1 \quad\quad\quad\quad\quad\quad\quad\quad\quad\quad i \in op$

$$\alpha_{ij} = 1, \quad \beta_{ij} = 1/2 \begin{cases} i \in cl, j \in op \\ i \in op, j \in cl \end{cases}$$

$$\alpha_{ij} = 1/2, \quad \beta_{ij} = 1/2 \quad i \in op, j \in op, i \neq j$$

$$\hat{F}_i = \hat{h} + \sum_{j \in cl} (2\hat{J}_j - \hat{K}_j) + \sum_{j \in op} (\hat{J}_j - \hat{K}_j/2) = \hat{F}_{cl} \quad i \in cl \tag{III.45}$$

$$\hat{F}_i = \frac{1}{2}\hat{h} + \sum_{j \in cl} (\hat{J}_j - \hat{K}_j/2) + \sum_{j \in op} (\hat{J}_j - \hat{K}_j)/2$$
$$= \frac{1}{2}\left(\hat{F}_{cl} - \frac{1}{2}\hat{K}_{op}\right), \quad i \in op \tag{III.46}$$

where

$$\hat{F}_{cl} = \hat{h} + \hat{G}_T \tag{III.47}$$

and

$$\hat{G}_T = \sum_{j \in M} n_j (\hat{J}_j - \hat{K}_j/2) \tag{III.48}$$

where n_j is the occupation number:

$$n_j = 2 \quad \text{if } j \in cl$$
$$n_j = 1 \quad \text{if } j \in op \tag{III.49}$$

$$\hat{K}_{op} = \sum_{j \in op} \hat{K}_j$$

(iii) $\omega_i = 2$, $\alpha_{ij} = 2$, $\beta_{ij} = 1$ $\quad i,j \in cl$

$\omega_i = 1$ $\quad\quad\quad\quad\quad\quad\quad\quad\quad i \in op\alpha$ or $op\beta$

$\quad\quad\quad\quad \alpha_{ij} = 1$, $\beta_{ij} = 1/2$ $\quad i \in cl, j \in op\alpha$ or $op\beta$

$$\alpha_{ij} = 1/2, \quad \beta_{ij} = 1/2 \quad \begin{cases} i \in op\alpha, j \in op\alpha \\ \text{or } i \in op\beta, j \in op\beta \end{cases} \quad\quad (III.50)$$

$$\alpha_{ij} = 1/2, \quad \beta_{ij} = 0 \quad \begin{cases} i \in op\alpha, j \in op\beta \\ i \in op\beta, j \in op\alpha \end{cases}$$

$$\hat{F}_i = \hat{h} + \sum_{j \in cl}(2\hat{J}_j - \hat{K}_j) + \sum_{\substack{j \in op\alpha \\ \text{or } op\beta}}(\hat{J}_j - \hat{K}_j/2) = \hat{h} + \hat{G}_T = \hat{F}_{cl} \quad i \in cl$$

$$\hat{F}_i = \hat{h}/2 + \sum_{j \in cl}(\hat{J}_j - \hat{K}_j/2) + \sum_{j \in op\alpha}(\hat{J}_j - \hat{K}_j)/2 + \sum_{j \in op\beta}\hat{J}_j/2$$

$$= \frac{1}{2}\left[\hat{F}_{cl} + \frac{1}{2}(\hat{K}_\beta - \hat{K}_\alpha)\right] \quad\quad\quad i \in op\alpha \quad\quad (III.51)$$

$$\hat{F}_i = \hat{h}/2 + \sum_{j \in cl}(\hat{J}_j - \hat{K}_j/2) + \sum_{j \in op\beta}(\hat{J}_j - \hat{K}_j)/2 + \sum_{j \in op\alpha}\hat{J}_j/2$$

$$= \frac{1}{2}\left[\hat{F}_{cl} - \frac{1}{2}(\hat{K}_\beta - \hat{K}_\alpha)\right] \quad\quad\quad i \in op\beta \quad\quad (III.52)$$

The Hartree–Fock equation (equation III.38) can be written in the form

$$\hat{F}_i|i\rangle = \sum_{j \in M}\varepsilon_{ij}|i\rangle = \sum_{j \in M}|j\rangle\langle j|\hat{F}_i|i\rangle = \hat{P}_M\hat{F}_i|i\rangle \quad\quad (III.53)$$

where $\hat{P}_M$ is the projector operator on the occupied space:

$$\hat{P}_M = \sum_{j \in M}|j\rangle\langle j| \quad\quad (III.54)$$

Let us consider further the case (ii) of closed shell plus open shell with parallel spins, $i \in cl$ and $k \in op$. Then, $\hat{F}_i = \hat{F}_{cl}$ and $\hat{F}_k = \hat{F}_{op}$ and equation III.43 takes the form

$$\varepsilon_{ik} = \langle i|\hat{F}_{cl}|k\rangle = \langle i|\hat{F}_{op}|k\rangle \quad\quad (III.55)$$

$$\hat{F}_{op} = \frac{1}{2}(\hat{h} + \hat{G}_T - \hat{K}_{op}/2) = \frac{1}{2}(\hat{F}_{cl} - \hat{K}_{op}/2) \quad\quad (III.56)$$

Therefore

$$2\varepsilon_{ik} = \langle i|\hat{F}_{cl} - \hat{K}_{op}/2|k\rangle = 2\langle i|\hat{F}_{op}|k\rangle \quad\quad (III.57)$$

and using equation III.55, one obtains

$$\varepsilon_{ik} = -\frac{1}{2} \langle i|\hat{K}_{op}|k\rangle = \langle i|\hat{h} + \hat{G}_T|k\rangle = \langle i|\hat{F}_{cl}|k\rangle \qquad \text{(III.58)}$$

The Hartree–Fock equation (equation III.53) can be written as

$$i \in cl: \quad \hat{F}_{cl}|i\rangle = \sum_{j \in cl} |j\rangle \langle j|\hat{F}_{cl}|i\rangle + \sum_{l \in op} |l\rangle \langle l|\hat{F}_{cl}|i\rangle \qquad \text{(III.59)}$$

$$k \in op: \quad \hat{F}_{op}|k\rangle = \sum_{j \in cl} |j\rangle \langle j|\hat{F}_{op}|k\rangle + \sum_{l \in op} |l\rangle \langle l|\hat{F}_{op}|k\rangle \qquad \text{(III.60)}$$

or, alternatively, as

$$\begin{aligned}(\hat{F}_{cl} - \hat{P}_{op}\hat{F}_{cl})|i\rangle &= (\hat{I} - \hat{P}_{op})\hat{F}_{cl}|i\rangle \\ &= (\hat{I} - \hat{P}_{op})(\hat{h} + \hat{G}_T)|i\rangle = \sum_{j \in cl} \varepsilon_{ij}|j\rangle \end{aligned} \qquad \text{(III.61)}$$

$$\begin{aligned}(\hat{F}_{op} - \hat{P}_{cl}\hat{F}_{op})|k\rangle &= (\hat{I} - \hat{P}_{cl})\hat{F}_{op}|k\rangle \\ &= (\hat{I} - \hat{P}_{cl})\frac{1}{2}(\hat{h} + \hat{G}_T - \hat{K}_{op}/2)|k\rangle \\ &= \sum_{l \in op} \varepsilon_{kl}|l\rangle \end{aligned} \qquad \text{(III.62)}$$

One of the operators defined in this way acts on a closed-shell orbital to yield a linear combination of closed-shell orbitals, while the other operator acts on an open-shell orbital $|k\rangle$ to give a linear combination of open-shell orbitals. Combining equations III.58 and III.61, one obtains

$$(\hat{h} + \hat{G}_T + \hat{P}_{op}\hat{K}_{op}/2)|i\rangle = \sum_{j \in cl} \varepsilon_{ij}|j\rangle \qquad \text{(III.63)}$$

Combining equations III.58 and III.62, one obtains

$$\frac{1}{2}(\hat{h} + \hat{G}_T - \hat{K}_{op}/2 + \hat{P}_{cl}\hat{K}_{op})|k\rangle = \sum_{l \in op} \varepsilon_{kl}|l\rangle \qquad \text{(III.64)}$$

These are the coupled Hartree–Fock equations for the open-shell case with parallel spins.

Introducing a new operator $\hat{F}'$,

$$\begin{aligned}\hat{F}' &= \hat{h} + \hat{G}_T - \hat{K}_{op} - \hat{P}_{cl}\hat{K}_{op}\hat{P}_{cl} - \hat{P}_{op}\hat{K}_{op}\hat{P}_{op}/2 \\ &\quad + (\hat{P}_{cl} + \hat{P}_{op}/2)\hat{K}_{op} + \hat{K}_{op}(\hat{P}_{cl} + \hat{P}_{op}) \\ &= \hat{F}_{cl} + \frac{1}{2}[\hat{P}_{cl}\hat{K}_{op}\hat{P}_{op} + \hat{P}_{op}\hat{K}_{op}\hat{P}_{cl} - \\ &\quad (\hat{P}_{op} + \hat{P}_v)\hat{K}_{op}(\hat{P}_{op} + \hat{P}_v) - \hat{P}_v\hat{K}_{op}\hat{P}_v] \end{aligned} \qquad \text{(III.65)}$$

where

$$\hat{P}_v = \hat{I} - \hat{P}_{cl} - \hat{P}_{op} \qquad (III.66)$$

it follows that only one equation is to be solved:

$$\hat{F}'|i\rangle = \sum_j \varepsilon'_{ij}|j\rangle$$

$$\varepsilon'_{ij} = \varepsilon_{ij} \quad \text{if } i,j \in cl$$

$$\varepsilon'_{ij} = 2\varepsilon_{ij} \quad \text{if } i,j \in op$$

$$\varepsilon'_{ij} = 0 \quad \text{otherwise.} \qquad (III.67)$$

Since $\hat{F}'$ is Hermitian, the matrix ε'_{ij} can be diagonalized, leading to eigenequations. These have to be solved iteratively because of the dependence of the $\hat{K}_j$ and $\hat{J}_j$ on the MOs.

The configuration interaction method. In this method, the wave function of the j-th state is a linear combination of symmetry- and spin-adapted Slater determinants

$$|\Psi_j\rangle = \sum_K |\phi_K\rangle C_{Kj} = \sum_K |\phi_K\rangle \langle \phi_K|\Psi_j\rangle \qquad (III.68)$$

For normal molecules, usually one spin-adapted configuration (symmetry-adapted function) is dominant. The expansion of the wave function for a given state j can then be written as

$$|\Psi_j\rangle = C_{0j}|\phi_0\rangle + \sum_{r\to s} C_{r\to s;j}|\phi_{r\to s}\rangle + \frac{1}{2}\sum_{\substack{r\to s \\ t\to u}} C_{r\to s, t\to u;j}|\phi_{r\to s, t\to u}\rangle + \cdots \qquad (III.69)$$

The second and third terms in the expansion represent singly and doubly excited configurations. We shall introduce a short-hand notation for the second term, $\Sigma_M C_M |\phi_M\rangle$, and the third term, $\frac{1}{2}\Sigma_D C_D |\phi_D\rangle$, where the symbols M and D indicate single and double excitations from $|\phi_0\rangle$, respectively. The dots indicate that the expansion continues; in practice, it is nearly always truncated at some point.

For the energy expectation value, we have quite generally

$$E_j = \langle \Psi_j|\hat{H}|\Psi_j\rangle = \sum_K \sum_L C_{Kj} C_{Lj} H_{KL} \qquad (III.70)$$

where

$$H_{KL} = \langle \phi_K|\hat{H}|\phi_L\rangle \qquad (III.71)$$

From the variational principle, we have

$$\sum_K C_{Kj} H_{KL} = E_j C_{Lj} \qquad \text{(III.72)}$$

$$|H_{KL} - E\delta_{KL}| = 0$$

If $|\phi_0\rangle$ in equation III.69 is a closed-shell ground-state HF configuration, the matrix elements $\langle\phi_0|\hat{H}|\phi_{r\to s}\rangle$ vanish,

$$\langle\phi_0|\hat{H}|\phi_{r\to s}\rangle = \langle r(1)|\hat{F}_{cl}|s(1)\rangle = 0 \qquad \text{(III.73)}$$

since

$$\langle s|\hat{F}_{cl}|r\rangle = \sum_{j \in M} \varepsilon_{rj} \langle s|j\rangle = 0 \qquad \text{(III.74)}$$

because s is not among the occupied orbitals. This result is known as *Brillouin's theorem*. It is also valid for certain open-shell cases (generalized Brillouin's theorem).

Equation III.72 can be written

$$\begin{vmatrix} H_{00} - E & H_{01} & H_{02} & H_{03} & \ldots & H_{0L} \\ H_{10} & H_{11} - E & H_{12} & H_{13} & \ldots & H_{1L} \\ \ldots & \ldots & \ldots & \ldots & \ldots & \ldots \\ H_{L0} & H_{L1} & \ldots & \ldots & \ldots & H_{LL} - E \end{vmatrix} = 0 \qquad \text{(III.75)}$$

Expanding the determinant with respect to the first row yields

$$(H_{00} - E_j)M_{00} + \sum_{K>0} H_{0K} M_{0K} = 0 \qquad \text{(III.76)}$$

where M_{00} and M_{0K} are the corresponding minors.

Assuming that E_j is close to H_{00} and that $H_{LL} - E_j$ for arbitrary $L > j$ is relatively large, it is possible to write

$$(H_{00} - E_j) \prod_{L=1} (H_{LL} - H_{00}) - \sum_K H_{0K}^2 \prod_{L \neq K} (H_{LL} - H_{00}) = 0 \qquad \text{(III.77)}$$

$$E_j = H_{00} + \sum_K \frac{H_{0K}^2}{(H_{KK} - H_{00})} = H_{00} + \sum_D \frac{H_{0D}^2}{(H_{DD} - H_{00})} \qquad \text{(III.78)}$$

where $H_{0K} \neq 0$ only if K is doubly excited. This is the second-order Möller–Plesset approximation.

Configuration selection rules for truncated CI. In actual calculations, the CI expansion usually needs to be truncated, and the selection of the configurations to be kept is discussed next. The lowering of the energy of the Hartree–Fock configuration due to the addition of a doubly excited configuration is obtained from

$$\begin{vmatrix} H_{00} - \varepsilon & H_{0K} \\ H_{0K} & H_{KK} - \varepsilon \end{vmatrix} = 0 \quad \text{(III.79)}$$

$$\varepsilon = \frac{H_{KK} + H_{00}}{2} - \frac{1}{2}\sqrt{(H_{KK} - H_{00})^2 + 4H_{0K}^2} \quad \text{(III.80)}$$

If $|H_{KK} - H_{00}|^2 \gg 4H_{0K}$, we obtain

$$\varepsilon \simeq H_{00} - \frac{H_{0K}^2}{(H_{KK} - H_{00})} \quad \text{(III.81)}$$

The individual contribution to the correlation energy produced by the doubly excited configuration is $\varepsilon - H_{00}$. If the individual contributions are small, the scheme is additive and therefore equivalent to the Möller–Plesset perturbation approach. Of course, the additivity is only approximate, and this perturbative scheme may easily overestimate the correlation energy. On the other hand, if the individual selection according to energy criterion is used for selecting the configurations to be used in the variational CI calculation, the interaction among the selected excited configurations will be allowed for. Then, the variational principle guarantees that the correlation energy is not overestimated.

Since in photochemical calculations it is often necessary to perform computations in the region of an avoided crossing and for biradicaloid geometries, more than one configuration often dominates the CI expansion for a given state. It is therefore important to be able to consider several reference configurations on equal footing and to use all single and double excitations with respect to them.

This is done in the "multi-reference single and double excitation" configuration interaction method (MRD-CI). The wave function is written in terms of the reference configurations R and the others, R':

$$|\Psi_j\rangle = \sum_{U \in R} C_{U,j}|\phi_U\rangle + \sum_{V \in R'} C_{V,j}|\phi_V\rangle \quad \text{(III.82)}$$

The selection of the configurations that do not belong to the reference set is done by examining the energy lowering of all reference configurations due to the separate addition of each individual singly or doubly excited configurations ϕ_V. The following secular determinant has to be solved:

$$\begin{vmatrix} \mathbf{H}(R) - E\mathbf{I} & \mathbf{H}_{RV} \\ \mathbf{H}_{VR}^\dagger & H_{VV} - E \end{vmatrix} = 0 \quad \text{(III.83)}$$

where $\mathbf{H}(R)$ is Hamiltonian matrix for the reference set R, and $\mathbf{H}_{RV}$ is the column vector of the matrix elements between configuration ϕ_V and configurations from the reference set R. The E_R obtained by solving the problem for the reference set only, $|\mathbf{H}(R) - E_R\mathbf{I}| = 0$, is compared with the E obtained by solving equation III.83.

The selection of the individual configurations according to the energy lowering criterion can be introduced by choosing a threshold T. In this manner, all the configurations that contribute less than the value of T are discarded.

If the configuration space is very large and only a small fraction of all the configurations can be selected, an extrapolation procedure to the full MRD-CI space is usually applied. According to this scheme, the selection of configurations can be made for more than one solution, and consequently several states of the same symmetry can be considered as the roots of the same secular problem. Since this is truncated CI, the choice of the one-electron functions (MOs) that are used to build the configurations is very important because a certain choice of MOs might be more suitable for the description of one state and less suitable for others that are obtained as roots of the secular equation. An example is covalent versus zwitterionic states.

Demonstration of the size consistency effect. Truncated CI expansions suffer from an important shortcoming, which we shall now illustrate. Consider a system of n noninteracting two-electron, two-center homonuclear subsystems (e.g., n H_2). The ground configuration $|\phi_0\rangle$ is the Slater determinant built from the occupied MOs of all subsystems. Doubly excited configurations belong to individual subsystems

$$|\Psi\rangle = |\phi_0\rangle + \sum_I C_I |\phi_I \begin{pmatrix} \bar{1} \rightarrow \bar{2} \\ 1 \rightarrow 2 \end{pmatrix} \rangle \qquad (III.84)$$

and the CI matrix takes the form

$$\begin{vmatrix} -E_c & K_{12} & \cdots & K_{12} \\ K_{12} & \Delta - E_c & \cdots & 0 \\ \vdots & \vdots & & \vdots \\ K_{12} & 0 & \cdots & \Delta - E_c \end{vmatrix} = 0 \qquad (III.85)$$

where $\Delta = 2(h_{22} - h_{11}) + (J_{22} - J_{11})$ and $\langle \phi_0 | \hat{H} | \phi_0 \rangle$ has been taken for energy zero. The solution is

$$-E_c(\Delta - E_c)^n - n(\Delta - E_c)^{n-1} K_{12}^2 = 0 \qquad (III.86)$$

$$E_c = \frac{\Delta}{2} - \sqrt{\frac{\Delta^2}{4} + nK_{12}^2} \qquad (III.87)$$

and, for large n, goes to

$$E_c \equiv E_c(D-CI) \simeq -\sqrt{n}\, K_{12} \qquad (III.88)$$

The exact solution, beyond D-CI, is

$$^nE_c \equiv n\left(\frac{\Delta}{2} - \sqrt{\frac{\Delta^2}{4} + K_{12}^2}\right) \qquad (III.89)$$

It is clear that for large n,

$$\lim_{n\to\infty} \frac{E_c(\text{D-CI})}{n} = 0 \tag{III.90}$$

which is wrong. Even for small n there is a difference between $^n E_c$ and $E_c(\text{D-CI})$. Therefore, corrections to the D-CI result need to be introduced (e.g., the Davidson correction).

Direct CI. The Hartree–Fock approximation does not consider at all the correlation of the motions of electrons with opposite spins and considers, in part, only the correlation of the motions of electrons with parallel spins. The pair correlation can be well described by the inclusion of configurations resulting from single and double replacements in the reference configuration. Evidently, the number of these configurations is very large. For large CI expansions containing more than a few thousand configurations, the construction of the CI matrix H and its straightforward diagonalization become impractical. Instead of solving the secular problem

$$\sum_L (H_{KL} - E\, S_{KL}) C_L = 0 \tag{III.91}$$

where

$$H_{KL} = \langle \phi_K | \hat{H} | \phi_L \rangle$$

and

$$S_{KL} = \langle \phi_K | \phi_L \rangle$$

by diagonalization, it can be solved in an iterative manner. The basic idea is to construct the auxiliary vectors

$$\boldsymbol{\sigma}^{(m)} = \mathbf{H}\, \mathbf{C}^{(m-1)} \tag{III.92}$$

where $\boldsymbol{\sigma}^{(m)}$ is calculated in the m-th step by multiplying the solution $\mathbf{C}^{(m-1)}$ of the $(m-1)$st step with the Hamiltonian matrix H. The connection between $\boldsymbol{\sigma}^{(m)}$ and $\mathbf{C}^{(m)}$ should be as simple as possible.

The key feature of the relation III.92 is that only one row of the matrix H is used for the calculation of an element $\sigma_i^{(m)}$ of the vector $\boldsymbol{\sigma}^{(m)}$. The individual Hamiltonian matrix elements H_{KL} are linear combinations of integrals over MOs, such as $(ab|cd)$. The contribution to the change $\Delta\sigma_i^{(m)}$ in the i-th element of the vector $\boldsymbol{\sigma}^{(m-1)}$ in the m-th step provided by a particular integral $(ab|cd)$ has the form

$$A(ab|cd) C_j^{(m-1)} \tag{III.93}$$

where A is determined solely by the nature of the spin coupling in the configurations involved.

Two closely related procedures for the construction of the vectors $\mathbf{C}^{(m)}$ from the auxiliary vectors $\boldsymbol{\sigma}^{(m)}$ are standard:

(i) *The perturbative approach*. The Hamiltonian is written in the form

$$\hat{H} = \hat{H}_0 + \hat{V} \tag{III.94}$$

and the wave function and energy are expanded in the usual way:

$$\Psi = \sum_n \Psi^{(n)} \tag{III.95}$$

$$E = \sum_n \varepsilon_n \tag{III.96}$$

The substitution of equations III.95 and III.96 into the eigenequation of the Hamiltonian III.94 leads to the following expression for the m-th iterative step:

$$(E_0 - \hat{H}_0)\Psi^{(m)} = V\Psi^{(m-1)} - \sum_{n=0}^{m-1} \varepsilon_{m-n}\Psi^{(n)} \tag{III.97}$$

An appropriate choice of $\hat{H}_0$ which allows for a simple calculation of the inverse $(E_0 - \hat{H}_0)^{-1}$ is

$$\hat{H}_0 = \sum_{K=1}^{P} |\phi_K\rangle\langle\phi_K|\hat{H}|\phi_K\rangle\langle\phi_K| \tag{III.98}$$

where P is the number of configurations. The component form of equation III.97 then takes the simple form

$$C_K^{(m)} = \frac{1}{E_0 - H_{KK}} \left[\sum_{L=1}^{P} V_{KL} C_L^{(m-1)} - \sum_{n=1}^{m-1} \varepsilon_{m-n} C_K^{(n)} \right] \tag{III.99}$$

The sums in the bracket on the right-hand side of equation III.99 have the form III.92. The right-hand sides of the equations for the perturbation energies III.96,

$$\varepsilon_{2m-1} = \sum_{K=1}^{P} C_K^{(m-1)} \sum_{L=1}^{P} V_{KL} C_L^{(m-1)} - \sum_{n=1}^{m-1} \sum_{n'=1}^{m-1} \varepsilon_{2m-1-n-n'} \sum_K C_K^{(n)} C_K^{(n')} \tag{III.100}$$

$$\varepsilon_{2m} = \sum_{K=1}^{P} C_K^{(m)} \sum_{L=1}^{P} V_{KL} C_L^{(m-1)} - \sum_{n=1}^{m} \sum_{n'=1}^{m-1} \varepsilon_{2m-n-n'} \sum_K C_K^{(n)} C_K^{(n')} \tag{III.101}$$

also have the desired form III.92. The computation is continued until convergence.

Other choices of $\hat{H}_0$ are possible, such as

$$\hat{H}_0 = \sum_K |\phi_K\rangle \langle\phi_K|\hat{F}_{\phi_K}|\phi_K\rangle \tag{III.102}$$

and yield faster convergence, but the above choice III.98 has a better theoretical justification.

(ii) *The partitioning approach.* The equation to be solved has the following form:

$$\mathbf{AC} = 0 \tag{III.103}$$

Let us partition the matrix $\mathbf{A}$:

$$(\mathbf{A}' + \mathbf{A}'')\mathbf{C} = 0 \tag{III.104}$$

so that

$$\mathbf{A}'\mathbf{C}^{(m-1)} + \mathbf{A}''\mathbf{C}^{(m)} = 0$$

or $\tag{III.105}$

$$\mathbf{C}^{(m)} = -(\mathbf{A}'')^{-1}\mathbf{A}'\mathbf{C}^{(m-1)}$$

$\mathbf{A}''$ should be diagonal in order to be inverted easily. A convenient choice is to include the diagonal terms of the matrix $(\mathbf{H} - E\mathbf{I})$ in $\mathbf{A}''$; consequently,

$$\begin{aligned} A''_{KL} &= (H_{KK} - E)\delta_{KL} \\ A'_{KL} &= H_{KL}(1 - \delta_{KL}) \end{aligned} \tag{III.106}$$

The component form of equation III.105 is

$$C_K^{(m)} = \frac{1}{E^{(m-1)} - H_{KK}} \left[\sum_{L=1}^{P} H_{KL} C_L^{(m-1)} - H_{KK} C_K^{(m-1)} \right] \tag{III.107}$$

where $E^{(m-1)}$ is the energy calculated in the $(m-1)$st step,

$$E^{(m-1)} = \sum_{K,L=1}^{P} C_K^{(m-1)} H_{KL} C_L^{(m-1)} \bigg/ \sum_{K=1}^{P} |C_K^{(m-1)}|^2 \tag{III.108}$$

The contributions on the right-hand side of equation III.107 have again the desired form III.92.

In both the perturbative (i) and the partitioning (ii) schemes, simple algebraic expressions, III.99 and III.107, respectively, for the auxiliary vectors $\sigma^{(m)}$ permit a solution of the secular problem III.91. Both of these direct CI procedures determine the CI expansion coefficients directly from molecular integrals, avoiding the need to keep all elements of the Hamiltonian matrix in computer memory.

The generalized valence bond (GVB) method. This method is particularly suitable for the description of bond dissociation processes. Consider a closed-shell singlet wave function in a minimum AO basis set describing a chemical bond between equivalent atoms A and B by means of delocalized normalized MO, $|b\rangle = [2(1 + S_{AB})]^{-1/2} (|A\rangle + |B\rangle)$ (Chapter 4):

$$\hat{\mathscr{A}} b(1)b(2)\alpha(1)\beta(2) = b(1)b(2)\Sigma(1,2) \qquad \text{(III.109)}$$

$$\Sigma(1,2) = \frac{1}{\sqrt{2}} [\alpha(1)\beta(2) - \alpha(2)\beta(1)] \qquad \text{(III.110)}$$

After complete bond dissociation the minimum AO basis set description is

$$\hat{\mathscr{A}} A(1)B(2)\Sigma(1,2) = \frac{1}{\sqrt{2}} [A(1)B(2) + A(2)B(1)]\Sigma(1,2) \qquad \text{(III.111)}$$

where A is localized on atom A and B is localized on atom B, so that in the limit of infinite separation we obtain

$$\langle A|B \rangle = 0 \qquad \text{(III.112)}$$

A smooth transition between the two cases is offered by a GVB wave function for two electrons, based on GVB orbitals of the type $|a\rangle = C_A|A\rangle + C_B|B\rangle$:

$$\hat{\mathscr{A}} a(1) b(2) \Sigma(1,2) = \frac{1}{\sqrt{2(1 + S_{ab}^2)}} [a(1) b(2) + a(2) b(1)]\Sigma(1,2) \qquad \text{(III.113)}$$

where

$$\langle a | b \rangle = S_{ab} \neq 0 \qquad \text{(III.114)}$$

and the orbitals $|a\rangle, |b\rangle$ gradually change from the delocalized $|a\rangle, |b\rangle$ to the localized AOs $|A\rangle, |B\rangle$ as the bond dissociates, allowing for a variation in the covalent and the ionic character of the bond in the process.

The GVB wave function for a larger number of electrons is

$$\psi_{GVB} = \hat{\mathscr{A}} [a_1(1) b_1(2)\Sigma(1,2)\cdots a_m(2m-1) b_m(2m)$$
$$\times \Sigma(2m-1, 2m) \cdots a_{m+1}(2m+1)\alpha(2m+1) \cdots a_{m+m'}(N)\alpha(N)] \qquad \text{(III.115)}$$

$$\langle a_j | b_k \rangle = \delta_{jk} S_j \qquad \text{(III.116)}$$

where m is the number of singlet-coupled electron pairs and m' is the number of unpaired electrons:

$$N = 2m + m' \qquad \text{(III.117)}$$

Note that

$$E_{GVB} = \frac{\langle \Psi_{GVB}|\hat{H}|\Psi_{GVB}\rangle}{\langle \Psi_{GVB}|\Psi_{GVB}\rangle} \quad (III.118)$$

does not have the appropriate form

$$E = \sum_j \omega_j h_j + \sum_{j,k} (\alpha_{jk} J_{jk} - \beta_{jk} K_{jk}) \quad (III.119)$$

because a_j and b_j are not orthogonal ($S_j \neq 0$). The transformation to orthogonal orbitals $\mathscr{A}_j$ and $\mathscr{B}_j$,

$$\mathscr{A}_j = \frac{1}{\sqrt{2(1-S_j)}}(a_j - b_j)$$

$$\mathscr{B}_j = \frac{1}{\sqrt{2(1+S_j)}}(a_j + b_j)$$

$$\langle \mathscr{A}|\mathscr{B}\rangle = 0$$

which has the inverse

$$a_j = \frac{1}{\sqrt{2}}[\sqrt{1+S_j}\,\mathscr{B}_j + \sqrt{1-S_j}\,\mathscr{A}_j]$$
$$b_j = \frac{1}{\sqrt{2}}[\sqrt{1+S_j}\,\mathscr{B}_j - \sqrt{1-S_j}\,\mathscr{A}_j] \quad (III.120)$$

yields the appropriate form for equation III.119.

After the transformation III.120, the GVB wave function takes the form

$$\Psi_{GVB} = \hat{\mathscr{A}}\left\{\prod_{j=1}^{m}[(1+S_j)\,b_j(2j-1)\,b_j(2j)\right.$$
$$- (1-S_j)\,a_j(2j-1)\,a_j(2j)]\Sigma(2j-1,2j)\,a_{m+1}(2m+1)$$
$$\times \alpha(2m+1)\cdots a_{m+m'}(N)\alpha(N)] \quad (III.121)$$

Natural orbitals. A density matrix is defined for every molecular state as

$$\Gamma^{(N)}(1,2,\ldots,N|1',2',\ldots,N')$$
$$= |\Psi(1,2,\ldots,N)\rangle\langle\Psi(1',2',\ldots,N')| \quad (III.122)$$

The use of the unprimed and primed labels for the coordinates of each electron indicates that they can have different values in the $|ket\rangle$ and the $\langle bra|$ symbol. A reduced one-electron density matrix is obtained by integration over the coordinates of all electrons but one:

$$\Gamma^{(1)}(1,1') \equiv \Gamma(1,1')$$
$$= N \int \psi(1,2,\ldots,N)\psi^*(1',2,\ldots,N)\,dx_2\cdots dx_N \quad (III.123)$$

It is Hermitian

$$\Gamma(1,1') = \Gamma^*(1',1) \quad (III.124)$$

and can be expanded in terms of an arbitrary orthogonal set of one-electron basis functions:

$$\Gamma(1,1') = \sum_{j,k} |j(1)\rangle \gamma_{jk}(1,1')\langle k(1')| = \mathbf{j}(1)\Gamma(1,1')\mathbf{k}(1') \quad (III.125)$$

The diagonalization of this density matrix $\Gamma(1,1')$ yields the eigenvectors (natural orbitals, NOs, $|s\rangle$) and their eigenvalues (occupation numbers n_s):

$$\mathbf{U}^+ \, \Gamma \, \mathbf{U} = \mathbf{n}, \quad [\mathbf{n}]_{st} = \delta_{st} n_s \quad (III.126)$$

The natural orbitals $|s\rangle$ can be expressed in terms of the basis set orbitals $|j\rangle$:

$$|s\rangle = \sum_j |j\rangle U_{js}$$
$$|s\rangle = |\mathbf{j}\rangle \mathbf{U} \quad (III.127)$$

In the NO basis, the reduced one-electron density matrix Γ is diagonal:

$$\Gamma(1,1') = \sum_s n_s |s(1)\rangle \langle s(1')| \quad (III.128)$$

In this basis, all one-electron observables can be written as a sum containing a contribution from each NO:

$$\langle \Psi|\hat{P}|\Psi\rangle = \sum_s n_s \langle s|\hat{p}|s\rangle \quad (III.129)$$

For additional information and further detail on *ab initio* computational methods, see: H. F. Schaefer III, Ed., *Modern Theoretical Chemistry, Vol. 3: Methods of Electronic Structural Theory*, Plenum Press, New York, 1977; P. Čársky and M. Urban, *Ab Initio Calculations*, Springer Verlag, Berlin, 1980; A. Szabo and N. S. Ostlund, *Modern Quantum Chemistry. Introduction to Advanced Electronic Structure Theory*, Macmillan, New York, 1982; W. J. Hehre, L. Radom, P. R. Schleyer, and J. A. Pople, *Ab Initio Molecular Orbital Theory*, Wiley, New York, 1986; T. Clark, *A Handbook of Computational Chemistry*, Wiley, New York, 1985.

A brief and informative survey of basis sets is found in E. R. Davidson and D. Feller, *Chem. Rev.* **86,** 681 (1986).

INDEX

Ab initio calculations, 462
  acetaldehyde, 379
  acrolein, 367
  acroleinimine, 360, 363, 365
  acroleiniminium, 320, 322
  aminoborane, 220
  ammonia–borane adduct, 207
  aziridine, 409
  bromine, 396
  1,3-butadiene, 256, 303, 308
  carbene, 412
  carbonyl addition to olefins, 397
  cycloaddition of two ethylenes, 271
  cyclobutadiene, 247
    heterocyclic analogs, 247
  cyclobutene, 256
  cyclopropane, 202
  diazene, 369, 373
  diazirine, 412, 414, 415
  diazomethane, 392, 414, 415
  digermane, 201
  disrotatory cyclobutene–butadiene interconversion, 256
  ethylene, 208, 213, 397
    dimerization, 271
    in the field of charge, 167, 222
    minimum basis set, 158
    non-Born–Oppenheimer coupling elements, 68, 69
    pyramidalized, 212
  formaldehyde, 397, 403
  formaldimine, 355, 356, 357, 358
  formaldiminium, 218
  formyl radical, 379
  H_2:
    accurate, 156
    in the field of charge, 164
    minimum basis set, 148, 149, 150, 153, 432
      polarized, 162, 163, 165
  H_4:
    linear, 288
    rectangular, 232
    square, 234
    trapezoidal, 241
  H abstraction from methane by formaldehyde, 403
  heterocyclic analogs of cyclobutadiene, 247
  hydrazoic acid, 395
  hydrogen 3,2 shift in propene, 262
  lithium hydride, 166
  methane, 403
  methanol, 377
  methods, 421, 442
  methylammonium, 205
  methyl bromide, 381
  methyl peroxide, 383
  Norrish I, 379
  oxaziridine, 410
  oxirane, 259, 409
  propene, 215, 262
  protonated methylamine, 205
  push–pull perturbed cyclobutadienes, 247
  trimethylene biradical, 203
  vinyl bromide, 385
Absorption, 56, 105. *See also* Excitation; Transition
Abstraction of hydrogen atom, 405. *See also* Atom or ion transfer
  by azaaromatics, 404
  by carbonyl, 400, 404, 405, 407, 416
  by cycloalkene, 296, 300
  by imines, 404
Abstraction of proton, *see* Atom or ion transfer
Acceptor, *see* Electron transfer; Energy transfer; Exciplexes; Push–pull substitution; TICT (twisted internal charge-transfer) states
Acetaldehyde, 378
Acetone methylimine, 12, 13, 14, 15
Acid–base equilibria, 98. *See also* Atom or ion transfer
Acrolein, 365
  C=C twist, 367
Acroleinimine, 359. *See also* Cis–trans and syn–anti isomerization
  C—C twist, 364
  C=C twist, 359
  CNH inversion, 363
  C=N twist, 362
Acroleiniminium, 320. *See also* Cis–trans and syn–anti isomerization
  C=C twist, 320
  C=N twist, 322
Active AOs, 112
Addition reaction, *see also* Atom or ion transfer; Cycloaddition
  carbonyl to olefins (Paterno–Büchi), 397
  nucleophilic, to C=N$^+$, 319
  photohydration, 297
  radical, to C=O and C=N, 400
  radical and ionic, to olefins, 296, 300
Adiabatic reaction, *see* Reaction
Alcohols, *see* Methanol
Alkenes, radical and ionic addition, 296, 300. *See also* Atom or ion transfer; 1,3-Butadiene; Cis–trans and syn–anti isomerization; Cycloaddition; Electrocyclic

463

Alkenes, radical and ionic addition (*Continued*)
  reaction; Ethylene; Propene;
  Pseudosigmatropic shift; Sigmatropic shift;
  Stilbene; Styrene
Alkyl halides, *see also* Vinyl bromide
  dissociation of methyl bromide, 381
  dissociation of trifluoromethyl chloride, 119
"Allowed" reaction, *see* Reaction
Allyl anion, 258
Allylic bond cleavage, 292, 318
Allylsulfonium cation, 318
Allyltrimethylsilane, 126
Alternant systems, 161
Aminoborane, 177, 219
Ammonia–borane adduct, 206
Angular momentum quenching, 439
Anilinium cation, 133
Antarafacial participation, 225, 260, 264
Anthracene, 333
Antiaromatic cyclic orbital array, 226
AO, *see* Orbital
AO basis set, 29, 442
Aromatics, photosubstitution, 297, 302
Arrhenius equation, 70
Atomic orbital (AO):
  active, 112
  correlation, 113
  definition, 29, 421, 442
  labels, 114
  phase, 114
Atom or ion transfer, 296, 346, 398. *See also*
    Correlation diagrams
  to alkenes, 296
  to charged double bonds, 319
  to a double bond, 296, 399
  hydrogen atom, 297, 400, 404, 405
  intermolecular, 400
  intramolecular, 405
  to a lone pair, 98, 325, 398
  to a lone pair on a double bond, 400
  photohydration, 297
  proton, 98, 325, 398, 400, 405
Aufbau principle, 34
Autoionization, 10, 86
Azaallyl, 320, 361
Azaaromatics, 404
Azides, 391
Aziridine, 285, 416
  C—N bond-breaking, 408
  electrocyclic ring opening, 248, 258
Azirine, 416
Azo compounds, *see* Diazene

B^-=N^+ bond isomerization, aminoborane, 219.
    See also Cis–trans and syn–anti
    isomerization; TICT (twisted internal
    charge-transfer) states
Back electron transfer, *see* Electron transfer
Barrier, 70, 139
Barrier-free reaction, *see* Reaction
Basis set, *see* Orbital
Bathorhodopsin, 324
Benzene dianion, 177
Benzene dication, 177

Benzylammonium cation, 138
Benzylic bond cleavage, 138, 139, 292, 294, 295,
    319. *See also* "Interacting subunits"
    correlation diagrams
9,9′-Bianthryl, 341, 347. *See also* TICT (twisted
    internal charge-transfer) states
Biradical, 169, 223. *See also* Biradicaloid;
    Occupancy of natural orbitals; Two-electron
    two-orbital model
  axial, 178, 437
  1,3 biradical, 184, 202
  energies, 436
  Hamiltonian matrix, 172, 176, 434
  in Norrish II reaction, 407
  pair, 176, 437
  perfect, 175, 436
  S_0–S_1 splitting, 178
  S_0–T_1 splitting, 178
  spin–orbit coupling, 198
  wave functions, 176, 436
Biradicaloid, 179, 223, 438. *See also* Biradical;
    Dissociation of single bond; Two-electron
    two-orbital model
  charge-transfer, 186, 190, 192, 220
  critically heterosymmetric, 112, 186
  density matrix, 438
  electronegativity difference parameter δ, 181,
    438
  energies, 182, 184
  Hamiltonian matrix, 172, 182, 184, 187, 434
  heterosymmetric, 184, 440
  homosymmetric, 182, 440
  interaction parameter γ, 181, 438
  natural orbital occupancy in, 175, 180, 184,
    185, 438, 441
  nonsymmetric, 187, 441
  from oxirane and aziridine, 258
  spin–orbit coupling in, 198
  strongly heterosymmetric, 186
  sudden polarization, 185, 187
  wave functions, 180, 182, 184, 440
  weakly heterosymmetric, 186
Bistability, 219, 323
Bitopic, 143. *See also* Topicity
Bond, *see also* Pi bond; Sigma bond
  charged, 205, 217
  covalent, 200, 206
  dative, 206, 219
  order, 41
Bond cleavage, *see* Cis–trans and syn–anti
    isomerization; Dissociation of double bond;
    Dissociation of single bond
Bond-shift reaction, *see* Pseudosigmatropic shift;
    Sigmatropic shift
Borane–ammonia adduct, 206
Born–Oppenheimer approximation, 8, 19, 48
Brillouin's theorem, 35, 454
Bromine, molecular, 395
1,3-Butadiene, 302. *See also* Cis–trans and syn–
    anti isomerization; Cyclobutene
  comparison with ethylene, 307
  cycloaddition, 123, 128
  electrocyclic ring closure, 248
  electronic states, 305, 306

planar, 249, 304
singlet isomerization, 309
triplet isomerization, 308
twisted, 305, 314
VB-like structures, 306

C_2, 11
C=C bond cleavage, in ketene, 386
C=C bond isomerization, 192. *See also* Cis–trans and syn–anti isomerization
  acrolein, 365
  acroleinimine, 359
  acroleiniminium, 320
  butadiene, 302
    in S_1 state, 309
    in T_1 state, 308
  ethylene, 207, 213
  ethylene in the field of a charge, 167, 222
  polyenes, 309
  propene, 215
  stilbene, 312
  styrene, 310
C=N bond cleavage, in diazomethane, 391
C=N bond isomerization, *see also* Cis–trans and syn–anti isomerization
  acroleinimine, 362
  acroleiniminium, 322
  formaldimine, 350
  formaldiminium, 217
Caldwell equation, 271, 286
Carbene, 386, 391, 413
  dimerization to twisted ethylene, 209
  from hydrogen shift in twisted ethylene, 214
Carbonyl, 401. *See also* Abstraction of hydrogen atom; Addition reaction; Diazene; Norrish I; Norrish II
Charge–bond-order matrix, 41
Charge density, *see* Electron charge density
Charged perimeter, 226, 258
Charge-resonance interaction, 240, 278
Charge transfer, 22. *See also* Charge translocation; Electron transfer; Excimer; Exciplexes; Marcus equation
  absorption, 78, 329
  biradicaloid, 186, 190, 192, 220, 338
  charge recombination, 88
  charge separation, 88
  complex, 78, 107, 328
  states, 240
  transition, 45
  TICT (twisted internal charge-transfer), 220
  VB structures, 237
Charge translocation, 88, 219. *See also* Charge transfer
  acroleiniminium, 320
  formaldiminium, 217
  rhodopsin, 325
Cheletropic reaction, *see* Diazirine
Chemiexcitation, *see* Excitation
Chemiluminiscence, 22, 63
Chromophore, 43
CI, *see* Configuration interaction (CI)
CIDNP, chemically induced dynamic nuclear polarization, 299, 347

Cis–trans and syn–anti isomerization, 192, 346, 415. *See also* Correlation diagrams; TICT (twisted internal charge-transfer) states; Two-electron two-orbital model
  acrolein, 367
  acroleinimine, 359, 362
  acroleiniminium, 320, 322
  aminoborane, 219
  butadiene, 302, 308, 309
  charged pi bond, 217, 320, 322
  covalent pi bond, 206, 215
  dative pi bond, 219
  diazene, 369
  diene, 302
  enone, 365
  ethylene, 207, 213
  formaldimine, 350, 354
  formaldiminium, 217
  intersystem crossing, 198, 214, 309, 314
  polyenes, 309
  propene, 215
  rhodopsin, 324
  stilbene, 312
  styrene, 310
Classification of MOs and states, 43, 48, 49
α-Cleavage, *see* Diazene; Norrish I
"Closs rule," 85
Complex, *see* Charge transfer; Energy transfer
Concerted reaction, *see* Reaction
Configuration, 32
  characteristic, 132
  charge separated, 342
  charge transfer, 279, 330. *See also* Charge transfer, biradicaloid
  in correlation diagrams, 127
  in diazene, 371
  doubly excited, 33
  energy, 35
    in H_2, 431
  locally excited, 279, 330. *See also* "Interacting subsystems" correlation diagrams; TICT (twisted internal charge-transfer) states
  reference, 32
  selection, 34, 454
  singly excited, 33
  state function, 32
  subsystem, 238
Configuration correlation diagrams:
  cycloaddition:
    2 + 2, two ethylenes, 266
    2 + 4, Diels–Alder, 128
  definition, 127, 129, 130
  dissociation:
    of C=C bond in ketene, 388
    of N=N bond in hydrazoic acid, 394
  fragmentation of diazirine, 411
  ring closure, pleiadiene, 257
Configuration interaction (CI), 34, 430, 453. *See also* Two-electron two-orbital model
  direct, 457
    partitioning, 459
    perturbative, 458
  full, 34, 229
  multi-reference, 34, 455

Configuration interaction (CI) (*Continued*)
  MRD-CI, 455
  single-reference, 34
  3 × 3, 153, 169, 433
Configuration state function, 32
Conical intersection, 22, 51, 66, 140, 168, 215, 218, 222, 356, 391, 398, 402
Conrotatory, 248. *See also* Electrocyclic reaction
Contact ion pair, *see* Ion pairs and free ions
Contour map, 12
Contracted basis set, *see* Orbital
Correlation diagrams, 112, 140. *See also* Configuration correlation diagrams; "Interacting subunits" correlation diagrams; Molecular orbital correlation diagrams; State correlation diagrams; Valence-bond correlation diagrams
  "allowedness," 127, 131
  AO, 113
  atom or ion transfer:
    to alkenes, 298, 299
    to carbonyl, 401, 404
    to styrene, 301
  cis–trans and syn–anti isomerization:
    diazene, 370
    double bond, 117, 121, 192
    ethylene, 117, 122
    formaldimine, 351
    push–pull substituted ethylene, 136
    stilbene, 313
    styrene, 311
  cycloaddition:
    2 + 2, two ethylenes, 265, 266
    2 + 4, Diels–Alder, 123, 128
    non-concerted, two ethylenes, 316
  dissociation:
    of allylic or benzylic bond, 138, 139, 294, 295, 318
    of Br—Br bond, 396
    of C—Br bond in vinyl bromide, 384
    of C—C bond in acetaldehyde, 379
    of C—Cl bond in trifluoromethyl chloride, 119
    of C—N bond in methylamine, 375
    of C—N$^+$ bond in anilinium, 133, 134
    of C—N$^+$ bond in benzylammonium, 138, 139
    of C—O bond in methanol, 377
    of C—S$^+$ bond in allylsulfonium, 318
    of C=C bond in ketene, 387, 388, 390
    of charged single bond, 133, 134, 138, 139, 318
    of lone pair carrying single bond, 119
    of N=N bond in hydrazoic acid, 393, 394
    of single bond, 116, 120, 191
  donor–acceptor pair, 332
  "forbiddenness," 127, 128, 131
  fragmentation of diazirine, 411
  H$_2$ + D$_2$ ⇌ 2HD, 229
  hydrogen abstraction, *see* Correlation diagrams, atom or ion transfer
  MO crossing:
    abnormal, 130, 132, 257, 387, 411
    normal, 129
  natural, 135
  Norrish I, 379
  ring closure and opening:
    butadiene–cyclobutene:
      conrotatory, 252, 253
      disrotatory, 249, 251, 253
      non-concerted, 315
    naphthalene–dewarnaphthalene, 101
    pleiadiene, disrotatory, 132, 257
    sigmatropic shift, 1,3-:
      with inversion, in allyltrimethylsilane, 126
      without inversion, in propene, 125
    syn–anti isomerization, *see* Correlation diagrams, cis–trans and syn–anti isomerization
  TICT (twisted internal charge-transfer), 339, 341
  valence angle variation:
    in formaldimine, 353
    in trimethylene biradical, 204
Correlation energy, 34
Correlation of twisted ethylene with two carbenes, 209
Coulomb integral, 37, 283, 333, 424, 430, 433
Coupling, *see* Dexter energy transfer mechanism; Exchange integral; Förster; Hyperfine interactions; Nonadiabatic coupling; Spin–orbit coupling
Covalent, 32, 161. *See also* State, electronic
Critical heterosymmetry, 186
Critical transfer distance R_0, *see* Förster
Crossing, *see also* Conical intersection, Correlation diagrams
  configuration, 130
  strongly avoided, *see* Strong coupling
  weakly avoided, *see* Weak coupling
Cyanine dyes, 323, 346
Cyano-substituted benzenes, 333
Cycloaddition, *see also* Correlation diagrams
  concerted, 264, 285
    Caldwell equation, 271
    regiochemistry, 267, 268, 272
    singlet, 267
    stereochemistry, 267, 272
    triplet, 267
    two nonpolar bonds, 226
    two polar bonds, 244
    2$_s$ + 2$_s$, two ethylenes, 265
  cycloreversion, 270
  electronic states, 240
  nonconcerted, 314
Cycloalkene, 300
Cyclobutadiene, 177
  heterocyclic analogs, 247
  push–pull substituted, 247, 285
Cyclobutene:
  concerted ring opening, 248
  conrotatory, 253, 255
  disrotatory, 248, 253, 285
  nonconcerted ring opening, 314
Cycloheptatrienide anion, 177
Cyclopentadienyl cation, 177
Cyclopentenide anion, 260
Cyclopropane, 202

Cyclopropyl anion, 258
Cycloreversion, see Cycloaddition

D* value, 334
Debye–Einstein–Stokes equation, 80
Density matrix, 41, 438, 461
Density of states, 55, 89
De-Rydbergization, 47, 140, 156, 376, 378
1,4-Dewarnaphtalene, 100, 108
Dewar rules for pericyclic reaction, 226
Dexter energy transfer mechanism, 84. See also Triplet–triplet, annihilation
Diabatic, see Nonadiabatic coupling; Reaction
Diazene, 367, 416
  α-cleavage, 382
  MO correlation diagram, 370
  molecular orbitals, 368
  N—H bond dissociation, 382
  NNH inversion, 372
  N=N twist, 369
  planar states, 367
Diazirine, 416
  cheletropic fragmentation, 413
  ring opening, 414
Diazomethane, 416
  fragmentation, 391
  ring closure, 414
Diels–Alder, see Correlation diagrams; Cycloaddition
Diethylaniline, exciplex with anthracene, 333
Diffusion, 107
  rotational, 80
  translational, 81
Digermane, 200
1,2-Dihydrocyclobut[a]acenaphthylene, 257
Diphenylmethane dyes, 323
Dipole–dipole mechanism of energy transfer, see Förster
Dipole moment, see also Sudden polarization
  twisted acroleinimine, 360, 363
  twisted propene, 217
Direct CI, see Configuration interaction (CI)
Direct reaction, see Reaction
Disrotatory, 248. See also Electrocyclic reaction
Dissociation of double bond, 374. See also Cis-trans and syn-anti isomerization; Correlation diagrams
  diazomethane, 391
  ethylene, 209
  hexatopic, 392
  hydrazoic acid, 392
  ketene, 386
  pentatopic 386, 391
Dissociation of single bond, 189, 200, 205, 206, 373. See also Correlation diagrams; Two-electron two-orbital model
  allylic bond cleavage, 292, 318
  benzylic bond cleavage, 138, 139, 294, 295, 319
  bitopic:
    allylic bond, 292, 318
    benzylic bond, 138, 139, 292, 319, 381
    charged bond, 133, 134, 138, 139, 205, 318, 346

  covalent bond, 200, 292
  dative bond, 206
  B^-—N^+ in borane–ammonia adduct, 206
  bromine, 395
  C—Br in methyl bromide, 380
  C—Br in vinyl bromide, 384
  C—C in acetaldehyde, 378
  C—N in methylamine, 374
  C—N^+ in methylammonium, 205
  C—O in methanol, 376
  α-cleavage in diazene, 382
  cyclopropane, 202
  Ge—Ge in digermane, 200
  hexatopic, Br—Br, 395
  hydrogen, 148, 156
  O—O in methyl peroxide, 382
  pentatopic, C—Br, 384
  tetratopic:
    C—Br, 380
    N—H, 382
    O—O, 382
  tritopic:
    C—C, 378
    C—N, 374
    C—O, 376
  Norrish I, 378
Donor, see Electron transfer; Energy transfer; Exciplexes; Push–pull substitution; TICT (twisted internal charge-transfer) states
Donor–acceptor pairs, 328. See also Charge transfer, complex; Exciplexes; Ion pairs and free ions; TICT (twisted internal charge-transfer) states
Dot–dot VB structure, 32, 145, 159, 161, 235, 422
Double bond, see Alkenes; Carbonyl; Diazene; Diazirine; Imine; Iminium
Doublet state, 26, 62
Double zeta basis set, see Orbital
Dyes, see Cyanine dyes; Diphenylmethane dyes; Triphenylmethane dyes
Dynamic spin polarization, 209, 238

Efficiency, 96, 101, 102
Electrochemiluminescence, 80
Electrocyclic reaction, 248
  aziridine, 248, 258
  1,3-butadiene, 248
  cyclobutene, 248
  diazirine, 414
  diazomethane, 414
  heterocycle-producing, 248, 258, 286
  iminium ylide, 248, 258
  oxirane, 258
  oxonium ylide, 258
  pleiadiene, 257
  polycyclic systems, 256
  stereochemistry, 256
Electron affinity, 35, 40, 283, 330, 333, 339, 341
Electron charge density:
  in an MO, 37
  repulsion, 37
Electronegativity parameter δ, 172, 181, 224, 433, 438. See also Two-electron two-orbital model

Electronic transfer processes, *see* Electron transfer; Energy transfer
Electronic wave function, *see* Wave function
Electron transfer, 22, 86, 91, 107. *See also* Exciplexes; TICT (twisted internal charge-transfer) states; Tunneling
 adiabatic, 86
 back transfer, 92, 338
 charge recombination, 87
 charge separation, 88
 charge shift, 88
 charge translocation, 88
 distance dependence, 89
 electron translocation, 88
 hole translocation, 88
 Marcus equation, 91
 nonadiabatic, 86
 nonradiative, 88, 92
 photoinduced, 88, 92
 radiative, 88, 92
 in a radical pair, 299, 405
 spontaneous, 88
El Sayed's rule, 77
Emission, *see also* Fluorescence; Phosphorescence
 from higher excited state, 75
 stimulated, 75
Energy:
 configuration, 35
 correlation, 34
 gap law, 76, 77
 hydrogen molecule:
  MO model, 155, 429
  VB model, 155, 421
 one-electron, 155, 430
 orbital, 35, 155, 431
 singlet, 41
 triplet, 40
Energy transfer, 22, 78, 81, 107. *See also* Triplet–triplet, annihilation; Triplet–triplet, quenching
 Dexter mechanism, 84
 Förster mechanism, 83
 intramolecular, 85
 non-vertical, 83
 to oxygen, 85
 radiative, 82
 singlet–singlet, 79
 through spacer, 81
 triplet–triplet, 78
 trivial, 82
Environmental effects, *see* Heavy atom effect; Solvent effects
Epoxide, *see* Oxirane
Equipotential surfaces, 12, 240
Ethylene, 121, 128, 157, 207, 223, 265, 316. *See also Ab initio* calculations; Biradical; Biradicaloid; Cis–trans and syn–anti isomerization; Cycloaddition; Two-electron two-orbital model
 in external field, 167, 222
 nonadiabatic coupling, 67
 path to methylcarbene, 214
 push–pull substituted, 136

 pyramidalized, 181, 212
 Rydberg states, 209
 twisted, 117, 118, 177, 181, 207, 214
 wave functions, 210
Euler equations, 448
Exchange integral:
 MO, 38, 430, 433
 VB, 154, 160, 425
Exchange mechanism of energy transfer, *see* Dexter energy transfer mechanism
Excimer, 274, 286, 344
 laser, 75
 minimum, 243, 266, 268, 273, 282
 singlet, 277, 282
 triplet, 267, 277, 284
 wave function, 283
Exciplexes, 92, 286, 328, 347
 singlet, 329
 solvent effects, 334
 triplet, 334
 wave function, 333
Excitation, 49, 56, 105
 by charge-transfer absorption, 78
 chemical, 63, 80
 density, 62, 78
 energy, 39
 by energy transfer, 78
 by ion recombination, 79, 337
 by radical recombination, 80
 singlet–singlet, 60
 singlet–triplet, 61
 by slow electrons, 62
 transfer, *see* Energy transfer
 two-photon, 62
Exciton interaction, 237, 240, 276
Eyring equation, 70

Fermi golden rule, 55
Fluorescence, 73, 105
 delayed:
  E type, 76
  P type, 86
 excimer, 75, 266, 274, 282
 exciplex, 92, 331, 333, 335
"Forbidden" reaction, *see* Reaction
Forces on nuclei, *see* Potential energy surfaces, gradient
Formaldehyde, 397, 403
Formaldimine, 350
 CNH inversion, 352, 354
 pyramidalization, 358
 S_1–S_0 touching, 356
 twisting, 351, 354
Formaldiminium:
 S_0–S_1 touching, 218
 twisting, 217
Formyl radical, 379
Förster:
 cycle, 99
 energy transfer mechanism, 83, 278
Four-electron four-orbital system, *see also* Orbital, array, cyclic; Pericyclic reaction
 cyclic, 226, 285
 dot–dot (covalent) states, 235

exact solution, 227
hole–pair (zwitterionic) states, 236
large and small squares, 237
linear, 287, 345
MO picture, 228
MO–VB dichotomy, square, 237
pericyclic minimum, 227
rectangular geometries, 228, 238
square geometries, 231
states in MO picture, 229
3 × 3 CI, 229
trapezoidal geometries, 240
$2_s + 2_s$ cycloaddition path, 232
VB picture, 235
Franck–Condon factor, 58
Franck–Condon principle, 56, 105
in electron transfer, 88, 89
in energy transfer, 83
Frontier MO theory, 124
Funnel, see Conical intersection
Furans, 260

Gaussian AO, see AO basis set
Generalized valence bond (GVB) method, 459
Gradient, see Potential energy surfaces
GVB, see Generalized valence bond (GVB) method

H_2, see also Two-electron two-orbital model
accurate calculations, 156
energy expression, 155, 425, 428
in external field, 164
MO method, 32, 148, 429
MO–VB comparison, 153
polarized, 162
Rydberg states, 156
valence-bond (VB) method, 31, 151, 421
H_4, see Four-electron four-orbital system
Half-wave oxidation potential, 330, 333, 335
Half-wave reduction potential, 330, 333, 335
Ham effect, 75
Hamiltonian operator:
for H_2, 423
for H_2^+, 429
spin–orbit coupling, 42
Hartree–Fock equations, 449
Hartree–Fock operator, 33, 449
Heavy atom effect, 42, 50, 61, 74, 77, 198, 200
Heterocycles, three-membered ring, 285, 407
fragmentation, 413
ring-opening, 258, 408
Heterosymmetric, see Biradicaloid
Hexatopic, 391. See also Topicity
Hole–pair VB structure, 32, 145, 159, 161, 236, 422
Homosymmetric, see Biradicaloid
Hot excited state reaction, see Reaction
Hot ground state reaction, see Reaction
Hot reaction, see Reaction
Hückel AO array, 226
Hund's rule, 27, 179
Hybrid electron repulsion integral, 427, 433
Hydrazoic acid, 392, 416
Hydrogen abstraction, see Abstraction of hydrogen atom

Hydrogen atom transfer, see Abstraction of hydrogen atom
Hydrogen bonding:
effect on Norrish II reaction, 407
in proton transfer reaction, 398
Hyperfine interactions, 27, 80, 85, 86, 87, 92, 108, 198, 298
Hyperspace, see Potential energy surfaces
Hypersurface, see Potential energy surfaces

Imine, see also Acetone methylimine; Acroleinimine; Formaldimine
abstraction of hydrogen, 404
twisting, 350, 359, 362
Iminium, see also Reaction; Rhodopsin
charge translocation, 320
cis–trans isomerization, 217, 322
nucleophilic addition, 319
twisted, 179
ylide, 248, 258
"Interacting subunits" correlation diagrams, 140
definition, 135
dissociation:
of benzylic or allylic bond, 294, 295
of C—N^+ bond in benzylammonium, 138, 139
of C—S^+ bond in allylsulfonium, 318
ring closure and opening:
butadiene–cyclobutene, 253
pleiadiene, 257
twisting:
of push–pull substituted ethylene, 136
of stilbene, 313
of styrene, 311
Interaction parameter γ, 172, 181, 433, 438. See also Two-electron two-orbital model
Intermolecular, see Atom or ion transfer; Electron transfer; Energy transfer; Excitation; Jumps between surfaces; Vibrational relaxation
Internal conversion, 64, 76
Intersystem crossing, 28, 64, 76, 92, 93, 106
Intramolecular, see Atom or ion transfer; Electron transfer; Energy transfer; Excitation; Jumps between surfaces; Vibrational relaxation
Inversion on a migrating group, see Reaction
Inversion on nitrogen, 352, 354, 363, 372
Iodine atom laser, 381
Ionic VB structures, see Hole–pair VB structure
Ionization potential, 35, 40, 46, 283, 330, 333, 339, 341
Ion pairs and free ions, 92, 189, 269, 328, 336. See also Exciplexes; Hole–pair VB structure
Ion recombination, 79, 87, 92, 93, 267, 268, 337, 345
Ion transfer, see Atom or ion transfer
Isomerization, see Reaction

Jablonski diagram, 51
Jumps between surfaces, 19, 28, 55, 105. See also Photophysical processes; Potential energy surfaces
radiationless, 19, 28, 76

Jumps between surfaces radiative
(*Continued*)
  chemiexcitation, 63, 337
  "direct" reaction, 66
  electron transfer, 88
  energy transfer, 78, 82
  internal conversion, 64, 76
  intersystem crossing, 26, 64, 76
  triplet–triplet annihilation, 85
  radiative, 19, 56, 73
    absorption:
      in charge transfer complexes, 78
      singlet–singlet, 60
      singlet–triplet, 61
    electron transfer, 88
    energy transfer, 82
    environmental effects, 75
    fluorescence, 73
    phosphorescence, 73
    stimulated emission, 75

Ketene, 386, 416
Kinetics, 95, 102, 108
Koopmans' theorem, 35, 431
Kramers' equation, 73, 106. *See also* Solvent effects

Landau–Zener jump probability, 66
Lifetime, 96
Lithium hydride, 166
Local excitation, *see* State, electronic
Lone pair orbital, *see* Orbital
Loose ion pair, *see* Ion pairs and free ions

Magnetic field effects, 27, 74, 80, 85, 86, 93, 108, 347
Marcus equation, 89, 91
Marcus inverted region, 91, 92, 93
Mechanism, 95
Memory effects, 20, 219
Methane, 403
Methanol, 416
  C—O bond dissociation, 376
  Rydberg states, 378
Methylamine, 374
Methylammonium, 205
Methyl bromide, 380
Methyl peroxide, 382
Minima, 109, 139
  biradicaloid, 111
  excimer, 111, 268, 273
  exciplex, 111, 328
  excited state, 110
  pericyclic, 228, 232, 269
  spectroscopic, 110
Minimum basis set, *see* Orbital
MO, *see* Molecular orbital (MO)
Möbius AO array, 226
MO crossing, *see* Correlation diagrams
Molecular orbital (MO), 32, 429. *See also* Correlation diagrams; Dissociation of single bond; Four-electron four-orbital system; Self-consistent field (SCF) method; Two-electron two-orbital model
  canonical, 32, 173, 182
  charge density, 37
  correlation, *see* Correlation diagrams
  energy in H_2, 431
  following, 120, 140
  in H_2, 144, 155, 429
  in H_4, 228, 247
  MO–VB comparison, 154, 231
  MO–VB transformation, 143, 146, 153
  need for CI, 150, 231, 237
  nodes, 120, 124
  in subunit, 275
Molecular orbital correlation diagrams:
  cycloaddition:
    2 + 2, two ethylenes, 265
    2 + 4, Diels–Alder, 123, 128
  definition, 119, 129, 130, 135
  dissociation:
    of Br—Br bond, 396
    of C=C bond in ketene, 387
    of C—N^+ bond in anilinium, 133
    of C—N^+ bond in benzylammonium, 138
    of N=N bond in hydrazoic acid, 393
    of sigma bond, 121
  fragmentation of diazirine, 411
  $H_2 + D_2 \rightleftarrows 2HD$, 229
  ring closure and opening:
    butadiene–cyclobutene:
      conrotatory, 252, 253
      disrotatory, 251, 253
    pleiadiene, 132, 257
  sigmatropic shift, 1,3-:
    with inversion, in allyltrimethylsilane, 126
    without inversion, in propene, 125
  twisting:
    of diazene, 370
    of ethylene, 122
    of push–pull substituted ethylene, 136
  valence angle variation in trimethylene biradical, 204
Molecular transport, *see* Diffusion
Möller–Plesset second-order approximation, 454
Motion (nuclear) on a surface, 15. *See also* Potential energy surfaces
MP2 approximation, *see* Möller–Plesset second-order approximation
MRD-CI, 455
Multiplicity:
  mixing, 26, 28, 43, 198
  spin, 24
Multi-reference configuration interaction, 34, 455

N=N bond isomerization, 369
Natural lifetime, *see* Radiative lifetime
Natural orbital, *see* Orbital
NEER principle, 346, 364, 365
Nitrone, cyclization, 412
Nodal properties, in MO correlation, 124
Nonadiabatic coupling, 19, 66, 106
Nonadiabatic reaction, *see* Reaction
Nonconcerted reaction:
  cycloaddition and cycloreversion, 314
  cyclobutene opening, 314

H$_2$ + H$_2$ linear path, 287
H$_3$ + H linear path, 290
Nonequilibration of excited rotamers, see NEER principle
Nonsymmetric, see Biradicaloid
Nonvertical excitation, see Energy transfer; Transition
Norbornene, 300
Norrish I, 378, 416
Norrish II, 407, 416
Nuclear configuration space, 9
Nuclear motion, 13. See also Jumps between surfaces
  on a single surface, 15
  thermal activation, 18
  thermal bath, 16
  tunneling, 17
  vibrational wave functions, 17
Nuclear wave function. See Wave function
Nucleophilic addition to C=N$^+$, 319

Occupancy of natural orbitals, 175, 180, 184, 185, 438, 441
Olefin, see Alkenes
One-electron energies, in H$_2$, 155, 430
Onium salts:
  bond cleavage, 133, 134, 205, 318
  nucleophilic attack on C=N$^+$, 319
Orbital:
  AO basis set, 29, 442, 462
  array:
    cyclic, 226
    cyclic charged, 178
    cyclic uncharged, 178
    linear, 287
  atomic (AO), see Atomic orbital (AO)
  canonical, 32, 173, 182
  delocalized, 145, 147
  energy, 35
  Gaussian, 30, 442
  localized, 145, 147
  lone pair, 44
  molecular (MO), see Molecular orbital (MO)
  natural (NO), 41, 175, 438, 461. See also Biradical; Biradicaloid; Two-electron two-orbital model
  pi, 44. See also Pi bond
  sigma, 44. See also Sigma bond
  Slater, 30, 442
  transformation, 173, 434
Overlap:
  charge density, 39
  integral, 39
Oxaziridine, O—N bond breaking, 411, 416
Oxetane, Paterno–Büchi reaction, 397
Oxidation potential, see Half-wave oxidation potential
Oxirane, 285, 416
  C—O bond-breaking, 408
  electrocyclic ring opening, 258
Oxonium ylide, interconversion with oxirane, 258

Paterno–Büchi reaction, 397, 416
Pauli matrices, 24
Pauli principle, 26
Pentadienide anion, 260
Pentatopic, 384. See also Topicity
Perfect biradicals, see Biradical
Pericyclic intermediate, see Pericyclic minimum
Pericyclic minimum, 247, 271
Pericyclic reaction, see Cycloaddition; Diazirine, cheletropic fragmentation; Electrocyclic reaction; Four-electron four-orbital system; H$_4$; Orbital, array, cyclic; Ring opening and closure; Sigmatropic shift
Permutation of nuclei, 10, 49
Peroxide, see Methyl peroxide
Phosphorescence, 73, 92
Photochemical processes, 50, 55, 105
  adiabatic reaction, 65, 70
  chemiexcitation, 63
  "direct" reaction, 66
  electron transfer, 86, 91
  hot excited-state reaction, 65
  hot ground-state reaction, 67, 94
  ion recombination, 79
  triplet–triplet annihilation, 267, 269
  tunneling, 17, 70, 100, 106
Photodetachment, 86, 318
Photohydration of alkenes, 300
Photoionization, 86
Photophysical processes, 50, 55, 105
  absorption, 56, 78
  diffusion, 80
  energy transfer, 78, 82
  fluorescence, 73
  internal conversion, 64, 76
  intersystem crossing, 64, 76
  phosphorescence, 73
  stimulated emission, 75
  triplet–triplet annihilation, 85
  vibrational relaxation, 65, 94
Photoprotonation of alkenes and arenes, 300
Phototautomerization, 405, 416
Phthalic anhydride, 333
Pi bond, see also Cis–trans and syn–anti isomerization; Dissociation of double bond; Two-electron two-orbital model
  charged, 192, 217
  comparison with sigma bond, 160
  dative, 192, 219
  in field of charge, 222
  MO view, 157
  non-polar, 192, 206
  polar, 165, 192
  VB view, 159
Pi orbital, see Orbital
Pleiadiene, 132, 257
Polarization functions, see AO basis set
Polar structures, see Hole–pair VB structure; Dot–dot VB structure
Polyenes, 309. See also 1,3-Butadiene
Potential energy surfaces, 10, 15, 48, 139. See also Jumps between surfaces; Two-electron two-orbital model

Potential energy surfaces (*Continued*)
   coupling, 20
   crossing and touching, 11, 140. *See also*
      Conical intersection
   gradient, 16
Propene, *see also* Biradicaloid; Two-electron
      two-orbital model
   dipole moment, 217
   pseudosigmatropic 1,2-shift, 261
   sigmatropic 1,3-shift, 125, 261
   twisting, 215
   wave functions, 216
Protonated azastilbenes, 323
Protonated methylamine, 205
Protonated Schiff base, *see* Iminium
Proton transfer, *see* Atom or ion transfer
Proton translocation, 325
Pseudosigmatropic shift, 261, 263, 286
Psoralen, 5, 48
Push–pull substitution:
   cycloaddition regiochemistry, 272
   cyclobutadiene, 247, 285
   ethylene, 136
Pyramidalization at a double bond, 212, 359. *See also* Sudden polarization
Pyrroles, 260

Quantum defect, 46
Quantum yield, 95, 104
Quartet state, 26, 62
Quenching, 77
   by oxygen, 85, 86

Radiation trapping, 82
Radiative energy transfer, 82
Radiative lifetime, 74
Radiative rate constant, 74
Radical:
   ions, 92, 93
   multiplicity, 26, 62
   recombination, 80, 108
Rate constants, 95, 97, 101, 102
Reaction, *see also* Addition reaction; Atom or
      ion transfer; Cis–trans and syn–anti
      isomerization; Cycloaddition; Dissociation
      of double bond; Dissociation of single
      bond; Electrocyclic reaction; Internal
      conversion; Potential energy surfaces;
      Pseudosigmatropic shift; Ring opening and
      closure; Sigmatropic shift; Topicity
   abstraction, *see* Atom or ion transfer
   addition, *see* Addition reaction
   adiabatic, 19, 65, 70, 100, 106, 260, 270, 285,
      286, 346
   aromatic photosubstitution, 297, 302
   atom transfer, *see* Atom or ion transfer
   barrier-free, 65, 344
   bitopic, 143. *See also* Topicity
   bond cleavage, *see* Cis–trans and syn–anti
      isomerization; Dissociation of double bond;
      Dissociation of single bond
   bond shift, 214, 260. *See also*
      Pseudosigmatropic shift; Sigmatropic shift
      center, 112
   concerted, 226
      in triplet state, 101, 248, 255, 258, 260
   cycloreversion, 263, 270. *See also*
      Cycloaddition
   diazirine fragmentation, 413
   direct, 66
   hexatopic, 391. *See also* Topicity
   hot excited-state, 65, 100, 107, 258
   hot ground-state, 67, 94, 107
   hydrogen bonding effects, 407
   inversion, in bond shift, 126, 261, 263
   ion transfer, *see* Atom or ion transfer
   nonadiabatic, *see* Conical intersection; Jumps
      between surfaces
   nonconcerted, 287, 314
   Paterno–Büchi, 397
   path, "allowed" and "forbidden,", 124, 126,
      127, 128, 226, 246, 254
   pentatopic, 384. *See also* Topicity
   pericyclic, 225
   photohydration, *see* Addition reaction
   pi bond twisting, *see* Cis–trans and syn–anti
      isomerization
   protonation, *see* Atom or ion transfer
   proton translocation, 325
   region, 112
   sigmatropic, *see* Sigmatropic shift
   singlet–triplet comparison, 248, 255, 260, 263,
      284, 290, 314, 315, 389, 398, 403
   solvent viscosity effects, 73, 344
   tetratopic, 380. *See also* Topicity
   tritopic, 374. *See also* Topicity
   upper triplet, 258
Recombination, *see* Ion recombination; Radical,
      recombination
Reduction potential, *see* Half-wave reduction
      potential
Regiochemistry, in cycloaddition, 272
Rehm–Weller equation, 330, 347
Reorganization energy, 89
Resonance, 32, 116, 423
Resonance integral, *see* Interaction parameter γ
Rhodopsin, 324, 346
Ring opening and closure, *see also* Correlation
      diagrams
   concerted, *see* Electrocyclic reaction
   non-concerted, 314
   three-membered ring:
      aziridine, 408
      diazirine, 414
      oxaziridine, 411
      oxirane, 408
   in triplet biradicals, 200
Rotational diffusion, *see* Diffusion
Rydberg constant, 46
Rydbergization, *see* De-Rydbergization
Rydberg orbitals, 46. *See also* AO basis set
Rydberg states, *see* State, electronic

S_0–S_1 touching, *see* Conical intersection
Salem diagrams, 119. *See also* Valence-bond
      correlation diagrams
Sandros equation, 85
Schiff base, *see* Imine
   protonated, *see* Iminium

Schrödinger equation (time-independent):
  electronic, 10
  nuclear, 17
Self-consistent field (SCF) method, 33, 443
  closed plus open shell, 445, 446
  closed shell, 445
Self-quenching, 77
Shear viscosity, 80
Sigma bond, 189. *See also* Dissociation of single bond; Two-electron two-orbital model
  charged, 205, 318
  comparison with pi bond, 160
  dative (donor–acceptor), 206, 338
  MO picture, 148, 162
  MO–VB comparison, 154
  nonpolar, 200
  polar, 162, 165
  VB picture, 151, 164
Sigma orbital, *see* Orbital
Sigmatropic shift, 125, 214, 261, 262, 286. *See also* Correlation diagrams
Singlet:
  configuration energy, 41
  energy, in H_2, 428
  ion pair, 92, 93
  oxygen, 85, 86
  spin function, 36, 143
  state, 25
Singlet–singlet:
  absorption, 60
  energy transfer, 79
Singlet–triplet:
  absorption, 61
  splitting, 41, 45, 46, 49, 78, 154, 197, 334, 339, 402
Size consistency, 456
Slater AO, *see* Orbital
Slater determinant, 32, 422, 443
Solvated electrons, 86
Solvent effects, *see also* Exciplexes; Thermal bath; TICT (twisted internal charge-transfer) states
  polarization energy, 336, 343
  transition moments, 75
  viscosity, 73, 80, 106, 344
Solvent-separated ion pairs, *see* Ion pairs and free ions
Spectral overlap, 83
Spin, electron, 24
  angular momentum, 25
  doublet and quartet, 26
  multiplicity, 24
  singlet and triplet, 26, 36, 143, 144, 434
  weakly interacting spins, 29
Spinorbital, 443
Spin–orbit coupling, 27, 28, 42, 49, 87, 106, 198. *See also* Two-electron two-orbital model
  in biradicaloids, 198, 223
  in a C—Br bond, 381
  Hamiltonian, 42
  in a O—O bond, 382
  in a twisted pi bond, 198, 214, 309, 314
Split-valence basis set, *see* Orbital

Stark effect, 437, 438
State, electronic, *see also* Correlation diagrams; Potential energy surfaces; Two-electron two-orbital model
  of a bond, 148, 154, 156, 160, 165, 195
    MO picture, 148, 157, 162
    VB picture, 151, 159, 164
  of carbonyl, VB description, 401
  charge-transfer excited, 22, 78, 329, 339
  classification, 43
  covalent, 161
  of an excimer, 275
  of an exciplex, 330
  locally excited, 22, 205, 311, 312, 339
  $n\pi^*$, 44
  order, at biradicaloid geometries, 197, 223, 233, 247
  $\pi\pi^*$, 44
  Rydberg, 46, 49, 108, 156, 197, 209, 375, 378
  singlet, 25
  TICT (twisted internal charge-transfer), 186, 220, 338, 347
  triplet, 25
  of two bonds:
    MO picture, 231
    non-polar, 240
    polar, 244
    subsystem configuration picture, 238
    in terms of subunits, 238
    VB picture, 235
  valence excited, 43
  zwitterionic, 161
State, spin, *see* Spin, electron
State, vibrational, 17
State correlation diagrams:
  atom or ion transfer:
    to alkene, 298, 299
    to styrene, 301
  cycloaddition:
    non-concerted, two ethylenes, 316
    2 + 2, two ethylenes, 266
    2 + 4, Diels–Alder, 128
  definition, 115, 127, 129, 130, 137
  dissociation:
    of benzylic or allylic bond, 139, 294, 295, 318
    of C—Br bond in vinyl bromide, 384
    of C—C bond in acetaldehyde, 379
    of C—Cl bond in trifluoromethyl chloride, 119
    of C—N bond in methylamine, 375
    of C—N^+ bond in anilinium, 134
    of C—N^+ bond in benzylammonium, 139
    of C—O bond in methanol, 377
    of C—S^+ bond in allylsulfonium, 318
    of C=C bond in ketene, 388, 390
    of N=N bond in hydrazoic acid, 394
    of single bond, 116, 120, 191
  donor–acceptor pair, 332
  fragmentation of diazirine, 411
  hydrogen or proton abstraction by a carbonyl, 401, 404
  ring closure and opening:

State correlation diagrams (*Continued*)
  butadiene–cyclobutene:
    conrotatory, 253
    disrotatory, 249, 253
    non-concerted, 315
  pleiadiene, 257
  TICT (twisted internal charge-transfer), 339, 341
  twisting:
    of double bond, 192
    of ethylene, 118
    of formaldimine, 351
    of stilbene, 313
    of styrene, 311
  valence angle variation in formaldimine, 353
Stern–Volmer equation, 98
Stilbene, 312
Strickler–Berg equation, 74
Strong coupling, 20. *See also* Potential energy surfaces
Structure, valence-bond (VB), 31, 144, 164, 421
  dot–dot, 152, 159, 422
  hole–pair, 152, 422
  pi-bond, 159
  resonance (mixing), 423
  sigma bond, 151
  VB-like, 306
  zwitterionic, 152
Styrene:
  atom or ion transfer to, 301
  cis–trans isomerization, 310
Sudden polarization, 185, 187, 216, 223, 308, 342, 439
Supermolecule, 13
Suprafacial participation, 225, 260, 264
Surface, *see* Potential energy surfaces
Symmetry, 49, 120
Syn-anti isomerization, *see* Cis–trans and syn–anti isomerization

Temperature effects, 76, 77
Tetramethylene biradical, 177, 316, 345
Tetramethyleneethane, 177
Tetratopic, 380. *See also* Topicity
Thermal bath, 16
Thiophenes, 260
Through-bond interaction, 204
Through-space interaction, 204
TICT (twisted internal charge-transfer) states, 186, 220, 338, 347. *See also* Charge transfer; State, electronic
  solvent effects, 221, 340
Topicity, 348, 373, 415, 416
  bitopic, 143
  hexatopic, 391
  pentatopic, 384
  tetratopic, 380
  tritopic, 374
Trajectories, 16, 106
Transfer, *see* Abstraction of hydrogen atom; Atom or ion transfer; Charge transfer; Electron transfer; Energy transfer
Transition:
  assignment, 47
  charge density, 39, 82
  Franck–Condon allowed, 58
  Franck–Condon forbidden, 58
  moment, 19, 56, 73, 88, 105
  singlet–singlet, 60
  singlet–triplet, 61
  solvent effects, 75
  state theory, 70, 106
  two-photon, 62
  vertical and "nonvertical", 58
Translational diffusion, *see* Diffusion
Transport, molecular, *see* Diffusion
Trifluoromethyl chloride, 119
Trimethylene biradical, 177, 183, 184, 202, 262
Triphenylmethane dyes, 323
Triple complex, 329
Triple zeta basis set, *see* AO basis set
Triplet:
  characteristics, 48
  configuration energy, 40
  energy, in H_2, 425, 431
  ion pair, 92, 93
  spin functions, 36, 143, 144, 434
  state, 25
Triplet–triplet:
  absorption, 63
  annihilation, 85, 107, 267, 269
  energy transfer, 78
  quenching, 86, 267
Tritopic, 349, 374. *See also* Topicity
Tunneling:
  electron, 87, 88
  nuclear, 17, 100, 106
Twisting, *see* Cis–trans and syn–anti isomerization
Two-electron two-orbital model, 143, 169, 223, 433, 441. *See also* Biradical; Biradicaloid
  axial biradicals, 178
  biradicaloids, 179, 438
  electronegativity difference parameter δ, 172, 181, 433, 438
  energy, 193
  Hamiltonian matrix, 172, 176, 182, 184, 187
  heterosymmetric biradicaloids, 184
  homosymmetric biradicaloids, 182
  interaction parameter γ, 172, 181, 433, 438
  MO description, 143, 148, 157
  MO–VB comparison, 154
  MO–VB transformation, 143, 146
  natural orbital occupancy, 175, 180, 184, 185, 438
  need for configuration interaction, 150, 153
  non-orthogonal basis, 193
  non-polar pi bond (twisting), 157, 159, 192
  non-polar sigma bond (dissociation), 148, 151, 189
  non-symmetric biradicaloids, 187
  orbital transformations, 173, 434
  orthogonal basis, 433
  pair biradicals, 176
  perfect biradicals, 175, 436
  polar pi bond (twisting), 165, 192
  polar sigma bond (dissociation), 161, 164, 165, 189

potential energy surfaces, 194
shortcomings, 196
sigma–pi comparison, 160
spin–orbit coupling, 198
VB description, 144, 151, 159
wave functions, 147, 149, 151, 157, 159, 162, 164, 180, 182, 184
Two-photon absorption, 62, 63
Two-photon photochemistry, 258

Uncharged perimeter, 226

Valence-bond (VB) method. *See also* Correlation diagrams; Dissociation of single bond; Four-electron four-orbital system; Two-electron two-orbital model, 31, 421
  carbonyl group, 401
  energy in H_2, 155
  GVB, 459
  in H_2, 151
  in H_4, 235, 287, 290
  MO–VB comparison, 154, 231
  MO–VB transformation, 143, 146, 153
  picture of cycloaddition, 235
  picture of electrocyclic reaction, 249
  structure, *see* Structure, valence-bond
Valence-bond correlation diagrams:
  atom or ion transfer to alkene, 298, 299
  cycloaddition, non-concerted, two ethylenes, 316
  definition, 115
  dissociation:
    of C—Br bond in vinyl bromide, 384
    of C—C bond in acetaldehyde, 379
    of C—Cl bond in trifluoromethyl chloride, 119
    of C—N bond in methylamine, 375
    of C—O bond in methanol, 377
    of single bond, 116
  hydrogen or proton abstraction by a carbonyl, 401
  ring closure and opening, butadiene–cyclobutene:
    disrotatory, 249
    non-concerted, 315

twisting:
  of ethylene, 118
  of formaldimine, 351
valence angle variation in formaldimine, 353
Valence excited states, *see* State, electronic
VB, *see* Correlation diagrams; Generalized valence bond (GVB) method; Valence-bond (VB) method
Vertical excitation, *see* Energy transfer; Franck–Condon principle; Transition
Vibrational relaxation:
  intermolecular, 65
  intramolecular (IVR), 65
Vibrations, 17
Vinyl bromide, 384
Virtual MO, 32
Viscosity, 73, 80, 106, 344
Vision, 326
Vitiligo, 3, 5, 48

Wave function, *see also Ab initio* calculations; Biradical; Biradicaloid; Correlation diagrams; Four-electron four-orbital system; Molecular orbital; Pi bond; Sigma bond; State, electronic; Two-electron two-orbital model; Valence-bond (VB) method
  electronic, 10, 49
  space part, 29
  spin part, 23
  excimer, 283
  exciplex, 333
  vibrational, 17, 48
Wavelength dependence, 104, 258
Wave packet, 16, 106
Weak coupling, 20. *See also* Potential energy surfaces
Woodward–Hoffmann rules, 226

Zeeman effect, 437, 438
Zero-field splitting, 27
Zimmerman rules for pericyclic reaction, 226
Zwitterionic, 32, 161, 422. *See also* State, electronic